Nanoparticles in Plant Biotic Stress Management

Masudulla Khan • Jen-Tsung Chen
Editors

Nanoparticles in Plant Biotic Stress Management

Springer

Editors
Masudulla Khan
Department of Botany
Aligarh Muslim University
Aligarh, India

Jen-Tsung Chen
Department of Life Sciences
National University of Kaohsiung
Kaohsiung, Taiwan

ISBN 978-981-97-0850-5   ISBN 978-981-97-0851-2 (eBook)
https://doi.org/10.1007/978-981-97-0851-2

This Springer imprint is published by the registered company Springer Nature Singapore Pte Ltd.
The registered company address is: 152 Beach Road, #21-01/04 Gateway East, Singapore 189721, Singapore

Dedicated to my beloved parents
Mr. Zaffar Ullah Khan
Mrs. Fatima Begum

Preface

Over the decades, the world has continued to face critical issues of an increasing human population and climate change, which threaten food security globally, and fortunately, new technologies based on the topic of plant sciences are upgraded fast and expected to overcome the crisis through the enhancement of agricultural production. Among these technologies, engineered nanomaterials/nanoparticles and associated emerging plant nanotechnology have been proven to have promising advantages when applied to a range of management in agriculture.

Each year, a huge loss in agricultural production can be caused by plant diseases that consequently damage global food security. However, the management of most diseases in crops remains a challenging task, and therefore, over the years, crop protection has been considered one of the most crucial research fields of plant sciences. In addition, when faced with global climate change, particularly the rising temperature, the negative impact of plant diseases can be more serious. Therefore, there is a high demand to organize strategies to make crops climate smart or resilient. In recent years, plant nanotechnology has gotten increasing attention on its remarkable potential to be applied in crop protection for releasing the potential of defense machinery and subsequently alleviating biotic stress caused by diverse pests and pathogens in the scenario of climate change.

This book collects a series of summaries organized by experts in nanotechnology to provide a systematic literature review and comprehensive discussion on the potential of nanomaterials/nanoparticles in combating plant biotic stress caused by bacteria, fungi, viruses, pests, pathogens, etc. The leading chapter presents fundamental and advanced methods for the management of crop diseases and then introduces a range of applications using nanomaterials in detecting diseases, targeted delivering pesticides, and enhancing tolerance to diseases. The following chapters provide insights into the molecular and physiological interactions involving phytohormones and cell signaling networks of some critical types of nanoparticles including AgNPs, MgO, SiO_2, and ZnO with plants and discuss their roles in disease management and accompanying issues of biosafety and phytotoxic behavior of nanomaterials, particularly at the stages of seed germination, vegetative growth, and fruit set.

The integration of green chemistry into plant nanotechnology gives considerable advances in the process of synthesis to produce a range of green synthesis nanoparticles using metallic and metalloid materials, which may ensure a more eco-friendly nanotechnology with biocompatible and biodegradable properties when applied in the sustainable management of plant diseases. Additionally, researchers have developed strategies based on the application of natural antimicrobials such as secondary metabolites, peptides, and enzymes that are encapsulated in stable, controlled-release organic-based nanoparticles in sustainably combating plant biotic stress. In brief, these achievements through the use of engineered nanoparticles as beneficial agents, elicitors, or carriers can highly support smart and sustainable agriculture in the future.

Based on the intensive exploration of literature, this book provides an in-depth analysis of the interaction of nanoparticles with crops when faced with major biotic stressors either under field conditions or in in vitro clonal propagation. For instance, seed priming and spraying using zinc oxide and copper oxide nanoparticles were found to be effective in the mitigation of plant diseases caused by significant species of bacteria, fungi, and parasitic nematodes. Interestingly, a frontier achievement is to use carbon-based nanomaterials such as graphene oxide not only to promote plant growth and alter plant metabolisms but also to induce immunity for better plant health under the attack of pathogens.

By providing a complete set of methods and applications using nanoparticles for the sustainable management of plant diseases, this book supports the UN's Sustainable Development Goals (SDGs), particularly, SDG2: Zero Hunger, which aims to end hunger and achieve food security for all. Undoubtedly, this book is an ideal reference for students and young scientists to quickly overview such a critical topic and for researchers, professors, and experts to efficiently gain summaries of each subtopic in plant nanotechnology with an emphasis on biotic stress management. Not only for plant pathologists, this book also provides valuable information for cross-field scientists, including the research fields of plant physiology, plant cell biology, plant molecular biology, plant biochemistry, plant biotechnology, plant tissue culture, plant stress, materials science and engineering, green chemistry, applied microbiology, and so on. As book editors, we would like to thank all the contributors for their insightful chapters. During the period of book organizing, the instruction and timely assistance from the publisher are very grateful.

Aligarh, Uttar Pradesh, India Masudulla Khan
Kaohsiung, Taiwan Jen-Tsung Chen

Acknowledgements

The current book, "Nanoparticles in Plant Biotic Stress Management," was developed to offer a one-pot remedy for controlling plant diseases brought on by pathogens and pests. As a result, most of the chapters were drawn from the works of erudite scientists, researchers, and academics to create a worthwhile and educational book. This book provides a quick overview of the various nanomaterials' roles in the sustainable management of plant diseases caused by pathogens and plant parasitic nematodes. As a result, the chapters have been created to ensure that consistency and coherence are maintained throughout and that the researchers have access to the most information possible on this subject. The most recent information regarding the management of plant diseases by the use of different nanomaterials has been compiled.

Furthermore, the Editorial Board would like to express its sincere gratitude to all of the authors for their contribution in this book. Without their help, this book would have remained the editors' vision alone.

Furthermore, editors extend hearty thanks to our beloved Prof. Mohammad Gulrez, Hon'ble Vice Chancellor, Aligarh Muslim University, and Prof. Naima Khatoon, Principal, Women's College, Aligarh Muslim University, Aligarh for being a source of inspiration.

I would like to express heartfelt gratitude to my thesis supervisor, Prof. Zaki Anwar Siddiqui for his guidance. Prof. Moinuddin, Dr. Fauzia Maushin, Dr. Amir Raina and Dr. Anushi A. Jahan of Women's College, Aligarh Muslim University, Aligarh, India, receive special thanks for providing critical suggestion.

I, Dr. Masudulla Khan would like to extend my deepest and heartfelt gratitude to my wife Dr. Aiman Parveen for her support.

I shall be failing in my duty if I don't thank my lovely family members K.U. Khan, A.U. Khan, M.U. Khan, S.U. Khan, Tanveer, Masarrat, Mohsina, Shazia, Shaista, Hiba, Saad, Ahmad, Sara for their selfless affection and motivation.

Without the unending help, prayers, and encouragement of their elders and youth in both happy and sad times, the editors could not have finished this endeavor. We hope that our attempts to guide readers toward a higher level of plant science will be successful.

Aligarh, Uttar Pradesh, India Masudulla Khan

Contents

Chapter 1
Nanomaterials for the Management of Crop Diseases: Methods and Applications

Manoharan Rajesh, Kempanna Sushmitha, Ganesan Megha, Ravichandran Sneha, Arockia Doss Cible, Mani Manoj, Manavalan Murugan, and Arumugam Vijaya Anand

Abstract Nanotechnology has emerged as a promising avenue for revolutionizing crop disease management. Nanomaterials have become a promising tool for the management of diseases in crops, offering innovative solutions to combat various agricultural challenges. This chapter provides an overview of the methods and applications of nanomaterials in crop disease management. The methods encompass the synthesis, characterization, and functionalization of nanomaterials tailored for specific crop diseases. Nanomaterials, due to their unique properties, offer innovative solutions for disease control in crops including nanoparticles and nanocomposites, with an emphasis on eco-friendly approaches. Additionally, we discuss the application of nanomaterials in disease detection, targeted delivery of pesticides, and enhancing plant resistance mechanisms. The potential benefits, challenges, and environmental implications of nanomaterial use in agriculture are also explored. This abstract highlights the significant strides made in harnessing nanotechnology for sustainable crop protection and disease management, shedding light on its potential to revolutionize modern agriculture.

Keywords Nanomaterials · Crop disease resistance · Crop disease protection · Crop disease control · Crop disease management · Sustainable agriculture

M. Rajesh (✉) · K. Sushmitha · G. Megha
Department of Biochemistry, Bharathiar University, Coimbatore, Tamil Nadu, India

R. Sneha · A. D. Cible · M. Manoj · A. V. Anand
Medical Genetics and Epigenetics Laboratory, Department of Human Genetics and Molecular Biology, Bharathiar University, Coimbatore, Tamil Nadu, India

M. Murugan
Department of Biomedical Engineering, Noorul Islam Centre for Higher Education, Noorul Islam College of Engineering, Kanyakumari, Tamil Nadu, India

M. Khan, J.-T. Chen (eds.), *Nanoparticles in Plant Biotic Stress Management*,
https://doi.org/10.1007/978-981-97-0851-2_1

1.1 Small Wonders, Big Impact: Nano-biosensors in Crop Disease Management

Nanotechnology is an emerging technology that has applications in many industries, including health (Anand et al. 2021; Rengarajan et al. 2022; Varalakshmi et al. 2022; Gunasangkaran et al. 2022), food safety and packaging (Pushparaj et al. 2022), agriculture and green synthesis (Jaison et al. 2023; Pushparaj et al. 2023), and other applications (Pavithran et al. 2020; Vadivelu et al. 2020). Nanotechnology has its roots in the discipline of material science. This technology has the potential to replace conventional medical treatment procedures, albeit further research is necessary before drawing definitive conclusions. Although nanoparticles can be produced chemically or physically, they prefer green synthesis, which makes use of plants, algae, and other organisms to manufacture nanoparticles that can be used in a range of fields (Thomas et al. 2022; Muthukrishnan et al. 2022; Srimurugan et al. 2022; Paramasivam et al. 2023).

Nano-biosensors, the intersection of nanotechnology and biotechnology, represent a groundbreaking advancement in the field of sensor technology. This technology offers a new dimension in plant disease diagnostic systems by providing non-destructive, minimally invasive, economical, and easy-to-use systems with improved detection limit, sensitivity, specificity, and in situ detection of plant pathogens (Kashyap et al. 2019). A nanosensor is a device with at least one nanoscale dimension that modulates a signal when it senses nanoscale objects or events. Nanosensors can operate independently or as active components of a large device. Nanosensors have been around for just over a decade, but their utility has already been demonstrated in industries as diverse as home security, medical, imaging, food marketing, environmental protection, and many others (Mahbub and Hoque 2020).

The journey of nanosensor evolution unfolds through history, starting with mechanical sensors in 1994, followed by an optical breakthrough in 1996 and optical fiber in 1998. Polymer beads and nanoelectrodes appeared in 1999, while variations appeared in the 2000s. The following years brought photonic explorers for biomedical use with biologically localized embedding, quantum dot, ion channel, and nano-strip sensors by 2002. By 2004, nanogap nanoparticles, magnetic nanoparticles, and Förster resonance energy transfer-based methods were in use. In 2005, advances were made in semiconductor, piezoelectric, and DNA sensors and localized surface plasmon resonance. Nanofiber sensors have marked a transformative period and are revolutionizing the properties of nanosensors (Lim and Ramakrishna 2006).

These nano-biosensors designed for disease detection in plants operate through the integration of biological recognition elements and nanomaterial transducers. These sensors play a pivotal role in identifying the presence of specific molecules closely linked with plant pathogens. The mechanism involves a series of intricate steps: Firstly, a bio-recognition element, which could be antibodies, enzymes, or nucleic acids, is employed. This element selectively binds to target molecules produced by the pathogens responsible for plant diseases. Secondly, nanomaterials such as carbon nanotubes or nanoparticles act as transducers. When the bio-recognition element interacts with the target molecule, it sets off a cascade of events leading to discernible alterations in the nanomaterial's characteristics. These changes manifest as modifications in electrical conductivity or optical properties, which constitute detectable signals. To amplify the sensitivity of detection, many nano-biosensors incorporate signal amplification strategies. Enzyme-catalyzed reactions or techniques harnessing surface plasmon resonance are often utilized for this purpose. The generated signal is then transduced into a readable format—for instance, electrical changes are quantified, while optical sensors might detect shifts in fluorescence intensity or color. This transduced signal subsequently undergoes thorough analysis, translating it into meaningful data. This critical information is meticulously assessed to deduce both the presence and concentration of the target pathogen. This streamlined process enables nano-biosensors to provide high sensitivity, swift detection, and real-time monitoring capabilities. These sensors can be conveniently deployed directly in the field, ushering in early detection and enabling prompt interventions to curtail further pathogenic proliferation. Figure 1.1 highlights the mechanism of nano-biosensors for identifying plant disease.

The below classification helps us identify the best nano-biosensor design for particular disease detection applications by taking into account elements like sensitivity, selectivity, usability, and compatibility with clinical settings (Fig. 1.2).

Utilizing nanosensors, the agricultural industry may benefit from the capabilities of nanotechnology. It can enhance disease control and therapy and enrich comprehension of host-parasite molecular associations. The detection of diverse pathogens, fertilizers, and soil moisture levels is also possible through nano-biosensors. By utilizing this technology, crop protection products can be reduced, nutrient loss can be lessened, and optimal nutrient management practices can allow for an increase in yield (Kaushal and Wani 2017).

Incorporating novel nanomaterials enhances the existing biochemical assays or bioassays and provides substantial improvement in sensitivity and selectivity. Meanwhile, rapid and site diagnosis of plant diseases has been facilitated, particularly in resource-deprived settings, by nanostructure-supportive non-invasive detection tools coupled with mobile visual equipment, e.g., smartphones, while long-term

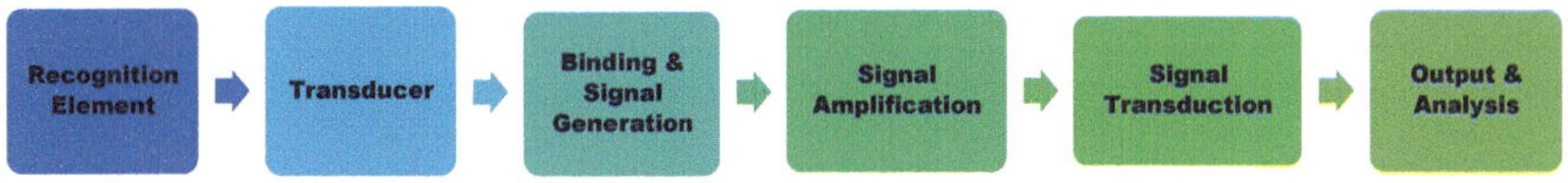

Fig. 1.1 Mechanism of nano-biosensors for recognizing plant diseases

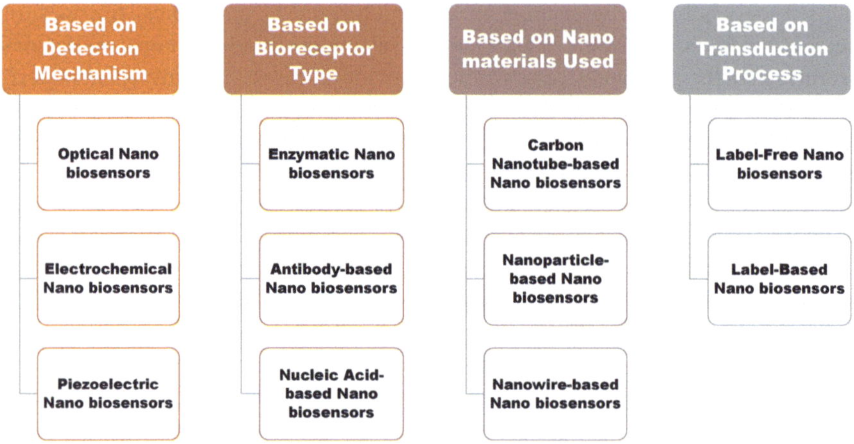

Fig. 1.2 Classification of nano-biosensors

monitoring of plant health conditions is made possible through a system of nanostructure-enabled non-invasive devices (Li et al. 2020). Sensors find application in agriculture encompassing areas like heavy metal ions, pollutants, microorganism load, and pathogens as well as the quickness with which temperature, traceability, or humidity is monitored is one of the application areas for sensors in agriculture (Johnson et al. 2021).

The application of nano-biosensors in plant disease detection is extremely promising in the future (Table 1.1). These sophisticated sensors have the potential to revolutionize agriculture by providing real-time, highly sensitive monitoring of plant health. It may detect pathogens at the molecular level by combining nanotechnology and biological identification elements, allowing for early and accurate illness diagnosis. This proactive strategy can result in more successful disease management techniques, more efficient use of resources such as pesticides and fertilizers, and higher crop yields. Nano-biosensors have the potential to greatly minimize economic losses and environmental problems associated with plant diseases by giving farmers timely information about the health of their plants, thus playing a critical role in guaranteeing global food security.

Table 1.1 Fabrication of biosensors for different pathogens

S. no.	Constituents of nano-biosensors	Sensor fabrication	Against pathogen	Detection mechanism	Sensitivity	Sample type	References
1.	Silicon nanoparticles (Rubpy-IgG)	Surface immunological functionalization	*Xanthomonas campestris*	Fluorescence quenching	7.5 ng	Cruciferous vegetables	Mondal and Jain (2021)
2.	Gold nanoparticles (ssDNA)	Surface functionalization	*Ralstonia solanacearum*	Colorimetry	25 µg/mL	Tomato, pepper, eggplant, and Irish potato	Mondal and Jain (2021)
3.	Gold nanoparticles	Surface functionalization	Aflatoxins	Surface-enhanced Raman scattering	5×10^{-5} M for aflatoxin B1 ~ 10^2 cfu	Maize, peanuts	Li et al. (2020)
4.	Cadmium quantum dots	Surface immunological functionalization	*Citrus tristeza*	Förster resonance energy transfer	220 ng/mL	Citrus trees	Li et al. (2020)
5.	Cadmium quantum dots (Rdk)	Surface immunological functionalization	*Citrus tristeza*	Förster resonance energy transfer	35–62 nM	Citrus trees	Li et al. (2020)
6.	Titanium dioxide nanotubes	Surface functionalization	Aflatoxin B_1 on acetylcholinesterase activity	Amperometric analysis using conventional three-electrode cell	0.33 nM	Detection of aflatoxins	Yuan et al. (2018)
7.	Gold nanoparticles (ssDNA)	Nanoparticle functionalization and DNA hybridization	*Ralstonia solanacearum*	Colorimetric detection	25 ng	Potato	Khaledian et al. (2017)
8.	Gold nanoparticles (ssDNA)	Nanoparticle functionalization and DNA hybridization	*Pseudomonas syringae*	Electrochemistry	214 pM	*Arabidopsis thaliana*	Lau et al. (2017)
9.	*Cadmium selenide* (polyethyleneimine quantum dots)	Surface functionalization	*Trichoplusia ni*	Fluorescence	pH range of 2.13–9.34	*Arabidopsis thaliana*	Koo et al. (2015)
10.	Carbon dots	Surface functionalization	*Fusarium avenaceum*	Fluorescence	50 nM	*Punica granatum*	Kasibabu et al. (2015)

(continued)

Table 1.1 (continued)

S. no.	Constituents of nano-biosensors	Sensor fabrication	Against pathogen	Detection mechanism	Sensitivity	Sample type	References
11.	Gold nanoparticles (ssDNA)	Nanoparticle functionalization and DNA hybridization	*Phytophthora ramorum* and *Phytophthora lateralis*	Surface-enhanced Raman scattering	N/A	Rhododendron leaves	Yüksel et al. (2015)
12.	Zinc oxide nanoparticles (chitosan nanocomposite membrane)	Single-stranded DNA probe	*Trichoderma harzianum*	Electrochemistry	1.0×10^{-19} mol/L	Detection of fungus in soil	Siddiquee et al. (2014)
13.	Gold nanoparticles	Immunosensor	*Pantoea stewartii*	Enzyme-linked immunosorbent assay	7.8×10^{-3} cfu/mL	Corn seed soak	Zhao et al. (2014)
14.	Titanium dioxide and stannic oxide nanoparticles	Surface functionalization	*p*-Ethylguaiacol	Electrochemistry	0.5 ppm	Strawberries	Fang et al. (2014)
15.	Gold nanorod	Antibody-functionalized gold nanorod Fiber optic particle plasmon resonance sensor	*Cymbidium mosaic virus* and *Odontoglossum ringspot virus*	Fiber optic particle plasmon resonance	4.8×10^{11} g/mL (*Cymbidium mosaic virus*) 4.2×10^{11} g/mL (*Odontoglossum ringspot virus*)	Orchids	Lin et al. (2014)
16.	Quantum dots	Surface functionalization DNA probe	*Ganoderma boninense*	Fluorescence resonance energy transfer	3.55×10^{-9} M	Oil palm	Mohd Bakhori et al. (2013)
17.	Cadmium-telluride quantum dots	Surface immunological functionalization	*Phytoplasma aurantifolia*	Fluorescence	5 *Ca. P. aurantifolia*/μL	Lime trees	Rad et al. (2012)
18.	Polypyrrole nanoribbon	Surface immunological functionalization	Cucumber mosaic virus	Lithographically patterned nanowire electrodeposition	10 ng/mL	Cucumbers	Chartuprayoon et al. (2013)

1.2 Precision Pest Control: Nanopesticides and Nano-insecticides Leading the Way in Crop Disease Mitigation

Globally, plant disease poses a significant obstacle for agriculturists in crop production, leading to estimated annual losses of 30–40% (Flood 2010). Various chemical choices are employed in agriculture to prevent plant diseases. While these chemicals offer advantages, they can also have detrimental effects on beneficial organisms that aid plants and pose risks to human health. So, in the quest for more potent and eco-friendly pest control solutions, integrating nanotechnology and agriculture has given rise to nano-insecticides and nanopesticides. Both of these have undoubtedly spurred innovation in the agriculture field. By exploiting the unique properties of nanoparticles, both of these aim to address some of the obstructions associated with traditional conventional pesticides, ushering in a new era of sustainable pest management with minimal or no harm to the environment (Wang et al. 2022).

The need for a shift from conventional pesticides and insecticides to nanotechnology-incorporated strategy is the need for this hour because of the risks it poses for both human health and non-target organisms. But nanopesticides and nano-insecticides, on the other hand, align with the pursuit of sustainable and secure agricultural practices. Also, they resist pests and insects by delivering the active compounds with heightened efficiency by controlled release with the application of lower doses itself, thereby fostering a more resource-efficient approach (Kumari et al. 2020). Nanotechnology is being driven in the pesticide industry by the need to decrease the amount of pesticides needed for crop protection. This is achieved by enhancing solubility, controlled release, precise delivery, better sticking ability, improved effectiveness and stability of active ingredients in the environment (Kah et al. 2018). They have the potential to take on various formulations such as emulsions, suspensions, polymer sheets, and gels. Alternatively, they can serve as capsules constructed from materials like silica, chitosan, sodium alginate, or polyethylene glycol (PEG) for encapsulating chemical substances (Shahzad and Manzoor 2021).

In 1997, a study explored the potential of nanospheres to enhance the transportation of a novel insecticide to plants. While nanosphere formulations exhibited reduced speed and release compared to a reference, their diminutive size led to enhanced penetration and systemic distribution of the insecticide within plants, despite the absence of controlled release (Boehm et al. 2003). Nanopesticides can be manufactured using two methods: (1) creating pesticides directly on the nanoscale and (2) loading the active pesticide components into nanocarriers. There are two main types of nanopesticides. The first type includes nanopesticides made from metals like silver, copper, and titanium. These have nanoparticles of the metal as the active ingredient without any carriers. The second type involves nanopesticides where the active ingredients are enclosed in nanocarriers such as polymers, clays, zein nanoparticles, or in the form of emulsions or liposomes (Kannan et al. 2023).

Efficient methods to address pest issues have been discovered through the utilization of nanoparticles. Imidacloprid-loaded sodium alginate nanoparticles controlled leafhoppers (Kumar et al. 2014). Nanogels composed of chitosan and cashew gum incorporating *Lippia sidoides* oil were made that demonstrated to have effective larvicide efficacy through slow, sustained release with optimized formulations (Abreu et al. 2012). PEGylated acephate nanoparticles acted as an alternative pesticide against *Spodoptera litura*. Lambda-cyhalothrin/silver nanoparticles were 37 times more efficient against *Spodoptera littoralis*. Novaluron nanoparticles caused 92% mortality in *Spodoptera littoralis*. Silver nanoparticles were effective against *Heterodera sacchari*. Nanocalcium protected crops from *Aonidiella aurantii* and *Bactrocera dorsalis*. Diacetyl hydrazine-based nanoformulations had potent anti-pest effects on *Spodoptera litura*. An emulsion cross-linking method was employed to produce sodium alginate nanoparticles loaded with imidacloprid. These nanoparticles demonstrated their ability to effectively combat sucking pests, particularly leafhoppers, displaying insecticidal properties (Kannan et al. 2023). In the forthcoming years, utilizing nanopesticides and nano-insecticides on estuarine plants can be studied, given their pronounced utility. Further research on environmental fate is required, alongside policy development, outreach to farmers, and innovation of new formulations with standardized protocols to foster a greater embrace of nano-empowered agriculture (Table 1.2).

Table 1.2 Synthesis of nanopesticides against different pathogens

S. no.	Nanoparticles	Against pest/ insect	Host	Impact	References
1.	Metal organic PCN-777 nanoparticles with avermectin	*Spodoptera litura*	Cabbage	Significant effectiveness in managing indoor insect populations and achieving successful control of pests in field conditions, coupled with the intelligent and responsive manner of releasing the pesticide	Liu et al. (2023)
2.	Nano-suspensions using leaves extracts of *Azadirachta indica* and flowers extracts of *Chrysanthemum coronarium*	*Tribolium castaneum* and *Rhyzopertha dominica*	Stored products	Both pests experienced a complete mortality rate of 100% when both plant extracts and its corresponding nano-suspensions were used at 100% concentrations. However, this combination with led to a greatest mortality rate of 100% in just 72 h	Hazafa et al. (2022)

(continued)

Table 1.2 (continued)

S. no.	Nanoparticles	Against pest/ insect	Host	Impact	References
3.	Neem oil-based nano-emulsion with natural adjuvant (*Prosopis juliflora* and *Cymbopogon citratus*)	*Bemisia tabaci*	*Solanum melongena*	Reduced levels of pesticide residues and the inclusion of natural adjuvants enhanced both the reliability and effectiveness of these biopesticides	Iqbal et al. (2022)
4.	Silica nanoparticles	*Fusarium oxysporum* f. sp. *niveum*	*Citrullus lanatus*	Enhanced plant growth, disease suppression (*fusarium* wilt disease) and remarkably increased fruit yield	Kang et al. (2021)
5.	Copper oxide nanoparticles	*Fusarium oxysporum* f. sp. chrysanthemi	Chrysanthemum	Increased average dry biomass and reduced disease severity	Elmer et al. (2021)
6.	Silver nanoparticles	*Meloidogyne incognita*	*Solanum nigrum*	Abnormalities include distorted embryonic development (tadpole stage), excessive growth, irregular division (gastrula stage), distinct endoderm-ectoderm cell variations, and noticeable paralyzed larvae	Fouda et al. (2020)
7.	Silver nanoparticles	*Meloidogyne graminicola*	*Oryza sativa*	Dose-dependent effectiveness in significantly reducing root gallings; progressively decreased galling as doses increased	Baronia et al. (2020)
8.	Copper nanoparticles and zinc nanoparticles	*Botrytis cinerea* and *Colletotrichum* spp.	*Prunus domestica*	Restrained fungal mycelial expansion, demonstrating potential variations in their mechanism, displayed escalated toxicity during spore germination	Malandrakis et al. (2019)

(continued)

Table 1.2 (continued)

S. no.	Nanoparticles	Against pest/ insect	Host	Impact	References
9.	$Cu_3(PO_4)_2{\cdot}3H_2O$ nanosheets	*Fusarium oxysporum* f. sp. *niveum*	*Citrullus lanatus*	Notable disease suppression effects and potential yield improvements	Borgatta et al. (2018)
10.	Citrus peel essential oil nanoformulation	*Tuta absoluta*	Tomato	Eggs experienced a postponed impact, decrease in the count of larvae that successfully developed into adults, disturbing influence on insect development	Campolo et al. (2017)
11.	Nanostructured alumina	*Sitophilus oryzae*	Stored products	Charged nanoparticles adhere to the insect's outer layer via triboelectric forces, absorbing its wax covering and causing dehydration. Additionally, ingesting these nanoparticles increased mortality in the affected insects	Stadler et al. (2017)
12.	Mesoporous alumina nanoparticles	*Fusarium oxysporum*	Tomato	Increased growth parameters by around twofold, reduced damage from root rot, and enhanced photosynthesis process	Shenashen et al. (2017)
13.	Green silver nanoparticles	*Meloidogyne javanica*	*Solanum melongena*	Reduced the population of *Meloidogyne javanica* second-stage juveniles in soil; beneficial effects on the growth and alterations in the DNA profile	Abdellatif et al. (2016)

(continued)

Table 1.2 (continued)

S. no.	Nanoparticles	Against pest/ insect	Host	Impact	References
14.	Silica nanoparticles	*Callosobruchus maculatus*	Pulse seeds of *Macrotyloma uniflorum*, *Vigna radiata*, *Cajanus cajan*, *Vigna mungo*, *Vigna unguiculata*, and *Cicer arietinum*	Notable decrease in egg laying; damaged cuticles, lead to water loss, dehydration, and insect mortality	Arumugam et al. (2016)
15.	Titanium dioxide nanoparticles and zinc oxide nanoparticles	*Agrius convolvuli*	Sweet potato leaf powder	Reduced both testis weight and sperm bundle count; led to nanoparticle-triggered vacuoles impacting chromatin condensation; harmful effects on insect spermatogenesis	Kubo-Irie et al. (2015)
16.	Myristic acid-chitosan nanogels encumbered with *Cuminum cyminum* essential oil	*Sitophilus granarius* and *Tribolium confusum*	Stored products	54% *Sitophilus granarius* mortality occurred in 10 days, while *Tribolium confusum* showed 54% mortality after 20 days of storage	Ziaee et al. (2014)
17.	Pyridalyl nanocapsule suspension	*Helicoverpa armigera*	Tomato fruit	Facilitated penetration of nanoparticles via the digestive tracts' epithelial lining and enhanced access to capillaries, encouraging widespread circulation; modifies protein tertiary structure, resulting in impaired function and eventual demise of the insect	Saini et al. (2014)

(continued)

Table 1.2 (continued)

S. no.	Nanoparticles	Against pest/ insect	Host	Impact	References
18.	Copper nanoparticles	*Phytophthora infestans*	*Lycopersicon esculentum*	Demonstrated greater efficacy than commercial agrochemicals, with no plant-harming effects	Giannousi et al. (2013)
19.	Nanogel of methyl eugenol pheromone	*Bactrocera dorsalis* Hendel	*Psidium guajava*	Highest trap catches occurred in the initial 3 weeks; yielded abundant harvest and crops without any damage; a positive safety record, rendering it highly suitable for controlling pests across various types of crops	Bhagat et al. (2013)
20.	Zinc oxide nanoparticles	*Botrytis cinerea* and *Penicillium expansum*	*Botrytis cinerea* (table grapes) *Penicillium expansum* (stored apples and pears)	Elevated nucleic acid and carbohydrate bands in *Botrytis cinerea*, signifying altered cell functions and stress response, and decreased bands related to carbohydrates, lipids, and proteins, resulting in complete inhibition of fungal growth found by Raman spectroscopy	He et al. (2011)

1.3 Different Types of Nanomaterials Used for Crop Protection

The population has increased by more than four times over the last century. An agricultural food supply for an unprecedentedly growing population exacerbates the situation (Habeeba 2022). Moreover, plant diseases and pests considerably reduce crop output, with annual losses estimated to be between 20% and 40% on a global scale (Worrall et al. 2018). In addition to that, the establishment and spread of new pathogenic races are the constant problems because chemical pest management is expensive and ineffective. Researchers have recently proposed nanomaterials as a viable replacement for the current techniques of plant disease management and crop production (Malukani et al. 2021). Because they have properties that extend shelf life, increase the solubility of least-harmful and water-soluble pesticides, and improve site-specific penetration into the target insect. They protect crops by themselves and serve as a vehicle for insecticides (Hayles et al. 2017).

At the moment, there are numerous nanomaterials here, including metalloids, metallic oxide, non-metals, carbon nanomaterials, and functionalized dendrimer, liposome, and quantum dot forms (Elmer and White 2018). However, compared to humans, its application in disease control is less widespread in plants. Metal, metal oxide, and carbon nanoparticles were used in the majority of the studies (Li et al. 2018). The silver nanoparticles are highly reactive, effective in bactericidal and fungicidal effects, and efficiently penetrate microbial cells at lower concentrations, resulting in microbial control (Lamsa et al. 2011). For example, the accumulation of silver impairs respiration and metabolism while also generating reactive oxygen species (ROS), which results in damaged *Magnaporthe grisea*, which causes the rice blast illness in *Oryza sativa* (Akter 2019).

Fungi toxicity experiments reveal that *zinc oxide nanoparticles* are more toxic to fungal species than $ZnSO_4$, while copper nanoparticles are more fungi-toxic than $CuSO_4$. Especially the silver *nanoparticles* suppress grey mold symptoms on plum fruit, inhibiting disease development by suppressing the symptoms of this disease (Malandrakis et al. 2019). When compared to bulk $CuSO_4$ and $MgSO_4$ treatments, copper oxide nanoparticles and magnesium oxide nanoparticles significantly decreased the incidence of brown rot disease to 71.2% and 69.4%, respectively. They improved enzyme productivity, chlorophyll content, and potato plant yield. Lipid peroxidation and ultrastructural studies revealed that nanomechanical forces severely harmed the bacterial cytomembrane (Rabea et al. 2023).

Silica is crucial for monocot plants, providing biotic and abiotic stress tolerance. It modulates host resistance to pathogens and is essential for silica nanoparticles to enhance plant resistance. Silica-mediated defense mechanisms against fungi increase phenolic compounds and enzymes like chitinases, peroxidases, and polyphenol oxidases (Suriyaprabha et al. 2014). Against infections like *Fusarium solani*,

Neofusicoccum sp., and *Fusarium oxysporum*, copper nanoparticles have antifungal properties. The phytotoxicity of clove oil reduces the growth of maize seeds, while copper nanoparticles grafted with essential oils improve plant defense systems by directly disrupting pathogen membranes and perhaps preventing *Bipolaris maydis* leaf blight (Dorjee et al. 2023).

Chitosan, a natural polymer, has been shown to protect hosts from fungal infections, boost defense-related enzyme activity in *Arabidopsis* and tomatoes, and regulate β-1,3-glucanase and chitinase gene expression. However, its insolubility in aqueous media limits its widespread application with low antifungal properties. Chitosan's chelating property makes it a suitable biopolymer for improved stability, solubility, and antifungal activity (Sarkar et al. 2022). For instance, silver and chitosan-loaded salicylic acid stimulates secondary metabolites in the cassava plant to increase plant protection against leaf spot disease (Hoang et al. 2022). The blast illness was delayed, and the induction of ROS and peroxidase activity in the leaves was facilitated by the use of copper chitosan in finger millet (Sathiyabama and Manikandan 2016).

Fusarium graminearum and *Fusarium poae* are significantly inhibited by single-walled carbon nanotubes, multi-walled carbon nanotubes, fullerene (C_{60}), and reduced graphene oxide (Hao et al. 2017). Additionally, they combat the chlorotic yellow virus illness by reducing cucurbit chlorotic yellows virus symptoms and preventing viral replication (Al-Zaban et al. 2022). Few studies have been conducted using dendrimers as a carrier system and extracted essential oils. Hydrophobic substances have a common propensity to become less volatile when combined with polyamidoamine (PAMAM). PAMAMG4.0's abundant positively charged surface NH_2 groups make it easier to engage with and break down negatively charged microbial membranes. Oil/PAMAMG4.0 as a result completely interacts with the fungus cell membrane and easily penetrates the cell layer, killing fungi cells (Thanh et al. 2019).

When combined with protective agents, Cymoxanil (CYM) is a powerful therapeutic agent in agriculture. However, due to its great sensitivity and quick degradation, it cannot be used in fields. These problems are addressed, and efficacy is increased by using nanocarriers like liposomes. Due to their simple manufacture, monodisperse makeup, and biocompatibility, liposomes are well-researched nanocarriers for drug loading. However, due to stability concerns and oxidation/hydrolysis, traditional liposomes with phospholipids are not suited for agricultural uses. The formation of large unilamellar non-phospholipid liposomes with high specific surface area and stability in the aqueous phase holds a promise as potential drug carriers in agriculture (Zhang et al. 2021).

Graphene-copper atomic or nanometric materials complement each other, improving functions and properties and enhancing interrelation. By examining physiological and metabolic changes in response to biotic stress, graphene-copper nanocomposites and functionalized graphene nanomaterials can promote resistance to *Fusarium oxysporium f.sp. lycopersici* (Fol) and aid in the growth and development of tomato plants affected by the "vascular wilt" disease (Cota-Ungson et al. 2023). According to the findings, the graphene oxide-silver nanocomposite showed a nearly fourfold improvement over pure silver *nanoparticles* (Liang et al. 2017). Eugenol oil nanoemulsions are stable oil-in-water emulsions with broad-spectrum activity against viruses, fungi, and bacteria. They inhibit radial growth, sporulation, and pigmentation of pathogens while reducing mold growth and mycotoxins produced by mold. Studies have shown that eugenol alone can reduce mold growth and mycotoxin production (Abd-Elsalam and Khokhlov 2015).

Researchers face challenges in studying nanomaterials in different plant and disease systems, requiring separate studies for each. Chemists must develop innovative nanocomposites and formulations to combat hetero−/homo-aggregation and agglomeration. Nanotechnology's disease suppression tactics are crucial for global food security. Undoubtedly, in the upcoming years, the design and synthesis of multifunctional nanocomposites will be quite appealing. For instance, it is feasible to develop composites with antimicrobial activity that improve plant growth and activate the plant immune system by co-loading pesticides, micronutrients, and immune elicitors. This multimodal action mechanism may be detrimental to the emergence of pathogen resistance while also being highly efficient in the suppression of disease (Fu et al. 2020). Table 1.3 represents the different types of nanomaterials used against different kinds of pathogens.

Table 1.3 Different types of nanomaterials used against different pathogens

Type of nanomaterials		Name of crop	Disease	Pathogen	Effect of nanoparticles	Concentration	References
Metal	Silver nanoparticle	*Beta vulgaris*	Soft rot disease	*Pectobacterium carotovorum*	Direct penetration of silver into the cytoplasm of pathogen inactivated the enzyme that leads to cell death	100 ppm	Ghazy et al. (2021)
		Cucumber and pumpkin	Powdery mildew	*Golovinomyces cichoracearum* or *Sphaerotheca cusca*	Inhibit respiration and metabolism to cause physical damage of microbe and also intercalating the DNA prevent proliferation	100 ppm	Lamsa et al. (2011)
		Oryza sativa	Rice blast disease	*Magnaporthe grisea*	Accumulation of silver decreases the metabolism and respiration and also produces ROS that leads to cell damage	100 ppm	Akter (2019)
		Pepper	Pepper mild mold	Pepper mild mold virus	Increase the phenolic compound and build up the total soluble protein	400 μg/L	Elbeshehy et al. (2022)
		Banana (*Musa* sp.)	Banana bunchy top	Virus	Inducing the change in chlorophyll a and b and carotenoids and also reducing the replication of BBTV	50 ppm	Mahfouze et al. (2020)
	Essential oil-grafted copper nanoparticle	Maize	Maydis leaf blight	*Bipolaris maydis* and *Rhizoctonia solani* f. sp. *saskii*	Optimal concentration of nanoparticle inhibits the pathogen activity	500 mg/L	Dorjee et al. (2023)

Chitosan	Chitosan nanoparticle	Chilli	*Alternaria* leaf spot disease	Fungi	Inducing the different defense enzyme along with total flavonoid and phenol content	–	Sarkar et al. (2022)
	Chitosan-encumbered silver and salicylic acid	Cassava plant	Leaf spot	*Alternaria* sp.	One of the best methods to improve plant immunity is to stimulate the secondary metabolites	200–400 ppm	Hoang et al. (2022)
	Chitosan nanoparticle	*Solanum lycopersicum*	Wilt disease	*Fusarium. oxysporum* f. sp. *lycopersici*	Delaying the symptom of wilt disease	–	Sathiyabama and Charles (2015)
	Copper chitosan	Soybean	Bacterial pustule disease	*Xanthomonas axonopodis*	Inhibiting the growth of bacteria by increasing the concentration of copper	–	Swati and Joshi (2020)
		Finger millet	Blast disease	*Pyricularia grisea*	Delayed expression of blast symptoms by inducing the ROS and peroxidase in activity leaves	1 mg/mL	Sathiyabama and Manikandan (2016)
		Maize	Post-flowering stalk rot disease	*Fusarium verticillioides*	Inhibiting the various metabolic reaction by interaction with the surface area of fungi and bacteria	0.1% and 0.14%	(Choudhary et al. 2017)
	Chitosan nanoparticle	*Oryza sativa*	Sheath blight pathogen	*Rhizoctonia solani*	Activating the defense enzyme like peroxidase, phenylalanine ammonia lyase, and chitinase	–	Divya et al. (2020)
Beta D glucan nanoparticle	Beta D glucan nanoparticle	Turmeric crop	Rhizome rot disease	*Pythium aphanidermatum*	Rapid activation of various defense enzymes which is present in leaves and rhizome	–	Anusuya and Sathiyabama (2015)

(continued)

Table 1.3 (continued)

Type of nanomaterials		Name of crop	Disease	Pathogen	Effect of nanoparticles	Concentration	References
Metal oxide	Zinc oxide, copper oxide, silver nanoparticles	*Prunus domestica* fruit	Gray mold disease	*Botrytis cinerea*	Suppressing the symptom of gray mold disease	162 and 310 μg/mL	Malandrakis et al. (2019)
	Fabricated mesoporous alumina nanoparticle	*Solanum lycopersicum*	Tomato root rot disease	*Fusarium oxysporum*	Increase the surface-to-surface ratio, active surface site, and open channel pore	–	Shenashen et al. (2017)
	Magnesium oxide		Bacterial wilt disease	Tomato ralstonia	Through induced resistance	–	Imada et al. (2016)
	Copper oxide and magnesium oxide	*Solanum tuberosum*	Potato brown rot disease	*Ralstonia solanacearum*	Increasing the total chlorophyll content and enzyme efficiency	3 mg/mL	Rabea et al. (2023)
	Zinc oxide	Strawberry	Gray mold	*Botrytis cinerea*	Nanoparticle activated by UV, produce the ROS, and delayed the spoilage	–	Hashemi and Ahmadzadeh (2023)
		Coffee	*Erythricium salmonicolor* fungi caused by pink disease	*Erythricium salmonicolor*	Inhibit the fungal growth	6 mmol/L	Arciniegas-Grijalba et al. (2017)
	Copper oxide	*Cucumis sativus*	Root rot disease	*Fusarium solani*	Plasmolysis the mycelia as well as shrinking, collapsing the *Fusarium solani* and also increasing the activity of defense enzyme	–	Kamel et al. (2022)
	Zinc compound (zinc sulfate, $ZnSO_4$, and zinc oxide)	*Triticum aestivum*	Fusarium head blight	*Fusarium graminearum*	Reducing fungi and deoxynivalenol by using lower concentration of zinc oxide nanoparticle	100 mM	Savi et al. (2015)
Non-metal	Nano silica	Maize	Stem and ear root disease	*Fusarium oxysporum*, *Aspergillus niger*	Enhance accumulation of phenolic compound and defense enzyme like chitinase and peroxidase	10 kg/ha	Suriyaprabha et al. (2014)

Nanocomposite	Graphene copper nanocomposite	*Solanum lycopersicum*	Vascular wilt disease	*Fusarium oxysporum* f. sp. *lycopersici*	Control the pathogen by activating the photosynthesis pigments, synthesis of secondary metabolites, and antioxidant defense mechanism	–	Cota-Ungson et al. (2023)
	Graphene oxide/ silver nanocomposite	*Oryza sativa*	Leaf blight	*Xanthomonas oryzae*	Inducing the production of ROS and inhibiting the DNA replication of bacteria		Liang et al. (2017)
	Fullerene C_{60} and *Eucommia ulmoides* (Chinese rubber tree)	Wheat seeds	*Fusarium* head blight	*Fusarium graminearum*	Photo-selective inhibition of fungus by blocking the UV light	2%	Wei et al. (2021)
	Chitosan and their nanocomposite	*Cicer arietinum*	*Fusarium* wilt (seed-borne disease)	*Fusarium oxysporum*	Inhibiting the mycelial growth	100 μg/mL	Kaur et al. (2018)

(continued)

Table 1.3 (continued)

Type of nanomaterials		Name of crop	Disease	Pathogen	Effect of nanoparticles	Concentration	References
Carbon-based nanomaterial	**Graphene** Graphene oxide and fungicide	Cereals	*Fusarium* head blight disease	*Fusarium graminearum*	Inhibiting the mycelial growth, biomass, and spore germination	–	Wang et al. (2021)
	DNA Directed silver nanoparticle grown on graphene oxide	*Solanum lycopersicum*	Bacterial spot	*Xanthomonas perforans*	Improving the stability and antibacterial activity by stronger absorption property	–	Ocsoy et al. (2013)
	Carbon nanotubes	*Solanum lycopersicum*	Stem fruit, leaf blight	*Alternaria solani*	Induce the antioxidant defense system in plants	100 mg/L	González-García et al. (2021)
	Fullerene Fullerene and metal oxide	Rose petals	Gray mold disease	*Botrytis cinerea*	Efficacy to suppress the *Botrytis cinerea* at lower concentration	50 mg/L	Hao et al. (2017)
	Fullerene C_{60}	*Nicotiana benthamiana*	Chlorotic yellow virus disease	Cucurbit chlorotic yellow virus	Suppression of CCYV symptom via inhibition of viral replication	100 mg/L	Al-Zaban et al. (2022)
Nanoemulsion	Herbal nanoemulsion (peppermint oil)	*Solanum lycopersicum*	Early blight disease	*Alternaria solani*	Nanoemulsion-treated plant able to adopt the stress condition by upregulated defense gene expression	–	Pandey et al. (2020)
	Eugenol oil nanoemulsion	Cotton seed (*Gossypium* sp.)	*Fusarium* wilt infection	*Fusarium oxysporum* f. sp. *vasinfectum*	Destruction of fungal structure	–	Abd-Elsalam and Khokhlov (2015)
	Nanoemulsion (cinnamon oil)	Strawberry fruits	*Rhizopus* rot and gray mold	*Botrytis cinerea* and *Rhizopus stolonifer*	Increasing cellular absorption and reducing adverse effect	0.2% (500–1000 μL)	Yousef et al. (2019)
Dendrimer	*Origanum majorana* essential oil into a PAMAM G_{40} dendrimer	Tomato	Tomato late blight	*Phytophthora infestans*	Completely interacts with the fungus cell membrane and easily penetrates into the cell layer, killing fungi cells	5000 ppm	Thanh et al. (2019)

1.4 Revolutionizing Crop Protection Through Nanoparticle-Enhanced Gene Delivery

Agricultural improvement for many years has benefited greatly from gene transfer between plant species. From non-cultivated plants, beneficial features have been transmitted to agricultural varieties, such as tolerance to disease, insects, and stress. The sources from which genetic information can be gathered for crop development have been substantially expanded by recombinant DNA techniques. Recombinant DNA-based gene transfer technologies are already in use for some crop species and are being developed for others. Crop improvement will be aided by the combined employment of both conventional and modern techniques for plant genetic manipulation (Goodman et al. 1987).

There is a possibility to genetically change plants using nanoparticle-mediated plant transformation. The use of nanoparticle technology in plant pathology focuses on problems with plant-pathogen interactions and provides cutting-edge crop protection techniques. Fullerenes with functionalized carbon nanotubes, a set of instruments made possible by bucky balls with bio-recognition capabilities, makes it possible to examine biochemical processes and alter individual molecules in living cells. Studies of this kind can give information on diseases that are likely to cause gene damage so that DNA-safe therapies might be investigated. During germination, carbon nanotubes act as a delivery system for desirable chemicals that can shield seeds against illness. Since it encourages growth, the plant won't experience any harmful or toxic effects from it. Mesoporous silica nanoparticles aid in the delivery of drugs and DNA to solitary plant cells. Mesoporous silica nanoparticles have a coating of chemicals that act as carriers for the genes that are given to the plants. The plant takes the particles through the cell walls when the coating is present. It was discovered that mannosylated polyethylenimine-DNA complexes combined with mesoporous silica nanoparticles increased transfection effectiveness via receptor-mediated endocytosis via mannose receptors. These findings suggest that future use of mannosylated polyethylenimine and mesoporous silica nanoparticles as a potential gene carrier to antigen-presenting cells is possible. As vectors for delivering genes into plant cells, liposomes may have a number of benefits. Liposomes may offer several advantages as vectors for gene delivery including the ability to specifically target plant cells such as protoplasts through plasmodesmata with improved delivery of enclosed DNA via membrane fusion, and safeguarding nucleic acids from degradation by nucleases (Srilatha 2011).

In various in vitro experiments, nanoparticles have demonstrated their capacity to facilitate gene transfer without external assistance. For instance, PAMAM dendrimer nanoparticles were employed to deliver GFP plasmids to *Agrostis stolonifera*. This delivery was passive in nature, resulting in approximately 48.5% of cells displaying transient expression of the target gene (Pasupathy et al. 2008). Another method involved calcium phosphate nanoparticles transporting GUS plasmids to *Brassica juncea* hypocotyl explants. This passive delivery approach yielded an impressive 80.7% efficiency in stable transformation (Naqvi et al. 2012).

Furthermore, organically functionalized carbon nanotubes were utilized to carry YFP plasmids to both *Nicotiana tabacum* protoplasts and leaf explants. This technique led to both transient and stable expressions of the target gene (Burlaka et al. 2015).

Moving beyond the confines of laboratory settings, in vivo studies revealed the potential of nanoparticles for gene transfer without the need for external aids. Organically functionalized mesoporous silica nanoparticles were employed to deliver mCherry plasmids to *Arabidopsis thaliana* roots using passive methods. Remarkably, this approach resulted in 46.5% transient expression efficiency (Chang et al. 2013). Notably, PAMAM dendrimer nanoparticles were again utilized, this time to transport double-stranded DNA or RNA interference to *A. thaliana* roots. This passive method of gene delivery induced developmental gene silencing, leading to systemic phenotypic changes (Liu et al. 2014). Additionally, polymer-functionalized carbon nanotubes were harnessed to transport GFP plasmids and siRNA for transgenic GFP silencing in leaves of *Eruca sativa*, *Nicotiana benthamiana*, and *Triticum aestivum*. The approach achieved a transient silencing efficiency of 95%, coupled with transient expression in mature leaves. These diverse experiments collectively underscore the potential of nanoparticles in various gene transfer scenarios, showcasing their ability to facilitate targeted gene expression and manipulation in plants (Demirer et al. 2018).

The creation of nanoformulations designed to transport biomolecules like siRNA, defense proteins, and protease inhibitors holds the potential to introduce an innovative approach for managing pest infestations. The assessment of how soybean trypsin inhibitor, a protease inhibitor, interacts with silica nanoparticles was conducted, aiming to subsequently deliver these nanoparticles to tomato plants with a specific goal of combating the insect pest *Helicoverpa armigera*. The utilization of soybean trypsin inhibitor-loaded nanoparticles demonstrated the ability to impede the growth of *Helicoverpa armigera* larvae, thus showcasing their deterrent effect (Bapat et al. 2020). The prospects of nanoparticle-mediated gene transfer for crop protection are highly promising and poised to bring transformative changes to agriculture. This technology offers the potential for precise and targeted delivery of genetic material to plants through genome editing enabling the integration of beneficial traits such as disease resistance and stress tolerance with greater accuracy and efficiency (Lv et al. 2020). By reducing the need for chemical pesticides and fertilizers, nanoparticle-based gene delivery could lead to a decreased environmental impact and improved soil health.

This approach also enables rapid responses to emerging pest and disease challenges, ensuring faster adaptation to evolving threats. Moreover, the customization of nanoparticle-mediated gene delivery to specific crop varieties and local environmental conditions holds the promise of tailored and resilient agricultural practices. As multifunctional nanoparticles that carry multiple genes emerge, crops could gain comprehensive protection against a range of stressors. This technology's potential impact extends to address food security challenges in developing countries and fostering sustainable agricultural practices while requiring careful consideration of ethical and regulatory aspects. In essence, nanoparticle-mediated gene transfer represents a pivotal advancement in crop protection, offering a path toward more efficient, resilient, and environmentally conscious agriculture.

1.5 Nanoparticles as Biostimulant: A Platform in Agro-environment

Nanoparticles are being explored for their potential as biostimulants in agriculture and medicine. Research is ongoing to ensure they do not harm the environment, plants, or humans. Regulations around the employment of nanoparticles in multiple sectors are evolving to address safety and efficacy concerns. Different types of nanoparticles (e.g., metallic, polymeric) have varying effects and applications, requiring specific research approaches. Nanoparticles hold a promise in enhancing sustainability in agriculture and improving ecological systems. Continued research and development will refine their applications and address potential risks. The field of nanoparticles in biostimulation is continuously evolving, so it's important to stay updated with the latest research and regulations in your area of interest (Tolisano and Del Buono 2023).

One of the most significant advantages of nanoparticles in agriculture is their ability to encapsulate and deliver nutrients and fertilizers to plants more efficiently. Traditional methods of nutrient application often result in wastage due to leaching and volatilization. Nanoparticles, however, can protect nutrients, ensuring they are delivered directly to the plant roots (Tortella et al. 2023). This targeted approach reduces the environmental impact and enhances nutrient uptake, ultimately leading to improved crop yields. Environmental stressors such as drought, salinity, and diseases pose significant challenges to agriculture. Nanoparticles can play an integral role in enhancing plant stress tolerance. They can act as carriers for stress-alleviating compounds, triggering a plant's natural defense mechanisms (Vargas-Hernandez et al. 2020). This can help crops withstand adverse conditions and maintain their productivity, even under challenging circumstances. Nanoparticles can stimulate plant growth by promoting cell division and enhancing photosynthesis (Chernikova et al. 2021). These particles can be engineered to release growth-promoting substances slowly, providing a continuous and controlled supply to the plants. Consequently, crops treated with nanoparticle-based biostimulants often exhibit increased biomass and improved overall health (Juárez-Maldonado et al. 2019). Agriculture has long relied on chemical pesticides to combat pests and diseases. However, excessive pesticide use has unfortunate consequences on the agricultural environment and human health.

1.6 Metallic Nanoparticles as Biostimulant in Crop Growth and Development in Agriculture

Several studies have shown nanoparticles as biostimulants in the relief of plants and crop defects and disease, for instance, the conceptual research leads an idea of the union of biostimulants and zinc nanoparticles. They are competent in cadmium remediation in maize and also additionally showed synergistic effect in *in silico* analysis (Tanveer et al. 2023). The utilization of nanoparticles such as iron and zinc accompanied by biostimulants gives rise to the upregulation in nutrient mobility by raising the aerial zones mineral content (Agüero-Esparza et al. 2022). The metallic nanoparticles such as ZnO, Fe_3O_4, and SiO_2 nanoparticles enhanced the activity of catalase and peroxidase antioxidant enzymes (Cai et al. 2019, 2020). Cucumber (*Cucumis sativus*) plants tested with SiO_2 nanoparticles showed an upregulated expression of the pox and pal genes. When cucumber plants were treated with nickel oxide nanoparticles, pod gene expression increased (Elsharkawy and Derbalah 2019). Tomato plants treated with silver nanoparticles increase the antioxidant activity in enzymes like peroxidases and polyphenol oxidase (Noha et al. 2018).

Transitioning to metallic nanoparticles such as TiO_2, ZnO, Fe_3O_4, and CeO_2, they impact soil enzyme activities and alter soil bacterial communities (You et al. 2018). CeO_2 nanoparticles were found to compromise soil fertility (Priester et al. 2012). The production of metallic nanoparticles through biogenesis has become more profitable and eco-friendly, involving bacteria, fungi, and plants (Peralta-Videa et al. 2016). Metallic nanoparticles can implement in seedlings to prevent virus contamination (Elbeshehy et al. 2015). Therefore, metallic nanoparticles are suggested that they can be implemented in budding early-stage plants, either to crop seeds or during seedlings as a preventive control measure (Sánchez-López et al. 2020). Based on the properties, types, and characteristics of nanoparticles and the plant's condition, nanoparticles as plant biostimulants can have either a eustress (positive) or distress (negative) impact (Du Jardin 2015; Vazquez-Hernandez et al. 2019; Rastogi et al. 2017). Demonstration studies on good physical condition tobacco plants treated with SiO_2, ZnO, and Fe_2O_3 nanoparticles showed increased tobacco growth (Cai et al. 2019, 2020). This biostimulant effect can induce eustress in both healthy and stressed plants (Derbalah and Elsharkawy 2019).

Infected *Cucumis sativus* plants in treatment with nickel oxide nanoparticles via foliar sprays such as kelp, compost and weed tea, herbal tea, fish emulsion, and soil drench exhibited a high rise in raw-fresh and dry mass (weights) and increased growth in the many numbers of leaves (Hao et al. 2018). The comparison study in turnip mosaic virus-infested tobacco plants treated with TiO_2 and FeO_3 nanoparticles in the proportion of 50 mg/L showed an increase in dry and fresh weight, but in the case of 200 mg/L, there is no potential effect in untreated plants (Derbalah and Elsharkawy 2019). Administration of silver nanoparticles to potato virus-infected tubers improved graded parameters in comparison to the untreated infected plants, possibly due to resistance induction or due to nanoparticles' impact on virus entry (El-Shazly et al. 2017). Unaffected *Nicotiana benthamiana* plants treated with ZnO and SiO_2

nanoparticles upregulated salicylic acid-inducible pathogenesis-related genes PR1 and PR2, similar to the effect of Fe_3O_4 nanoparticles. *Cucumis sativus* treated with SiO_2 nanoparticles and inoculated with *Papaya ringspot virus* induced the expression of PR1, and cucumber plants increased PR1 expression after 2 and 4 days post-infection (dpi) with *Cytomegalovirus*, possibly due to the upregulation of salicylic acid synthesis (Cai et al. 2020). Nanoparticles represent a promising frontier in agriculture and medicine, offering innovative solutions to enhance sustainability and improve human health. Ongoing research and development efforts are refining their applications and ensuring their safe and effective use. As the field continues to evolve, the potential for nanoparticles to transform these industries remains both exciting and hopeful.

1.7 Innovation in Organic Pest Control Methods for Sustainable Agriculture

Plants are usually attacked by many pathogens which suppress their growth and productivity. In recent years, plant microbial diseases have played a crucial role in the world economy, and it is the most common economic loss that was experienced by all nations making it an important issue (Khan and Rizvi 2014). As the population grows, the cultivation field areas are also growing so the management and protection of crops are essential. For almost 1200 years, silver and silver salts have been used as disinfectants by human beings; it is a nontoxic element that is a disinfectant for almost 650 different microbes (Shang et al. 2020; Ismail et al. 2017). In the past years, nanoparticles started to play a significant role in plant disease management, and the use of nanoparticles is a novel and fancy approach that has been widely used nowadays. The most simple and obvious application of nanoparticles is inducing the nanoparticles directly into the soil which not only plays a role in disease management but also plays a significant role in mineral fixation in the soil. Under flooded conditions, this nanomaterial can also be used as a carrier molecule which can effectively release the ingredients of pesticides under a controlled environment (Abd-Elsalam and Prasad 2018).

With lower doses nanoparticles can show great efficacy toward bacteria, viruses, fungi, etc. In the management of plant disease, nanoparticles are used in two ways: disease detection and pathogen suppression. In terms of pathogen suppression, nanoparticles can be packed with pathogen-suppressive chemicals which can directly act on microorganisms. In disease detection, the nanosensor or nanoparticle-based biosensor is used to detect and study the pathogens which will lead to the precise treatment of disease management. It could also be used to monitor crop health, and some other type of nanomaterial can also be used to remove the soil contaminants from the soil. Some other types of nanoparticles have been used to trigger enzymatic reactions by using nano-engineered molecules that can bind to the specific chemicals in the soil (Joshi et al. 2019).

Nanoforms of the metal have various applications; till now 15 nanoforms have been reported in the supervision of plant disease. From that $Na_2B_4O_7$ and $CuSO_4$ are found to be more effective than any other forms. Researchers found that the nanoparticle of zinc and manganese prevents charcoal-rot diseases in sunflower (Khan et al. 2019). Another example, using silver nanoparticles, is that a lot of researchers have recommended silver nanoparticles as an antimicrobial activity against a multiple and wide range of plant pathogens. Scientists are effectively evolving a nano-sized silver *nanoparticle* composition against numerous phytopathogens, *Pseudomonas syringae*, *Pythium* sp., *Xanthomonas campestris*, etc. This composition was found to be effective in managing powdery mildew in pumpkins at the lowest concentration. Additionally, inclusive work done by researchers provides a proof for managing different kinds of plant pathogens like gray mold in strawberries and *Pseudomonas syringae* in tomato. At this moment, it is especially meaningful to know the truth that silver *nanoparticles* exist neither from dissimilar sources nor by using chemical/physical methods (Mishra and Singh 2015).

The way nanoparticles suppress pathogens is very much similar to pesticides. If appropriate nanoparticles are induced in the soil, the effectiveness of non-target molecules is high. For example, the nanoparticles can enhance the function of nitrogenous bacteria and mineral-solubilizing microorganisms. Quantum dots are small nanomaterials used as perceptive tools to detect the specific biological markers in the different cells. Since they are restricted in all three dimensions, it is a promising tool for cell labeling, tracking, and imaging. Nanoparticles possess highly unique properties, for example, colloidal iron nanoparticles can be used as a catalyst in organic reactions (Ismail et al. 2017).

1.8 Conclusion

Nanotechnology has emerged as one of the potential technologies in life sciences. With the increasing challenges in agriculture for crop disease management, new findings are always needed to increase crop productivity to meet the increasing demand for food crops all over the world. The application of nanotechnology by fabricating nanomaterials and nanoparticles for plant disease management may lead to further investigation to improve crop productivity. The entire world is focusing on synthesizing nanopesticides, nanosensors, multifunctional nanocomposites, and engineering nanomaterials for improved crop productivity. Most of the research grants are successful in laboratories which might be released soon to the public and will have a tremendous impact on future agriculture.

Author Contributions MR, KS, GM, RS, ADC, MM, and MM did the conceptualization and writing. MR and AVA did material collection and editing.

Data Availability The authors confirm that the data supporting the study's findings are incorporated in the book/chapter.

Declaration of Competing Interest The authors declare no conflict of interest.

Acknowledgments All of the authors would like to state their gratitude to their various universities and institutions for their invaluable assistance.

References

Abdellatif KF, Abdelfattah RH, El-Ansary MSM (2016) Green nanoparticles engineering on root-knot nematode infecting eggplants and their effect on plant DNA modification. Iran J Biotechnol 14(4):250

Abd-Elsalam KA, Khokhlov AR (2015) Eugenol oil nanoemulsion: antifungal activity against *Fusarium oxysporum* f. sp. vasinfectum and phytotoxicity on cottonseeds. Appl Nanosci 5:255–265

Abd-Elsalam KA, Prasad R (eds) (2018) Nanobiotechnology applications in plant protection. Springer

Abreu FO, Oliveira EF, Paula HC, de Paula RC (2012) Chitosan/cashew gum nanogels for essential oil encapsulation. Carbohydr Polym 89(4):1277–1282

Agüero-Esparza M, Villalobos-Cano O, Sánchez E, Pérez-Álvarez S, Sida-Arreola JP, Palacio-Márquez A, Ramírez-Estrada CA (2022) Effectiveness of foliar application of biostimulants and nanoparticles on growth, nitrogen assimilation and nutritional content in green bean. Not Sci Biol 14(3):11261–11261

Akter R (2019) Efficacy of silver nanoparticles against rice blast disease and farmers perception about its management in Bangladesh

Al-Zaban MI, Alhag SK, Dablool AS, Ahmed AE, Alghamdi S, Ali B, Al-Saeed FA, Saleem MH, Poczai P (2022) Manufactured nano-objects confer viral protection against cucurbit chlorotic yellows virus (CCYV) infecting *nicotiana benthamiana*. Microorganisms 10(9):1837

Anand K, Chandrasekaran B, Gupta G, Dureja H, Singh SK, Gulati M, Chellappan DK, Balasubramanian B, Femeela I, Arumugam VA, Dua K (2021) Biosynthetic exosome nanoparticles isolation, characterization, and their diagnostic and therapeutic applications. In: Advanced drug delivery systems in the management of cancer, vol 1. Academic Press, pp 373–385

Anusuya S, Sathiyabama M (2015) β-d-Glucan nanoparticle pre-treatment induce resistance against *Pythium aphanidermatum* infection in turmeric. Int J Biol Macromol 74:278–282

Arciniegas-Grijalba PA, Patiño-Portela MC, Mosquera-Sánchez LP, Guerrero-Vargas JA, Rodríguez-Páez JE (2017) ZnO nanoparticles (ZnO-NPs) and their antifungal activity against coffee fungus *Erythricium salmonicolor*. Appl Nanosci 7:225–241

Arumugam G, Velayutham V, Shanmugavel S, Sundaram J (2016) Efficacy of nanostructured silica as a stored pulse protector against the infestation of bruchid beetle, *Callosobruchus maculatus* (Coleoptera: Bruchidae). Appl Nanosci 6:445–450

Bapat G, Zinjarde S, Tamhane V (2020) Evaluation of silica nanoparticle mediated delivery of protease inhibitor in tomato plants and its effect on insect pest Helicoverpa armigera. Colloids Surf B Biointerfaces 193:111079

Baronia R, Kumar P, Singh SP, Walia RK (2020) Silver nanoparticles as a potential nematicide against. J Nematol 52(1):1–9

Bhagat D, Samanta SK, Bhattacharya S (2013) Efficient management of fruit pests by *Pheromone nanogels*. Sci Rep 3(1):1294

Boehm AL, Martinon I, Zerrouk R, Rump E, Fessi H (2003) Nanoprecipitation technique for the encapsulation of agrochemical active ingredients. J Microencapsul 20(4):433–441

Borgatta J, Ma C, Hudson-Smith N, Elmer W, Plaza Perez CD, De La Torre-Roche R et al (2018) Copper based nanomaterials suppress root fungal disease in watermelon (*Citrullus lanatus*):

role of particle morphology, composition and dissolution behavior. ACS Sustain Chem Eng 6(11):14847–14856

Burlaka OM, Pirko YV, Yemets AI, Blume YB (2015) Plant genetic transformation using carbon nanotubes for DNA delivery. Cytol Genet 49(6):349–357

Cai L, Liu C, Fan G, Liu C, Sun X (2019) Preventing viral disease by ZnONPs through directly deactivating TMV and activating plant immunity in *Nicotiana benthamiana*. Environ Sci Nano 6(12):3653–3669

Cai L, Cai L, Jia H, Liu C, Wang D, Sun X (2020) Foliar exposure of Fe3O4 nanoparticles on *Nicotiana benthamiana*: evidence for nanoparticles uptake, plant growth promoter and defense response elicitor against plant virus. J Hazard Mater 393:122415

Campolo O, Cherif A, Ricupero M, Siscaro G, Grissa-Lebdi K, Russo A, Palmeri V (2017) Citrus peel essential oil nanoformulations to control the tomato borer, *Tuta absoluta*: chemical properties and biological activity. Sci Rep 7(1):13036

Chang FP, Kuang LY, Huang CA, Jane WN, Hung Y, Yue-ie CH, Mou CY (2013) A simple plant gene delivery system using mesoporous silica nanoparticles as carriers. J Mater Chem B 1(39):5279–5287

Chartuprayoon N, Rheem Y, Ng JC, Nam J, Chen W, Myung NV (2013) Polypyrrole Nanoribbon based chemi resistive immune sensors for viral plant pathogen detection. Anal Methods 5(14):3497–3502

Chernikova O, Mazhaysky Y, Buryak S, Seregina T, Ampleeva L (2021) Comparative analysis of the use of biostimulants on the main types of soil. Agron Res 19:711

Choudhary MK, Joshi A, Sharma SS, Saharan V (2017) Effect of laboratory synthesized Cu-Chitosan nanocomposites on control of PFSR disease of Maize caused by *Fusarium verticillioids*. Int J Curr Microbiol Appl Sci 6:1656–1664

Cota-Ungson D, González-García Y, Cadenas-Pliego G, Alpuche-Solís ÁG, Benavides-Mendoza A, Juárez-Maldonado A (2023) Graphene–Cu nanocomposites induce tolerance against *Fusarium oxysporum*, increase antioxidant activity, and decrease stress in tomato plants. Plants 12(12):2270

Demirer GS, Chang R, Zhang H, Chio L, Landry MP (2018) Nanoparticle-guided biomolecule delivery for transgene expression and gene silencing in mature plants. Biophys J 114(3):217a

Derbalah ASH, Elsharkawy MM (2019) A new strategy to control Cucumber mosaic virus using fabricated NiO-nanostructures. J Biotechnol 306:134–141

Divya K, Thampi M, Vijayan S, Varghese S, Jisha MS (2020) Induction of defence response in *Oryza sativa* L. against *Rhizoctonia solani* (Kuhn) by chitosan nanoparticles. Microb Pathog 149:104525

Dorjee L, Gogoi R, Kamil D, Kumar R, Mondal TK, Pattanayak S, Gurung B (2023) Essential oil-grafted copper nanoparticles as a potential next-generation fungicide for holistic disease management in maize. Front Microbiol 14:1204512

Du Jardin P (2015) Plant biostimulants: definition, concept, main categories and regulation. Sci Hortic 196:3–14

Elbeshehy EK, Elazzazy AM, Aggelis G (2015) Silver nanoparticles synthesis mediated by new isolates of Bacillus spp., nanoparticle characterization and their activity against Bean Yellow Mosaic Virus and human pathogens. Front Microbiol 6:453

Elbeshehy EK, Hassan WM, Baeshen AA (2022) Controlling pepper mild mottle virus (PMMoV) infection in pepper seedlings by use of chemically synthetic silver nanoparticles. Molecules 28(1):139

Elmer W, White JC (2018) The future of nanotechnology in plant pathology. Annu Rev Phytopathol 56:111–133

Elmer WH, Zuverza-Mena N, Triplett LR, Roberts EL, Silady RA, White JC (2021) Foliar application of copper oxide nanoparticles suppresses *fusarium* wilt development on *chrysanthemum*. Environ Sci Technol 55(15):10805–10810

Elsharkawy MM, Derbalah A (2019) Antiviral activity of titanium dioxide nanostructures as a control strategy for broad bean strain virus in faba bean. Pest Manag Sci 75(3):828–834

El-Shazly MA, Attia YA, Kabil FF, Anis E, Hazman M (2017) Inhibitory effects of salicylic acid and silver nanoparticles on potato virus Y-infected potato plants in Egypt. Middle East J Agric Res 6(3):835–848
Fang Y, Umasankar Y, Ramasamy RP (2014) Electrochemical detection of p-ethylguaiacol, a fungi infected fruit volatile using metal oxide nanoparticles. Analyst 139(15):3804–3810
Flood J (2010) The importance of plant health to food security. Food Secur 2(3):215–231
Fouda MM, Abdelsalam NR, IMA G, Hanfy AE, Othman SI, Zaitoun AF, El-Naggar M (2020) Utilization of High throughput microcrystalline cellulose decorated silver nanoparticles as an eco-nematicide on root-knot nematodes. Colloids Surf B Biointerfaces 188:110805
Fu L, Wang Z, Dhankher OP, Xing B (2020) Nanotechnology as a new sustainable approach for controlling crop diseases and increasing agricultural production. J Exp Bot 71(2):507–519
Ghazy NA, Abd El-Hafez OA, El-Bakery AM, El-Geddawy DI (2021) Impact of silver nanoparticles and two biological treatments to control soft rot disease in sugar beet (*Beta vulgaris* L). Egypt J Biol Pest Control 31:1–12
Giannousi K, Avramidis I, Dendrinou-Samara C (2013) Synthesis, characterization and evaluation of copper based nanoparticles as agrochemicals against *Phytophthora infestans*. RSC Adv 3(44):21743–21752
González-García Y, Cadenas-Pliego G, Alpuche-Solís ÁG, Cabrera RI, Juárez-Maldonado A (2021) Carbon nanotubes decrease the negative impact of *Alternaria solani* in tomato crop. Nanomaterials 11(5):1080
Goodman RM, Hauptli H, Crossway A, Knauf VC (1987) Gene transfer in crop improvement. Science 236(4797):48–54
Gunasangkaran G, Ravi AK, Arumugam VA, Muthukrishnan S (2022) Preparation, characterization, and anticancer efficacy of chitosan, chitosan encapsulated piperine and probiotics (*Lactobacillus plantarum* (MTCC-1407), and *Lactobacillus rhamnosus* (MTCC-1423)) nanoparticles. BioNanoScience 12(2):527–539
Habeeba SS (2022) Nanotechnology in crop protection: a review. Res J Sci Technol 14(3):177–182
Hao Y, Cao X, Ma C, Zhang Z, Zhao N, Ali A, Hou T, Xiang Z, Zhuang J, Wu S, Xing B (2017) Potential applications and antifungal activities of engineered nanomaterials against graymold disease agent Botrytis cinerea on rose petals. Front Plant Sci 8:1332
Hao Y, Yuan W, Ma C, White JC, Zhang Z, Adeel M, Xing B (2018) Engineered nanomaterials suppress Turnip mosaic virus infection in tobacco (*Nicotiana benthamiana*). Environ Sci Nano 5(7):1685–1693
Hashemi S, Ahmadzadeh M (2023) The effect of zinc oxide, copper, and silver nanoparticles synthesized by the green method for controlling strawberry gray mold fungus, B. cinerea pers. J Plant Sci Phytopathol 7:50
Hayles J, Johnson L, Worthley C, Losic D (2017) Nanopesticides: a review of current research and perspectives. In: New pesticides and soil sensors. Academic Press, pp 193–225
Hazafa A, Jahan N, Zia MA, Rahman KU, Sagheer M, Naeem M (2022) Evaluation and optimization of nanosuspensions of *Chrysanthemum coronarium* and *Azadirachta indica* using response surface methodology for pest management. Chemosphere 292:133411
He L, Liu Y, Mustapha A, Lin M (2011) Antifungal activity of zinc oxide nanoparticles against *Botrytis cinerea* and *Penicillium expansum*. Microbiol Res 166(3):207–215
Hoang NH, Le Thanh T, Thepbandit W, Treekoon J, Saengchan C, Sangpueak R, Papathoti NK, Kamkaew A, Buensanteai N (2022) Efficacy of chitosan nanoparticle loaded-salicylic acid and-silver on management of cassava leaf spot disease. Polymers 14(4):660
Imada K, Sakai S, Kajihara H, Tanaka S, Ito S (2016) Magnesium oxide nanoparticles induce systemic resistance in tomato against bacterial wilt disease. Plant Pathol 65(4):551–560
Iqbal N, Hazra DK, Purkait A, Agrawal A, Kumar J (2022) Bioengineering of neemnano-formulation with adjuvant for better adhesion over applied surface to give long term insect control. Colloids Surf B Biointerfaces 209:112176
Ismail M, Prasad R, Ibrahim AI, Ahmed AI (2017) Modern prospects of nanotechnology in plant pathology. In: Nanotechnology: an agricultural paradigm. Springer, pp 305–317

Jaison JP, Balasubramanian B, Gangwar J, James N, Pappuswamy M, Anand AV, Al-Dhabi NA, Valan Arasu M, Liu WC, Sebastian JK (2023) Green synthesis of bioinspired nanoparticles mediated from plant extracts of Asteraceae family for potential biological applications. Antibiotics 12(3):543

Johnson MS, Sajeev S, Nair RS (2021) Role of nanosensors in agriculture. In: 2021 International conference on computational intelligence and knowledge economy (ICCIKE), Mar. IEEE, pp 58–63

Joshi H, Choudhary P, Mundra SL (2019) Future prospects of nanotechnology in agriculture. Int J Chem Stud 7(2):957–963

Juárez-Maldonado A, Ortega-Ortíz H, Morales-Díaz AB, González-Morales S, Morelos-Moreno Á, Cabrera-De la Fuente M et al (2019) Nanoparticles and nanomaterials as plant biostimulants. Int J Mol Sci 20(1):162

Kah M, Kookana RS, Gogos A, Bucheli TD (2018) A critical evaluation of nanopesticides and nanofertilizers against their conventional analogues. Nat Nanotechnol 13(8):677–684

Kamel SM, Elgobashy SF, Omara RI, Derbalah AS, Abdelfatah M, El-Shaer A, Al-Askar AA, Abdelkhalek A, Abd-Elsalam KA, Essa T, Kamran M (2022) Antifungal activity of copper oxide nanoparticles against root rot disease in cucumber. J Fungi 8(9):911

Kang H, Elmer W, Shen Y, Zuverza-Mena N, Ma C, Botella P, Haynes CL (2021) Silica nanoparticle dissolution rate controls the suppression of *fusarium* wilt of watermelon (*Citrullus lanatus*). Environ Sci Technol 55(20):13513–13522

Kannan M, Bojan N, Swaminathan J, Zicarelli G, Hemalatha D, Zhang Y, Faggio C (2023) Nanopesticides in agricultural pest management and their environmental risks: a review. Int J Environ Sci Technol 20:1–26

Kashyap PL, Kumar S, Jasrotia P, Singh DP, Singh GP (2019) Nanosensors for plant disease diagnosis: current understanding and future perspectives. In: Nanoscience for sustainable agriculture. Springer, pp 189–205

Kasibabu BSB, D'Souza SL, Jha S, Singhal RK, Basu H, Kailasa SK (2015) One-step synthesis of fluorescent carbon dots for imaging bacterial and fungal cells. Anal Methods 7(6):2373–2378

Kaur P, Duhan JS, Thakur R (2018) Comparative pot studies of chitosan and chitosan-metal nanocomposites as nano-agrochemicals against fusarium wilt of chickpea (Cicerarietinum L.). Biocatal Agric Biotechnol 14:466–471

Kaushal M, Wani SP (2017) Nanosensors: frontiers in precision agriculture. In: Nanotechnology: an agricultural paradigm. Springer, pp 279–291

Khaledian S, Nikkhah M, Shams-bakhsh M, Hoseinzadeh S (2017) A sensitive biosensor based on gold nanoparticles to detect *Ralstonia solanacearum* in soil. J Gen Plant Pathol 83:231–239

Khan MR, Rizvi TF (2014) Nanotechnology: scope and application in plant disease management. Plant Pathol J 13(3):214–231

Khan MR, Ahamad F, Rizvi TF (2019) Effect of nanoparticles on plant pathogens. In: Advances in phytonanotechnology. Academic Press, pp 215–240

Koo Y, Wang J, Zhang Q, Zhu H, Chehab EW, Colvin VL, Braam J (2015) Fluorescence reports intact quantum dot uptake into roots and translocation to leaves of Arabidopsis thaliana and subsequent ingestion by insect herbivores. Environ Sci Technol 49(1):626–632

Kubo-Irie M, Shimoda M, Sato A, Shida K, Yamaguchi T, Mohri H, Irie M (2015) Effect of nanoparticles injected into larvae on spermatogenesis in the pupal testis of the sweet potato hornworm, *Agrius convolvuli* (L.). Fundam Toxicol Sci 2(1):1–8

Kumar S, Bhanjana G, Sharma A, Sidhu MC, Dilbaghi N (2014) Synthesis, characterization and on field evaluation of pesticide loaded sodium alginate nanoparticles. Carbohydr Polym 101:1061–1067

Kumari M, Tomar S, Singh D, Solankey SS (2020) Nanopesticides: a new paradigm in crop protection, vol 20, pp 104–109

Lamsa K, Kim SW, Jung JH, Kim YS, Kim KS, Lee YS (2011) Inhibition effects of silver nanoparticles against powdery mildews on cucumber and pumpkin. Mycobiology 39(1):26–32

Lau H, Wu H, Wee EJ, Trau M, Wang Y, Botella JR (2017) Specific and sensitive isothermal electrochemical biosensor for plant pathogen DNA detection with colloidal gold nanoparticles as probes. Sci Rep 7(1):38896

Li H, Huang J, Lu F, Liu Y, Song Y, Sun Y, Zhong J, Huang H, Wang Y, Li S, Lifshitz Y (2018) Impacts of carbon dots on rice plants: boosting the growth and improving the disease resistance. ACS Appl Bio Mater 1(3):663–672

Li Z, Yu T, Paul R, Fan J, Yang Y, Wei Q (2020) Agricultural nanodiagnostics for plant diseases: recent advances and challenges. Nanoscale Adv 2(8):3083–3094

Liang Y, Yang D, Cui J (2017) A graphene oxide/silver nanoparticle composite as a novel agricultural antibacterial agent against *Xanthomonas oryzae pv. oryzae* for crop disease management. New J Chem 41(22):13692–13699

Lim TC, Ramakrishna S (2006) A conceptual review of nanosensors. Z Nat A 61(7–8):402–412

Lin HY, Huang CH, Lu SH, Kuo IT, Chau LK (2014) Direct detection of orchid viruses using nanorod-based fiber optic particle plasmon resonance immunosensor. Biosens Bioelectron 51:371–378

Liu S, Liu Y, Yang X, Tong C, Edwards D, Parkin IA, Paterson AH (2014) The Brassica oleracea genome reveals the asymmetrical evolution of polyploid genomes. Nat Commun 5(1):3930

Liu J, Xu D, Xu G, Li X, Du X (2023) Smart controlled-release avermectin nanopesticides based on metal–organic frameworks with large pores for enhanced insecticidal efficacy. Available at SSRN: 4510431

Lv Z, Jiang R, Chen J, Chen W (2020) Nanoparticle-mediated gene transformation strategies for plant genetic engineering. Plant J 104(4):880–891

Mahbub T, Hoque ME (2020) Introduction to nanomaterials and nanomanufacturing for nanosensors. In: Nanofabrication for smart nanosensor applications. Elsevier, pp 1–20

Mahfouze HA, El-Dougdoug NK, Mahfouze SA (2020) Virucidal activity of silver nanoparticles against Banana bunchy top virus (BBTV) in banana plants. Bull Natl Res Cent 44(1):1–11

Malandrakis AA, Kavroulakis N, Chrysikopoulos CV (2019) Use of copper, silver and zinc nanoparticles against foliar and soil-borne plant pathogens. Sci Total Environ 670:292–299

Malukani KK, Gaur N, Patel HK (2021) Application of nanobiotechnology in agriculture: novel strategy for food security. In: Omics technologies for sustainable agriculture and global food security, vol II. Springer, pp 247–265

Mishra S, Singh HB (2015) Biosynthesized silver nanoparticles as a nano weapon against phytopathogens: exploring their scope and potential in agriculture. Appl Microbiol Biotechnol 99(3):1097–1107

Mohd Bakhori N, Yusof NA, Abdullah AH, Hussein MZ (2013) Development of a fluorescence resonance energy transfer (FRET)-based DNA biosensor for detection of synthetic oligonucleotide of *Ganoderma boninense*. Biosensors 3(4):419–428

Mondal K, Jain AG (2021) Precision agriculture with nano-based biosensors

Muthukrishnan S, Anand AV, Palanisamy K, Gunasangkaran G, Ravi AK, Balasubramanian B (2022) Novel organic and inorganic nanoparticles as a targeted drug delivery vehicle in cancer treatment. In: Emerging nanomaterials for advanced technologies, vol 22. Springer, pp 117–161

Naqvi S, Maitra AN, Abdin MZ, Akmal MD, Arora I, Samim MD (2012) Calcium phosphate nanoparticle mediated genetic transformation in plants. J Mater Chem 22(8):3500–3507

Noha K, Bondok AM, El-Dougdoug KA (2018) Evaluation of silver nanoparticles as antiviral agent against ToMV and PVY in tomato plants. Sciences 8(1):100–111

Ocsoy I, Paret ML, Ocsoy MA, Kunwar S, Chen T, You M, Tan W (2013) Nanotechnology in plant disease management: DNA-directed silver nanoparticles on graphene oxide as an antibacterial against *Xanthomonas perforans*. ACS Nano 7(10):8972–8980

Pandey S, Giri VP, Tripathi A, Kumari M, Narayan S, Bhattacharya A, Srivastava S, Mishra A (2020) Early blight disease management by herbal nanoemulsion in *Solanum lycopersicum* with bio-protective manner. Ind Crop Prod 150:112421

Paramasivam D, Balasubramanian B, Suresh R, Kumaravelu J, Vellingiri MM, Liu WC, Meyyazhagan A, Alanazi AM, Rengasamy KR, Arumugam VA (2023) One-pot synthesis of silver nanoparticles derived from aqueous leaf extract of *Ageratum conyzoides* and their biological efficacy. Antibiotics 12(4):688

Pasupathy K, Lin S, Hu Q, Luo H, Ke PC (2008) Direct plant gene delivery with a poly (amidoamine) dendrimer. Biotechnol J 3(8):1078–1082

Pavithran S, Pappuswamy M, Annadurai Y, Armugam VA, Periyaswamy T (2020) Green Synthesis of copper nanoparticles, characterization and their applications. J Appl Life Sci Int 23(7):10–24

Peralta-Videa JR, Huang Y, Parsons JG, Zhao L, Lopez-Moreno L, Hernandez-Viezcas JA, Gardea-Torresdey JL (2016) Plant-based green synthesis of metallic nanoparticles: scientific curiosity or a realistic alternative to chemical synthesis? Nanotechnol Environ Eng 1:1–29

Priester JH, Ge Y, Mielke RE, Horst AM, Moritz SC, Espinosa K, Holden PA (2012) Soybean susceptibility to manufactured nanomaterials with evidence for food quality and soil fertility interruption. Proc Natl Acad Sci 109(37):E2451–E2456

Pushparaj K, Liu WC, Meyyazhagan A, Orlacchio A, Pappusamy M, Vadivalagan C, Robert AA, Arumugam VA, Kamyab H, Klemeš JJ, Khademi T (2022) Nano-from nature to nurture: a comprehensive review on facets, trends, perspectives and sustainability of nanotechnology in the food sector. Energy 240:122732

Pushparaj K, Balasubramanian B, Kandasamy Y, Arumugam VA, Kaliannan D, Arumugam M, Alodaini HA, Hatamleh AA, Pappuswamy M, Meyyazhagan A (2023) Green synthesis, characterization of silver nanoparticles using aqueous leaf extracts of *Solanum melongena* and *in vitro* evaluation of antibacterial, pesticidal and anticancer activity in human MDA-MB-231 breast cancer cell lines. J King Saud Univ Sci 35(5):102663

Rabea A, Naeem E, Balabel NM, Daigham GE (2023) Management of potato brown rot disease using chemically synthesized CuO-NPs and MgO-NPs. Bot Stud 64(1):1–20

Rad F, Mohsenifar A, Tabatabaei M, Safarnejad MR, Shahryari F, Safarpour H, Fotokian M (2012) Detection of *Candidatus Phytoplasma aurantifolia* with a quantum dots fret-based biosensor. J Plant Pathol 94:525–534

Rastogi A, Zivcak M, Sytar O, Kalaji HM, He X, Mbarki S, Brestic M (2017) Impact of metal and metal oxide nanoparticles on plant: a critical review. Front Chem 5:78

Rengarajan RL, Rathinam A, Suganthy N, Balamuralikrishnan B, Anand AV, Velayuthaprabhu S (2022) Perspectives of metals and metal oxide nanoparticles for antimicrobial consequence—an overview. In: Emerging nanomaterials for advanced technologies, vol 22. Springer, pp 397–418

Saini P, Gopal M, Kumar R, Srivastava C (2014) Development of pyridalyl nanocapsule suspension for efficient management of tomato fruit and shoot borer (*Helicoverpa armigera*). J Environ Sci Health B 49(5):344–351

Sánchez-López E, Gomes D, Esteruelas G, Bonilla L, Lopez-Machado AL, Galindo R, Souto EB (2020) Metal-based nanoparticles as antimicrobial agents: an overview. Nanomaterials 10(2):292

Sarkar A, Chakraborty N, Acharya K (2022) Chitosan nanoparticles mitigate Alternaria leaf spot disease of chilli in nitric oxide dependent way. Plant Physiol Biochem 180:64–73

Sathiyabama M, Charles RE (2015) Fungal cell wall polymer based nanoparticles in protection of tomato plants from wilt disease caused by *Fusarium oxysporum f.* sp. *lycopersici*. Carbohydr Polym 133:400–407

Sathiyabama M, Manikandan A (2016) Chitosan nanoparticle induced defense responses in finger millet plants against blast disease caused by *Pyricularia grisea* (Cke.) Sacc. Carbohydr Polym 154:241–246

Savi GD, Piacentini KC, de Souza SR, Costa ME, Santos CM, Scussel VM (2015) Efficacy of zinc compounds in controlling Fusarium head blight and deoxynivalenol formation in wheat (*Triticum aestivum L.*). Int J Food Microbiol 205:98–104

Shahzad K, Manzoor F (2021) Nanoformulations and their mode of action in insects: a review of biological interactions. Drug Chem Toxicol 44(1):1–11

Shang H, Zhou Z, Wu X, Li X, Xu Y (2020) Sunlight-induced synthesis of non-target biosafety silver nanoparticles for the control of rice bacterial diseases. Nanomaterials 10(10):2007

Shenashen M, Derbalah A, Hamza A, Mohamed A, El Safty S (2017) Antifungal activity of fabricated mesoporous alumina nanoparticles against root rot disease of tomato caused by *Fusarium oxysporium*. Pest Manag Sci 73(6):1121–1126

Siddiquee S, Rovina K, Yusof NA, Rodrigues KF, Suryani S (2014) Nanoparticle-enhanced electrochemical biosensor with DNA immobilization and hybridization of *Trichoderma harzianum* gene. Sens Biosens Res 2:16–22
Srilatha B (2011) Nanotechnology in agriculture. J Nanomed Nanotechnol 2(7):1000123
Srimurugan S, Ravi AK, Arumugam VA, Muthukrishnan S (2022) Biosynthesis of silver nanoparticles using *Citrus hystrix* leaf extract and evaluation of its anticancer efficacy against HeLa cell line. Drug Dev Ind Pharm 48(9):480–490
Stadler T, LópezGarcía GP, Gitto JG, Buteler M (2017) Nanostructured alumina: biocidal properties and mechanism of action of a novel insecticide powder. Bull Insectol 70:17
Suriyaprabha R, Karunakaran G, Kavitha K, Yuvakkumar R, Rajendran V, Kannan N (2014) Application of silica nanoparticles in maize to enhance fungal resistance. IET Nanobiotechnol 8(3):133–137
Swati KJAJ, Joshi A (2020) Cu-chitosan nanoparticle induced plant growth and antibacterial activity against bacterial pustule disease in soybean [Glycine max (L.)]. J Pharmacogn Phytochem 9:450–455
Tanveer Y, Jahangir S, Shah ZA, Yasmin H, Nosheen A, Hassan MN, Ahmad P (2023) Zinc oxide nanoparticles mediated biostimulant impact on cadmium detoxification and *in silico* analysis of zinc oxide-cadmium networks in *Zea mays* L. regulome. Environ Pollut 316:120641
Thanh VM, Bui LM, Bach LG, Nguyen NT, Thi HL, Hoang Thi TT (2019) Origanummajorana L. essential oil-associated polymeric nanodendrimer for antifungal activity against *Phytophthora infestans*. Materials 12(9):1446
Thomas S, Gunasangkaran G, Arumugam VA, Muthukrishnan S (2022) Synthesis and characterization of zinc oxide nanoparticles of *solanum nigrum* and its anticancer activity via the induction of apoptosis in cervical cancer. Biol Trace Elem Res 200(6):2684–2697
Tolisano C, Del Buono D (2023) Biobased: biostimulants and biogenic nanoparticles enter the scene. Sci Total Environ 885:163912
Tortella G, Rubilar O, Pieretti JC, Fincheira P, de Melo Santana B, Fernández-Baldo MA, Seabra AB (2023) Nanoparticles as a promising strategy to mitigate biotic stress in agriculture. Antibiotics 12(2):338
Vadivelu B, Meyyazhagan A, Palanisamy S, Anand V, Chelliapan KK (2020) Synthesis of silver nanoparticles from fish scale extract of *Cyprinus carpio* and its decolorization activity of textile dyes. J Environ Treat Technol 8:870–874
Varalakshmi B, Karpagam T, Anand AV, Balamuralikrishnan B (2022) Nanoscale smart drug delivery systems and techniques of drug loading to nanoarchitectures. In: Emerging nanomaterials for advanced technologies. Springer, pp 29–82
Vargas-Hernandez M, Macias-Bobadilla I, Guevara-Gonzalez RG, Rico-Garcia E, Ocampo-Velazquez RV, Avila-Juarez L, Torres-Pacheco I (2020) Nanoparticles as potential antivirals in agriculture. Agriculture 10(10):444
Vazquez-Hernandez C, Feregrino-Perez AA, Perez-Ramirez I, Ocampo-Velazquez RV, Rico-García E, Torres-Pacheco I, Guevara-Gonzalez RG (2019) Controlled elicitation increases steviol glycosides (SGs) content and gene expression-associated to biosynthesis of SGs in *Stevia rebaudiana* B. cv. Morita II. Ind Crop Prod 139:111479
Wang X, Peng F, Cheng C, Chen L, Shi X, Gao X, Li J (2021) Synergistic antifungal activity of graphene oxide and fungicides against *Fusarium* head blight *in vitro* and *in vivo*. Nanomaterials 11(9):2393
Wang D, Saleh NB, Byro A, Zepp R, Sahle-Demessie E, Luxton TP, Su C (2022) Nano-enabled pesticides for sustainable agriculture and global food security. Nat Nanotechnol 17(4):347–360
Wei X, Yang X, Ma A, Zhang J, Peng P, Dong J (2021) Fullerene C60/*Eucommia ulmoides* rubber nanocomposite films with excellent UV-blocking performance for controlling wheat scab fungus. Ind Crop Prod 162:113301
Worrall EA, Hamid A, Mody KT, Mitter N, Pappu HR (2018) Nanotechnology for plant disease management. Agronomy 8(12):285

You T, Liu D, Chen J, Yang Z, Dou R, Gao X, Wang L (2018) Effects of metal oxide nanoparticles on soil enzyme activities and bacterial communities in two different soil types. J Soils Sediments 18:211–221

Yousef N, Niloufar M, Elena P (2019) Antipathogenic effects of emulsion and nanoemulsion of cinnamon essential oil against Rhizopus rot and grey mold on strawberry fruits. Foods Raw Mater 7(1):210–216

Yuan Q, He C, Mo R, He L, Zhou C, Hong P, Li C (2018) Detection of AFB1 via TiO2 nanotubes/Au nanoparticles/enzyme photoelectrochemical biosensor. Coatings 8(3):90

Yüksel S, Schwenkbier L, Pollok S, Weber K, Cialla-May D, Popp J (2015) Label-free detection of Phytophthoraramorum using surface-enhanced Raman spectroscopy. Analyst 140(21):7254–7262

Zhang Z, Yang J, Yang Q, Tian G, Cui ZK (2021) Fabrication of non-phospholipid liposomal nanocarrier for sustained-release of the fungicide cymoxanil. Front Mol Biosci 8:627817

Zhao Y, Liu L, Kong D, Kuang H, Wang L, Xu C (2014) Dual amplified electrochemical immunosensor for highly sensitive detection of *Pantoea stewartiis busp.* stewartii. ACS Appl Mater Interfaces 6(23):21178–21183

Ziaee M, Moharramipour S, Mohsenifar A (2014) MA-chitosan nanogel loaded with *Cuminum cyminum* essential oil for efficient management of two stored product beetle pests. J Pest Sci 87:691–699

Chapter 2
Nanoparticles: A Novel Promising Strategy for the Management of Plant Disease-Causing Pathogens

Amara Rafi, Syazwan Afif Mohd Zobir, and Khairulmazmi Ahmad

Abstract Plant disease control is in high demand around the world to protect agricultural production against pathogens that can infect a wide range of crops. However, effective disease protection from viruses, fungi, and bacteria in crops is still one of the major challenges in agriculture. Nanotechnology has been well-recognized as a novel strategy for combating a range of pathogens that cause diseases in plants. Nanoparticles such as ZnO, SiO_2, MgO, and AgNPs were demonstrated to play a vital role in plant protection. Among these, for instance, ZnO nanoparticles were used to inhibit the growth of *Penicillium expansum*, *Fusarium graminearum*, *F. oxysporum*, and *Pseudomonas aeruginosa* and consequently reduce their negative impacts. In recent years, more and more literature revealed that nanotechnology could serve as a promising tool for the suppression of various diseases caused by plant pathogenic microorganisms, and this will help immensely toward global food security as well as increase income to the farmers. This chapter aims to summarize the use of nanoparticles to control different diseases in plants; their mode of action and uptake; interactions between nanoparticles, pathogens, and plants; toxicological impacts; and additionally, the limitations were comprehensively discussed.

Keywords Antimicrobial mechanism · Disease control · Nanomaterials · Nanoparticles · Plant diseases · Toxicity

A. Rafi · S. A. M. Zobir · K. Ahmad (✉)
Sustainable Agronomy and Crop Protection, Institute of Plantation Studies, Universiti Putra Malaysia, Serdang, Selangor, Malaysia

Department of Plant Protection, Faulty of Agriculture, Universiti Putra Malaysia, Serdang, Selangor, Malaysia
e-mail: khairulmazmi@upm.edu.my

M. Khan, J.-T. Chen (eds.), *Nanoparticles in Plant Biotic Stress Management*,
https://doi.org/10.1007/978-981-97-0851-2_2

2.1 Introduction

By 2050, it is expected that there will be 9.8 billion people on the planet. Food and water will be in greater demand, particularly in developing countries where the incomes are expected to be raised significantly (Islam and Karim 2019). The global population at risk of hunger is projected to increase from 2010 to 2050, as the total demand for food is anticipated to rise from 35% to 56%. The impact of pests and diseases on the global food supply is substantial. Plant diseases led to a 16% reduction in global crop yields between 2001 and 2003 (Moore et al. 2011). Developing countries experienced a 25% yield reduction due to phytopathogens, with fungi being the most prevalent (42%), followed by bacteria (27%) and viruses (18%) (Khan and Sharma 2020). Plant diseases targeting cultivated crops can cause partial or complete death and significant yield loss even before harvesting. Plant diseases alone account for approximately $220 billion in global crop losses annually. Agrochemicals have been heavily relied upon in commercial agriculture to shield crops from pathogens and pests (Ladner et al. 2008). However, continuous and unregulated use of pesticides and fungicides has led to resistance among pests and pathogens, resulting in health risks (Agrios 2005). Recent studies have indicated serious threats to crops from pathogen resistance to many synthetic fungicides. By adopting environmentally friendly nanoparticle materials that enhance pesticide efficiency at lower dosages and reduced application frequency, nanotechnology offers a solution to support agriculture and mitigate the environmental challenges posed by synthetic pesticides, chemical fertilizers, and plant disease management (Patel et al. 2014; Singh et al. 2010). While synthetic pesticides have historically been employed on a micron scale, there is a growing interest in transforming these pesticides to the nanoscale (Bhattacharyya et al. 2016). Coined and popularized by Eric Drexler, "nanotechnology" provides the means to revolutionize biological systems (Gopal et al. 2011) and offers numerous opportunities for advancement in the agricultural sector (Fortina et al. 2005). Diverse industries, encompassing pathogen diagnostics, packaging technology, and agribusiness, have widely embraced nanotechnology (Gopal et al. 2011; Chhipa and Joshi 2016; Nel et al. 2006; Oliveira et al. 2016). "Nanoparticles" (NPs) or "nanoscale particles" denote molecular clusters measuring 1–100 nm in size (Perez-de-Luque and Diego 2009; Raliya and Tarafdar 2013). Their diminutive size imparts a wide array of physicochemical attributes. Additionally, their high surface energy and significant surface-to-volume ratio augment their reactivity and biochemical activity (Roco 2003; Nel et al. 2006). NPs can be synthesized using various physical, chemical, and biological methods (Singh et al. 2016, 2017), and their interactions with plants can yield multifarious effects (Nel et al. 2006; Dubchak et al. 2010; Singh et al. 2016; Binns 2021). The integration of nanobiotechnology tools into agriculture has gained momentum, as NPs contribute to enhanced plant growth, development, productivity, and resilience against abiotic and biotic stressors. Notably, Tripathi et al. (2015) revealed plants' natural NP production under normal growth conditions, while (Tripathi et al. (2015) and Shweta et al. (2017) underscored the potential for NPs to boost plant metabolism. Utilizing nanoparticles as nano-vehicles can enhance material efficiency, resulting in reduced usage. Additionally, nanoparticles could heighten the sensitivity of plant disease detection in diagnostic tools (Tan et al. 2017). Nanotechnology's significance is pervasive across various fields such as food and agricultural sciences, materials science, food processing, product development, and food safety (Tripathi et al. 2015). The incorporation of nanotechnology techniques into integrated plant disease management systems holds a promise for enhancing crop yields and mitigating the impact of diseases. Numerous pathogens pose a direct threat to plant health, affecting food and crop security amidst poverty and hunger. Phytopathogens inflict substantial damage on crops, during both developmental and storage phases (Binns 2021). The integrated management of plant diseases is vital for safeguarding food supplies. Examples such as potato late blight, rice blast, citrus canker, soybean cyst nematode, and chestnut blight underscore the severity of plant diseases. The complex interplay of crops and environmental factors makes disease management exceptionally challenging. Various methods have been employed to control diseases, including resistant varieties, crop rotation, pathogen-free seedlings, optimal cultivation practices, pesticides, and moisture regulation. Nanotechnology has emerged as a potent tool in addressing this challenge, particularly with the utilization of nanoparticles as pesticide alternatives or enhancers in pesticide formulation (Tripathi et al. 2015). This chapter presents a comprehensive overview of nanoparticles, encompassing their types, applications, uptake in plants, and their potential to curb pathogen infections in plants.

2.2 Nanoparticles (NPs): An Overview

The Greek word "nanos," which signifies very small or dwarf, is where the word "nano" originates. Nanoparticles are defined as elements between 1 and 100 nm in size. Nanoclusters have a narrow size range and at least one dimension between 1 and 10 nm. The properties of conventional products will alter when they are made from nanoparticles because they have a larger surface area (Roy et al. 2021). Nanotechnology is concerned with materials that are smaller in size, larger surfaces, and bulk materials that can integrate nanoparticles (Fig. 2.1).

2.2.1 Definition and Applications of Nanoparticles in the Past

Nanoparticles, often denoted as larger groups with diameters spanning from several nanometers to several hundreds of nanometers, have a rich history that extends beyond the emergence of nanotechnology. Ancient civilizations had unknowingly engaged with and utilized various nanoscale objects. For instance, in ancient Egypt, the popular practice of black hair dyeing was thought to involve plant-based

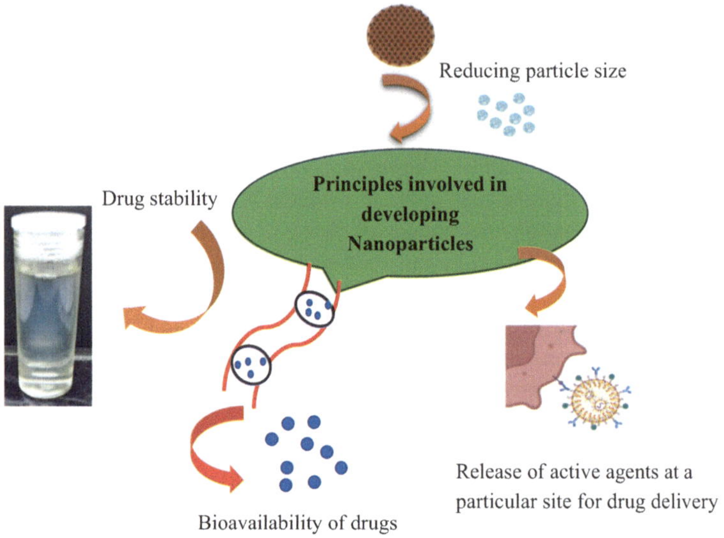

Fig. 2.1 Basic principles involved in developing nanoparticles

materials like henna. Recent studies on hair samples from Egyptian graves have revealed that lime, lead oxide, and water paste were used to dye hair, inadvertently creating galenite (lead sulfide, PbS) nanoparticles during the process. These tiny PbS nanoparticles interacted with sulfur in hair keratin, resulting in even and consistent dyeing (Giraldo et al. 2014; Sudha et al. 2018; Binns 2021). A renowned historical illustration of early nanotechnology application can be seen in the Lycurgus Cup from ancient Rome (fourth century CE). This extraordinary cup showcases exceptional optical characteristics, displaying varying colors based on the illumination source. When lit from within, it transforms into a crimson hue. Recent analysis has disclosed the presence of Au and Ag nanoparticles within the cup, measuring between 50 and 100 nm. These nanoparticles are responsible for the cup's captivating color-changing phenomenon due to the excitation of electrons through plasmon resonance (Shweta et al. 2017; Joudeh and Linke 2022). Traces of ancient nanotechnology practices have also been uncovered in civilizations such as Mesopotamia, Ancient India, and the Maya, highlighting the historical origins of nanotechnology processes (Tripathi et al. 2015; Shweta et al. 2017). In the modern context, the International Organization for Standardization (ISO) defines nanoparticles as nano-objects that encompass all their external dimensions within the nanoscale, without distinct variations between their longest and shortest axes. In instances where dimensions significantly deviate (typically more than threefold), alternative terms like nanofibers or nanoplates might be favored over the term "NPs." Nanoparticles exhibit an array of shapes, sizes, and configurations, encompassing irregular forms, as well as spherical, cylindrical, conical, tubular, spiraled, and various others (Montanarella and Kovalenko 2022; Bassett 2007). Their sizes range from 1 to 100 nm. When nanoparticles measure less than 1 nm, they are commonly referred to as atom clusters.

2.2.2 A Nanoscale Analysis of Materials' Behavior

Increased relative surface area and quantum size are the main effects on the properties of the materials which greatly differ from their bulk form. Properties like reactivity, strength, and electrical characteristics can be altered or improved because of these variables.

Relative Surface Area

A greater percentage of atoms are discovered at the particle's surface compared to its interior as it gets smaller. At a diameter of 10 nm, 20% of the particle's roughly 30,000 atoms are located on its surface; at a diameter of 5 nm, the percentage rises to 40%; and at a diameter of 1 nm, almost all of the atoms are located on the surface. As a result of having fewer immediate neighbors than atoms inside the material, the surface atoms have more unsaturated bonds. They are responsible for the particle surface's increased sensitivity. Consequently, in contrast to larger particles, nanoparticles exhibit significantly greater surface area per unit mass. This implies that a given mass of substance presented in nanoparticle structures will display notably heightened reactivity in comparison to an equivalent mass of material comprised of larger particles. This increased reactivity arises from surface-mediated processes such as growth and catalytic chemical reactions (Montanarella and Kovalenko 2022; de Mello Donega 2014).

Quantum Size

In the realm of nanoscale, particularly in its smaller reaches, the influence of quantum effects becomes more pronounced in dictating the characteristics of substances. This impact extends to the optical, electrical, and magnetic attributes of materials, shaping their behavior.

2.2.3 Types of Nanoparticles

NPs are often divided into three categories, namely, organic, carbon-based, and inorganic (Tekle et al. 2008).

Organic Nanoparticles

The most well-known examples of this class include protein complexes like ferritin, dendrimers, liposomes, and micelles. These nanoparticles are normally non-toxic, and biodegradable as in the case of liposomes having a hollow center. Organic NPs are susceptible to electromagnetic and thermal radiation, including heat and light (Schmalz et al. 2017). Additionally, because they are frequently created through non-covalent intermolecular interactions, they are more brittle by nature and provide a pathway for excretion from the body (Guerra et al. 2018). The potential application of organic nanoparticles (NPs) is contingent upon several elements, encompassing composition, surface structure, stability, loading capacity, and more. Presently, the primary utilization of organic NPs revolves around targeted drug delivery (Guerra et al. 2018; Xin et al. 2017).

Carbon-Based Nanoparticles

Carbon-based nanoparticles (NPs) (Khan et al. 2019; Chenthamara et al. 2019) consist entirely of carbon atoms. Among these, recognized members include fullerenes, carbon black nanoparticles, and carbon quantum dots. Fullerenes, categorized as carbon compounds, adopt a symmetrical closed-cage structure. Alongside the well-known C60 fullerenes, which feature an elliptical arrangement of 60 carbon atoms (Das et al. 2020; Georgakilas et al. 2015), descriptions also extend to C70 and C540 fullerenes (Das et al. 2020; Khan et al. 2019; Spinato et al. 2016). Resembling clusters of grapes, carbon black NPs form densely fused spherical aggregates (Jović et al. 2020). Within the realm of carbon nanoparticles, those measuring below 10 nm in diameter are identified as carbon quantum dots (CQDs) (Anwar et al. 2019). Carbon-based nanoparticles seamlessly integrate the distinctive attributes of sp2-hybridized carbon bonds with exceptional physicochemical characteristics at the nanoscale. These carbon-based nanoparticles exhibit remarkable versatility across various domains, including drug delivery (Spinato et al. 2016; Tao et al. 2017), energy storage, bioimaging, photovoltaic devices, and environmental sensing—facilitating tasks such as monitoring microbial ecosystems and detecting microbial pathogens (Bourassa et al. 2019; Cho et al. 2013). This adaptability is rooted in their distinctive amalgamation of traits, including electrical conductivity, durability, electron affinity, optical capabilities, thermal conductivity, and sorption capacity (Cho et al. 2013; Tyszczuk-Rotko et al. 2016). Diverse manifestations of carbon-based nanoparticles encompass nanodiamonds and carbon nano-onions, which hold a value in fields like tissue engineering and pharmaceutical delivery due to their inherent low toxicity and compatibility with biological systems (Montanarella and Kovalenko 2022; Guerra et al. 2018).

Inorganic Nanoparticles

Non-biogenic and carbon-free nanoparticles form a distinct category, encompassing materials like metal, ceramic, and semiconductor nanoparticles. Among these, metal nanoparticles (MNPs) stand out, composed solely of metal precursors and existing in monometallic (Tao et al. 2017), bimetallic (Tao et al. 2017), or polymetallic (Bourassa et al. 2019) variations. Bimetallic nanoparticles can be generated through alloys or structured arrangements like the core-shell configuration (Tao et al. 2017), harnessing the unique properties of localized surface plasmon resonance for exceptional optical and electrical traits (Montanarella and Kovalenko 2022). Certain metal nanoparticles also exhibit distinct biological, magnetic, and thermal characteristics (Binns 2021), rendering them pivotal components for crafting nanodevices spanning physics, chemistry, biology, medicine, and pharmacology (Cho et al. 2013). The primary group of nanomaterials involves nanoparticles made from metals, referred to as pure metal nanoparticles. These include metals like silver, copper, gold, titanium, platinum, zinc, magnesium, iron, and alginate nanoparticles. Another category encompasses nanoparticles made from metal oxides, such as zinc oxide and titanium dioxide, among others. Another distinct class consists of nanomaterials that involve metals, metal oxides, and doped variations of both. Metal sulfide and metal-organic framework (MOF) nanomaterials have attracted significant attention due to their promising attributes and potential applications in various biological fields. Examples include nanoparticles like AgS, CuS, and FeS, as well as MOFs based on zinc, copper, and manganese, which have demonstrated applications in drug delivery and antibacterial functions. Notably, silver nanoparticles exhibit exceptional antibacterial properties against viruses, bacteria, and other pathogens, making them incredibly valuable. Among all these materials, nanoparticles of this type are undoubtedly the most widely utilized.

2.2.4 Metal-Based Nanoparticles in Plant Disease Management

In order to combat the current situation due to plant diseases, a different approach adoption through nanotechnology for plant disease control needs to be implemented intensively. Currently, nanofungicides such as AgNPs, CuNPs, TiO_2, chitosan nanoparticles, and nanoemulsions are being used in the agricultural sectors (Fig. 2.2). The antifungal, antibacterial, and antiviral qualities of modern nanoparticles are exploited in the agriculture industry (Tyszczuk-Rotko et al. 2016; Elmer et al. 2018).

2.2.5 Uses of Nanoparticles

Nano-derived compounds for agricultural production are essential for improving plant growth development and crop yield. They are favored by farmers over traditional pesticides and methods due to several characteristics, including their small particle size, portability, ease of handling, long-term storage, high effectiveness, and non-toxicity. Worldwide commercialization of nanotechnology is now expanding significantly:

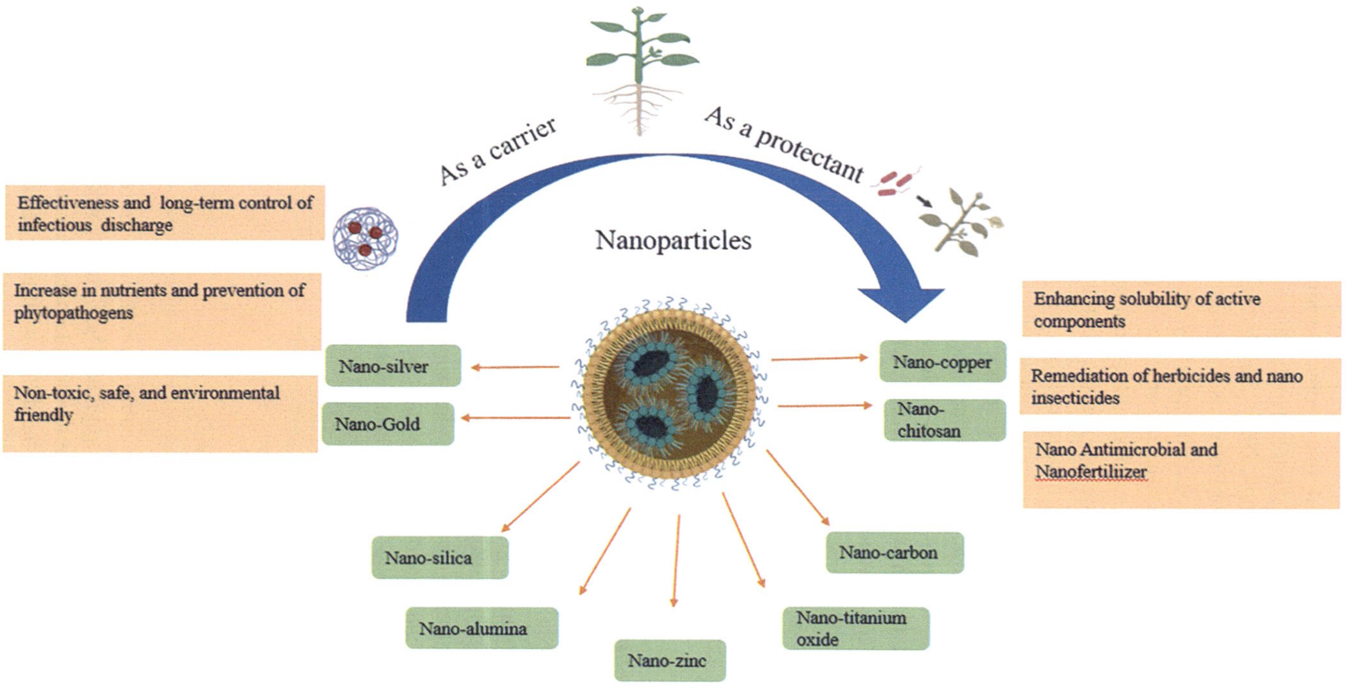

Fig. 2.2 Nanoparticles as a carrier and protectant were used in managing plant diseases in the field

1. The nanoparticles (NPs) that have been most extensively applied in plant science and have demonstrated efficacy in shaping plant growth and development are Ag, Au, Si, Ti, and Zn. While several other types of NPs also show potential outcomes with plants, additional studies are required to fully understand the processes of absorption, movement, and metabolic consequences. Multiple research endeavors have indicated that NPs impart beneficial impacts on both the structure and functioning of plants.
2. According to the existing records, gold nanoparticles (AuNPs) have a significant role in enhancing plant growth. For instance, in *Arabidopsis thaliana*, AuNPs exhibited the capacity to enhance antioxidant protective mechanisms and support seed germination. Furthermore, the impact of AuNPs extended to the control of microRNAs (miRNAs), which oversee a range of plant physiological, morphological, and metabolic processes. By utilizing AuNPs coated with DNA, successful integration of the neomycin phosphotransferase II gene into the soybean genome was accomplished (Rico et al. 2011).
3. Numerous research endeavors have been dedicated to exploring the impact of metal and metal oxide nanoparticles (NPs) on plant growth and development, especially when contrasted with their respective larger NPs or ionic forms. A specific study underscored that spinach (*Spinacia oleracea*) cultivated using a 2.5% solution of TiO_2NPs (rutile) exhibited accelerated germination and increased vitality in comparison to plants subjected to conventional bulk TiO_2-treated seeds (Abdel-Aziz et al. 2016; Liu et al. 2016). Similarly, in canola seedlings, the application of TiO_2NPs was observed to enhance germination and promote the growth of both the radicle and plumule (Kim et al. 2017).
4. Photosynthesis characterizes the process by which plants absorb light energy, converting it into chemical energy through the participation of vital enzymes. The potential improvement in the photosynthetic rate arises as a possibility with the integration of nanotechnology, facilitated by the intervention of nanoparticles (NPs). The influence of NPs on photosynthesis manifests through their interaction with the pivotal enzyme Rubisco, a cornerstone of photosynthesis recognized globally as the most abundant protein. Research led by Liu et al. (2016) and Banerjee and Kole (2016) unveiled the capacity of SiO_2NPs to amplify the functionality of carbonic anhydrase, an enzyme accountable for converting CO_2 into bicarbonate. This augmentation subsequently culminates in heightened concentrations of photosynthetic pigments, consequently fostering an elevation in the photosynthetic rate.
5. In pearl millet, ZnNPs prompt an increase in the activity of acid phosphatase, alkaline phosphatase, phytase, and dehydrogenase in comparison to bulk-Zn metals (Tao et al. 2017; Abdel-Aziz et al. 2016). In the context of peroxidase and dehydrogenase activities, MWCNTs facilitate a favorable response (Banerjee and Kole 2016). The supplementation of MS media with ZnONPs yields positive outcomes concerning proline synthesis activity, along with SOD, catalase, and peroxidase (POX), thereby enabling plants to recover more effectively under biotic stress conditions (Guerra et al. 2018; Khan et al. 2019; Jović et al. 2020).

2.2.6 *Uptake and Modes of Applications of Nanoparticles*

Uptake of Nanoparticles

The manner of exposure plays a pivotal role in determining how nanoparticles (NPs) are taken up by plants. When NPs are applied to leaves, they directly settle on the above-ground portions of the plant. In contrast, in soil-based applications, NPs gradually navigate through soil pores, propelled by factors like water movement and physical pressure, ultimately approaching the root system. Upon encountering root systems or leaf surfaces, a range of forces, including electrostatic, hydrophobic, and van der Waals interactions, contribute to their adherence to plant surfaces (Tekle et al. 2008). After this interaction, wrapping or docking occurs on the surface, facilitating their entry into neighboring cells (Yang et al. 2015). While NPs often enter the leaves primarily through the stomata and occasionally through the trichomes, their uptake in underground sections is largely facilitated by physiologically active lateral roots and root hairs (Chichiriccò and Poma 2015). Some plants can hinder NP uptake by releasing mucilage, which binds NPs to the root surface (Geisler-Lee et al. 2014). Once NPs enter the roots and leaves, they diffuse from one cell to another and traverse the xylem or phloem tissues, respectively (Anjum et al. 2013). The research underscores that NPs can permeate and navigate among cells through various mechanisms, encompassing cell wall openings, endocytosis, plasmodesmata, and ion transporters. It is important to note that the mechanisms governing the movement and entry of NPs into plant cells are significantly influenced by the size of the NPs.

Cell Wall Pores

Regardless of the plant part, the primary hurdle for NPs is the cell wall (in the epidermis, an extra layer known as the cuticle, mainly composed of lipid substances, overlays the outer periclinal walls). Navigating this barrier is a prerequisite for NPs to access the cell. The dimensions of the wall's pores control the passage of molecules. Pore dimensions typically span from 1.6 to 4.6 nm, with variations influenced by the plant species and the developmental stage of the cells. Measured dimensions display variability: 3.5–3.8 nm for *Raphanus sativus* hair cells and fibers, 3.8–4 nm for *Acer pseudoplatanus* cells, and 4.5–5.2 nm for isolated cells within the palisade parenchyma leaves of *Xanthium strumarium* and *Commelina communis* (Hu et al. 2020). Although studies have documented pore sizes ranging from 2 to 20 nm, the upper limit of 20 nm has been achieved only under specific wall treatment conditions (Banerjee et al. 2019).

Endocytosis

The process of endocytosis involves the cellular uptake of a substance by engulfing it. Plant absorption of nanoparticles (NPs) is believed to follow a more prevalent mechanism. Upon direct contact with root or leaf surfaces, particles adhere to these surfaces, a phenomenon often termed wrapping or docking (Anjum et al. 2013; Banerjee et al. 2019). To facilitate effective wrapping, the cell surface should exhibit a certain level of curvature. This adherence occurs through a combination of interactions such as hydrophobic contacts, electrostatic forces, van der Waals attractions, and/or binding to ligand receptors (Tripathi et al. 2017). Moreover, these forces initiate the process of endocytosis, where intracellular vesicles form and are subsequently engulfed by the plasma membrane (Tilney et al. 1991; Milewska-Hendel et al. 2016). While endocytosis is feasible for nanoparticles of various sizes, the most rapid movement has been observed for particles falling within the size range of 20–50 nm (Erwee and Goodwin 1985; Nair et al. 2012).

Plasmodesmata

Plasmodesmata offer cells an intricate intracellular path of cytoplasmic bridges. Endomembrane continuity between cells is achieved by the endoplasmic reticulum (ER), which connects the ER of adjacent cells and forms a tube that passes through the plasmodesmata. The plasmodesmata serve as tiny channels through which different chemicals and water can pass through. They are cylinder-shaped and around 40 nm broad structurally (Faulkner et al. 2009). For NPs larger than 10 nm (Tao et al. 2017) that cannot fit through the pores in the cell wall, the plasmodesmata may act as an alternative route of transportation (Demirer et al. 2019). As a result, the plasmodesmata can carry 40 nm particles. They may also allow for the passage of larger particles.

However, the precise mechanism regulating the movement of larger particles is elusive. To transport larger particles, several chemicals have been involved to widen the plasmodesmata channels efficiently (Tighe-Neira et al. 2018). To permit the passage of bigger particles and molecules, plasmodesmata's outer aperture may widen (Tani and Barrington 2005). For instance, by adjusting the diameter of the passage, plant virus particles can pass through the plasmodesmata (Tani and Barrington 2005). Similarly, the plasmodesmata may also allow the entry of larger NPs by altering their breadth due to physical pressure generated by a concentration gradient (Sun et al. 2020). Larger particles, however, are thought to move far more slowly.

Ion Transporters

Transport carrier proteins and ion transporters could also contribute to the traversal of NPs across the cell wall and plasma membrane (Sun et al. 2020). The significance of ion transporters in facilitating NP transport within the root cells has been highlighted in a study conducted by Raliya et al. (2016).

2.2.7 Modes of Applications

The two main routes for the uptake, transport, and accumulation of nanoparticles are as follows:

Foliar Spray of Nanoparticles

In agriculture, foliar-applied NPs are frequently utilized as biosensors, post-harvest preservatives, fertilizers, herbicides, and nutritional supplements for plants (Sun et al. 2020; El-Batal et al. 2016). Because plants use soil-applied fertilizers infrequently due to rainfall and adsorption on soil complexes, chemical fertilizers are applied more frequently, which causes eutrophication (Banerjee et al. 2019). Plants develop poorly when fertilizers have limited solubility because they are unable to absorb adequate trace elements. Foliar NPs have been shown in studies to have antimicrobial properties. Due to its antibacterial characteristics, silver can be a key ingredient in spray treatments that are used to treat different diseases of plants caused by fungi and molds (Li et al. 2017). A number of biologically generated NPs, such as silver NPs produced by bacteria isolated from soil, are utilized as fungicides on a large scale (Sun et al. 2020). These NPs also have high antibacterial activity (Kim et al. 2017). Inorganic metal-based NPs, including CuZn bimetallic nanoparticles, are frequently utilized as insecticides and antibacterial agents since they also have antibacterial characteristics (Khan et al. 2019).

Root Application

The application of nanoparticles to roots has been extensively investigated in contrast to their foliar application. When nanoparticles are applied to roots, they pass through the cell wall pores and the porous cuticle region using a passive mechanism. Several factors contribute to the complexity of the nanoparticle uptake in plants via root application, including nanoparticle size, surface modifications (such as positive and negative charges), physical attributes, exposure conditions, plant species, growth stage, root health, and rhizospheric processes (Mishra et al. 2017). Within the root environment, nanoparticles encounter a range of challenges, involving both physiological barriers like root hairs and external elements such as root-associated bacteria, mycorrhizae, mucin/exudates, and the cuticle of the root periderm. The process of nanoparticles entering the cell membrane involves interactions with these barriers through mechanisms like fluid-phase endocytosis, passive diffusion, or mechanical penetration. This leads to their accumulation within the mucilage, cuticle, and cell wall. Microorganisms in the rhizosphere contribute to nanoparticle transport, facilitated by actions like mucilage exudation or biomineralization (El-Batal et al. 2016). After entering through root pores, nanoparticles navigate the plant cell membrane, initiating diffusion through two distinct pathways: (1) traveling through the space between the cell wall and plasma membrane and (2) moving through the intercellular space, bypassing direct cell membrane entry. The apoplast and symplast, which connect cell walls and intercellular gaps, along with protoplasmic connections facilitated by ion channels, are recognized as crucial routes when nanoparticles approach the root epidermis (Sun et al. 2020; Antonoglou et al. 2018). The apoplastic route plays a key role in transporting nanoparticles to above-ground plant segments and significantly contributes to their radial movement within plant tissues. As nanoparticles penetrate the central cylinder, they can also be transported through the xylem, utilizing the transpiration stream to reach aerial parts. However, the Casparian strip, mainly associated with the symplastic pathway, poses an obstacle for nanoparticles attempting to take the apoplastic route from the roots to the xylem. Comprising cell wall components bonded to the vascular system and sealed by lipophilic hydrocarbons, this strip acts as a barrier, particularly for larger nanoparticles (Antonoglou et al. 2018; Lv et al. 2019). The symplastic pathway, involved in water and nutrient transport through phloem sieve tube components, enables nanoparticles to move to non-photosynthetic tissues and organs. Within this route, nanoparticles enter the cytoplasm or neighboring cells through plasmodesmata, specialized structures connecting the cytoplasm of adjacent cells and facilitating intercellular communication (Fig. 2.3).

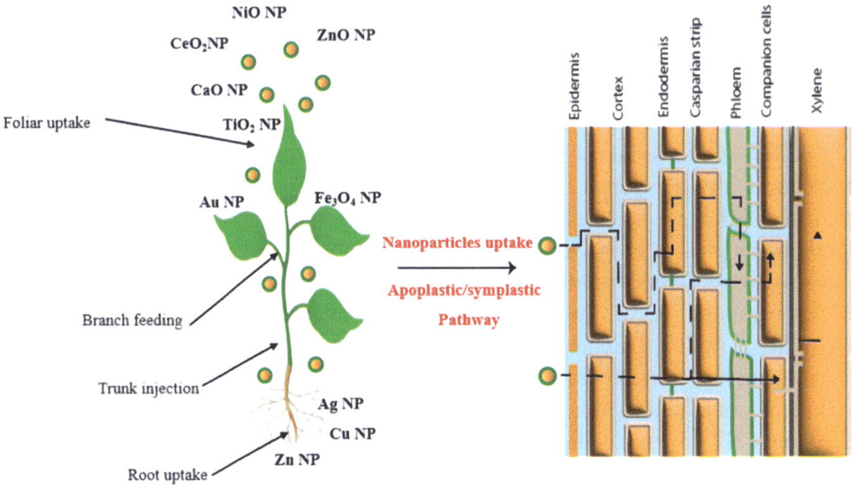

Fig. 2.3 Uptake of nanoparticles and their translocation within the root cells of the plant

2.2.8 Interactions Between NPs and Plants

The external covering of the plant continues to shield plant sections from outside elements. NPs, on the other hand, are smaller and can more easily pass through the cell membrane of plants, which is nearly permeable to macromolecules in the 3.5–20 nm range. Through the formation of vesicles around the transport component or endocytosis, NPs can enter plant cells through cytoplasmic membrane proteins, ionic channels, or both. Root hairs in plant roots allow for the passage of solutes through the tips, whereas suberin renders the remaining surface of the root waterproof. These holes make it simple for NPs to be transported and absorbed. It is therefore reported that NPs may enter the leaves through the stomatal holes. The presence of stomata on the dorsal side of the leaves, however, makes it more difficult for airborne particles to enter the leaf (Lv et al. 2019; Schwab et al. 2016). Stomata can be easily penetrated by NPs smaller than 43 nm (El-Batal et al. 2016; Schwab et al. 2016).

2.3 Nanoparticles' Interaction with Plant Pathogens

There is a suggestion that bacteria can encounter natural nanoparticles (NPs) without experiencing adverse effects. Much of the research has concentrated on the potential medical applications of NPs, particularly their antimicrobial attributes. Only a limited number of studies have explored how CuO or ZnO (Chichiriccò and Poma 2015); CeO_2, Fe_3O_4, or SnO_2 (Chichiriccò and Poma 2015; Rousk et al. 2012); and Ag-based NPs (Vittori Antisari et al. 2013) can impact microbial communities. For instance, CuONPs inhibit pyoverdine levels in *Pseudomonas chlororaphis* O6 cells, while ZnONPs increase them (Rousk et al. 2012). The primary metabolic pathways of bacteria exhibit varying alterations upon exposure to Ag, CuO, and ZnONPs (Chichiriccò and Poma 2015; Rousk et al. 2012; Vittori Antisari et al. 2013). The antibacterial role of AgNPs has long been recognized within the scientific community (Pawlett et al. 2013; Prasad and Swamy 2013). However, their antimicrobial efficacy, tied to the concentration of free Ag^+ ions, is mitigated by components present in the soil, leading to divergent antimicrobial behavior in field conditions due to the formation of AgCl, which reduces toxicity. The interaction of carbon nanotubes (CNTs) with microbial populations remains incompletely understood. Some findings suggest that carbon nanomaterials adversely affect soil bacteria, influencing nutrient availability for microbes and potentially diminishing the toxicity of metabolites. While the precise mechanism remains unclear, it is postulated that CNTs interact with cell membranes, leading to the disruption of microbial cells through oxidative stress or physical damage to the membrane (Eichert et al. 2008; Schreiber 2005). The antibacterial properties of NPs have opened avenues for the development of nanotechnology-based formulations targeting plant pathogens (Dimkpa et al. 2012; Soenen et al. 2011). These NPs hold a promise as substitutes for conventional agrochemicals, indirectly fostering plant growth through disease management. It's important to note that many of these experiments have primarily been conducted in controlled bacteriological media, either liquid or solid, under optimal growth conditions. Nevertheless, it's reasonable to anticipate that the effects of NPs in environmental settings will be more pronounced due to nutrient limitations and suboptimal growth conditions.

2.3.1 Mechanisms of Action of Nanoparticles Against Plant Pathogens

The precise mechanisms by which NPs combat plant pathogens remain largely unclear. Among the processes identified so far are the generation of reactive oxygen species (ROS) and subsequent antioxidant degradation, disruption of membranes (including membrane integrity and permeability), genotoxicity, and inhibition of transporter gene expression (Prasad and Swamy 2013; Aziz et al. 2015; Ibrahim et al. 2020). These pathways often interconnect and collaborate to counteract various phytopathogens. As the charges on nanoparticles (NPs) diminish, an electrostatic attraction forms between negatively charged NPs and bacterial cell membranes, leading to NPs adhering to these membranes. The unique attributes of NP-influenced membranes, coupled with membrane depolarization, disrupt membrane permeability and respiratory activity, ultimately resulting in cell demise by compromising essential cellular components. This disruption of cellular structures triggers the release of enzymes, proteins, metabolites, and DNA from the cell. Furthermore, NPs are capable of generating irregular perforations in the cell walls of microorganisms, enabling their infiltration into both extracellular and intracellular compartments (Aziz et al. 2016; Masum et al. 2019). This impairment of the cell surface and membranes is observable through transmission electron microscopy (TEM). TEM images depict the distortion of cell walls, accompanied by swelling and the release of nuclear and cytoplasmic contents (Dinesh et al. 2012; Jackson et al. 2019; Prasad et al. 2017). Similar effects have been noted in other plant fungi, such as *Alternaria alternata* and *Trichosporon asahii*, in response to silver nanoparticles (AgNPs). Research by Prasad et al. (2017), Alghuthaymi et al. (2015), and Zhang et al. (2018) suggests that the toxicity of NPs may originate from the generation of reactive oxygen species (ROS). These free radicals possess the capability to induce harm to DNA, proteins, lipids, cell walls, and other essential biological components. DNA damage may manifest as deletions, mutations, single-strand breaks, double-strand breaks, and protein cross-linking (Jackson et al. 2019; Hajipour et al. 2012). Both the impairment of cell membranes and the production of ROS function as antimicrobial mechanisms (Lemire et al. 2013). In TEM analyses, evidence of cell membrane damage and the influence of AgNPs on *Azotobacter vinelandii* cells has been observed, while the generation of hydroxyl radicals due to AgNPs has been detected in bacterial cells through the use of electron spin resonance (Prasad et al. 2017; Lemire et al. 2013).

2.3.2 Nanoparticles in Plant Disease Management

Numerous studies in the realm of nanotechnology have employed nanoparticles (NPs) derived from metalloids, metallic oxides, nonmetals (both in single and composite forms), and carbon nanomaterials for the purpose of managing plant diseases. These various NPs encompass antifungal and antibacterial properties, with some even offering nutritional benefits to plants and enhancing their resilience against diseases. The initial exploration into NPs for plant disease management centered on silver (Ag) NPs. Researchers such as Park et al. (2016), Lamsal et al. (2011), and Kim et al. (2008) were pioneers in investigating the use of AgNPs against powdery mildew, achieving notable disease suppression. Notably, Lamsal et al. (2011) observed that at a concentration of 100 mg/mL, disease control was comparable to traditional fungicide treatment, and even a curative response was evident. Further studies have since corroborated the efficacy of AgNPs against bacteria (Ocsoy et al. 2013), nematodes (Abdellatif et al. 2016; Ardakani 2013; Cromwell et al. 2014; Nassar 2016), and fungal diseases (Aguilar-Méndez et al. 2011; Gajbhiye et al. 2009; Jo et al. 2009; Jung et al. 2010; Mallaiah 2015). Given its established history as a contact bactericide/fungicide in plant disease management (Chenthamara et al. 2019; Evans et al. 2007; Kanhed et al. 2014), the application of copper (Cu) NPs emerged as another logical avenue. However, serious investigation into the potential of CuNPs for disease management only gained momentum in 2013 (Evans et al. 2007). This exploration involved comparing foliar applications of CuONPs, Cu_2O, and Cu/Cu_2O composites with conventionally registered Cu fungicides in controlling late blight disease caused by *Phytophthora infestans* (Table 2.1). Interestingly, CuONPs at concentrations of 150–340 mg/mL positively influenced plant performance. Copper is an essential element for plants, critical for both growth and disease control. Key proteins linked to host defense, such as plastocyanins, peroxidases, and multi-Cu oxidases, rely on copper and are generated in response to pathogen infections (Kanhed et al. 2014). Furthermore, NP Si has garnered increased attention across various systems due to its recognized impact on enhancing plant health (Giannousi et al. 2013).

Numerous works have reported that ZnNPs exhibit antibacterial activity against bacteria (Giannousi et al. 2013; Graham et al. 2016), a variety of fungi (Indhumathy and Mala 2013; He et al. 2011), and viral infections (Hao et al. 2018). However, there aren't many field studies demonstrating a real decrease in disease. On rose leaves, NP of Zn has been used to prevent bacterial leaf spots (He et al. 2011). Sugar beet's *Cercospora* leaf blight (Dubchak et al. 2010) and sorghum's *Fusarium graminearum* (Wani and Shah 2012; Paret et al. 2013a) are two examples. For usage in preventing citrus canker incited by *Xanthomonas citri* subsp. *citri*, one Zn-based nanoproduct (Zinkicide) is now going through the registration procedure (Joudeh and Linke 2022; Hu et al. 2020). In field tests, classic cuprous oxide (Cu_2O) and cuprous oxide/zinc oxide (Cu_2O/ZnO) bactericides were less effective at reducing the incidence of the disease. Additionally, grapefruit melanose (*Diaporthe citri*) and citrus scab (*Elsinoe fawcettii*) are successfully controlled with Zinkicide treatments.

Table 2.1 The use of nanoparticles for effective management of plant fungal diseases

NPs	Particle size (nm)	Concentration used	Pathogen	Crop diseases	Affected crops	Effects	References
Ag	~20	0.12–10 ppm	*Fusarium culmorum*	*Fusarium* head blight disease (FHB)	Wheat, barley, rye, corn, triticale, canary seed, and various forage grasses	Impeded the growth of the fungus	Kasprowicz et al. (2010)
Ag	10–50	5, 10, and 15 mg/L	*Alternaria alternata*	Black rot, black spot	Fruits and vegetables	Strong antifungal action	Kasprowicz et al. (2010), Kim et al. (2012)
Ag	10–50	5, 10, and 15 mg/L	*Sclerotinia sclerotiorum*	White mold	Legumes, sunflowers, canola, most vegetables, tobacco, many flowering bedding plants, and stone fruits	Strong antifungal action	Lamsal et al. (2010)
Ag	10–50	5, 10, and 15 mg/L	*Curvularia lunata*	*Curvularia* leaf spot	Maize	Strong antifungal action	Lamsal et al. (2010)
Ag	10–50	5, 10, and 15 mg/L	*Macrophomina phaseolina*	Damping off, seedling blight, collar rot, stem rot, charcoal rot, basal stem rot, and root rot	Peanut, cabbage, pepper, chickpea, soybean, sunflower, sweet potato, alfalfa, sesame, potato, sorghum, wheat, and corn	Strong antifungal action	Kim et al. (2012)
Ag	10–50	10 and 15 mg/L	*Botrytis cinerea*	Gray mold disease	Mainly dicotyledonous plant species	Reduction in the rate of microbial growth	Jo et al. (2009)
Ag	10–50	5, 10, and 15 mg/L	*Rhizoctonia solani*	Sheath blight	Rice	Antifungal activity	Krishnaraj et al. (2012)
$AgNO_3$	20–30	25–200 ppm	*Bipolaris sorokiniana*	Black point, root rot, crown rot, and spot blotch	Wheat	Suppressed colony formation	Jo et al. (2009)

(continued)

Table 2.1 (continued)

NPs	Particle size (nm)	Concentration used	Pathogen	Crop diseases	Affected crops	Effects	References
AgCl	20–30	25–200 ppm	*Magnaporthe grisea*	Blast disease	Rice and other species of the Poaceae family	Suppressed colony formation	Krishnaraj et al. (2012)
Cu	30	2.6 and 4 wt%	*Micrococcus luteus*, *Issatchenkia orientalis*	Leafspot	*Mangifera indica*	Antifungal activity	Kanhed et al. (2014)
ZnO	58	1.2–2.9 μg/mL	*Aspergillus flavus*	Ear rot disease	Maize kernel	Inhibit growth	Jayaseelan et al. (2012)
ZnO	27	25, 50, and 100 μg/mL	*Fusarium oxysporum*	*Fusarium* wilt	Sweet potatoes, tomatoes, legumes, melons, and bananas	Efficacy in inhibiting fungal growth	Arciniegas-Grijalba et al. (2017)
SiO_2			*Blumeria graminis*	*Powdery mildew*	Grasses, cereals	Higher inhibition rate	Suriyaprabha et al. (2014)
CuO and Cu_2O	11–55	35, 54, 110, and 224 g/ml	*Phytophthora infestans*	*Late blight*	Potato	Antifungal activity	Kanhed et al. (2014)
CuO	20	250, 500, and 750 mg/kg	*Colletotrichum graminicola*, *Sclerotium rolfsii*	Anthracnose disease, southern blight	Maize, ornamental crops, vegetables, fruits	Suppresses the development of colonies	Kanhed et al. (2014)

It has been demonstrated that the usage of NP Al_2O_3 (Datnoff et al. 2007; Shenashen et al. 2017), MgO (Imada et al. 2016), NP TiO_2 (Boxi et al. 2016; De Filpo et al. 2013; Elmer and White 2016; Owolade et al. 2008; Paret et al. 2013b), NP S (Choudhury et al. 2010; Gogoi et al. 2013; Rao and Paria 2013), and NP CeO_2 (Adisa et al. 2018) is effective in reducing diseases in many systems (Table 2.2). The majority of research has been able to directly link NP amendment to altered host defense mechanisms.

Table 2.2 Nanoparticles as potential agents for managing plant bacterial diseases

Nanoparticles	Size	Pathogens	Effects	References
AgNPs	35–500 nm	*S. aureus*, *Proteus vulgaris*	Strong antibacterial activity	Patra and Baek (2017), Bryaskova et al. (2011)
AgNPs	14 nm	*E. coli*, *P. aeruginosa*	Sporicidal and bactericidal action	Safavi et al. (2011)
AgNPs	500 nm	*Listeria monocytogenes*, *Staphylococcus aureus*, *Escherichia coli*	Suppress the growth of bacteria	Patra and Baek (2017)
ZnONPs	13 nm	*S. aureus*	Inhibited bacterial growth	Jayaseelan et al. (2012)
ZnNPs	57.72 nm	*S. aureus*, *E. faecalis*, *S. pyogenes*, *P. aeruginosa*	Antibacterial activity	Jayaseelan et al. (2012)
ZnNPs	<100 nm	*Campylobacter jejuni*	Effective against bacteria (in vitro)	Tayel et al. (2011)
ZnNPs	–	*P. aeruginosa*, *K. pneumoniae*, *P. vulgaris*, *Bacillus megaterium*	The surface of the bacteria becomes tightly bound by stimulation and accumulates in biofilm, resulting in obvious cell damage	Tayel et al. (2011), Yousef and Danial (2012)
ZnNPs	15–32 nm	*Klebsiella pneumoniae*, *Staphylococcus epidermidis*, *Klebsiella aerogenes*, *Candida albicans*	Antibacterial effect	Tayel et al. (2011), Yousef and Danial (2012), Palanikumar et al. (2014)
CuNPs	–	*Xanthomonas axonopodis* pv. *punicae*	Completely damaged the cells	Mondal and Mani (2012)
CuNPs	50 nm	*Staphylococcus aureus*, *Listeria*	Antibacterial activity	Cioffi et al. (2004)

(continued)

Table 2.2 (continued)

Nanoparticles	Size	Pathogens	Effects	References
CuONPs	20–27 nm	*P. aeruginosa*, *B. subtilis*, *S. aureus*, *E. coli*	Antibacterial activity	Azam et al. (2012)
CuONPs	5–10 nm	*S. aureus*	Good antimicrobial properties	Jadhav et al. (2011)
CuONPs	20–30 nm	*B. subtilis*, *S. aureus*	Extremely toxic	Jadhav et al. (2011), Baek and An (2011), Agarwala et al. (2014)
CuONPs	25–30 nm	*K. pneumoniae*, *Pseudomonas* spp.	Pathogens (MRSA and *E. coli*) that produce biofilms have been significantly reduced	Azam et al. (2012)
α-Fe_2O_3NPs	2–540 nm	*Pseudomonas aeruginosa*	Enhanced growth of microorganisms and the establishment of biofilm structures	Borcherding et al. (2014)
Maghemite (Fe_2O_3) NPs	25–30 nm	*Klebsiella pneumoniae*, *Pseudomonas* spp., *Proteus mirabilis*	Antibacterial activity	Al-Hazmi et al. (2012)
Magnesium oxide (MgO) nanowires	6 nm	*Bacillus* sp., *E. coli*	Strong bacteriostatic action	Al-Hazmi et al. (2012)
Nickel oxide (NiO) NPs	10–20 nm	*B. subtilis* and *S. aureus*	Extremely toxic	Agarwala et al. (2014)
NiONPs	20–30 nm	*Bacillus subtilis*, *P. aeruginosa*	Antibacterial activity	Agarwala et al. (2014), Borcherding et al. (2014)
Silica-silver NP	20 nm	*Xanthomonas campestris*, *Azotobacter chroococcum*, *Pseudomonas syringe*	Inhibiting results	Park et al. (2006)
CaONPs	15–180 nm	*Lactobacillus plantarum*	Substantial bactericidal action	Tang et al. (2013)
CaONPs	16 nm	*Candida tropicalis*, *P. aeruginosa*, *Staphylococcus epidermidis*	Gram-positive strains are more susceptible to antibacterial action than Gram-negative ones	Roy et al. (2013)
CeO_2NPs engineered	6–40 nm	*Bacillus subtilis*	Reduced the growth of *B. subtilis* and *E. coli*	Pelletier et al. (2010)
Chromium (III) oxide (Cr_2O_3) NPs	65 nm	*P. aeruginosa*, *E. coli*	Bacterial inhibition	Rakesh et al. (2013)
	79 nm	*P. aeruginosa*, *E. coli*	Bacterial inhibition	Rakesh et al. (2013)

Table 2.2 (continued)

Nanoparticles	Size	Pathogens	Effects	References
TiO_2NPs	<50 nm	*Methicillin-resistant Staphylococcus aureus*	Antibacterial effects	Jesline et al. (2015), Behera et al. (2012)
Iron oxide (IO) NPs	10–120 nm	*S. aureus*, *S. epidermis*, *B. subtilis*, *Bacillus licheniformis*, *Brevibacillus brevis*, *Vibrio cholerae*, *Pseudomonas aeruginosa*	Medium level of antibacterial activity	Borcherding et al. (2014)

2.3.3 Roles of Nanoparticles in Plant Viral Disease Management

Plant viruses have a significant detrimental impact on agriculture. It is critical to take significant action to combat viral infections (Shafie et al. 2018). Approximately 50% of newly emerging plant diseases are thought to be caused by viruses, and agriculture plays a significant role in the spread of these diseases (Jain 2014). Over 700 different crop species have been found to be infected by over 900 different plant viral species (El-Sawy et al. 2017; El-Dougdoug et al. 2018). The severity of a virus' effect on a plant depends on the stage of infection, habitat, strain, type of host, rate of reproduction, and concentration (Elazzazy et al. 2017). Viral infections can manifest as mottling, drying of leaves, distortion of fruit, necrosis, stunting of plants, and even mortality (Shafie et al. 2018; El-Dougdoug et al. 2018) in conditions that are favorable to the virus. Because diseases are difficult to treat and viruses have no known treatments, the primary alternatives for disease control are to prevent or avoid infection (El-Sawy et al. 2017; El-Dougdoug et al. 2018; Elazzazy et al. 2017). Currently, novel antiviral chemicals for use in animals and plants have been discovered in nanoscale materials (Shafie et al. 2018; Elazzazy et al. 2017; Elbeshehy et al. 2015). Nanoparticles enhance the plant's disease resistance through the activation of plant defense mechanisms or the inactivation of microorganisms, demonstrating promising antiviral effects against a range of plant viruses in various NP-based assays (Table 2.3).

Table 2.3 The use of nanoparticles (NPs) for combatting plant viral diseases

Nanoparticles (NPs)	Size (nm)	Target pathogen	Host plant	Effects	References
Silver nanoparticles (AgNPs)	12.6 ± 5 nm	Tomato-spotted wilt virus	*Solanum tuberosum*	TSWV infection is decreased, and local lesions are prevented	Shafie et al. (2018)
AgNPs	10–20 nm	Sunn-hemp rosette virus	*Cyamopsis tetragonoloba*	Total repression of decrease	Jain (2014)
AgNPs	12 nm	Potato virus Y	*Solanum tuberosum*	Development of viral resistance in the host plant	El-Dougdoug et al. (2018)
AgNPs	13 nm	Potato virus Y	*Solanum lycopersicum*	Decreased viral infection and disease severity	Elazzazy et al. (2017)
AgNPs	13 nm	Tomato mosaic virus	*Solanum lycopersicum*	Lessening of the disease's intensity	El-Dougdoug et al. (2018)
Graphene oxide-silver NPs	3050 nm	Tomato bushy stunt virus	*Lactuca sativa*	Reduced virus prevalence, infection rate, and disease severity	Elazzazy et al. (2017)
AgNPs	77 nm	Bean yellow mosaic virus	*Vicia faba*	Reduction in the prevalence and severity of the disease as well as the viral concentration lesions on diseased leaves are less visible	Elbeshehy et al. (2015)
AuNPs		Barley yellow mosaic virus	*Hordeum vulgare*	In vitro viral particle dissociation	Jain (2014)
Schiff base nano-silver NPS	–	TMV	*Nicotiana tabacum*	TMV-induced infection decreases	El-Sawy et al. (2017), Elazzazy et al. (2017)
AuNPs	3.151–31.67 nm	Barley yellow dwarf virus—PAV	*Hordeum vulgare*	Destroys the gold barley yellow dwarf virus and gets rid of the risks of infection	Aref et al. (2012)
Zinc oxide NPs (ZnONPs)	18 nm	Tobacco mosaic virus	*Nicotiana benthamiana*	Slowing the rate of TMV invasion	Cai et al. (2019)

Table 2.3 (continued)

Nanoparticles (NPs)	Size (nm)	Target pathogen	Host plant	Effects	References
TiO_2NPs	20 nm	Turnip mosaic virus	*Nicotiana benthamiana*	Effectively reduces viral multiplication and infection	Elsharkaway and Derbalah (2018)
Cerium oxide NPs (CeONPs)		Tobacco mosaic virus	*Datura stramonium* and *Nicotiana tabacum*	Decreased viral symptoms	Eugene and Zholobak (2016)
Iron oxide NPs (Fe_3O_4)	0.19 nm	TMV	*N. benthamiana*	Upregulation of SA genes induces plant resistance and activates antioxidants against TMV	Eugene and Zholobak (2016)
Silicon dioxide NPs (SiO_2NPs)	–	Papaya ringspot virus	*Cucumis sativus*	Lessening of symptoms and viral accumulation	Elsharkawy and Mousa (2015)
SiO_2NPs	20 nm	Tobacco mosaic virus (TMV)	*Nicotiana tabacum*	Slowing down the rate of TMV invasion	Elsharkawy and Mousa (2015)
Silicon dioxide NPs (SiO_2NPs)	100 nm	Tomato yellow leaf curl virus (TYLCV)	*Solanum lycopersicum*	Reductions in TYLCV levels and symptom severity	Elsharkawy and Mousa (2015)
Iron oxide NPs (Fe_2O_3NPs)	40–100 nm	Turnip mosaic virus	*Nicotiana tabacum*	Effectively reduces viral multiplication and infection	El-Dougdoug et al. (2018)
Nickel oxide NPs (NiONPs)	15–20 nm	Cucumber mosaic virus	*Cucumis sativus*	Reduces CMV concentration and severity of the disease	Derbalah and Elsharkawy (2019)
Carbon nanotubes (CNTs)	–	TMV	*N. benthamiana*	TMV replication and movement are inhibited, and resistance is induced	Adeel et al. (2021)
ZnONPs	–	*Solanum melongena*	Cucumber mosaic virus	Disease incidence and severity were markedly reduced	El-Sawy et al. (2017)
Titanium dioxide NPs (TiO_2NPs)	~3–5 μm	Broad bean stain virus (BBSV)	Fabaceae	Significantly reduction in disease severity	Elbeshehy et al. (2015)

2.4 Toxicological Effect of Nanoparticles

In spite of the various applications of nanoparticles (NPs) across industries like agriculture and medicine, their potential toxicological impacts (Bahadar et al. 2016; Ibrahim 2013; Khlebtsov and Dykman 2011) demand a foundational understanding. NPs can inadvertently enter ecosystems through human activities, permeating water bodies, soils, and the atmosphere. Conversely, the controlled introduction of NPs occurs when they are deliberately employed for environmental treatment, being released into the soil or aquatic systems. While magnetic NPs offer benefits like compact size, potent reactivity, and substantial capacity, these very attributes might lead to cellular toxicity and harmful consequences uncommon in larger counterparts. Studies show that NPs can infiltrate organisms via inhalation or ingestion, traveling within the body to various organs and tissues, potentially inducing reactive toxic effects. However, investigations into the toxicological effects of magnetic NPs on plants remain scarce, despite studies addressing their impact on both animal and plant cells. Silver NPs, widely used in consumer goods, enter aquatic environments, releasing dissolved silver and triggering toxic repercussions in aquatic organisms such as bacteria, algae, fish, and daphnia (Khlebtsov and Dykman 2010). Although NPs have gained traction in biological applications and nanobiotechnology, the potential adverse outcomes of prolonged exposure to varying environmental NP concentrations on human health remain uncertain. Nevertheless, it's foreseeable that NPs' environmental influence will expand over time. Notably, NPs possess the ability to associate with protein concentrations, influenced by particle size, charge, curvature, shape, surface properties, functional groups, and free energy. This interaction may lead to protein unfolding, fibrillation, thiol cross-linking, and enzyme activity loss, exerting harmful biological effects. Additionally, some NPs can release hazardous ions when their thermodynamic characteristics prompt particle dissolution in suspension media or biological environments (Xia et al. 2008). In aquatic ecosystems, the behavior of NPs is influenced by the presence of organic matter and colloidal particles, which can affect their dispersion, aggregation, and ecotoxicity. Variables such as pH, salinity, and organic substances play roles in this regard, warranting comprehensive examination as part of ecotoxicological inquiries (Navarro et al. 2008). Numerous detrimental impacts of nanoparticles (NPs) on plants, animals, and the ecosystem have been recognized. Nanopesticides can accumulate on plant surfaces, inhibiting pollen germination, sealing stomata, and obstructing pollen tubes from penetrating stigma. NPs are believed to enter vascular tissues, potentially blocking nutrient and water flow. Exposure to nano-silver has demonstrated detrimental impacts on both plants and microorganisms, altering microbial community composition, biomass, and extracellular enzyme activity (Handy et al. 2008). Further research is essential to comprehend NPs' interaction with other environmental pollutants, their entry into the food chain, and subsequent bioaccumulation. NPs have the potential to infiltrate human systems and harm vulnerable organs. Although the use of chemically engineered nanomaterials is widely acknowledged as detrimental to the environment, green nanotechnology provides a solution by producing NPs from plant systems (Zhang et al. 2015; Maity and Pramanick 2020). This approach minimizes energy consumption, waste generation, and greenhouse gas emissions, aligning with sustainable practices and exhibiting reduced environmental impact (Maity and Pramanick 2020; Lv et al. 2019).

2.5 Conclusion and Future Perspectives

Nanotechnology and nanoparticle sciences have the potential to revolutionize the current technology employed in the control of plant diseases and can offer new solutions for future agricultural applications. The application of nanotechnology in disease diagnostics, disease control, or increasing overall plant health by immune elicitors will play a significant part in the creation of numerous new techniques to improve plant health. Increased bioavailability and efficiency of pesticides, when loaded onto nanoparticles, can minimize toxicity, enhanced shelf-life, and controlled distribution of active compounds which are just a few of the unheard-of benefits of developing nanopesticides. The most creative and promising solutions rely heavily on nanotechnology and nanoscale sciences. Nanoscale materials change their biological, physical, and chemical characteristics, and this could help greatly in the development of some pesticides and make it possible to be more effective to combat weeds, disease infections, and other pest outbreaks.

To effectively manage the controlled release of synthetic chemicals from nanocarriers, nanomaterials, nanotubes, and microcapsules can hold higher amounts of active pesticide components, compounds that stimulate host resistance, inhibitors of polyamine synthesis, and more. Enhancing the effectiveness of diagnostic probes for detecting plant diseases can be achieved by employing enzyme-based biosensors coated with nanoparticles such as Au, Ag, Co, and Ti. Concerns over biosafety and phytotoxic behavior of nanomaterials, particularly with regard to seed germination, seedling growth, and fruit setting, must be addressed carefully by the scientist. Although there are concerns about the various negative effects of nanomaterials on plants, animals, and the environment, the development of nanopesticides is one potential future option for increasing global agricultural production and lowering chemical inputs. The development and standardization of nanomaterial application techniques are also required in order to fully realize the benefits and promises of nanomaterials in the control of plant diseases in the field globally.

References

Abdel-Aziz HMM, Hasaneen MNA, Omer AM (2016) Nano chitosan-NPK fertilizer enhances the growth and productivity of wheat plants grown in sandy soil. Span J Agric Res 14:17

Abdellatif KF, Abdelfattah RH, El-Ansary MSM (2016) Green nanoparticles engineering on root-knot nematode infecting eggplants and their effect on plant DNA modification. Iran J Biotechnol 14:250–259

Adeel M, Farooq T, White JC, Hao Y, He Z, Rui Y (2021) Carbonbased nano-materials suppress tobacco mosaic virus (TMV) infection and induce resistance in Nicotiana benthamiana. J Hazard Mater 404:124167. https://doi.org/10.1016/j.jhazmat.2020.124167

Adisa IO, Pullagurala VLR, Rawat S, Hernandez-Viezcas JA, Dimkpa CO, Elmer WH, White JC, Peralta-Videa JR, Gardea-Torresdey JL (2018) Role of cerium compounds in Fusarium suppression and growth enhancement in tomato (Solanum lycopersicum). Environ Nano Sci, in press

Agarwala M, Choudhury B, Yadav RNS (2014) Comparative study of antibiofilm activity of copper oxide and iron oxide nanoparticles against multidrug resistant biofilm forming uropathogens. Indian J Microbiol 54(3):365–368

Agrios GN (2005) Plant pathology, 5th edn. Elsevier Academic Press, San Diego

Aguilar-Méndez MA, Martín-Martínez ES, Ortega-Arroyo L, Cobián-Portillo G, Sánchez-Espíndola E (2011) Synthesis and characterization of silver nanoparticles: effect on phytopathogen Colletotrichum gloesporioides. J Nanopart Res 13:2525–2532

Alghuthaymi MA, Almoammar H, Rai M, Said-Galiev E, Abd-Elsalam KA (2015) Myconanoparticles: synthesis and their role in phytopathogens management. Biotechnol Biotechnol Equip 29(2):221–236

Al-Hazmi F, Alnowaiser F, Al-Ghamdi AA, Al-Ghamdi AA, Aly MM, Al-Tuwirqi RM, El-Tantawy F (2012) A new large–scale synthesis of magnesium oxide nanowires: structural and antibacterial properties. Superlattice Microstruct 52(2):200–209

Anjum NA, Gill SS, Duarte AC, Pereira E, Ahmad I (2013) J Nanoparticle Res. https://doi.org/10.1007/s11051-013-1896-7

Antonoglou O, Moustaka J, Adamakis IS, Sperdouli I, Pantazaki AA, Moustakas M, Dendrinou-Samara C (2018) Nanobrass CuZn nanoparticles as foliar spray nonphytotoxic fungicides. ACS Appl Mater Interfaces 10(5):4450–4461

Anwar S, Ding H, Xu M, Hu X, Li Z, Wang J, Liu L, Jiang L, Wang D, Dong C (2019) Recent advances in synthesis, optical properties, and biomedical applications of carbon dots. ACS Appl Bio Mater 2:2317–2338

Arciniegas-Grijalba PA, Patiño-Portela MC, Mosquera-Sánchez LP, Guerrero-Vargas JA, Rodríguez-Páez JE (2017) ZnO nanoparticles (ZnO-NPs) and their antifungal activity against coffee fungus Erythricium salmonicolor. Appl Nanosci 7(5):225–241

Ardakani AS (2013) Toxicity of silver, titanium and silicon nanoparticles on the root-knot nematode, Meloidogyne incognita, and growth parameters of tomato. Nematology 15(6):671–677

Aref N, Alkubaisi N, Marraiki N, Hindi A (2012) Multi-functional effects of gold nano-particles inducing plant virus resistance crops. In: Proceedings of the 5th annual world congress of industrial biotechnology-2012, Xi'an, China, 25 Apr 2012

Azam A, Ahmed AS, Oves M, Khan MS, Memic A (2012) Size-dependent antimicrobial properties of CuO nanoparticles against gram-positive and-negative bacterial strains. Int J Nanomedicine 7:3527–3535

Aziz N, Faraz M, Pandey R, Sakir M, Fatma T, Varma A, Barman I, Prasad R (2015) Facile algae derived route to biogenic silver nanoparticles: synthesis, antibacterial and photocatalytic properties. Langmuir 31:11605–11612. https://doi.org/10.1021/acs.langmuir.5b03081

Aziz N, Pandey R, Barman I, Prasad R (2016) Leveraging the attributes of Mucor hiemalis-derived silver nanoparticles for a synergistic broad-spectrum antimicrobial platform. Front Microbiol 7:1984. https://doi.org/10.3389/fmicb.2016.01984

Baek YW, An YJ (2011) Microbial toxicity of metal oxide nanoparticles (CuO, NiO, ZnO, and Sb2O3) to Escherichia coli, Bacillus subtilis, and Streptococcus aureus. Sci Total Environ 409(8):1603–1608

Bahadar H, Maqbool F, Niaz K, Abdollahi M (2016) Toxicity of nanoparticles and an overview of current experimental models. Iran Biomed J 20:1–11. https://doi.org/10.7508/ibj.2016.01.001

Banerjee J, Kole C (2016) Plant nanotechnology: an overview on concepts, strategies, and tools. In: Kole C, Kumar D, Khodakovskaya M (eds) Plant nanotechnology. Springer, Cham, pp 1–14

Banerjee K, Pramanik P, Maity A, Joshi DC, Wani SH, Krishnan P (2019) Chapter 4: Methods of using nanomaterials to plant systems and their delivery to plants (mode of entry, uptake, translocation, accumulation, biotransformation and barriers). In: Ghorbanpour M, Wani SH (eds) Advances in phytonanotechnology. Academic Press, Cambridge, MA, pp 123–152

Bassett RK (2007) To the digital age: research labs, start-up companies, and the rise of MOS technology. Johns Hopkins University Press

Behera SS, Patra JK, Pramanik K, Panda N, Thatoi H (2012) Characterization and evaluation of antibacterial activities of chemically synthesized iron oxide nanoparticles. World J Nano Sci Eng 2(4):196

Bhattacharyya A, Duraisamy P, Govindarajan M, Buhroo AA, Prasad R (2016) Nano-biofungicides: emerging trend in insect pest control. In: Advances and applications through fungal nanobiotechnology. Springer, pp 307–319

Binns C (2021) Introduction to nanoscience and nanotechnology. John Wiley & Sons

Borcherding J, Baltrusaitis J, Chen H, Stebounova L, Wu CM, Rubasinghege G, Mudunkotuwa IA, Caraballo JC, Zabner J, Grassian VH, Comellas AP (2014) Iron oxide nanoparticles induce Pseudomonas aeruginosa growth, induce biofilm formation, and inhibit antimicrobial peptide function. Environ Sci Nano 1(2):123–132

Bourassa DJ, Kerna NA, Desantis M (2019) Will nanocarbon onion-like fullerenes (NOLFs) play a decisive role in the future of molecular medicine. J Nanomed Nanosci 3:JNAN-152

Boxi SS, Mukherjee K, Paria S (2016) Ag doped hollow TiO2 nanoparticles as an effective green fungicide against Fusarium solani and Venturia inaequalis phytopathogens. Nanotechnology 27:085103. https://doi.org/10.1088/0957-4484/27/8/085103

Bryaskova R, Pencheva D, Nikolov S, Kantardjiev T (2011) Synthesis and comparative study on the antimicrobial activity of hybrid materials based on silver nanoparticles (AgNps) stabilized by polyvinylpyrrolidone (PVP). J Chem Biol 4:185–191

Cai L, Liu C, Fan G, Liu C, Sun X (2019) Preventing viral disease by ZnONPs through directly deactivating TMV and activating the plant immunity in Nicotiana benthamiana. Environ Sci Nano 6:3653

Chenthamara D et al (2019) Therapeutic efficacy of nanoparticles and routes of administration. Biomater Res 23(1):20

Chhipa H, Joshi P (2016) Nanofertilisers, nanopesticides and nanosensors in agriculture. Nanosci Food Agric 1:247–282

Chichiriccò G, Poma A (2015) Penetration and toxicity of nanomaterials in higher plants. Nanomaterials 5(2):851–873

Cho EJ, Holback H, Liu KC, Abouelmagd SA, Park J, Yeo Y (2013) Nanoparticle characterization: state of the art, challenges, and emerging technologies. Mol Pharm 10:2093

Choudhury SR, Nair KK, Kumar R, Gogoi R, Srivastava C, Gopal M, Subhramanyam BS, Devakumar C, Goswamiet A (2010) Nanosulfur: a potent fungicide against food pathogen, Aspergillus niger. AIP Conf Proc 1276:154–157

Cioffi N, Torsi L, Ditaranto N (2004) Antifungal activity of polymer-based copper nanocomposite coatings. Appl Phys Lett 85(12):2417–2419

Cromwell WA, Yang J, Starr JL, Jo YK (2014) Nematicidal effects of silver nanoparticles on root-knot nematode in bermudagrass. J Nematol 46(3):261

Das L et al (2020) Synthesis of hybrid hydrogel nano-polymer composite using graphene oxide, chitosan and PVA and its application in waste water treatment. Environ Technol Innov 18:100664

Datnoff LE, Rodrigues FA, Seebold KW (2007) Silicon and plant disease, pp 233–246

De Filpo G, Palermo AM, Rachiele F, Nicoletta FP (2013) Preventing fungal growth in wood by titanium dioxide nanoparticles. Int Biodeterior Biodegrad 85:217–222

de Mello Donega C (2014) In: de Mello Donegá C (ed) Nanoparticles. Springer, Berlin. https://doi.org/10.1007/978-3-662-44823-6

Demirer GS, Zhang H, Matos JL, Goh NS, Cunningham FJ, Sung Y, Chang R, Aditham AJ, Chio L, Cho MJ, Staskawicz B, Landry MP (2019) Nat Nanotechnol 14:456–464

Derbalah A, Elsharkawy M (2019) A new strategy to control cucumber mosaic virus using fabricated NiO-nanostructures. J Biotechnol 306:134

Dimkpa CO, McLean JE, Latta DE, Manangón E, Britt DW, Johnson WP et al (2012) CuO and ZnO nanoparticles: phytotoxicity, metal speciation, and induction of oxidative stress in sand-grown wheat. J Nanopart Res 14:1–15

Dinesh R, Anandaraj M, Srinivasan V, Hamza S (2012) Engineered nanoparticles in the soil and their potential implications to microbial activity. Geoderma 173–174:19–27. https://doi.org/10.1016/j.geoderma.2011.12.018

Dubchak S, Ogar A, Mietelski JW, Turnau K (2010) Influence of silver and titanium nanoparticles on arbuscular mycorrhiza colonization and accumulation of radiocaesium in Helianthus annuus. Span J Agric Res 8:S103–S108

Eichert T, Kurtz A, Steiner U, Goldbach HE (2008) Size exclusion limits and lateral heterogeneity of the stomatal foliar uptake pathway for aqueous solutes and water-suspended nanoparticles. Physiol Plant 134(1):151–160

Elazzazy A, Elbeshehy E, Betiha M (2017) In vitro assessment of activity of graphene silver composite sheets against multidrug-resistant bacteria and tomato bushy stunt virus. Trop J Pharm Res 16:2705–2711

El-Batal AI, El-Lateef Gharib FA, Ghazi SM, Hegazi AZ, El Hafz AGMA (2016) Physiological responses of two varieties of common bean (Phaseolus vulgaris L.) to foliar application of silver nanoparticles. Nanomater Nanotechnol 6:13

Elbeshehy EKF, Elazzazy AM, Aggelis G (2015) Silver nanoparticles synthesis mediated by new isolates of Bacillus spp., nanoparticle characterization and their activity against bean yellow mosaic virus and human pathogens. Front Microbiol 6:453

El-Dougdoug NK, Bondok AM, El-Dougdoug KA (2018) Evaluation of silver nanoparticles as antiviral agent against ToMV and PVY in tomato plants. Middle East J Appl Sci 8:100–111

Elmer WH, White J (2016) Nanoparticles of CuO improves growth of eggplant and tomato in disease infested soils. Environ Sci Nano 3:1072–1079

Elmer W, Ma C, White J (2018) Nanoparticles for plant disease management. Curr Opin Environ Sci Health 6:66–70

El-Sawy MM, Elsharkawy MM, Abass JM, Kasem MH (2017) Antiviral activity of 2-nitromethyl phenol, zinc nanoparticles and seaweed extract against cucumber mosaic virus (CMV) in eggplant. J Virol Antivir Res 6:189. https://doi.org/10.4172/2324-8955.1000173

Elsharkaway M, Derbalah A (2018) Antiviral activity of titanium dioxide nanostructures as a control strategy for broad bean strain virus in faba bean: control of broad bean stain virus by TiO2 nanostructures. Pest Manag Sci 75:828–834. https://doi.org/10.1002/ps.5185

Elsharkawy MM, Mousa KM (2015) Induction of systemic resistance against Papaya ring spot virus (PRSV) and its vector Myzus persicae by Penicillium simplicissimum GP17–2 and silica (SiO2) nanopowder. Int J Pest Manag 61:353–358

Erwee M, Goodwin P (1985) Planta 163:9–19

Eugene K, Zholobak N (2016) Antiviral activity of cerium dioxide nanoparticles on tobacco mosaic virus model. In: Proceedings of the topical issues of new drugs development, Kharkiv, Ukraine, Apr 2016

Evans I, Solberg E, Huber DM (2007) Copper and plant disease. In: Datnoff LE, Elmer WH, Huber DN (eds) Mineral nutrition and plant disease. APS Press, St. Paul, MN, pp 177–188

Faulkner CR, Blackman LM, Collings DA, Cordwell SJ, Overall RL (2009) Eur J Cell Biol 88:357–369

Fortina P, Kricka LJ, Surrey S, Grodzinski P (2005) Nanobiotechnology: the promise and reality of new approaches to molecular recognition. Trends Biotechnol 23(4):168–173

Gajbhiye M, Kesharwani J, Ingle A, Gade A, Rai M (2009) Fungus-mediated synthesis of silver nanoparticles and their activity against pathogenic fungi in combination with fluconazole. Nanomed Nanotechnol Biol Med 5(4):382–386

Geisler-Lee J, Brooks M, Gerfen JJ, Wang Q, Fotis C, Sparer A, Ma X, Berg R, Geisler M (2014) Nanomaterials 4:301–318

Georgakilas V, Perman JA, Tucek J, Zboril R (2015) Broad family of carbon nanoallotropes: classification, chemistry, and applications of fullerenes, carbon dots, nanotubes, graphene, nanodiamonds, and combined superstructures. Chem Rev 115:4744–4822

Giannousi K, Avramidis I, Dendrinou-Samara C (2013) Synthesis, characterization and evaluation of copper based nanoparticles as agrochemicals against Phytophthora infestans. RSC Adv 3(44):21743–21752

Giraldo JP, Landry MP, Faltermeier SM, McNicholas TP, Iverson NM, Boghossian AA, Reuel NF, Hilmer AJ, Sen F, Brew JA, Strano MS (2014) Plant nanobionics approach to augment photosynthesis and biochemical sensing. Nat Mater 13:400–408

Gogoi R, Singh PK, Kumar R, Nair KK, Alam I, Srivastava C, Yadav S, Gopal M, Roy SC, Goswamiet A (2013) Suitability of nano-sulphur for biorational management of powdery mildew of okra (Abelmoschus esculentus Moench) caused by Erysiphe cichoracearum. J Plant Pathol Microbiol 4(4):71–175

Gopal MA, Gogoi RO, Srivastava CH, Kumar RA, Singh PK, Nair KK, Yadav SA, Goswami AR (2011) Nanotechnology and its application in plant protection. In: Plant pathology in India: vision 2030. Indian Phytopathological Society, Indian Agricultural Research Institute, New Delhi, pp 224–232

Graham JH, Johnson EG, Myers ME, Young M, Rajasekaran P, Das S, Santra S (2016) Potential of nano-formulated zinc oxide for control of citrus canker on grapefruit trees. Plant Dis 100(12):2442–2447

Guerra FD, Attia MF, Whitehead DC, Alexis F (2018) Nanotechnology for environmental remediation: materials and applications. Molecules 23(7):1760

Hajipour MJ, Fromm KM, Ashkarran AA, de Aberasturi DJ, de Larramendi IR, Rojo T et al (2012) Antibacterial properties of nanoparticles. Trends Biotechnol 30(10):499–511

Handy RD, von der Kammer F, Lead JR, Hassellov M, Owen R, Crane M (2008) The ecotoxicology and chemistry of manufactured nanoparticles. Ecotoxicology 17:287–314. https://doi.org/10.1007/s10646-008-0199-8

Hao Y, Yuan W, Ma C, White J, Zhang Z, Adeel M, Zhou T, Yukui R, Xing B (2018) Engineered nanomaterials suppress turnip mosaic virus infection in tobacco (Nicotiana benthamiana). Environ Sci Nano 5:1685. https://doi.org/10.1039/C8EN00014J

He L, Liu Y, Mustapha A, Lin M (2011) Antifungal activity of zinc oxide nanoparticles against Botrytis cinerea and Penicillium expansum. Microbiol Res 166(3):207–215

Hu P, An J, Faulkner MM, Wu H, Li Z, Tian X, Giraldo JP (2020) Nanoparticle charge and size control foliar delivery efficiency to plant cells and organelles. ACS Nano 14:7970–7986

Ibrahim KS (2013) Carbon nanotubes-properties and applications: a review. Carbon Lett 14:131–144. https://doi.org/10.5714/CL.2013.14.3.131

Ibrahim E, Zhang M, Zhang Y, Hossain A, Qiu W, Chen Y et al (2020) Green-synthesization of silver nanoparticles using endophytic bacteria isolated from garlic and its antifungal activity against wheat Fusarium head blight pathogen Fusarium graminearum. Nanomaterials 10(2):219

Imada K, Sakai S, Kajihara H, Tanaka S, Ito S (2016) Magnesium oxide nanoparticles induce systemic resistance in tomato against bacterial wilt disease. Plant Pathol 65(4):551–560

Indhumathy M, Mala R (2013) Photocatalytic activity of zinc sulphate nano material on phytopathogens. Int J Agric Environ Biotechnol 6(4S):737–743

Islam SMF, Karim Z (2019) Chapter 4: World's demand for food and water: the consequences of climate change. In: Farahani MHDA, Vatanpour V, Taheri AH (eds) Desalination-challenges and opportunities. IntechOpen, London, pp 57–84

Jackson KK, Ochoo L, Maingi JM, Swaleh S, Njue W (2019) Green sonochemical synthesis of silver nanoparticles using Bridelia micrantha extract and evaluation of their antibacterial activity. Afr J Pure Appl Chem 13(3):34–42

Jadhav S, Gaikwad S, Nimse M, Rajbhoj A (2011) Copper oxide nanoparticles: synthesis, characterization and their antibacterial activity. J Clust Sci 22(2):121–129

Jain D (2014) Green synthesis of silver nanoparticles and their application in plant virus inhibition. J Mycol Plant Pathol 44:21–24
Jayaseelan C, Rahuman AA, Kirthi AV, Marimuthu S, Santhoshkumar T, Bagavan A, Gaurav K, Karthik L, Rao KB (2012) Novel microbial route to synthesize ZnO nanoparticles using Aeromonas hydrophila and their activity against pathogenic bacteria and fungi. Spectrochim Acta A Mol Biomol Spectrosc 90:78–84
Jesline A, John NP, Narayanan PM, Vani C, Murugan S (2015) Antimicrobial activity of zinc and titanium dioxide nanoparticles against biofilm-producing methicillin-resistant Staphylococcus aureus. Appl Nanosci 5(2):157–162
Jo YK, Kim BH, Jung G (2009) Antifungal activity of silver ions and nanoparticles on phytopathogenic fungi. Plant Dis 93(10):1037–1043
Joudeh N, Linke D (2022) Nanoparticle classification, physicochemical properties, characterization, and applications: a comprehensive review for biologists. J Nanobiotechnol 20(1):262
Jović D, Jaćević V, Kuča K, Borišev I, Mrdjanovic J, Petrovic D et al (2020) The puzzling potential of carbon nanomaterials: general properties, application, and toxicity. Nanomaterials 10(8):1508
Jung JH, Kim SW, Min JS, Kim YJ, Lamsal K, Kim KS, Lee YS (2010) The effect of nano-silver liquid against the white rot of the green onion caused by Sclerotium cepivorum. Mycobiology 38:39–45
Kanhed P, Birla S, Gaikwad S, Gade A, Seabra AB, Rubilar O, Duran N, Rai M (2014) In vitro antifungal efficacy of copper nanoparticles against selected crop pathogenic fungi. Mater Lett 115:13–17
Kasprowicz MJ, Kozioł M, Gorczyca A (2010) The effect of silver nanoparticles on phytopathogenic spores of Fusarium culmorum. Can J Microbiol 56(3):247–253
Khan MR, Sharma RK (2020) Fusarium-nematode wilt disease complexes, etiology and mechanism of development. Indian Phytopathol 73:615–628
Khan I, Saeed K, Khan I (2019) Nanoparticles: properties, applications and toxicities. Arab J Chem 12(7):908–931
Khlebtsov NG, Dykman LA (2010) Optical properties and biomedical applications of plasmonic nanoparticles. J Quant Spectrosc Radiat Transf 111:1–35. https://doi.org/10.1016/j.jqsrt.2009.07.012
Khlebtsov N, Dykman L (2011) Biodistribution and toxicity of engineered gold nanoparticles: a review of in vitro and in vivo studies. Chem Soc Rev 40:1647–1671. https://doi.org/10.1039/C0CS00018C
Kim HS, Kang HS, Chu GJ, Byun HS (2008) Antifungal effectiveness of nanosilver colloid against rose powdery mildew in greenhouses. Solid State Phenom 135:15–18
Kim SW, Jung JH, Lamsal K, Kim YS, Min JS, Lee YS (2012) Antifungal effects of silver nanoparticles (AgNPs) against various plant pathogenic fungi. Mycobiology 40(1):53–58
Kim DH, Gopal J, Sivanesan I (2017) Nanomaterials in plant tissue culture: the disclosed and undisclosed. RSC Adv 7:36492–36505
Krishnaraj C, Ramachandran R, Mohan K, Kalaichelvan PT (2012) Optimization for rapid synthesis of silver nanoparticles and its effect on phytopathogenic fungi. Spectrochim Acta A Mol Biomol Spectrosc 93:95–99
Ladner DC, Tchounwou PB, Lawrence GW (2008) Evaluation of the effect of ecologic on root knot nematode, Meloidogyne incognita, and tomato plant, Lycopersicon esculenum. Int J Environ Res Public Health 5:104–110
Lamsal K, Sang-Woo K, Jung JH, Kim YS, Kim KU, Lee YS (2010) Inhibition effects of silver nanoparticles against powdery mildews on cucumber and pumpkin. Microbiology 39(1):26–32
Lamsal K, Kim SW, Jung JH, Kim YS, Kim KS, Lee YS (2011) Inhibition effects of silver nanoparticles against powdery mildews on cucumber and pumpkin. Mycobiology 39:26
Lemire JA, Harrison JJ, Turner RJ (2013) Antimicrobial activity of metals: mechanisms, molecular targets and applications. Nat Rev Microbiol 11(6):371–384

Li CC, Dang F, Li M, Zhu M, Zhong H, Hintelmann H, Zhou DM (2017) Effects of exposure pathways on the accumulation and phytotoxicity of silver nanoparticles in soybean and rice. Nanotoxicology 11(5):699–709

Liu RH, Kang YH, Pei L, Wan SQ, Liu SP, Liu SH (2016) Use of a new controlled-loss-fertilizer to reduce nitrogen losses during winter wheat cultivation in the Danjiangkou reservoir area of China. Commun Soil Sci Plant Anal 47:1137–1147

Lv J, Christie P, Zhang S (2019) Uptake, translocation, and transformation of metal-based nanoparticles in plants: recent advances and methodological challenges. Environ Sci Nano 6:41–59. https://doi.org/10.1039/C8EN00645H

Maity S, Pramanick K (2020) Perspectives and challenges of micro/nanoplastics-induced toxicity with special reference to phytotoxicity. Glob Chang Biol 6(26):3241–3250

Mallaiah B (2015) Integrate approaches for the management of Crossandra (Crossandra infundibuliformis L. nees) wilt caused by fusarium incarnatum (Desm.) Sacc. PhD thesis. Tamil Nadu Agric. Univ., Madurai

Masum MMI, Siddiqa MM, Ali KA, Zhang Y, Abdallah Y, Ibrahim E et al (2019) Biogenic synthesis of silver nanoparticles using Phyllanthus emblica fruit extract and its inhibitory action against the pathogen Acidovorax oryzae strain RS-2 of rice bacterial brown stripe. Front Microbiol 10:820

Milewska-Hendel A, Gawecki R, Zubko M, Stróż D, Kurczyńska E (2016) Acta Agrobot 69:1694

Mishra S, Singh BR, Naqvi AH, Singh HB (2017) Potential of biosynthesized silver nanoparticles using Stenotrophomonas sp. BHU-S7 (MTCC 5978) for management of soil-borne and foliar phytopathogens. Sci Rep 7:1–15

Mondal KK, Mani C (2012) Investigation of the antibacterial properties of nanocopper against Xanthomonas axonopodis pv. punicae, the incitant of pomegranate bacterial blight. Ann Microbiol 62:889

Montanarella F, Kovalenko MV (2022) Three millennia of nanocrystals. ACS Nano 16(4):5085–5102

Moore D, Robson GD, Trinci APJ (2011) 21st Century guidebook to fungi, 1st edn. Cambridge University Press, Cambridge, pp 367–391

Nair R, Mohamed MS, Gao W, Maekawa T, Yoshida Y, Ajayan PM, Kumar DS (2012) J Nanosci Nanotechnol 12:2212–2220

Nassar AM (2016) Effectiveness of silver nano-particles of extracts of Urtica urens (Urticaceae) against root-knot nematode Meloidogyne incognita. Asian J Nematol 2(5):14–19

Navarro E, Piccapietra F, Wagner B, Marconi F, Kaegi R, Odzak N, Sigg L, Behra R (2008) Toxicity of silver nanoparticles to Chlamydomonas reinhardtii. Environ Sci 42:8959

Nel A, Xia T, Madler L, Li N (2006) Toxic potential of materials at the nanolevel. Science 311:622–627

Ocsoy I, Paret ML, Ocsoy MA, Kunwar S, Chen T, You M, Tan W (2013) Nanotechnology in plant disease management: DNA-directed silver nanoparticles on graphene oxide as an antibacterial against Xanthomonas perforans. ACS Nano 7(10):8972–8980

Oliveira HC, Gomes BC, Pelegrino MT, Seabra AB (2016) Nitric oxide-releasing chitosan nanoparticles alleviate the effects of salt stress in maize plants. Nitric Oxide 61:10–19

Owolade OF, Ogunleti DO, Adenekan MO (2008) Titanium dioxide affects disease development and yield of edible cowpea. EJEAF Chem 7(50):2942–2947

Palanikumar L, Ramasamy SN, Balachandran C (2014) Size-dependent antimicrobial response of zinc oxide nanoparticles. IET Nanobiotechnol 8(2):111–117

Paret ML, Palmateer AJ, Knox GW (2013a) Evaluation of a light-activated nanoparticle formulation of titanium dioxide with zinc for management of bacterial leaf spot on rosa 14 294 'Noare'. HortScience 48(2):189–192

Paret ML, Vallad EG, Averett RD, Jones BJ, Olson MS (2013b) Photocatalysis: effect of light activated nanoscale formulations of TiO2 on Xanthomonas perforans, and control of bacterial spot of tomato. Phytopathology 103:228–236

Park HJ, Kim SH, Kim HJ, Choi SH (2006) A new composition of nanosized silica-silver for control of various plant diseases. Plant Pathol J 22(3):295–302
Park H-J, Kim SH, Kim HJ, Choi S-H (2016) A new composition of nanosized silica-silver for control of various plant diseases. Plant Pathol J 22:295–302
Patel N, Desai P, Patel N, Jha A, Gautam HK (2014) Agronanotechnology for plant fungal disease management: a review. Int J Curr Microbiol App Sci 3:71–84
Patra JK, Baek K-H (2017) Antibacterial activity and synergistic antibacterial potential of biosynthesized silver nanoparticles against foodborne pathogenic bacteria along with its anticandidal and antioxidant effects. Front Microbiol 8:167
Pawlett M, Ritz K, Dorey RA, Rocks S, Ramsden J, Harris JA (2013) The impact of zero-valent iron nanoparticles upon soil microbial communities is context dependent. Environ Sci Pollut Res 20:1041–1049
Pelletier DA, Suresh AK, Holton GA, McKeown CK, Wang W, Gu B, Mortensen NP, Allison DP, Joy DC, Allison MR, Brown SD (2010) Effects of engineered cerium oxide nanoparticles on bacterial growth and viability. Appl Environ Microbiol 76(24):7981–7989
Perez-de-Luque A, Diego R (2009) Nanotechnology for parasitic plant control. Pest Manag Sci 65:540–545
Prasad R, Swamy VS (2013) Antibacterial activity of silver nanoparticles synthesized by bark extract of Syzygium cumini. J Nanopart 2013:1. https://doi.org/10.1155/2013/431218
Prasad R, Gupta N, Kumar M, Kumar V, Wang S, Abd-Elsalam KA (2017) Nanomaterials act as plant defense mechanism
Rakesh S, Ananda S, Gowda NM (2013) Synthesis of chromium (III) oxide nanoparticles by electrochemical method and Mukia maderaspatana plant extract, characterization, KMnO4 decomposition and antibacterial study. Mod Res Catal 2:127–135
Raliya R, Tarafdar JC (2013) ZnO nanoparticle biosynthesis and its effect on phosphorous-mobilizing enzyme secretion and gum contents in cluster bean (Cyamopsistetragonoloba L.). Agric Res 2(1):48–57
Raliya R, Franke C, Chavalmane S, Nair R, Reed N, Biswas P (2016) Quantitative understanding of nanoparticle uptake in watermelon plants. Front Plant Sci 7:1288
Rao KJ, Paria S (2013) Use of sulfur nanoparticles as a green pesticide on Fusarium solani and Venturia inaequalis phytopathogens. RSC Adv 3(26):10471–10478
Rico CM, Majumdar S, Duarte-Gardea M, Peralta-Videa JR, Gardea-Torresdey JL (2011) Interaction of nanoparticles with edible plants and their possible implications in the food chain. J Agric Food Chem 59:3485–3498
Roco MC (2003) Broader societal issue of nanotechnology. J Nanopart Res 5:181–189
Rousk J, Ackermann K, Curling SF, Jones DL (2012) Comparative toxicity of nanoparticulate CuO and ZnO to soil bacterial communities. PLoS One 7(3):e34197
Roy A, Gauri SS, Bhattacharya M, Bhattacharya J (2013) Antimicrobial activity of CaO nanoparticles. J Biomed Nanotechnol 9(9):1570–1578
Roy A, Murthy HA, Ahmed HM, Islam MN, Prasad R (2021) Phytogenic synthesis of metal/metal oxide nanoparticles for degradation of dyes. J Renew Mater 10(7):1911
Safavi K, Mortazaeinezahad F, Esfahanizadeh M, Javad Asgari M (2011) In vitro antibacterial activity of nanomaterial for using in tobacco plants tissue culture. World Acad Sci Eng Technol 79:372–373
Schmalz G, Hickel R, van Landuyt KL, Reichl FX (2017) Nanoparticles in dentistry. Dent Mater 33(11):1298–1314
Schreiber L (2005) Polar paths of diffusion across plant cuticles: new evidence for an old hypothesis. Ann Bot 95:1069–1107
Schwab F, Zhai G, Kern M, Turner A, Schnoor JL, Wiesner MR (2016) Barriers, pathways and processes for uptake, translocation and accumulation of nanomaterials in plants–critical review. Nanotoxicology 10:257–278. https://doi.org/10.3109/17435390.2015.1048326
Shafie RM, Salama AM, Farroh KY (2018) Silver nanoparticles activity against tomato spotted wilt virus. Middle East J Agric Res 7:1251–1267

Shenashen M, Derbalah A, Hamza A, Mohamed A, El Safty S (2017) Antifungal activity of fabricated mesoporous alumina nanoparticles against root rot disease of tomato caused by Fusarium oxysporum. Pest Manag Sci 73(6):1121–1126

Shweta, Vishwakarma K, Sharma S, Narayan RP, Srivastava P, Khan AS, Dubey NK, Tripathi DK, Chauhan DK (2017) Plants and carbon nanotubes (CNTs) interface: present status and future prospects. In: Nanotechnology. Springer, Singapore, pp 317–340

Singh S, Singh M, Agrawal VV, Kumar A (2010) An attempt to develop surface plasmon resonance based immuno sensor for Karnal bunt (Tilletia indica) diagnosis based on the experience of nano-gold based lateral flow immune-dipstick test. Thin Solid Films 519:1156–1159

Singh S, Tripathi DK, Dubey NK, Chauhan DK (2016) Effects of nano-materials on seed germination and seedling growth: striking the slight balance between the concepts and controversies. Mater Focus 5(3):195–201

Singh S, Tripathi DK, Singh S, Sharma S, Dubey NK, Chauhan DK, Vaculık M (2017) Toxicity of aluminium on various levels of plant cells and organism: a review. Environ Exp Bot 137:177–193

Soenen SJ, Rivera-Gil P, Montenegro JM, Parak WJ, De Smedt SC, Braeckmans K (2011) Cellular toxicity of inorganic nanoparticles: common aspects and guidelines for improved nanotoxicity evaluation. Nano Today 6(5):446–465

Spinato C, Giust D, Vacchi IA, Ménard-Moyon C, Kostarelos K, Bianco A (2016) Different chemical strategies to aminate oxidised multi-walled carbon nanotubes for siRNA complexation and delivery. J Mater Chem B 4:431–441

Sudha PN, Sangeetha K, Vijayalakshmi K, Barhoum A (2018) Nanomaterials history, classification, unique properties, production and market. In: Emerging applications of nanoparticles and architecture nanostructures. Elsevier, pp 341–384

Sun H, Du W, Peng Q, Lv Z, Mao H, Kopittke PM (2020) Development of ZnO nanoparticles as an efficient Zn fertilizer: using synchrotron-based techniques and laser ablation to examine elemental distribution in wheat grain. J Agric Food Chem 68(18):5068–5075

Suriyaprabha R, Karunakaran G, Kavitha K, Yuvakkumar R, Rajendran V, Kannan N (2014) Application of silica nanoparticles in maize to enhance fungal resistance. IET Nanobiotechnol 8(3):133–137

Tan W, Du W, Barrios AC, Armendariz R, Zuverza-Mena N, Ji Z, Chang CH, Zink JI, HernandezViezcas JA, Peralta-Videa JR, Gardea-Torresdey JL (2017) Surface coating changes the physiological and biochemical impacts of nano-TiO2 in basil (Ocimum basilicum) plants. Environ Pollut 222:64–72

Tang ZX, Yu Z, Zhang ZL, Zhang XY, Pan QQ, Shi LE (2013) Sonication-assisted preparation of CaO nanoparticles for antibacterial agents. Quim Nova 36(7):933–936

Tani F, Barrington S (2005) Environ Pollut 138:538–547

Tao S, Zhu S, Feng T, Xia C, Song Y, Yang B (2017) The polymeric characteristics and photoluminescence mechanism in polymer carbon dots: a review. Mater Today Chem 6:13–25

Tayel AA, El-Tras WF, Moussa S, El-Baz AF, Mahrous H, Salem MF, Brimer L (2011) Antibacterial action of zinc oxide nanoparticles against foodborne pathogens. J Food Saf 31(2):211–218

Tekle C, Deurs BV, Sandvig K, Iversen T-G (2008) Cellular trafficking of quantum dot-ligand bioconjugates and their induction of changes in normal routing of unconjugated ligands. Nano Lett 8(7):1858–1865

Tighe-Neira R, Carmora E, Recio G, Nunes-Nesi A, Reyes-Diaz M, Alberdi M, Rengel Z, Inostroza-Blancheteau C (2018) Metallic nanoparticles influence the structure and function of the photosynthetic apparatus in plants. Plant Physiol Biochem 130:408–417

Tilney LG, Cooke TJ, Connelly PS, Tilney MS (1991) J Cell Biol 112:739–747

Tripathi DK, Singh VP, Prasad SM, Chauhan DK, Dubey NK (2015) Silicon nanoparticles (SiNp) alleviate chromium (VI) phytotoxicity in Pisum sativum (L.) seedlings. Plant Physiol Biochem 96:189–198

Tripathi DK, Singh S, Singh S, Pandey R, Singh VP, Sharma NC, Prasad SM, Dubey NK, Chauhan DK (2017) An overview on manufactured nanoparticles in plants: uptake, translocation, accumulation and phytotoxicity. Plant Physiol Biochem 110:2–12

Tyszczuk-Rotko K, Sadok I, Barczak M (2016) Thiol functionalized polysiloxanes modified by lead nanoparticles: synthesis, characterization and application for determination of trace concentrations of mercury(II). Microporous Mesoporous Mater 230:109–117

Vittori Antisari L, Carbone S, Gatti A, Vianello G, Nannipieri P (2013) Toxicity of metal oxide (CeO2, Fe3O4, SnO2) engineered nanoparticles on soil microbial biomass and their distribution in soil. Soil Biol Biochem 60:87–94

Wani AH, Shah MA (2012) A unique and profound effect of MgO and ZnO nanoparticles on some plant pathogenic fungi. J Appl Pharm Sci 2(3):40–44

Xia T, Kovochich M, Liong M, Madler L, Gilbert B, Shi H, Yeh JI, Zink JI, Nel AE (2008) Comparison of the mechanism of toxicity of zinc oxide and cerium oxide nanoparticles based on dissolution and oxidative stress properties. ACS Nano 2:2121–2134. https://doi.org/10.1021/nn800511k

Xin Y et al (2017) Recent progress on nanoparticle-based drug delivery systems for cancer therapy. Cancer Biol Med 14(3):228–241

Yang C, Powell CA, Duan Y, Shatters R, Zhang M (2015) Antimicrobial nanoemulsion formulation with improved penetration of foliar spray through citrus leaf cuticles to control citrus huanglongbing. PLoS One 10:e0133826. https://doi.org/10.1371/journal.pone.0133826

Yousef JM, Danial EN (2012) In vitro antibacterial activity and minimum inhibitory concentration of zinc oxide and nano-particle zinc oxide against pathogenic strains. J Health Sci 2(4):38–42

Zhang P, Ma Y, Zhang Z (2015) Interactions between engineered nanomaterials and plants phytotoxicity uptake translocation, and biotransformation. Springer International Publishing, Berlin

Zhang L, Wu L, Si Y, Shu K (2018) Size-dependent cytotoxicity of silver nanoparticles to Azotobacter vinelandii: growth inhibition, cell injury, oxidative stress and internalization. PLoS One 13(12):e0209020

Chapter 3
The Interaction of Nanoparticles with Phytopathogens and Plants: The Impact on Plant Health

Azhar U. Khan, Kim Min, and Mahboob Alam

Abstract There is a growing human population on Earth, and the demand for food and nourishment is also increasing. However, crop diseases cause food losses. Even though chemical pesticides improve yield, they harm both humans and the environment; therefore, new antimicrobial agents are required. In recent studies, nanoparticles have been found to be a valuable tool for reducing plant disease. The use of nanotechnologies in agriculture is therefore worthwhile. Nanotechnology's application in crop yield improvement and disease reduction is summarized in this chapter. It examines the use of nanotechnology in agriculture and the food industry to meet the growing human population's demand for food and nourishment. By allowing the manipulation of individual molecules and atoms, nanotechnology offers prospective solutions to problems such as crop diseases and food losses. Nanomaterials can be utilized as pesticides, fertilizers, and biosensors, as well as to improve nutrition, distribute bioactive components and micronutrients safely, and preserve food. The chapter also explores how nanotechnology has the potential to reduce pesticide use and enhance soil remediation while emphasizing the importance of conducting comprehensive research on the impacts of nanoparticles on plants, animals, and humans. The chapter discusses prospective research topics for enhancing nano-integrated agriculture systems and highlights nanotechnology applications in crop yield improvement and disease reduction.

Keywords Crop yield · Food security · Nanomaterials · Phyto-nanotechnology · Plant disease

---

A. U. Khan
School of Life and Basic Sciences, Department of Chemistry, SIILAS Jaipur National University, Jaipur, Rajasthan, India

K. Min · M. Alam (✉)
Department of Safety Engineering, Dongguk University, Gyeongju-si, Gyeongsangbuk-do, Republic of Korea

M. Khan, J.-T. Chen (eds.), *Nanoparticles in Plant Biotic Stress Management*,
https://doi.org/10.1007/978-981-97-0851-2_3

3.1 Introduction

Nanotechnology is a rapidly developing field with the potential to transform many industries, including agriculture. Nanomaterials, which have at least one dimension in the nanometer range, can be used to boost crop yields, increase nutrient uptake, and protect crops from pests and diseases (Shalaby et al. 2016). The development of nanofertilizers is one of the most promising applications of nanotechnology in agriculture. Nanofertilizers are made up of nanoparticles that release nutrients slowly and efficiently. This has the potential to reduce nutrient runoff and pollution while increasing crop yields. Nanoparticles can also be used to create more effective and environmentally friendly pesticides and herbicides. Nanosensors can detect pests and diseases early, allowing them to be treated before they cause significant damage (Siddiqui et al. 2015). Nanotechnology is still in its infancy, but it has the potential to revolutionize agriculture and help feed the world's growing population.

Carbon, metal, organic, and semiconductor nanoparticles are the four types of nanoparticles (Buzea and Pacheco 2017). Nanoparticles have unique optical, electrical, and magnetic properties, as well as a small size and a large surface area. Because of these properties, nanoparticles are promising candidates for a wide range of applications, including agricultural improvement. Nanoparticles have been shown to improve plant growth in a variety of ways. Silver nanoparticles, for example, have been shown to promote seed germination and seedling growth (Yin et al. 2012). Titanium nanoparticles have been shown to improve plant resistance to drought and salt stress (Abdel Latef et al. 2018). It has been shown that zinc oxide nanoparticles can increase photosynthesis and crop yields (Laware and Raskar 2014). A study by Rastogi et al. (2019) found that silica nanoparticles enhanced plant water and nutrient uptake. Plants produce antioxidants more effectively when carbon nanoparticles are introduced to them (Mohamed et al. 2018). A study found that boron nanoparticles enhance flowering and fruiting in plants (Goudar et al. 2018; Shireen et al. 2018). Studies found that plants are more resistant to pests and diseases when they are treated with gold nanoparticles (Shukla et al. 2015). Zeolite nanoparticles (Yılmaz et al. 2014) have been shown to improve the efficiency of nutrient uptake by plants. In recent years, nanotechnology-based plant viral disease detection kits and nanobiosensors have gained popularity due to their improved efficiency in detecting various viral diseases (Chaudhary et al. 2018). In the realm of agricultural enhancement, nanofertilizers have gained attention for surpassing the efficacy of traditional chemical fertilizers. Nanofertilizers are more effective at delivering nutrients to plants, and they can be used in smaller quantities, which reduces the risk of soil and water pollution. There are different types of exudates secreted by plants that attract microbes to the rhizosphere and soil. Various minerals of nano size can also be produced by exudates and microbial communities in soil, although this has not yet been extensively studied (Yu 2018). In conjunction with plant growth-promoting rhizobacteria (PGPR) organisms, nanoparticles have been reported to promote plant growth. Agriculture can use PGPR and nanotechnology to improve crops more efficiently than it can with conventional crop improvement techniques.

In 2020, around 720 and 811 million people faced chronic hunger. Globally, 149.2 million children under 5 suffer from stunting. The 2030 Agenda for Sustainable Development aims to achieve food security and promote sustainable agriculture. The new strategy to achieve this vision is through sustainability for better production, better nutrition, a better environment, and better life (World Health Organization 2021).

Cultivators mainly use toxic chemicals to prevent plant diseases. The increasing use of herbicides carries risks to the environment and human health. Around 2 million tons of pesticides are consumed worldwide per year. The use of these chemicals on a large scale gives rise to environmental deterioration, and the residues of pesticides in food cause harmful diseases like cancer and asthma in humans. DDT (dichlorodiphenyltrichloroethane) pesticide causes genotoxicity in humans (Cohn et al. 2007). 3,6-Dichloro-2-methoxy benzoic acid and 2,4-dichlorophenoxyacetic acid (2,4-D) and the herbicides belong to the auxinic group. It has been reported that they cause DNA damage in humans. Glyphosate is the herbicide and constitutes 60% of the herbicide sales (Atwood and Paisley-Jones 2017). In 1999, around 1 million deaths occured in the world due to pesticides. According to the World Health Organization, 3 million poisoning cases per year are caused by pesticides (WHO 2016). Herbicides cause acute and chronic toxicity. Many herbicides have chronic effects on human health. In addition, pesticides cause prostate cancer in farmers, decreased sperm count, endocrine disruption, or increased abnormalities in sperm (Weisenburger 1993; Kortekamp 2011).

There is a need to reduce the utilization of toxic pesticides in agriculture. Moreover, the toxicity of chemical pesticides has called for alternative measures to control plant pathogens. Among these alternatives, some of the most effective are nanomaterials. Management of plant disease, using nanomaterials is eco-friendly and host-targeted. Nanotechnology has the potential to revolutionize agriculture, crop yield, and food systems and has given birth to a new era of agronanotechnology. The word "nano" means the one-billionth part. Different types of nanomaterials, like zinc, copper, magnesium, titanium, gold, alginate, and silver, have been developed. Approximately 260,000–309,000 metric tons of nanoparticles were produced globally in 2010, and 225,060 metric tons to 585,000 metric tons were consumed worldwide from 2014 to 2019. Nanotechnology has the potential to revolutionize agriculture, but it is in the nascent stage and needs more research. The management of diseases and improving yield are key applications (Khan et al. 2019, 2021a).

3.2 Plant Growth-Promoting Rhizobacteria (PGPR) and Nanoparticles

Plant growth-promoting rhizobacteria (PGPR) are a diverse group of bacteria that live in the soil and promote plant growth in a variety of ways. They can directly improve plant growth by mobilizing nutrients, fixing nitrogen, and producing phytohormones (Glick 2012). They can also indirectly improve plant growth by protecting them from diseases and abiotic stresses (Beneduzi et al. 2012). The effectiveness of PGPR can vary depending on the environmental conditions, such as the climate, soil type, and the presence of other microorganisms (Timmusk et al. 2017). As a result, using various strains of PGPR at the same time can help ensure that there are always enough bacteria to promote plant growth (Mendis et al. 2018). Nanoparticles (NPs) are microscopic particles with numerous applications in agriculture. They can be used to feed plants, protect them from disease, and improve their adaptation to abiotic stress. Nanoparticles can have synergistic benefits when paired with PGPR, leading to larger increases in plant growth and health. NP, for example, can protect PGPR from adverse soil conditions, including heat and acidity. They can also improve plant nutrient uptake, which boosts plant growth. A promising new technique to increase agricultural productivity is the combination of PGPR with NP. It has the potential to reduce the need for chemical fertilizers and pesticides while also improving the health and resilience of crops (Pandey et al. 2012). By combining the synergistic effects of both NP and PGPR, a promising opportunity emerges to enhance plant yields and bolster their resistance to diseases.

3.3 Effect of Zinc Oxide Nanoparticles on Plant Pathogens and Plants

Zinc remains an indispensable micronutrient fundamental for the food of different living beings, spreading from plants and creatures to people. Inside the human hereditary code, there exists a plan for around 3000 proteins that consolidate zinc, each serving unmistakable primary and functional jobs (Process 2013). Across different classes of proteins—oxidoreductases, transferases, hydrolases, lyases, isomerases, and ligases—zinc expects the jobs of both a reactant and underlying co-factor (Mccall et al. 2000). These compound classes hold significant importance in the coordination of cell processes, whether as go-betweens for extracellular improvements or intracellular couriers (Maret 2017). The impact of zinc reaches out across various spaces, affecting cycles like photosynthesis (Wang et al. 2009), protein union (Obata and Umebayashi 1988), and the amalgamation of phytohormones like auxins and ABA (Atici et al. 2005). Furthermore, it contributes to sugar development (Mousavi 2011), film usefulness, and a stronghold against illnesses and abiotic stresses like dry season (Ma et al. 2017; Dang et al. 2008). In examples of deficiency, zinc lack can yield significant misfortunes in crop efficiency and, in extreme cases, lead to establish (Hafeez 2014). Shockingly, this lack can persevere in fields regardless of the obvious wealth of zinc (Sharma et al. 2013). To address this test, the use of zinc composts arises as the essential arrangement because of the moderately low solvency of zinc. Broadly utilized cures incorporate zinc sulfate and EDTA-Zn chelate (Cakmak and Kutman 2018). Khan and Siddiqui (2018) observe the effect of zinc oxide nanoparticles (ZnO-NPs) on root-knot nematode *Meloidogyne incognita*. ZnO-NPs at 200 ppm concentration showed the highest mortality of *M. incognita*.

Nanofertilizer formulations have shown improved performance compared to conventional fertilizers, as they release essential nutrients in a controlled manner. This, along with their small size and large surface area, enhances their effectiveness. Zinc oxide nanoparticles increase plant length, boost the antioxidant system of tomato plants, and also increase the photosynthetic pigments (Faizan et al. 2018). However, zinc oxide nanoparticles can also have adverse effects on plants. Treatment of tomato plants with zinc oxide nanoparticles negatively impacted both plant growth and chlorophyll content, which in turn affected the photosynthetic rate (Wang et al. 2018). A supernatant of zinc oxide nanoparticle suspensions containing Zn^{2+} was not found to affect tomato plant growth, indicating that the harmfulness is due to the zinc oxide nanoparticles themselves. Zinc oxide nanoparticles (ZnO-NPs) have been shown to improve plant defense responses by increasing the expression of antioxidant proteins (Parveen and Siddiqui 2021). Various zinc-solubilizing microorganisms have been identified that can solubilize zinc in soil, making it more available to plants. Kamran et al. (2017) found that *Pseudomonas putida* B6 could solubilize zinc in soil, and Mumtaz et al. (2017) showed that inoculation of maize with zinc-solubilizing bacteria resulted in improved plant growth. In addition to solubilizing zinc, some zinc-solubilizing bacteria can also produce zinc nanoparticles. Sultana et al. (2019) found that *Pseudomonas aeruginosa*, *Bacillus subtilis*, and *Azospirillum brasilense* could all produce zinc nanoparticles. Seed vaccination with PGPR and nano-zinc oxide significantly increased plant height, number of nodules per plant, grain yield, and grain weight. This was found in a study by Khoramdel (2016). ZnO nanoparticles also showed a dose-dependent increase in siderophore production by microbes. This was reported by Haris and Ahmad (2017). A safe and environmentally friendly dose of Zn nanoparticles along with PGPR could revolutionize the agricultural sector.

3.4 Effect of Silicon Nanoparticles on Plant Pathogens and Plants

Influencing development in a multitude of ways, silicon and silica nanoparticles significantly contribute to promoting plant growth. Through both direct and indirect methods, these nanoparticles penetrate the plants and affect their metabolic functions. A noteworthy example is the protective film silica nanoparticles that can create on the epidermal cell wall upon ingestion by plants. As noted by Strout et al. (2013) and Rajput et al. (2021), this strengthens the structural integrity of leaves and increases their resistance to nematodes and bacterial and parasitic infections. There are various advantages to forming a nanosilica layer using silica nanoparticles. Plants lose less water, which increases their ability to tolerate harsh circumstances such as high temperatures, drought, and humidity (Ashkavand et al. 2015).

Consequently, it enables plants to exhibit substantial growth and elevated yields even in adverse environmental conditions. Nanosilica oxide's influence on seed germination processes is particularly evident, as it has been shown to have a significant impact (Siddiqui and Al-whaibi 2014).

Silica nanoparticles offer a variety of applications in sustainable agriculture, including their use as nanopesticides, nanoherbicides, and nanofertilizers (Aa et al. 2016). According to Karunakaran et al.'s (2013) study, nanosilica can improve soil quality and benefit bacterial populations. Specifically, nanosilica produced from rice husks promoted microbial growth better than sodium silicate, which inhibited the formation of plant-beneficial rhizobacteria. The peculiar hydration properties of the nanosilica surface may be responsible for this result, since they may improve microbial surface interactions (Gordienko and Kurdish 2007). In another study by Rangaraj et al. (2014), nanosilica treatment demonstrated a significant increase in populations of phosphate-solubilizing bacteria and nitrogen-fixing bacteria. Silica nanoparticles indirectly acted as substrates for these beneficial microorganisms, further contributing to plant health and growth. Silica and PGPR have been recognized as potent agents for enhancing crop growth and development (Suriyaprabha et al. 2012; Ramprasad et al. 2015). Consequently, the combination of these factors holds the potential to replace conventional fertilizers and serve as an effective means for advancing biofertilizer technology.

The antifungal activity of silica and titanium dioxide (SiO_2 and TiO_2) nanoparticles against *Phomopsis vexans*, *Ralstonia solanacearum*, and plant inhibition of egg hatching by the parasitic nematode *Meloidogyne incognita*, also known as root-knot nematode, was investigated by Khan et al. (2022a). Both types of nanoparticles were shown to be efficient against *M. incognita*, but SiO_2 nanoparticles were more effective than TiO_2 nanoparticles. The most effective nanoparticle concentration is 0.20 mg/mL. As demonstrated in Fig. 3.1, the effect of nanoparticles on *M. incognita* was concentration dependent. Khan et al. (2022b) also studied the plant-mediated inhibition of *M. incognita* by synthetic silver nanoparticles (AgNPs). They discovered that the concentration of AgNP had a direct effect on *M. incognita*. Ahamad and Siddiqui (2021) investigated the inhibitory effect of three distinct nanoparticles on *M. incognita*: SiO_2, ZnO, and TiO_2. All three nanoparticles were found to be efficient against *M. incognita*; however, SiO_2 nanoparticles were the most effective. In summary, Khan et al. (2022a, b) and Ahamad and Siddiqui (2021) discovered that SiO_2, TiO_2, and AgNP nanoparticles are efficient against the plant-parasitic nematode *M. incognita*. Although SiO_2 nanoparticles appear to be the most effective, more research is needed to confirm this finding and determine the appropriate nanoparticle concentration for suppressing *M. incognita*.

SiO_2 and TiO_2 nanoparticles, according to Khan et al. (2022a), are antifungal against *Phomopsis vexans* and antibacterial against *Ralstonia solanacearum* and have an inhibitory effect on the hatching of the plant-parasitic worm *M. incognita*'s eggs. Both nanoparticles were found effective against *M. incognita*, but SiO_2 nanoparticles showed high mortality than TiO_2 nanoparticles. Among all three

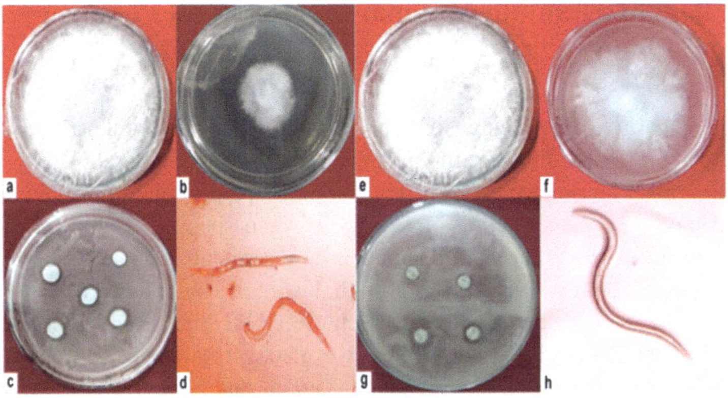

Fig. 3.1 Experiments on nanoparticle effects on *P. vexans* and *R. solanacearum* growth. Panels (**a**) and (**e**) show *P. vexans* growth on the PDA medium. Panel (**b**) displays *P. vexans* growth with 0.20 mg/mL SiO_2-NPs. Panel (**f**) exhibits *P. vexans* growth TiO_2-NPs. Inhibitory effects are in panels (**c**) and (**g**). Panel (**c**) shows SiO_2-NPs with *R. solanacearum*. Panel (**g**) displays TiO_2-NPs. The nanoparticle impact on *M. incognita* is in panels (**d**) and (**h**). Panel (**d**) illustrates the treatment of *M. incognita* J2 with SiO_2-NPs. Panel (**h**) demonstrates the treatment of *M. incognita* J2 with TiO_2-NPs (Khan et al. 2022a)

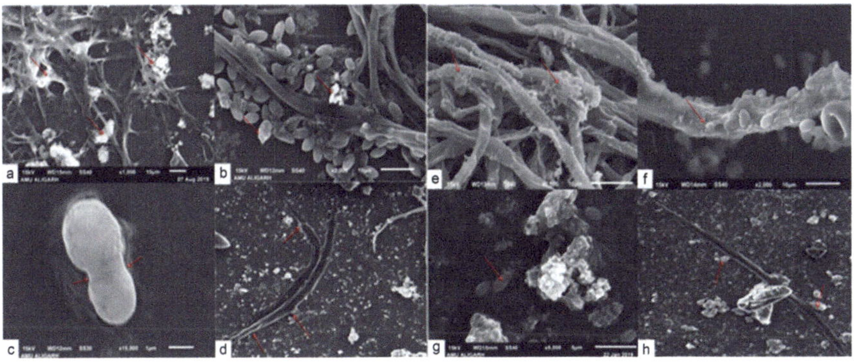

Fig. 3.2 SEM images of test pathogens subjected to SiO_2-NPs and TiO_2-NPs treatment. (**a, b**) *P. vexans* mycelium and conidia sprayed with SiO_2-NPs. (**c**) *R. solanacearum* treated with SiO_2-NPs. (**d**) *M. incognita* nematode treated with SiO_2-NPs. (**e, f**) *P. vexans* mycelium and conidia treated with TiO_2-NPs. (**g**) *R. solanacearum* treated with TiO_2-NPs. (**h**) *M. incognita* J2 nematode treated with TiO_2-NPs (Khan et al. 2022a)

concentrations (0.05, 0.10, and 0.20 mg/mL), the most effective concentration was 0.20 mg/mL. The effect of both nanoparticles on *M. incognita* was concentration-based (Figs. 3.1 and 3.2). Khan et al. (2020) reported that plant median-synthesized silver nanoparticles (AgNPs) have an inhibitory effect on plant pathogenic fungi and bacteria.

3.5 Effect of Titanium Nanoparticles on Plant Pathogens and Plants

Titania nanoparticles have extensive applications in the fields of agriculture and food industries. In agriculture, titania nanoparticles have been shown to have a favorable impact on plant biology. They can participate in plant nitrogen metabolism and modulate the signaling of reactive oxygen species (ROS) within plants (Nayana et al. 2020). When administered in low concentrations to either roots or leaves, nano-TiO_2 can stimulate crop growth, leading to heightened enzyme activities (Lyu et al. 2017). It can also enhance chlorophyll content and boost photosynthesis (Gao et al. 2008). This phenomenon can also facilitate nutrient uptake (Larue et al. 2012) and augment stress tolerance (Karami and Sepehri 2018), ultimately resulting in an overall improvement in crop yield. Titania nanoparticles have also been shown to have adhesive characteristics on bacteria. *Pseudomonas fluorescens* 5RL and *Pseudomonas putida* TVA8 exhibit pronounced nanophase adhesion when exposed to titania nanoparticles, as compared to conventional titania (Park et al. 2008). This can lead to a significant enhancement in the colonization of plant growth-promoting rhizobacteria (PGPR). When double inoculated with sol gel-synthesized titania nanoparticles, PGPR colonization can be increased by approximately 25%. The mechanism for this enhancement involves the creation of micro-niches around the roots that are distinct from the surrounding microbiome. This unique microenvironment allows beneficial bacteria to function as a cohesive unit, ultimately contributing to improved crop health. Application of titania nanoparticles can also result in the formation of larger and denser bacterial clusters compared to those produced by self-generated biofilm matrices (Timmusk et al. 2018). At the nanoscale interface between *Bacillus amyloliquefaciens* UCMB5113, a plant growth-promoting bacterium, and oilseed rape plants (*Brassica napus*), titania nanoparticles play a significant role. These nanoparticles can enhance the adhesion of beneficial bacteria to the roots of oilseed rape plants and also confer protection against the fungal pathogen, *Alternaria brassicae* (Timmusk et al. 2018; Palmqvist et al. 2015). The colonization process, facilitated by titania nanoparticles, is verified through various analytical techniques including SEM, EDS, CLSM, SDS-PAGE, and fluorescence measurements. The outcome is an increased bacterial biomass. By providing an effective platform for PGPR colonization on plants, titania nanoparticles emerge as potential agents for the development of PGPR-based solutions in sustainable agriculture (Timmusk et al. 2018; Palmqvist et al. 2015).

3.6 Effect of Silver Nanoparticles on Plant Pathogens and Plants

AgNPs, or silver nanoparticles, are frequently utilized in antimicrobial products. AgNPs have been shown by Radhakrishnan (2017) to be capable of efficiently preventing bacterial growth. Medical equipment (Gherasim et al. 2018), common products like sports socks and washing machines (Dargo et al. 2017), and agriculture (Babu et al. 2014) are just a few areas where AgNPs have been applied. AgNPs can, however, also have ecotoxicological impacts (Mao et al. 2018), so it's crucial to keep that in mind. According to Jasim et al. (2017), biosynthesized AgNPs can enhance fenugreek seedling growth. AgNPs produced biologically have been shown to be less hazardous than those produced chemically, according to Sharma et al. (2015).

Savithramma et al. (2012) found that AgNPs have the ability to aid in the penetration of seed coverings by water and nutrients, which can improve seed germination and seedling growth. Khan and Bano (2015) demonstrated that the presence of AgNPs can boost the activity of PGPR (plant growth-promoting rhizobacteria), which in turn can cause maize plants to experience greater root growth. In addition, the effects of phytohormones like abscisic acid (ABA), indole acetic acid (IAA), and squamous acid (GA) can be amplified by the presence of AgNPs, according to Khan and Bano (2015).

AgNPs can have a major impact on soil bacterial diversity even at low concentrations. Mishra and Sarma (2018) explored how fungal endophytes, such as taxol, podophyllotoxin, polyketides, terpenes, and peptides, might create significant plant secondary metabolites. According to Khan et al. (2021b), spraying of biosynthesized AgNPs can increase the plant length and photosynthetic pigments. The effect of AgNPs was dose based.

3.7 Effect of Gold Nanoparticles on Plant Pathogens and Plants

Gold nanoparticles have been investigated extensively and are routinely manufactured utilizing a number of processes, including chemical, biological, and physical approaches (Rashid et al. 2014; Herizchi et al. 2014; Shah et al. 2014). Because of their environmental friendliness and vast application, biological approaches are becoming increasingly popular (Raghuvanshi et al. 2017). For example, the plant growth-promoting *Bacillus thuringiensis* strain PG-4 has been proven to be a potent and less harmful agent for the biosynthesis of gold nanoparticles (Aljabali et al. 2018; Roshmi et al. 2015). Gold nanoparticles have also been demonstrated to improve plant growth in *Brassica juncea* and *Arabidopsis thaliana* (Rashid et al. 2014; Raghuvanshi et al. 2017; Pestovsky and Martínez-antonio 2017). Shukla et al. (2015) discovered that cultivating *Pseudomonas* and *Bacillus*, plant growth-promoting rhizobacteria (PGPR), in the presence of 6.25 μg/mL gold nanoparticles, increased their growth. Furthermore, Panichikkal et al. (2019) demonstrated that gold nanoparticles boosted indole-3-acetic acid (IAA) production in *P. monteilii*. IAA is a phytohormone that influences processes such as vascular tissue development, root initiation, blooming, fruit ripening, and leaf abscission in plants (Basheer 2017). The precise processes by which gold nanoparticles stimulate PGPR development and boost IAA synthesis remain unknown. One theory is that the nanoparticles bind to the bacterial cell surface, perhaps changing the cell's size and shape without causing harm (Shukla et al. 2015). According to Phenrat et al. (2009)), this could potentially promote bacterial growth.

3.8 Nanozeolites

One of the minerals that can be found in abundance in sedimentary rocks is called zeolite, and it is a crystalline aluminosilicate. Because it has a porous structure and a high cation exchange capacity (CEC), it is beneficial in a range of applications, one of which is agriculture. Specifically, it can absorb and release cations. For many years, Japanese farmers have employed zeolites to control soil moisture content and pH (Ramesh and Reddy 2011). Zeolites' porous structure allows them to absorb and store water, which aids in preventing soil water loss. Zeolites' CEC allows them to bind to cations such as potassium and calcium, which aids in soil fertility (Mahesh et al. 2018). Zeolites are also capable of transporting nutrients to plants. They can be employed as a carrier for slow-release fertilizers, ensuring that plants receive a consistent supply of nutrients throughout the growth season (Sangeetha and Baskar 2016). Zeolites can also be used to trap heavy metals that are toxic to plants. Nanozeolites are zeolites that are smaller than traditional zeolites. They have several advantages over traditional zeolites, including a higher CEC, which makes them more effective at transporting nutrients; a small particle size, which allows plants to absorb them; and a low bulk density, which makes them easy to transport and apply.

Nanozeolites have been found to improve agricultural production and nutrient utilization efficiency. They have also been demonstrated to help reduce nutrient leaching from the soil. Composites containing zeolites have also been demonstrated to offer several advantages over regular zeolites (Khati et al. 2018). These composites have a higher water retention capacity, which aids in preventing soil water loss. They also have higher equilibrium water content, allowing them to hold more water. This can be advantageous in arid climates. The ability of nanozeolites to increase organic carbon content in the soil while also successfully stabilizing micro- and macroaggregates outperforms that of bulk zeolites (Mirzaei et al. 2015). According to Lateef et al. (2016), it has been demonstrated that increasing agricultural productivity can be accomplished through the application of nanozeolites in conjunction with plant growth-promoting rhizobacteria (PGPR). Beneficial bacteria known as PGPR have been shown to stimulate increased plant development. When coupled with nanozeolites, PGPR has the potential to improve plant health by increasing nutrient uptake and lowering the prevalence of the disease.

3.9 Nano-carbon

The use of carbon nanoparticles in agriculture has elicited both good and negative responses. Carbon nanomaterials, according to Mukherjee et al. (2016), can be employed in agriculture for plant growth stimulation, plant protection, plant transformation, and nanodiagnostics. Al-Whaibi and Mohammad (2017) discovered that carbon nanoparticles can improve agricultural productivity and quality. However, Tong et al. (2007) found that nano-fullerenes (nC60) had little impact on the soil microbial community. It is possible that the application of PGPR (plant growth-promoting rhizobacteria) along with suitable nano-carbon or fullerenes could have a future application in agriculture.

3.10 Nanoboron

Boron functions as a minor trace element with a significant impact on the development of plants. Its role encompasses various critical processes, including cell division, elongation, nitrogen and carbohydrate metabolism, sugar transport, cytoskeletal proteins, ion movements (such as H^+, K^+, PO_4^{3-}, Rb^+, Ca^{2+}) across cellular membranes, and phenol metabolism and transportation (Shireen et al. 2018). In a related study, Maziah et al. (2010) observed that introducing plant growth-promoting rhizobacteria (PGPR), specifically *Bacillus sphaericus* UPMB10, into a modified MS medium enriched with boron, led to notable improvements in the growth and root biomass of banana plantlets, outperforming the control group. Notably, Goudar et al. (2018) found that the application of nanoboron nitride as a fertilizer contributed to increased seed and stalk yields in sunflower (*Helianthus annuus* L.). Harnessing PGPR to enhance boron uptake under boron-deficient conditions holds a considerable promise within the realm of agriculture, signifying a notable advancement. Furthermore, the integration of boron nanomaterials into PGPR formulations exhibits the potential for compelling applications.

3.11 Nano-chitosan

Chitosan, derived from the deacetylation of chitin, is a natural biomaterial. It exhibits the capacity to induce biotic and abiotic stress tolerance, control plant diseases, and foster growth in plants (Malerba and Cerana 2016). A noteworthy illustration of this is the employment of nano-chitosan alongside plant growth-promoting rhizobacteria (*Bacillus* spp.), which has been documented to heighten plant height, expand leaf area, bolster seed germination, and augment organic acid production as a response to stressors in maize (Khati et al. 2017). These observations propose that integrating natural biopolymers into the formulation of plant growth-promoting rhizobacteria could have a substantial influence on the technology surrounding these inoculants.

3.12 Nanoencapsulation

The realm of contemporary agriculture has been marked by a novel trend in development, namely, bio-nanoencapsulation employing beneficial microorganisms sourced from plants along with nanoparticles. This emerging approach presents a distinctive pathway for progress within the agricultural landscape. To fully harness its potential, it is imperative to grasp the potential toxicity stemming from the interaction of nanoparticles and microbes, as this knowledge forms the bedrock for optimizing their practical application. In this context, the utilization of nanoparticles derived from biopolymers such as proteins and carbohydrates emerges as a pivotal player, offering substantial contributions. A noteworthy current in agriculture involves the utilization of nanomaterials for delivering both macro- and micronutrients to plants, a trend highlighted by Jampílek and Kráľová (2017). In parallel, the concept of nanoencapsulation carries within it the capability to provide a safeguard for crops against a gamut of threats, encompassing disease-causing agents, insects, and other pests. This notion was underscored by Castañeda et al. (2014). Moreover, the study conducted by De Gregorio et al. (2017) demonstrated that the immobilization of *Pantoea agglomerans* and *Burkholderia caribensis* in nanofibers yielded no alteration in the viability or beneficial attributes inherent to these rhizobacteria. Instead, the immobilized PGPR (plant growth-promoting rhizobacteria) exhibited an association with amplified seed germination, as well as increased length and dry weight of soybean roots. Notably, the challenge of maintaining viable bacterial counts looms large when dealing with the application of PGPR in field conditions. Here, the integration of nanofibers and other nanomaterials holds a significant promise as an eco-friendly strategy for inoculum development. Within this framework, PGPR can exert their advantageous traits that foster plant growth, while the nanofibers play a protective role, shielding bacteria and seeds from local abiotic stresses.

3.13 Nanobiofertilizers with PGPR

Nanofertilizers have the potential to replace traditional fertilizers due to their many advantages. They can improve nutrient use efficiency, reduce soil toxicity, and mitigate the negative effects of over-application. This can lead to a reduction in the need for frequent reapplication (Qureshi et al. 2018). Mala et al. (2017) developed a slow-release nanofertilizer formulation that not only delivers nutrients but also contains plant growth-promoting rhizobacteria (PGPR). They combined nanofertilizers with neem cake, a byproduct of the neem tree (*Azadirachta indica*), and PGPR. This slow-release fertilizer mechanism was found to accelerate enzyme activity during germination, resulting in a higher seed vigor index (Mala et al. 2017). PGPR also helps plants use NPK fertilizer more efficiently. Therefore, the synergistic action of these components can lead to increased crop yields.

3.14 Conclusion

Nanomaterials could play an important role in crop production and improving yield. The combination of nanoparticles and PGPR (plant growth-promoting rhizobacteria) has the potential to revolutionize agriculture. Nanoparticles can help PGPR colonize plant roots more effectively, facilitate the synthesis of phytohormones and secondary metabolites, and ultimately amplify the overall effectiveness of rhizobacteria. There is a growing trend toward encapsulating PGPR within nanoparticles and formulating nanobiofertilizers. These environmentally friendly alternatives to conventional fertilization methods have the potential to supplant traditional agricultural practices. The application of nanotechnology is an environmentally sustainable, low-input technique that can play a vital role in mitigating plant stress and managing diseases. Anticipated advancements in this field promise to rejuvenate the agricultural sector through innovative technologies.

References

Aa E et al (2016) The silica-nano particles treatment of squash foliage and survival and development of Spodoptera littoralis (Bosid.) larvae. J Entomol Zool Stud 4(1):175–180

Abdel Latef AAH et al (2018) Titanium dioxide nanoparticles improve growth and enhance tolerance of broad bean plants under saline soil conditions. Land Degrad Dev 29(4):1065–1073. https://doi.org/10.1002/ldr.2780

Ahamad L, Siddiqui ZA (2021) Effects of silicon dioxide, zinc oxide and titanium dioxide nanoparticles on *Meloidogyne incognita, Alternaria dauci* and *Rhizoctonia solani* disease complex of carrot. Exp Parasitol 230:108176

Aljabali AAA et al (2018) Synthesis of gold nanoparticles using leaf extract of Ziziphus zizyphus and their antimicrobial activity. Nanomaterials 8:174. https://doi.org/10.3390/nano8030174

Al-Whaibi M, Mohammad F (2017) Chapter 2: Role of nanoparticles in plants. In: Nanotechnology and plant sciences. Springer. https://doi.org/10.1007/978-3-319-14502-0

Ashkavand P et al (2015) Effect of SiO2 nanoparticles on drought resistance in hawthorn seedlings. For Res Papers 76(4):350–359. https://doi.org/10.1515/frp-2015-0034

Atici Ö, Ağar G, Battal P (2005) Changes in phytohormone contents in chickpea seeds germinating under lead or zinc stress. Biol Plant 49(2):215–222. https://doi.org/10.1007/s10535-005-5222-9

Atwood D, Paisley-Jones C (2017) Pesticides industry sales and usage 2008–2012 market estimates

Babu MY et al (2014) Application of biosynthesized silver nanoparticles in agricultural and marine application of biosynthesized silver nanoparticles in agricultural and marine pest control. Curr Nanosci 10:374. https://doi.org/10.2174/157341371130966601O3

Basheer J (2017) Plant growth promoting potential of endophytic bacteria isolated from Piper nigrum plant growth promoting potential of endophytic bacteria isolated from Piper nigrum. Plant Growth Regul. https://doi.org/10.1007/s10725-013-9802-y

Beneduzi A, Ambrosini A, Passaglia LMP (2012) Plant growth-promoting rhizobacteria (PGPR): their potential as antagonists and biocontrol agents. Genet Mol Biol 4(35):1044–1051

Buzea C, Pacheco I (2017) Nanomaterials and their classification. In: EMR/ESR/EPR spectroscopy for characterization of nanomaterials. Springer, New Delhi, pp 3–45

Cakmak I, Kutman UB (2018) Agronomic biofortification of cereals with zinc: a review. Eur J Soil Sci 69(1):172–180. https://doi.org/10.1111/ejss.12437

Castañeda L et al (2014) Innovative Rice seed coating (Oryza Sativa) with polymer nanofibres and microparticles using the electrospinning method. J Res Updates Polym Sci 3(1):33–39. https://doi.org/10.6000/1929-5995.2014.03.01.5
Chaudhary V, Jangra S, Yadav NR (2018) Nanotechnology based approaches for detection and delivery of microRNA in healthcare and crop protection. J Nanobiotechnol 16(1):40
Cohn B, Wolff M, Cirillo P, Sholtz R (2007) DDT and breast cancer in young women: new data on the significance of age at exposure. Environ Health Perspect 115:1406–1412
Dang H et al (2008) Dominant chloramphenicol-resistant bacteria and resistance genes in coastal marine waters of Jiaozhou Bay, China. World J Microbiol Biotechnol 24(2):209–217. https://doi.org/10.1007/s11274-007-9458-8
Dargo H, Ayaliew A, Kassa H (2017) Synthesis paradigm and applications of silver nanoparticles (AgNPs), a review. Sustain Mater Technol 13:18–23. https://doi.org/10.1016/j.susmat.2017.08.001
De Gregorio PR, Michavila G, Muller LR, de Souza BC, Pomares MF, de Sa ELS, Pereira C, Vincent PA (2017) Beneficial rhizobacteria immobilized in nanofibers for potential application as soybean seed bioinoculants. PLoS One 12(5):e0176930
Faizan M, Faraz A, Yusuf M, Khan ST, Hayat S (2018) Zinc oxide nanoparticle-mediated changes in photosynthetic efficiency and antioxidant system of tomato plants. Photosynthetica 56:678–686
Gao F et al (2008) Was improvement of spinach growth by nano-TiO(2) treatment related to the changes of rubisco activase? Biometals 21:211. https://doi.org/10.1007/s10534-007-9110y
Gherasim O, Ficai A, Andronescu E (2018) Biomedical applications of silver nanoparticles: an up-to-date overview. Nanomaterials 8:681. https://doi.org/10.3390/nano8090681
Glick BR (2012) Plant growth-promoting bacteria: mechanisms and applications. Scientifica 2012:1
Gordienko AS, Kurdish I (2007) Surface electrical properties of Bacillus subtilis cells and the effect of interaction with silicon dioxide particles. Biophysics 52:217. https://doi.org/10.1134/S0006350907020121
Goudar KM et al (2018) Response of sunflower (Helianthus annuus L.) to nano boron nitride fertilization. Int J Chem Stud 6(5):2624–2630
Hafeez B (2014) Role of zinc in plant nutrition—a review. Am J Exp Agric 3(2):374–391. https://doi.org/10.9734/ajea/2013/2746
Haris Z, Ahmad I (2017) Impact of metal oxide nanoparticles on beneficial soil microorganisms and their secondary metabolites. Int J Life Sci Sci Res 3:1020–1030. https://doi.org/10.21276/ijlssr.2017.3.3.10
Herizchi R et al (2014) Current methods for synthesis of gold nanoparticles. Artif Cells Nanomed Biotechnol 44:596. https://doi.org/10.3109/21691401.2014.971807
Jampílek J, Kráľová K (2017) Nanomaterials for delivery of nutrients and growth-promoting compounds to plants. In: Nanotechnology. Springer, Singapore, pp 177–226
Jasim B et al (2017) Plant growth and diosgenin enhancement effect of silver nanoparticles in fenugreek (Trigonella foenum-graecum L.). Saudi Pharm J King Saud Univ 25(3):443–447. https://doi.org/10.1016/j.jsps.2016.09.012
Kamran S et al (2017) Contribution of zinc solubilizing bacteria in growth promotion and zinc content of wheat. Front Microbiol 8. https://doi.org/10.3389/fmicb.2017.02593
Karami A, Sepehri A (2018) Nano titanium dioxide and nitric oxide alleviate salt induced changes in seedling growth, physiological and photosynthesis attributes of barley. Zemdirbyste Agric 105(2):123–132. https://doi.org/10.13080/z-a.2018.105.016
Karunakaran G et al (2013) Effect of nanosilica and silicon sources on plant growth promoting rhizobacteria, soil nutrients and maize seed germination. IET Nanobiotechnol:1–8. https://doi.org/10.1049/iet-nbt.2012.0048
Khan N, Bano A (2015) Role of plant growth promoting Rhizobacteria and Ag-nano particle in the bioremediation of heavy metals and maize growth under municipal wastewater irrigation. Int J Phytoremediation 6514:211. https://doi.org/10.1080/15226514.2015.1064352

Khan M, Siddiqui ZA (2018) Zinc oxide nanoparticles for the management of *Ralstonia solanacearum*, *Phomopsis vexans* and *Meloidogyne incognita* incited disease complex of eggplant. Indian Phytopathol 71:355–364

Khan AU, Khan M, Malik N, Cho MH, Khan MM (2019) Recent progress of algae and blue–green algae-assisted synthesis of gold nanoparticles for various applications. Bioprocess Biosyst Eng 42:1–15

Khan M, Khan AU, Alam MJ, Park S, Alam M (2020) Biosynthesis of silver nanoparticles and its application against phytopathogenic bacterium and fungus. Int J Environ Anal Chem 100(12):1390–1401

Khan M, Khan AU, Hasan MA, Yadav KK, Pinto MMC, Malik N, Yadav VK, Khan AH, Islam S, Sharma GK et al (2021a) Agro-nanotechnology an emerging field: a novel sustainable approach for improving the plant growth by reducing biotic stress. Appl Sci 11:2282

Khan M, Khan AU, Moon IS, Falimban R et al (2021b) Synthesis of biogenic silver nanoparticles from the seed coat waste of pistachio (*Pistacia vera*) and their effect on the growth of eggplant. Nanotechnol Rev 10:1789–1800

Khan M, Siddiqui ZA, Parveen A, Khan AA, Moon IS, Alam M (2022a) Elucidating the role of silicon dioxide and titanium dioxide nanoparticles in mitigating the disease of the eggplant caused by *Phomopsis vexans, Ralstonia solanacearum,* and root-knot nematode *Meloidogyne incognita*. Nanotechnol Rev 11(1):1606–1619

Khan M, Khan AU, Rafatullah M, Alam M, Bogdanchikova N, Garibo D (2022b) Search for effective approaches to fight microorganisms causing high losses in agriculture: application of *P. lilacinum* metabolites and mycosynthesised silver nanoparticles. Biomol Ther 12:174

Khati P et al (2017) Nanochitosan supports growth of Zea mays and also maintains soil health following growth. 3 Biotech 7(1):1–9. https://doi.org/10.1007/s13205-017-0668-y

Khati P et al (2018) Effect of nanozeolite and plant growth promoting rhizobacteria on maize. 3 Biotech 8:141. https://doi.org/10.1007/s13205-018-1142-1

Khoramdel RSSS (2016) Effects of nano-zinc oxide and seed inoculation by plant growth promoting rhizobacteria (PGPR) on yield, yield components and grain filling period of soybean (Glycine max L.). Iran J Field Crop Res 13(4):738–753

Kortekamp A (2011) Herbicides and environment, 8 Jan. BoD–Books on Demand

Larue C et al (2012) Comparative uptake and impact of TiO 2 nanoparticles in wheat and rapeseed. J Toxicol Environ Health A 75:722. https://doi.org/10.1080/15287394.2012.689800

Lateef A et al (2016) Microporous and mesoporous materials synthesis and characterization of zeolite based nano e composite: an environment friendly slow release fertilizer. Microporous Mesoporous Mater 232:174–183. https://doi.org/10.1016/j.micromeso.2016.06.020

Laware SL, Raskar S (2014) Influence of zinc oxide nanoparticles on growth, flowering and seed productivity in onion. Int J Appl Microbiol 3(7):874–881

Lyu S et al (2017) Titanium as a beneficial element for crop production. Front Plant Sci 8:1–19. https://doi.org/10.3389/fpls.2017.00597

Ma D et al (2017) Physiological responses and yield of wheat plants in zinc-mediated alleviation of drought stress. Front Plant Sci 8:1–12. https://doi.org/10.3389/fpls.2017.00860

Mahesh M et al (2018) Zeolite farming: a sustainable agricultural prospective. Int J Appl Sci 7(5):2912–2924

Mala R et al (2017) Evaluation of Nano structured slow release fertilizer on the soil fertility, yield and nutritional profile of Vigna radiata. Recent Pat Nanotechnol 11(1):50–62. https://doi.org/10.2174/1872210510666160727093554

Malerba M, Cerana R (2016) Chitosan effects on plant systems. Int J Mol Sci 17(7):1–15. https://doi.org/10.3390/ijms17070996

Mao B et al (2018) Silver nanoparticles have lethal and sublethal adverse effects on development and longevity by inducing ROS-mediated stress responses. Sci Rep 500:1–16. https://doi.org/10.1038/s41598-018-20728-z

Maret W (2017) Zinc in cellular regulation: the nature and significance of "zinc signals". Int J Mol Sci 18. https://doi.org/10.3390/ijms18112285

Maziah M, Zuraida AR, Halimi MS, Zulkifli HS, Sreeramanan S (2010) Influence of boron on the growth and biochemical changes in plant growth promoting rhizobacteria (PGPR) inoculated banana plantlets. World J Microbiol Biotechnol 26(5):933–944

Mccall KA, Huang C, Fierke CA (2000) Zinc and health: current status and future directions. Function and mechanism of zinc metalloenzymes 1. Am Soc Nutr Sci 29(5):1437S–1446S

Mendis HC, Thomas VP, Schwientek P, Salamzade R, Chien JT, Waidyarathne P, Kloepper J, De La Fuente L (2018) Strain-specific quantification of root colonization by plant growth promoting rhizobacteria Bacillus firmus I-1582 and Bacillus amyloliquefaciens QST713 in non-sterile soil and field conditions. PLoS One 13(2):e0193119

Mirzaei M, Ali A, Safari A (2015) Aggregation stability and organic carbon fraction in a soil amended with some plant residues, nanozeolite, and natural zeolite. Int J Recycling Organic Waste Agric 4:11–22. https://doi.org/10.1007/s40093-014-0080-0

Mishra R, Sarma VV (2018) Secondary metabolite production by endophytic fungi: the gene clusters, nature, and expression in. In: Jha S (ed) Endophytes and secondary metabolites. Springer, Dordrecht, pp 1–16. https://doi.org/10.1007/978-3-319-76900-4_20-1

Mohamed MA, Hashim AF, Alghuthaymi MA, Abd-Elsalam KA (2018) Nano-carbon: plant growth promotion and protection. In: Nanobiotechnology applications in plant protection. Springer, Cham, pp 155–188

Mousavi SR (2011) Zinc in crop production and interaction with phosphorus. Aust J Basic Appl Sci 5(9):1503–1509. Available at: https://www.scopus.com/inward/record.uri?eid=2-s2.84155179110&partnerID=40&md5=17b9048c5e857bbd085b53ac7d590ab4

Mukherjee A et al (2016) Carbon nanomaterials in agriculture: a critical review. Front Plant Sci 7:1–16. https://doi.org/10.3389/fpls.2016.00172

Mumtaz MZ, Ahmad M, Jamil M, Hussain T (2017) Zinc solubilizing Bacillus spp. potential candidates for biofortification in maize. Microbiol Res 202:51–60

Nayana AR, Joseph BJ, Jose A, Radhakrishnan EK (2020) Nanotechnological advances with PGPR applications. In: Sustainable agriculture reviews 41, Nanotechnology for plant growth and development. Springer, pp 163–180

Obata H, Umebayashi M (1988) Effect of zinc deficiency on protein synthesis in cultured tobacco plant cells. Soil Sci Plant Nutr 34(3):351–357. https://doi.org/10.1080/00380768.1988.10415691

Palmqvist NGM et al (2015) Nano titania aided clustering and adhesion of beneficial bacteria to plant roots to enhance crop growth and stress management. Sci Rep 5:10146. https://doi.org/10.1038/srep10146

Pandey P, Bisht S, Sood A, Aeron A, Sharma GD, Maheshwari DK (2012) Consortium of plantgrowth-promoting bacteria: future perspective in agriculture. In: Bacteria in agrobiology: plant probiotics. Springer, Berlin, pp 185–200

Panichikkal J, Thomas R, John JC (2019) Biogenic gold nanoparticle supplementation to plant beneficial pseudomonas monteilii was found to enhance its plant probiotic effect. Curr Microbiol. https://doi.org/10.1007/s00284-019-01649-0

Park MR et al (2008) Influence of nanophase titania topography on bacterial attachment and metabolism. Int J Nanomedicine 3(4):497–504

Parveen A, Siddiqui ZA (2021) Zinc oxide nanoparticles affect growth, photosynthetic pigments, proline content and bacterial and fungal diseases of tomato. Arch Phytopathol Plant Protect 54:1–20

Pestovsky YS, Martínez-antonio A (2017) The use of nanoparticles and nanoformulations in agriculture. J Nanosci Nanotechnol 17(12):8699–8730. https://doi.org/10.1166/jnn.2017.15041

Phenrat T et al (2009) Partial oxidation ("aging") and surface modification decrease the toxicity of nanosized zerovalent. Environ Sci Technol 43:195. https://doi.org/10.1021/es801955n

Process DG (2013) Detonation gun process zinc biochemistry: from a single zinc enzyme to a key element of life. Am Soc Nutr Adv Nutr 1:82–91. https://doi.org/10.3945/an.112.003038.82

Qureshi A, Singh DK, Dwivedi S (2018) Nano-fertilizers: a novel way for enhancing nutrient use efficiency and crop productivity. Int J Curr Microbiol Appl Sci 7(2):3325–3335. https://doi.org/10.20546/ijcmas.2018.702.398

Radhakrishnan EK (2017) Microbially and phytofabricated AgNPs with different mode of bactericidal action were identified to have comparable potential for surface fabrication of central venous catheters to combat Staphylococcus aureus biofilm. J Photochem Photobiol B Biol 171:96. https://doi.org/10.1016/j.jphotobiol.2017.04.036

Raghuvanshi P et al (2017) Biosynthesis and characterization of gold nanoparticles from plant growth promoting rhizobacteria. Int J Chem Stud 5(5):525–532

Rajput VD, Minkina T, Feizi M, Kumari A, Khan M, Mandzhieva S et al (2021) Effects of silicon and silicon-based nanoparticles on rhizosphere microbiome, plant stress and growth. Biology 10:791

Ramesh K, Reddy DD (2011) Zeolites and their potential uses in agriculture. Adv Agron 113. https://doi.org/10.1016/B978-0-12-386473-4.00004-X

Ramprasad D, Sahoo D, Sreedhar B (2015) Plant growth promoting Rhizobacteria—an overview. Eur J Biotechnol Biosci 2(2):30–34

Rangaraj S et al (2014) Effect of silica nanoparticles on microbial biomass and silica availability in maize rhizosphere. Biotechnol Appl Biochem 61:668. https://doi.org/10.1002/bab.1191

Rashid R, Murtaza G, Zahra A (2014) Gold nanoparticles: synthesis and applications in drug. Trop J Pharm Res 13:1169–1177

Rastogi A et al (2019) Application of silicon nanoparticles in agriculture. 3 Biotech 9(3):1–11. https://doi.org/10.1007/s13205-019-1626-7

Roshmi T et al (2015) Effect of biofabricated gold nanoparticle-based antibiotic conjugates on minimum inhibitory concentration of bacterial isolates of clinical origin. Gold Bull 48:63–71. https://doi.org/10.1007/s13404-015-0162-4

Sangeetha C, Baskar P (2016) Zeolite and its potential uses in agriculture: a critical review. Agric Rev 37(2):101–108. https://doi.org/10.18805/ar.v0iof.9627

Savithramma N, Ankanna S, Bhumi G (2012) Effect of nanoparticles on seed germination and seedling growth of Boswellia ovalifoliolata an endemic and endangered medicinal tree taxon. Nano Vision 2(1):2

Shah M et al (2014) Gold nanoparticles: various methods of synthesis and antibacterial applications. Front Biosci (Landmark Ed) 1(1):1320–1344

Shalaby TA, Bayoumi Y, Abdalla N, Taha H, Alshaal T, Shehata S, Amer M, Domokos-Szabolcsy É, El-Ramady H (2016) Nanoparticles, soils, plants and sustainable agriculture. In: Nanoscience in food and agriculture 1. Springer, Cham, pp 283–312

Sharma A, Patni B, Shankhdhar D (2013) Zinc—an indispensable micronutrient. Physiol Mol Biol Plants 19:11–20. https://doi.org/10.1007/s12298-012-0139-1

Sharma D, Kanchi S, Bisetty K (2015) Biogenic synthesis of nanoparticles: a review. Arab J Chem. https://doi.org/10.1016/j.arabjc.2015.11.002

Shireen F et al (2018) Boron: functions and approaches to enhance its availability in plants for sustainable agriculture. Int J Mol Sci 19:95–98. https://doi.org/10.3390/ijms19071856

Shukla SK et al (2015) Prediction and validation of gold nanoparticles (GNPs) on plant growth promoting rhizobacteria (PGPR): a step toward development of nano-biofertilizers. Nanotechnol Rev 4(5):439–448. https://doi.org/10.1515/ntrev-2015-0036

Siddiqui MH, Al-whaibi MH (2014) Role of nano-SiO2 in germination of tomato (*Lycopersicum esculentum* seeds mill.). Saudi J Biol Sci King Saud Univ 21(1):13–17. https://doi.org/10.1016/j.sjbs.2013.04.005

Siddiqui MH, Al-Whaibi MH, Mohammad F (2015) Nanotechnology and plant sciences: nanoparticles and their impact on plants. Springer, pp 1–303. https://doi.org/10.1007/978-3-319-14502-0

Strout G et al (2013) Silica nanoparticles aid in structural leaf coloration in the Malaysian tropical rainforest understorey herb Mapania caudata. Ann Bot 112:1141–1148. https://doi.org/10.1093/aob/mct172

Sultana U, Desai S, Reddy G, Prasad TNVKV (2019) Zinc solubilizing plant growth promoting microbes produce zinc nanoparticles. bioRxiv. https://doi.org/10.1101/602219

Suriyaprabha R et al (2012) Growth and physiological responses of maize (Zea mays L.) to porous silica nanoparticles in soil. J Nanopart Res 14(12). https://doi.org/10.1007/S11051-012-1294-6

Timmusk S et al (2017) Perspectives and challenges of microbial application for crop improvement. Front Plant Sci 8:1–10. https://doi.org/10.3389/fpls.2017.00049

Timmusk S, Seisenbaeva G, Behers L (2018) Titania (TiO2) nanoparticles enhance the performance of growth-promoting rhizobacteria. Sci Rep 8:617. https://doi.org/10.1038/s41598-017-18939-x

Tong Z, Bischoff M, Nies L, Applegate B, Turco RF (2007) Impact of fullerene (C60) on a soil microbial community. Environ Sci Technol 41(8):2985–2991

Wang H, Liu RL, Jin JY (2009) Effects of zinc and soil moisture on photosynthetic rate and chlorophyll fluorescence parameters of maize. Biol Plant 53(1):191–194. https://doi.org/10.1007/s10535-009-0033-z

Wang XP et al (2018) Effects of zinc oxide nanoparticles on the growth, photosynthetic traits, and antioxidative enzymes in tomato plants. Biol Plant 62(4):801–808. https://doi.org/10.1007/s10535-018-0813-4

Weisenburger DD (1993) Human health effects of agrichemical use. Hum Pathol 24(6):571–576

WHO (2016) Public health impact of pesticides used in agriculture. World Health Organization

World Health Organization (WHO) (2021) The state of food security and nutrition in the world transforming food systems for food security, improved nutrition and affordable healthy diets for all, 12 Jul 2021. Food & Agriculture Organization

Yılmaz E, Sönmez İ, Demir H (2014) Effects of zeolite on seedling quality and nutrient contents of cucumber plant (Cucumis sativus L. cv. Mostar F1) grown in different mixtures of growing media. Commun Soil Sci Plant Anal 45(21):2767–2777

Yin L, Colman BP, McGill BM, Wright JP, Bernhardt ES (2012) Effects of silver nanoparticle exposure on germination and early growth of eleven wetland plants. PLoS One 7(10):e47674. https://doi.org/10.1371/journal.pone.0047674

Yu G (2018) Root exudates and microbial communities drive mineral dissolution and the formation of nano-size minerals in soils: implications for soil carbon storage. Root Biol 52. https://doi.org/10.1007/978-3-319-75910-4

Chapter 4
The Impact of Nanoparticles on Plant Growth, Development, and Stress Tolerance Through Regulating Phytohormones

Fadime Karabulut

Abstract To enhance the production, nutritional value of plant-based products, and plant physiology, nanoparticles (NPs) are frequently used in agriculture. However, it is still unclear how nanoparticles are absorbed and transported. It is demonstrated in agriculture that plants can absorb nanoparticles and affect their physiological processes. Significant limitations that have a negative impact on the metabolic and developmental processes of plants are biotic stress factors. However, these results have not always been encouraging partially because NPs have occasionally been found to hurt plants by the potential of toxicity. Depending on the number of NPs applied, the stage of plant growth, and environmental factors, the relief of stress brought on by such NPs varies greatly. The activation of the antioxidant defense system, hormonal regulation, and stress are just a few examples of the physiological and biochemical processes that NPs are thought to be regulatory molecules for plants that can modulate. Plant growth hormones are known to participate in a variety of signaling cascades and function as molecules that regulate stress when exposed to nanoparticles. Reactive oxygen species (ROS), which can be controlled with the help of nanoparticles, are produced more frequently when under stress. According to recent research, nanoparticles can strengthen the effects of stress by regulating phytohormone signaling. This section discusses the phytotoxicity of NPs, how it affects plant growth and cell structure, and the underlying mechanisms that lead to damage and the oxidative stress response. In addition, this study will be beneficial for ensuring the longevity of nanoparticles used in agricultural production and directing their rational application.

Keywords Nanoparticles · Plant physiology · Phytohormone · Biotic stress

F. Karabulut (✉)
Department of Biology, Faculty of Science, Firat University, Elazıg, Türkiye

M. Khan, J.-T. Chen (eds.), *Nanoparticles in Plant Biotic Stress Management*,
https://doi.org/10.1007/978-981-97-0851-2_4

4.1 Introduction

In agriculture, nanoparticles can be used for a variety of purposes, such as modifying plant genetics, controlling crop growth, and the release of agrochemicals (Grillo et al. 2021). Nanoparticles provide agrochemicals and different macromolecules necessary for plant growth, resulting in less exposure for effective use and environmental protection (Rajput et al. 2020). Additionally, compared to traditional methods, nanoparticle-mediated plant transformation technology makes it easier to genetically modify plants. Because they are so small, nanoparticles can transport foreign substances to plant cells while also preventing their deterioration (Gao et al. 2023). Furthermore, nanoparticles offer a fresh approach to protecting crops from specific agricultural issues (Singla et al. 2019).

Nanoparticles can be absorbed and transported by plants, which can affect a variety of factors including particle size, surface charge, concentration, exposure time, and plant species. Stomata, root hairs, and leaf surface cracks are just a few of the numerous ways that nanoparticles can get into a plant's system. Once inside the plant, the nanoparticles can either diffuse, bulk flow, or phloem load their way through the system. Previous studies have primarily used seed treatment, branch injection, root application, and foliar spraying to apply nanoparticles to plants. Plants can absorb nanoparticles, according to some studies (Khan et al. 2022). However, limited direct evidence of nanoparticle transport in plants exists (Rico et al. 2011). To create the best nanoparticles for agricultural use, it is essential to comprehend how agricultural nanoparticles are absorbed and transported by plants (Grillo et al. 2021). It is important to understand the mechanisms underlying the effects and bioaccumulation of agricultural nanoparticles in plants. It can also help clarify the biological safety of nanoparticles and their safe use (Khan et al. 2022). Although research has shown that nanoparticles can persist in edible plant parts and pass through the food chain to consumers of plant products, they are not covered in this report.

Two pathways for NP uptake and transport have been established, similar to non-nanoparticle nutrients (Ma et al. 2015). These pathways are either from the leaf to the root or from the root to the leaf and fruit. On the one hand, soil culture techniques are used to conduct root-to-fruit path studies (Rico et al. 2015). Additionally, reported NP displacement starts with mature tubers or seedlings (Ghodake et al. 2011). After being absorbed by the root's endodermis via apoplastic pathways, NPs can be transported to the vascular cylinder by symplastic pathways (Zhao et al. 2012). NPs can travel from the root of the plant to the leaf or fruit by passing through the xylem or phloem (Hernandez-Viezcas et al. 2013; Zhao et al. 2014). The routes from the leaves to the roots, however, entail NP entrapment and redistribution to the stems or NP penetration through the stomata (Hong et al. 2014).

The use of nano-biofertilizers improves nutrient uptake, photosynthesis, nutrient transport, and high photosynthate accumulation. Plant quality and productivity are improving as a result of rising crop growth. The use of nano-biofertilizer in agricultural products has been found to significantly increase crop growth by encapsulating silver and gold nanoparticles with a biofertilizer. Growth was seen in a variety of agricultural products when nanoparticles were applied directly to them as fertilizer (Dikshit et al. 2013). By encouraging the germination of the crops in the seedlings, PGPR increases the productivity of the plants. Additionally, some leguminous plants may be effective at boosting crop productivity (Rahman and Zhang 2018).

Continuous use of chemical fertilizers reduces the soil's fertility because it depletes the soil's yield content. The degradation of soil fertility is also significantly impacted by soil acidification (Rahman and Zhang 2018). Due to the significant correlations between low soil fertility and soil nutrient imbalance and decreased crop productivity, soil nutrient deficiencies, and decreased food nutritional value, farmers are in a difficult position (Monreal et al. 2016). Nano-biofertilizer offers a cost-effective, sustainable, and effective integrated nutrient management solution to address these drawbacks. It improves plants' ability to absorb and assimilate nutrients. By leaching or competing with other organisms, it also lowers nutrient losses in the soil (Janmohammadi et al. 2016). The bioorganic elements of nano-biofertilizer (PGPR) aid in phosphate dissolution, nitrogen fixation, and soil nutrient richness restoration. The biofertilizer's nanoparticle coating aids in the synchronized slow release of nutrients in response to crop demand (Morales-Dıaz et al. 2017).

The chapter discusses the behavior of agricultural nanoparticles in plant physiology as well as the uptake and transport of phytohormones and nanoparticles that affect their synthesis.

4.2 Nanoparticles' Impact on Plant Growth

Nanoparticles can affect plant growth in a variety of ways, depending on the type of plant, application technique, dose, and type of nanoparticle (Chen et al. 2015; Nair 2016). Moderate-to-low levels of general phytotoxicity have been observed in many studies on nanoparticles in terrestrial plant species (Servin and White 2016). The excessive generation of ROS, binding interactions, dissolution/release of toxic ions, and oxidation of biomolecules via the redox cycle are all ways that nanoparticles can cause phytotoxicity. According to a study, the size and concentration of silver nanoparticles have an impact on the toxicity of the effects they have on rice seedlings. As silver nanoparticle concentration rises from 0.1 to 1000 mg/L and size rises from 20 to 150 nm, the effect of inhibition on seedling growth and germination increases (Thuesombat et al. 2014). Nanoparticles' dosage and size have a significant impact on their reactivity, toxicity, and behavior (Wang et al. 2016). The other study (Dimkpa et al. 2013) examined how Zn in ZnO nanoparticles affected plant growth and how readily it was absorbed by maize plants. In conclusion, ZnO nanoparticles were a stimulatory effect on maize growth and nutrition, root activity (dehydrogenase), and photosynthetic pigments. Their phytotoxicity may be significantly influenced by nanoparticles (Dimkpa et al. 2013).

Carbon nanotube-treated tomato seeds germinate more quickly and produce more plant biomass. The seeds germinate more quickly due to increased water uptake by the seeds and the carbon nanotubes' ability to pierce the dense seed coat (Liu et al. 2015). By activating the pumpkin's antioxidant system while it is under NaCl stress, nano-SiO_2 promotes seed germination (Siddiqui et al. 2014). By increasing the transpiration rate, true photochemical efficiency, net photosynthesis rate, effective photochemical efficiency, gas exchange, stomatal conductivity, and chlorophyll fluorescence parameters, nano-SiO_2 promotes plant development and growth (Siddiqui et al. 2014). As opposed to that, titanium dioxide (TiO_2) nanoparticles encourage the production of carbohydrates in plants by speeding up photosynthesis (Wang et al. 2014). It promotes the growth of canola seedlings' roots, hair, and plants. TiO_2 nanoparticles regulate enzymes involved in nitrogen metabolism. These nanoparticles help plants absorb nitrate (Chen et al. 2014). These enzymes also help convert inorganic nitrogen to organic nitrogen (Jaberzadeh et al. 2013).

Physiological parameters, antioxidant activities, micro-macro element values, agricultural parameters, photosynthetic parameters, and food quality were observed in both ways of NP uptake to the plant (such as chlorophyll content, photosynthetic ratio, and chloroplast number), and differences were found (Rico et al. 2011). In a helical series of events, nanoparticles that have been exposed to the root epidermis' cell membrane and cell wall move from symplastic translocation to stellate leaves before entering the plant's vascular system (xylem). In spite of the fact that NP must pass through cell membrane pores in order to pass through the membrane unharmed, this suggests that nanomaterial uptake is size-specific (Rico et al. 2011). Passive integration is how NPs enter the stele from the endodermis apoplast. The soil, seeds, roots, and leaves are all improved and benefited by NPs (Fig. 4.1).

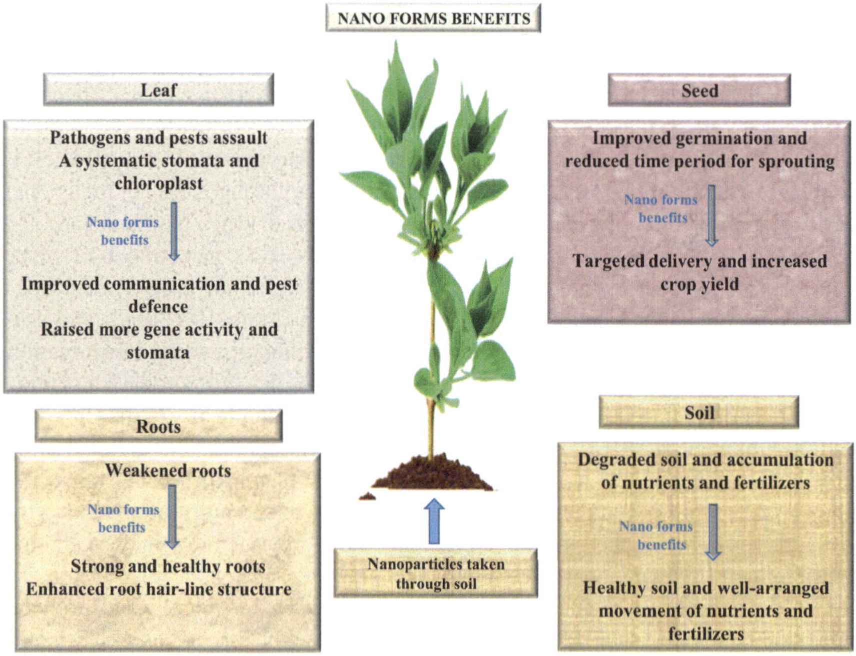

Fig. 4.1 Beneficial effects of nanoparticles on seeds, soil, roots, and leaves

4.3 Nanoparticle Transport in Biosolids

The ecosystem's capacity to operate as a living system is necessary to maintain soil health, water quality, and air quality. In addition, healthy soils support a wide variety of soil organisms that collaborate with plant roots to form advantageous symbiotic relationships, fight crop-destroying pests, and control plant diseases. Furthermore, it is necessary to enhance crop production and the recycling of essential plant nutrients, both of which will have a positive effect on the structure of the soil and ultimately increase its capacity to hold water and nutrients (Yang et al. 2014). The four main tasks must be carried out in order to maintain soil health (FAO 2008). These include maintaining the soil's structural integrity, carbon cycling, nutrient cycles, and disease and pest management. Each of these functions can be accounted for as a combination of different biological processes carried out by various soil organisms in interaction with one another while being influenced by the abiotic soil environment. Physical properties of the soil such as aeration, porosity, macro- and mesofauna, availability of nutrient mineralization, and soil organisms can also be impacted. Earthworms, for instance, can impact soil structure changes and nutrient availability. It offers data on the potential nutrient status of the soil, as well as estimates of biomass and microbial activity. Organic matter in the soil is the foundation of soil fertility. In contrast, the mineralization of soil organic matter can be significantly reduced when soil microorganisms are absent. Microbial activity thus contributes significantly to the health of the soil.

In addition to the numerous positive effects of biosolids, when applied in excess, it is harmful to the environment (McCall et al. 2015), which is why the government applies biosolids regularly. Biosolids are categorized into two classes based on their biological content: class A for partially detectable pathogens and class B for more detectable pathogens (McCall et al. 2015). Class A is defined as bagged and intended for sale or land application, as well as use in lawns, gardens, or other types of land. Other uses for class B exist (Lu et al. 2012). When regulating pharmaceuticals and nanoparticles, additional contaminants also need to be considered. The release of nanoparticles occurs as a result of applying biosolids to a field when they are used as organic fertilizers in agriculture (Kim et al. 2010). Although the process can remove some of the nanoparticles, the biosolids made from the sludge still contain a sizable amount of them (Rottman et al. 2012).

4.4 Nanoparticles Regulated the Biosynthesis and Signaling of Phytohormones in Plants

Plant growth regulators (auxin, gibberellins, cytokinin, abscisic acid (ABA), ethylene, brassinosteroids, salicylic acid (SA), and jasmonic acid (JA)) are examples of small endogenous signaling molecules (Sytar et al. 2019; Khan et al. 2020). Signal crosstalk is the study of the synergistic and antagonistic interactions between plant growth regulators and nanoparticles that primarily control the protective responses (or signals) of plants. These molecules, as well as other molecules that might adhere to the aforementioned general signaling pathway, start the subsequent round of signaling. Studies on the perception of stress signals by plants and the impact on plant growth regulators have been carried out in the last 10 years. Nanoparticles can modify the pool of available phytohormones, alter the biosynthesis of a specific phytohormone, and regulate the signaling process in plants, among other processes (Rasool 2022). However, depending on a variety of factors (Xie et al. 2020), the responses mediated by nanoparticles are probably dependent on plant species, the type of nanoparticles, exposure conditions, etc. For instance, *Momordica charantia* treated with nano-selenium was used for root development and embryo-like structure formations during callogenesis/organogenesis. Phytohormones with NO/H_2S signaling are an effect of nanoparticle-mediated changes (Rajaee Behbahani et al. 2020). They proposed investigating the phytohormonal changes mediated by nanoparticles and suggested that stem bending, inhibition of primary root growth, the emergence of adventitious roots, and inhibition of xylem tissue differentiation under nano-Se exposure indicate hormonal changes (especially ethylene and auxin).

Auxin is useful for controlling both plant growth regulation processes and stress regulation of nanoparticles (Xiong et al. 2021a). Tryptophan is converted to IAA (indole-3-acetic acid), which is the primary auxin biosynthesized by tryptophan aminotransferase from the *Arabidopsis*-1 (TAA1) family and the YUC (YUCCA) family of flavin monooxygenases in two separate chemical reactions. According to recent reports, auxin intraplant biosynthesis and transport are two processes that are regulated by nanoparticle exposure and can control plant growth. According to numerous earlier studies, auxin content rises when exposed to metallic nanoparticles (Jonapa-Hernandez et al. 2020; Zahedi et al. 2019). ZnO NP treatment has a detrimental effect on the concentration of IAA at the root tips in *Arabidopsis* (Vankova et al. 2017). Increased auxin accumulation was seen in TiO_2 nano-treated *Arabidopsis* root tips (Wei et al. 2020). Comparative transcriptome analysis was carried out to better understand the intricate physiological and metabolic processes at work in *Lactuca sativa* L. (Xiong et al. 2021b). Numerous genes related to auxin transport, hormone response, and stress response were found to be overexpressed, according to the findings. This study anticipated the importance of auxin levels in CuO NP-induced stress regulation.

Gibberellin controls a number of plant processes, including flowering, senescence, stem lengthening, and seed germination (Davies 2012). Numerous studies (Hao et al. 2016; Sheteiwy et al. 2017) have shown that nanoparticles regulate the endogenous gibberellin concentration in plants. The floral transition is essential for affecting plant yield and avoiding gibberellin-regulated abiotic stress (Boss et al. 2004). Abiotic stress is frequently identified by the presence of gibberellins. Rice plants under nitrogen (N) stress have an increase in gibberellin content (Wang et al. 2020a). Low N stress caused a significant change in GA content, while high N stress reduced roots' higher GA content by 83% when compared to normal medium N conditions.

The well-known stress phytohormone ABA plays numerous significant roles in helping plants adapt to and withstand abiotic stress (Vishwakarma et al. 2017). Nanoparticle exposure in plants can quickly have an impact on the ABA signaling pathway (Cai et al. 2020). It provides ABA signaling by helping ABA production (Syu et al. 2014). In maize seedlings exposed to La_2O_3 NPs, Yue et al. (2019) described a change in ABA signaling. The ABA content of maize seedlings exposed to La_2O_3 NPs was found to be higher in both the root and the shoots. The ABA receptor quickly recognized the high ABA concentration, activating the ABA signaling pathway, which in turn activated the antioxidant defense system, ultimately increasing stress tolerance (Sheteiwy et al. 2017). Endogenous brassinosteroids are altered in plants exposed to metallic nanoparticles. At the concentration of brassinosteroid, which prevented plant tolerance, thin-walled carbon nanotubes (CNTs) significantly decreased phytotoxicity in the roots and shoots of *O. sativa* L. (Hao et al. 2016).

Adenine derivatives are found in *cytokinins*, a group of plant hormones with isoprenoid or aromatic side chains (Haberer and Kieber 2002). The growth and development of plants that depend on them are greatly influenced by the inhibition of lateral root initiation and vascular differentiation. Furthermore, cytokinins regulate cell division, differentiate photomorphogenic cells in growing leaves and shoots, and prevent leaf senescence in addition to supporting tissues in the roots and controlling cell division (Hai et al. 2020). The cytokinin hormone regulates plant growth under various stress conditions (Azhar et al. 2021). Furthermore, it improved the function of cytokinin in metallic nanoparticles that cause abiotic stress. It's possible that the effects of nanoparticle stress were lessened by the higher cytokinin content even through the growth of the treated pepper plants. The cytokinin signaling pathway is altered by nanoparticle accumulation in the plant tissue and analysis of signaling phytohormones in *Arabidopsis* by exposure to various metallic nanoparticles (Azhar et al. 2021). The study also provided a proof that the positive effects of SiO_2 NPs on plant growth are caused by the upregulation of genes in the cytokinin pathway.

Ethylene, a gaseous hormone, is involved in a variety of physiological, biochemical, and molecular processes throughout a plant's life cycle and functions as a signaling molecule (Iqbal et al. 2017). Methionine is used to biosynthesize ethylene, which is then converted into ethylene using the reactions *S*-adenosyl-L-methionine, 1-aminocyclopropane-1-carboxylic acid synthase (ACS), 1-aminocyclopropane-1-carboxylic acid (ACC), and 1-aminocyclopropane-1-carboxylic acid oxidase. In plants exposed to nanoparticles, ethylene synthesis and signaling play important regulatory roles (Azhar et al. 2020). The transcript level of ACS produced in *T. undulata* leaves under in vitro conditions was decreased by AgNP application. Additionally, it was discovered that ZnO NPs control ethylene signaling-dependent biochemical processes in plants. In cucumbers exposed to Ag_2S-NPs, a notable increase in ethylene-related genes was also noted (Wang et al. 2017). Ethylene-related genes showed a significant increase in expression under ZnO NPs. SiO_2 NPs appear to have positive effects and ZnO NPs negative effects on *Arabidopsis* plant growth based on ethylene signaling.

Controlling plant immunity depends on the defense-related plant hormone *salicylic acid* (SA), which is produced by the shikimate pathway (Koo et al. 2020). Abiotic stress tolerance, plant disease resistance, fruit yield, DNA damage/repair, and seed germination are just a few of the additional benefits it can provide in addition to SA. It is crucial for regulation as well as activation of genes related to pathogenesis (PR). According to reports, nanoparticles caused a significant increase in the endogenous SA concentration in *A. thaliana* leaves (Vankova et al. 2017). Furthermore, some studies have discovered that SA signaling is crucial for stress reduction when exposed to nanoparticles (Cai et al. 2020; El-Shetehy et al. 2021). Fe_3O_4 NPs applied topically to *Nicotiana benthamiana* have been shown to increase plant biomass, the active plant antioxidant system, and plant resistance to TMV infection (Cai et al. 2020). The increase in endogenous SA content in tobacco plants was investigated (Peng et al. 2020). Therefore, it would seem that the SA signaling pathway contributes to the reduction of ZnO NP toxicity in plants.

The signaling molecule *jasmonic acid* (JA) controls a number of developmental processes, including seed germination, plant growth, pollen fertility, fruit ripening, senescence, and stress tolerance (Ghorbel et al. 2021). When under stress, JA helps with stomatal opening and closing, the antioxidant defense system, and amino acids (Wang et al. 2020b). Nanoparticle exposure affects the endogenous JA concentration in plants as a stressor. For instance, Vankova et al. (2017) discovered that *A. thaliana* tips and roots exposed to ZnO NPs were lower concentrations of JA. However, they did observe that when ZnO NP concentrations were low, JA in *A. thaliana* leaves increased, and when concentrations were high, it decreased. Kasote et al. (2019) demonstrated the activation of JA-mediated defense responses in watermelon seedlings following seed preparation with FeNPs. It has been discovered that in some situations with additional abiotic and biotic constraints and low nanoparticle concentration, stress-coping JA signaling is upregulated, which ultimately activates defense responses and has detrimental effects on plants (Hernandez-Hernandez et al. 2018). Furthermore, it was found that in response to a biotic restriction (fungal infection of *G. fujikuroi*), copper sulfide nanoparticles increased the synthesis of JA in *O. sativa* (Shang et al. 2020).

A class of phytohormones called *brassinosteroids* regulates a variety of biological functions in plants, including photosynthesis, antioxidant defense mechanisms, osmolyte accumulation, nitrogen metabolism, and plant-water interaction under both favorable and unfavorable conditions (Ahanger et al. 2018). Certain studies have shown that plants exposed to metallic nanoparticles change the level of endogenous brassinosteroids. Phytotoxicity in plant roots and shoots at the concentration of brassinosteroids was significantly reduced by thin-walled carbon nanotubes (CNTs) in *O. sativa* L., which inhibited plant tolerance (Hao et al. 2016).

4.5 Crosstalk Between Plant Hormones and Nanoparticles as a Method of Controlling Plant Stress

The physiological function of plants is directly correlated with the modulation of their hormone pools, and these signaling molecules are also closely correlated with the control of plant growth under biotic and abiotic stress (Xiao et al. 2021). Because of this, it's crucial to characterize plant hormone metabolism and signal transduction in order to fully understand how regulatory networks function when environmental stress is present. Plants evolved a defensive reaction under the stress condition that raised their tolerance threshold through a variety of plant signals. The presence and activity of plant hormones serve as a crucial indicator of plant toxicity (Yang et al. 2017). Just starting to be understood is the intricate crosstalk between phytohormones and nanoparticles. Plant stress responses are influenced by interactions between plant hormones and nanoparticles that can be either synergistic or antagonistic. Plant hormones, which are versatile regulators of plant growth and development, have a lot of potential but have not yet been fully explored in research on the interactions between plants and nanoparticles (Vinkovic et al. 2017). Plant growth may be impacted either favorably or unfavorably depending on the concentration of metallic nanoparticles present (Tripathi et al. 2021). The plant hormones auxin and cytokinin are primarily responsible for these effects. In response to exposure to metal oxide-based nanoparticles, Vankova et al. (2017) linked the hormonal profile of *Arabidopsis* with its physiological state. They observed the plant's reaction to ZnO NP exposure at various concentrations of auxin, cytokinins, ABA, SA, and JA. It was found that the concentration of cytokinins and cytokinin phosphatase, their active precursor, could be regulated in *Arabidopsis* up to a very low ZnO NP concentration. The interaction between various plant growth hormones and the nanomaterial (graphene oxide) may help control the growth of *Brassica napus* roots (Xie et al. 2020). The regulatory function of nanoparticles in plants under stressful circumstances has been extensively studied in recent decades. However, only a few studies have shown that using nanomaterials to modulate the synthesis and signaling of various phytohormones and other environmental constraints can improve them. Here, illustrate the control mechanisms used by nanoparticles to reduce plant stress responses. However, more research needs to be done to determine precisely how nanomaterials alter the expression of genes linked to phytohormones important for plant growth (Apel and Hirt 2004; Torres and Dangl 2005). Both biotic and abiotic stress factors cause disruption in the equilibrium between the production and removal of ROS, which is present under normal metabolic conditions (optimal levels of ethylene, brassinosteroids, ABA, SA, ROS, and JA, among other things). It makes it easier for different phytohormones to be synthesized. However, under unfavorable circumstances, plant hormones are in charge of the growth of plants by quickening the plant's defense mechanism and eliminating ROS (Xia et al. 2015). Wheat, rice, peanuts, and other crops may become weakened by pathogens and abiotic stress. In commercial cash crops, excessive ROS production has been shown to cause severe yield loss and toxicity (Abhinandan et al. 2018; Trinh et al. 2014). Kashyap et al. (2020) also reported on the possibility of stress reduction through nanotechnology. According to studies, the use of nanoparticle therapy to treat *Solanum lycopersicum* nanoparticles may increase the antioxidant defense system's activity and hormonal signaling, which together may be the primary regulators of the negative effects of stress on plants (Hernandez-Hernandez et al. 2018). They listed phytohormone signaling and synthesis changes as a result of nanoparticle-mediated stress reduction in plants. SiO_2 NP treatment was found to be a cost-effective, safe, and long-lasting alternative to increasing the defense activity against pathogenic bacteria in *A. thaliana* (El-Shetehy et al. 2021).

4.6 The Impact of Nanomaterials and Nanofertilizers on Plant Growth

4.6.1 Nanomaterials in Crop Growth, Seed Germination, and Quality Enrichment

The development of cost-effective nanotechnology application approaches for plant growth, improved seed germination, development, and environmental adaptation are all parts of the new scientific innovation platform known as nanoscience. The development of seedlings and the survival of the plant during its life cycle are two important outcomes of the germination of seeds, which is a crucial stage that supports population dynamics. The genetic make-up, environmental aspects, soil moisture, and fertility are also major determinants of seed germination (Manjaiah et al. 2019). The germination, development, and growth of the plant are positively impacted by the nanomaterials that have been used, according to numerous studies. Multi-walled carbon nanotubes (MWCNTs) have been found to improve seed germination in barley, maize, soybean, tomato, maize, wheat, garlic, and peanut (Joshi et al. 2018). Similarly, applying nano-zeolite, TiO_2, and SiO_2 to cultivated plants promotes seed germination (Manjaiah et al. 2019). The mechanisms by which nanomaterials can promote germination are still unknown, according to studies on them (Disfani et al. 2017). Other studies have shown that nanomaterials stimulate the enzymatic system, increasing water utilization and absorption capacity, which, by penetrating the seed coat, improves seedling growth and germination (Lu et al. 2002). Studies with TiO_2, MWCNTs, FeO, ZnFeCu-oxide, ZnO and hydroxyl fullerene nanomaterials have also shown that they improve crop quality in wheat, onion, potato, mung bean, tomato, soybean, spinach, and peanut crop types, as well as crop development and growth (Gao et al. 2011). Additionally, a study on bitter melon found that seed dressings containing fullerol increased fruit yield, number, and size by up to 128% (*Momordica charantia*), as well as stimulating the content of bioactive compounds cucurbitacin B, inulin, and lycopene in fruits (Kole et al. 2013). *Dracocephalum moldavica* plants' essential oil content was improved along with their planting density and agronomic qualities after receiving fertilizer containing nano-iron (Yousefzadeh and Sabaghnia 2016). These results showed how nanomaterials could be used to raise product quality and yield. The ability of nanomaterials to absorb more nutrients and water may help partially explain the mechanism promoting plant growth and enriched quality, despite the fact that this mechanism is still unknown. As a result of the increased enzymatic activity, root systems become stronger (Shojaei et al. 2019). Conventional bulk fertilizer loses nitrogen in just 30 days with an erratic release, which has a negative impact on crop growth and lowers plant nutrient efficiency (Kottegoda et al. 2011). Furthermore, the length of these seedlings' shoots was adversely impacted, and it was found that the concentration of the nanomaterial application affected how much the product grew (Disfani et al. 2017). It has been demonstrated that the way nanomaterials are applied affects how well plants grow (El-Feky et al. 2013). The amount of total chlorophyll, iron content, essential oil levels, total carbohydrates, plant height, leaves per plant, branch per plant, dry weight, and fresh weight can all be significantly increased when nano-Fe_3O_4 is applied to the leaves of *Ocimum basilicum* plants. These results were reported to be in comparison to soil.

4.6.2 Use of Nanofertilizers to Boost Plant Growth

The use of raw or processed plant products and plant growth is known to be influenced by the presence of nitrogen in structural and functional proteins. Additionally, crops are unable to utilize nitrogen effectively. Nitrogen losses can be minimized by using nanofertilizers. Nanoporous materials or nanotubes can be used to encapsulate nutrients like K, P, and N. These nutrients can also be given as NPs or emulsions, or they can be coated with thin protective polymer films (DeRosa et al. 2010). The study, which used commercial fertilizer as a control, was carried out in Sri Lanka on soils with varying levels of acidity (pH 4.2, 5.2, and 7). Investigated was the nitrogen released from a nanofertilizer made of hydroxyapatite (HA) NPs modified with urea and embedded in the soft wood of *Gliricidia sepium*. The urea-modified nano-HA fertilizer encapsulated with *G. sepium* demonstrated a sustained and slow release of nitrogen at three different pH values in comparison to the control. After 30 days of application, the commercial fertilizer's release was low and uneven after beginning with a heavy release (Kottegoda et al. 2011), whereas the nanofertilizer released more than 10 mg of nitrogen to the soil 60 days after application. Along with nanofertilizers, some NPs also have the ability to improve plant growth directly. ZnO NPs were found to improve root growth and the phytohormone indole acetic acid (IAA) in *Cicer arietinum* (chickpea) (Pandey et al. 2010). Additionally, carbon nanotubes (CNTs) have been shown to penetrate the seed coat and encourage plant growth and germination (Khodakovskaya et al. 2009). Tomato seeds were placed in the common Murashige and Skoog medium, which is based on agar, and exposed to various CNT concentrations (0, 10, 20, 40 g/mL). The results showed that while the seeds in the CNT-treated experiments began to grow on the third day, this was not the case for the seeds in the control experiment, which lacked CNTs. CNT-treated tomato seedlings had stronger stems and a nicer appearance. According to their findings, *Cicer arietinum* plants treated with CNT exhibited faster root, shoot, and branch growth rates. As a result, they discovered that plants can absorb more water when water-soluble CNTs are attached to cell walls (Tripathi et al. 2011). In order to maximize water absorption and transport and prevent water losses in agriculture, they came to this conclusion.

4.7 Phytotoxicity of Nanoparticles

Despite their many beneficial effects, the toxicity of nanoparticles in plants cannot be disregarded. In addition to encouraging plant oxidative stress and physically harming plants by causing stomatal closure as a result of nanoparticle accumulation, the toxicity of nanoparticles can harm plants in a number of other ways as well (Gao et al. 2023). High concentrations of NPs that are sprayed onto the leaves cause them to collect on the surface of the leaf. This will lead to stomatal occlusion, which will prevent gas exchange and photosynthesis in the plants. According to some studies, when Zn and Cu NP concentrations are too high, the rate of plant growth slows down, and the leaves start to turn yellow. *H. sativum* is poisoned by CuO NPs, which also harm the plant by delaying photosynthesis and inhibiting the growth of roots and stems (Chaplygin et al. 2021).

Current research on polymer-based nanoparticles focuses primarily on the toxicity of these materials when applied to plants. For instance, when exposed to plastic particles of 50 and 500 nm for 8 h, *Lepidium sativum* seeds' germination rates fell by 56% and 46%, respectively (Bosker et al. 2019). Following a 30-day exposure to PLA nanoparticles, the germination rate of ryegrass fell by 6% (Boots et al. 2019). However, after being exposed to PS NPs for 72 h, wheat seeds and onions showed no difference in the rate of germination (Boots et al. 2019). As a result, prolonged exposure of plants to polymer nanoparticles will reduce the germination rate, which is primarily brought on by the plants' decreased capacity to absorb nutrients as a result of the nanoparticles' pore-blocking effects. In addition to their impact on plant seed germination, polymer nanoparticles have also been shown in other studies to affect plant seedling growth (Kalcikova 2020). The polyethylene nanoparticles also caused the *Lemna minor* root's length to shorten. When *Arabidopsis thaliana* was exposed to polystyrene nanoparticles with different surface charges (PS-NH_2, PS-SO_3H), its fresh weight was decreased by an average of 50%. A 15% reduction in plant length and an average 30% reduction in root length were observed. The inhibition of *Arabidopsis thaliana* seedlings increased with nanoparticle concentration (Sun et al. 2020).

Nanoparticles are genotoxic in addition to making plants produce ROS and causing hormonal changes. Nanoparticles exert genotoxicity that can interfere with plant cell division by interacting with biological macromolecules like lipids and nuclei (Youssef and Elamawi 2020) (Fig. 4.2).

Plant germination and pigment content are not affected by Fe-coated zinc oxide nanoparticles (Sturikova et al. 2018).

Plant roots actively transport nutrients and NPs from the soil to their roots, which is thought to be a significant reservoir of NPs (Khan et al. 2021). After entering the plant through the roots, NPs travel to the upper parts of the plant where they can be extremely toxic before entering the root cortex and epidermis of cell walls. The natural fauna, such as bacteria, fungi, and nematodes, are negatively impacted by NPs, which are absorbed by soil systems (Khan et al. 2021). Nanotechnology has become a crucial tool for enhancing environmental safety, reducing energy consumption, and resolving a number of medical issues. It is regarded as a crucial strategy that can boost manufacturing output while lowering costs (Khan et al. 2015). Each method of synthesis for NPs has pros and cons, and they can be produced physically, chemically, biologically, or mechanically. The physical method of NP synthesis costs money, whereas chemical methods can have slow growth rates and environmental risks (Pandey and Jain 2020). Since the biological method of NP

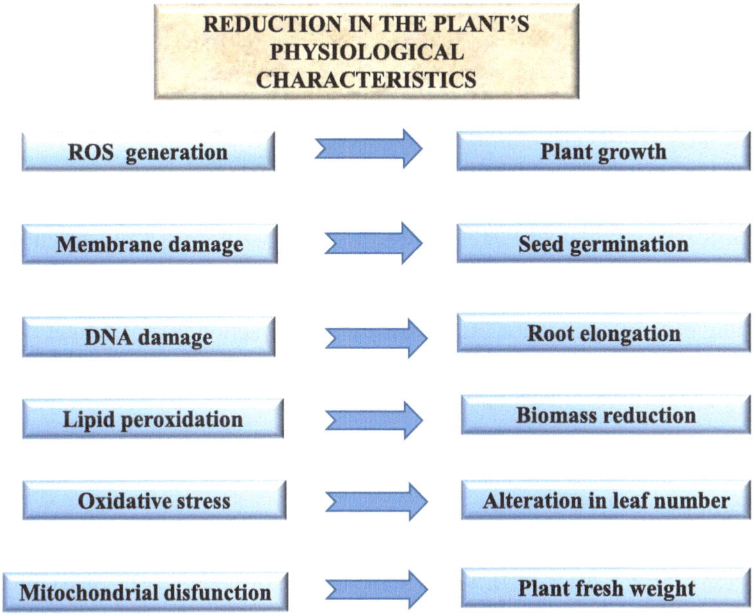

Fig. 4.2 Effect of ENM toxicity on plant physiology

synthesis is non-toxic and friendly to the environment, it is more widely accepted than conventional techniques (Patra and Baek 2014).

4.8 Conclusion and Future Remarks

For the development and sustainability of plant production in the future, it is essential to comprehend the control of phytohormone synthesis and signal transmission under stressful circumstances. In addition to discussing how nanoparticles affect plant growth, it also examines how plants absorb and transport them. When plant cells are under stress, nanoparticles may disrupt a variety of hormone biosynthesis and signaling pathways. Plant roots, primary roots, damaged areas, and the stomata and pores of the stratum corneum of the stem cell wall are where nanoparticles are absorbed by plant leaves. Both symplastic and apoplastic pathways are used to transport these nanoparticles inside the plant body as well as the xylem and phloem to move them between different plant tissues. The physical properties of nanoparticles, such as size, chemical composition, surface charge, and surface modification, influence the absorption and transport process. Additionally, while being transported inside a plant, nanoparticles have various effects on plant seed germination and growth. Different nanoparticles can harm plants at high concentrations while stimulating plant growth at low concentrations. Changes in the concentrations of plant hormones and even changes in gene expression in plants will result from the genotoxic effects of nanoparticles, such as chromosome aberration and the formation of micronuclei. Despite widespread confirmation of nanoparticles' negative effects on plants, little is understood about the mechanism underlying plant toxicity. Additionally, NPs can stunt plant growth and damage plant cells, which compromise the integrity of cell walls and result in membrane damage and cytoplasm contraction. It primarily depends on the number of NPs exposed. Lower concentrations of many NP species have been shown to support plant growth, while higher concentrations of NPs are likely to harm plants. As a result, the species of nanoparticles used and their concentration depend on how bio-based natural polymer nanoparticles are applied as an emerging technology to improve agricultural output, as does the timing of their application. To maximize its interaction in a particular environment, it should also be further investigated. As a result, these technologies aid in expected plant cultivation and production.

References

Abhinandan K, Skori L, Stanic M, Hickerson N, Jamshed M, Samuel MA (2018) Abiotic stress signaling in wheat–an inclusive overview of hormonal interactions during abiotic stress responses in wheat. Front Plant Sci 9:734

Ahanger MA, Ashraf M, Bajguz A, Ahmad P (2018) Brassinosteroids regulate growth in plants under stressful environments and crosstalk with other potential phytohormones. J Plant Growth Regul 37(4):1007–1024

Apel K, Hirt H (2004) Reactive oxygen species: metabolism, oxidative stress, and signal transduction. Annu Rev Plant Biol 55:373–399

Azhar W, Khan AR, Muhammad N et al (2020) Ethylene mediates CuO NP-induced ultrastructural changes and oxidative stress in *Arabidopsis thaliana* leaves. Environ Sci Nano 7(3):938–953

Azhar BJ, Noor A, Zulfiqar A et al (2021) Effect of ZnO, SiO2 and composite nanoparticles on *Arabidopsis thaliana* and involvement of ethylene and cytokinin signaling pathways. Pak J Bot 53(2):437–446

Boots B, Russell CW, Green DS (2019) Effects of microplastics in soil ecosystems: above and below ground. Environ Sci Technol 53:11496–11506

Bosker T, Bouwman LJ, Brun NR, Behrens P, Vijver MG (2019) Microplastics accumulate on pores in seed capsule and delay germination and root growth of the terrestrial vascular plant Lepidium sativum. Chemosphere 226:774–781

Boss PK, Bastow RM, Mylne JS, Dean C (2004) Multiple pathways in the decision to flower: enabling, promoting, and resetting. Plant Cell 16(Suppl 1):S18–S31

Cai L, Cai L, Jia H, Liu C, Wang D, Sun X (2020) Foliar exposure of Fe3O4 nanoparticles on *Nicotiana benthamiana*: evidence for nanoparticles uptake, plant growth promoter and defense response elicitor against plant virus. J Hazard Mater 393:122415

Chaplygin VA, Bauer TV, Sushkova SN et al (2021) The toxic effect of CuO of different dispersion degrees on the structure and ultrastructure of spring barley cells (Hordeum sativum distichum). Environ Geochem Health 43:1673–1687

Chen H, Seiber JN, Hotze M (2014) ACS select on nanotechnology in food and agriculture: a perspective on implications and applications. J Agric Food Chem 62:1209–1212

Chen G, Qiu J, Liu Y, Jiang R, Cai S, Liu Y, Zhu F, Zeng F, Luan T, Ouyang G (2015) Carbon nanotubes act as contaminant carriers and translocate within plants. Sci Rep 5:15682

Davies PJ (ed) (2012) Plant hormones and their role in plant growth and development. Springer Science & Business Media

DeRosa MC, Monreal C, Schnitzer M, Walsh R, Sultan Y (2010) Nanotechnology infertilizers. Nat Nanotechnol 5(2):91

Dikshit A, Shukla SK, Mishra RK (2013) Exploring nanomaterials with PGPR in current agricultural scenario. Lap Lambert Academic Publishing, Saarbrucken. ISBN: 978-3-659-36774-8

Dimkpa CO, McLean JE, Britt DW, Anderson AJ (2013) Antifungal activity of ZnO nanoparticles and their interactive effect with a biocontrol bacterium on growth antagonism of the plant pathogen *Fusarium graminearum*. Biometals 26:913–924

Disfani MN, Mikhak A, Kassaee MZ, Maghari A (2017) Effects of nano Fe/SiO2 fertilizers on germination and growth of barley and maize. Arch Agron Soil Sci 63:817–826

El-Feky SA, Mohammed MA, Khater MS, Osman YA, Elsherbini E (2013) Effect of magnetite nano-fertilizer on growth and yield of Ocimum basilicum L. Int J Indig Med Plants 46:1286–1293

El-Shetehy M, Moradi A, Maceroni M et al (2021) Silica nanoparticles enhance disease resistance in Arabidopsis plants. Nat Nanotechnol 16(3):344–353

FAO (2008) An international technical workshop. Investing in sustainable crop intensification. The case for improving soil health, 22–24 Jul 2008, Integrated crop management, vol 6. FAO, Rome

Gao J, Wang Y, Folta KM, Krishna V, Bai W, Indeglia P, Georgieva A, Nakamura H, Koopman B, Moudgil B (2011) Polyhydroxy fullerenes (fullerols or fullerenols): Beneficial effects on growth and lifespan in diverse biological models. PLoS One 6:e19976

Gao M, Chang J, Wang Z, Zhang H, Wang T (2023) Advances in transport and toxicity of nanoparticles in plants. J Nanobiotechnol 21:75

Ghodake G, Seo YD, Lee DS (2011) Hazardous phytotoxic nature of cobalt and zinc oxide nanoparticles assessed using Allium cepa. J Hazard Mater 186:952–955

Ghorbel M, Brini F, Sharma A, Landi M (2021) Role of jasmonic acid in plants: the molecular point of view. Plant Cell Rep 40(8):1471–1494

Grillo R, Mattos BD, Antunes DR, Forini MML, Monikh FA, Rojas OJ (2021) Foliage adhesion and interactions with particulate delivery systems for plant nanobionics and intelligent agriculture. Nano Today 37:101078

Haberer G, Kieber JJ (2002) Cytokinins. New insights into a classic phytohormone. Plant Physiol 128(2):354–362

Hai NN, Chuong NN, Tu NHC, Kisiala A, Hoang XLT, Thao NP (2020) Role and regulation of cytokinins in plant response to drought stress. Plants 9(4):422

Hao Y, Yu F, Lv R et al (2016) Carbon nanotubes filled with different ferromagnetic alloys affect the growth and development of rice seedlings by changing the C:n ratio and plant hormones concentrations. PLoS One 11(6):e0157264

Hernandez-Hernandez H, Juarez-Maldonado A, Benavides-Mendoza A, Ortega-Ortiz H, Cadenas-Pliego G, Sanchez-Aspeytia D, Gonzalez-Morales S (2018) Chitosan-PVA and copper nanoparticles improve growth and overexpress the SOD and JA genes in tomato plants under salt stress. Agronomy 8(9):175

Hernandez-Viezcas JA, Castillo-Michel H, Andrews JC, Cotte M, Rico C, Peralta-Videa JR, Gardea-Torresdey JL (2013) In situ synchrotron X-ray fluorescence mapping and speciation of CeO2 and ZnO nanoparticles in soil cultivated soy-bean (*Glycine max*). ACS Nano 7:1415–1423

Hong J, Peralta-Videa JR, Rico C, Sahi S, Viveros MN, Bartonjo J, Zhao LJ, Gardea-Torresdey JL (2014) Evidence of translocation and physiological impacts of foliar applied CeO2 nanoparticles on cucumber (Cucumis sativus) plants. Environ Sci Technol 48:4376–4385

Iqbal N, Khan NA, Ferrante A, Trivellini A, Francini A, Khan MIR (2017) Ethylene role in plant growth, development and senescence: interaction with other phytohormones. Front Plant Sci 8:475

Jaberzadeh A, Moaveni P, Moghadam HR, Zahedi H (2013) Influence of bulk and nanoparticles titanium foliar application on some agronomic traits, seed gluten and starch contents of wheat subjected to water deficit stress. Not Bot Horti Agrobot 41:201–207

Janmohammadi M, Navid A, Segherloo AE, Sabaghnia N (2016) Impact of nano-chelated micronutrients and biological fertilizers on growth performance and grain yield of maize under deficit irrigation condition. Biologija 62:134–147

Jonapa-Hernandez F, Gutierrez-Miceli F, Santos-Espinosa A, Ruíz-Lau N, Ruíz-Valdiviezo V, Valdez-Salas B, Gonzalez-Mendoza D (2020) Foliar application of green nanoparticles in *Annona muricata* L. plants and their effects in physiological and biochemical parameters. Biocatal Agric Biotechnol 28:101751

Joshi A, Kaur S, Dharamvir K, Nayyar H, Verma G (2018) Multi-walled carbon nanotubes applied through seed-priming influence early germination, root hair, growth and yield of bread wheat (Triticum aestivum L.). J Sci Food Agric 98:3148–3160

Kalcikova G (2020) Aquatic vascular plants—a forgotten piece of nature in microplastic research. Environ Pollut 262:114354

Kashyap PL, Kumar S, Jasrotia P, Singh DP, Singh GP (2020) Nanotechnology in wheat production and protection. In: Environmental nanotechnology, vol 4. Springer, Cham, pp 165–194

Kasote DM, Lee JH, Jayaprakasha GK, Patil BS (2019) Seed priming with iron oxide nanoparticles modulate antioxidant potential and defense-linked hormones in watermelon seedlings. ACS Sustain Chem Eng 7(5):5142–5151

Khan M, Naqvi AH, Ahmad M (2015) Comparative study of the cytotoxic and genotoxic potentials of zinc oxide and titanium dioxide nanoparticles. Toxicol Rep 2:765–774. https://doi.org/10.1016/j.toxrep.2015.02.004

Khan N, Bano A, Ali S, Babar M (2020) Crosstalk amongst phytohormones from planta and PGPR under biotic and abiotic stresses. Plant Growth Regul 90(2):189–203

Khan M, Khan MSA, Borah KK, Goswami Y, Hakeem KR, Chakrabartty I (2021) The potential exposure and hazards of metal-based nanoparticles on plants and environment, with special emphasis on ZnO NPs, TiO2 NPs, and AgNPs: a review. Environ Adv 6:100128. https://doi.org/10.1016/j.envadv.2021.100128

Khan I, Awan SA, Rizwan M, Hassan ZU, Akram MA, Tariq R, Brestic M, Xie W (2022) Nanoparticle's uptake and translocation mechanisms in plants via seed priming, foliar treatment, and root exposure: a review. Environ Sci Pollut Res 29:89823–89833

Khodakovskaya M, Dervishi E, Mahmood M, Xu Y, Li Z, Watanabe F, Biris AS (2009) Carbon nanotubes are able to penetrate plant seed coat and dramatically affect seed germination and plant growth. ACS Nano 3(10):3221–3227

Kim B, Park CS, Murayama M, Hochella MF (2010) Discovery and characterization of silver sulfide nanoparticles in final sewage sludge products. Environ Sci Technol 44:7509–7514

Kole C, Kole P, Randunu KM, Choudhary P, Podila R, Ke PC, Rao AM, Marcus RK (2013) Nanobiotechnology can boost crop production and quality: first evidence from increased plant biomass, fruit yield and phytomedicine content in bitter melon (*Momordica charantia*). BMC Biotechnol 13:37

Koo YM, Heo AY, Choi HW (2020) Salicylic acid as a safe plant protector and growth regulator. Plant Pathol J 36(1):1

Kottegoda N, Munaweera I, Madusanka N, Karunaratne V (2011) A green slow-release fertilizer composition based on urea-modified hydroxyapatite nanoparticles encapsulated wood. Curr Sci 101:73–78

Liu X, Wang F, Shi Z, Tong R, Shi X (2015) Bioavailability of Zn in ZnO nanoparticle-spiked soil and the implications to maize plants. J Nanopart Res 17:11

Lu C, Zhang C, Wen J, Wu G, Tao MX (2002) Research of the effect of nanometer materials on germination and growth enhancement of glycine max and its mechanism. Soybean Sci 21:168–171

Lu Q, He ZL, Stoffella PJ (2012) Land application of biosolids in the USA: a review. Appl Environ Soil Sci 2012:201462. https://doi.org/10.1155/2012/201462

Ma CX, White JC, Dhankher OP, Xing B (2015) Metal-based nanotoxicity and detoxification pathways in higher plants. Environ Sci Technol 49(12):7109–7122

Manjaiah KM, Mukhopadhyay R, Paul R, Datta SC, Kumararaja P, Sarkar B (2019) Clay minerals and zeolites for environmentally sustainable agriculture. In: Mariano M, Binoy S, Alessio L (eds) Modified clay and zeolite nanocomposite materials. Elsevier, Amsterdam, pp 309–329

McCall CA, Jordan KS, Habash MB, Dunfield KE (2015) Monitoring *Bacteroides spp.* markers, nutrients, metals and *Escherichia coli* in soil and leachate after land application of three types of municipal biosolids. Water Res 70:255–265

Monreal CM, De Rosa M, Mallubhotla SC, Bindraban PS, Dimkpa C (2016) Nanotechnologies for increasing the crop use efficiency of fertilizer-micronutrients. Biol Fertil Soils 52:423–437

Morales-Dıaz AB, Ortega-Ortız H, Juarez-Maldonado A, Cadenas-Pliego G, Gonzalez-Morales S, Benavides-Mendoza A (2017) Application of nanoelements in plant nutrition and its impact in ecosystems. Adv Nat Sci Nanosci Nanotechnol 8:013001, 13 pp

Nair R (2016) Effects of nanoparticles on plant growth and development. In: Plant nanotechnology: principles and practices. Springer, pp 95–118

Pandey G, Jain P (2020) Assessing the nanotechnology on the grounds of costs, benefits, and risks. beni-suef uni. J Basic App Sci 9(1):1–10. https://doi.org/10.1186/s43088-020-00085-5

Pandey AC, Sanjay SS, Yadav RS (2010) Application of ZnO nanoparticles in influencing the growth rate of Cicer arietinum. J Exp Nanosci 5(6):488–497

Patra JK, Baek KH (2014) Green nanobiotechnology: factors affecting synthesis and characterization techniques. J Nanomater 2014:417305. https://doi.org/10.1155/2014/417305

Peng D, Zhang Y, Li Q et al (2020) Exogenous application and endogenous elevation of salicylic acid levels by overexpressing a salicylic acid-binding protein 2 gene enhance nZnO tolerance of tobacco plants. Plant Soil 450(1):443–461

Rahman KM, Zhang D (2018) Effects of fertilizer broadcasting on the excessive use of inorganic fertilizers and environmental sustainability. Sustain For 10(3):759

Rajaee Behbahani S, Iranbakhsh A, Ebadi M, Majd A, Ardebili ZO (2020) Red elemental selenium nanoparticles mediated substantial variations in growth, tissue differentiation, metabolism, gene transcription, epigenetic cytosine DNA methylation, and callogenesis in bittermelon (*Momordica charantia*); an *in vitro* experiment. PLoS One 15(7):e0235556

Rajput V, Minkina T, Mazarji M, Shende S, Sushkova S, Mandzhieva S, Burachevskaya M, Chaplygin V, Singh A, Jatav H (2020) Accumulation of nanoparticles in the soil-plant systems and their effects on human health. Ann Agric Sci 65:137–143

Rasool N (2022) Plant hormones: role in alleviating biotic stress

Rico CM, Majumdar S, Duarte-Gardea M, Peralta-Videa JR, Gardea-Torresdey JL (2011) Interaction of nanoparticles with edible plants and their possible implications in the food chain. J Agric Food Chem 59:3485–3498

Rico CM, Peralta-Videa JR, Gardea-Torresdey JL (2015) Chemistry, biochemistry of nanoparticles, and their role in antioxidant defense system in plants. In: Nanotechnology and plant sciences. Springer International Publishing, pp 1–17

Rottman J, Shadman F, Sierra-Alvarez R (2012) Interactions of inorganic oxide nanoparticles with sewage biosolids. Water Sci Technol 66:1821–1827

Servin AD, White JC (2016) Nanotechnology in agriculture: Next steps for understanding engineered nanoparticle exposure and risk. Nano Impact 1:9–12

Shang H, Ma C, Li C, White JC, Polubesova T, Chefetz B, Xing B (2020) Copper sulfide nanoparticles suppress *Gibberella fujikuroi* infection in rice (*Oryza sativa* L.) by multiple mechanisms: contact-mortality, nutritional modulation and phytohormone regulation. Environ Sci Nano 7(9):2632–2643

Sheteiwy MS, Dong Q, An J et al (2017) Regulation of ZnO nanoparticles-induced physiological and molecular changes by seed priming with humic acid in *Oryza sativa* seedlings. Plant Growth Regul 83:27–41. https://doi.org/10.1007/s10725-017-0281-4

Shojaei TR, Salleh MAM, Tabatabaei M, Mobli H, Aghbashlo M, Rashid SA, Tan T (2019) Applications of nanotechnology and carbon nanoparticles in agriculture. In: Suraya AR, Raja NIRO, Mohd ZH (eds) Synthesis, technology and applications of carbon nanomaterials. Elsevier, Amsterdam, pp 247–277

Siddiqui MH, Al-Whaibi MH, Faisal M, Al Sahli AA (2014) Nano-silicon dioxide mitigates the adverse effects of salt stress on *Cucurbita pepo* L. Environ Toxicol Chem 33:2429–2437. https://doi.org/10.1002/etc.2697

Singla R, Kumari A, Yadav SK (2019) Impact of nanomaterials on plant physiology and functions. In: Husen A, Iqbal M (eds) Nanomaterials and plant potential. Springer International Publishing, Cham, pp 349–377

Sturikova H, Krystofova O, Huska D, Adam V (2018) Zinc, zinc nanoparticles and plants. J Hazard Mater 349:101–110

Sun X-D, Yuan X-Z, Jia Y, Feng L-J, Zhu F-P, Dong S-S, Liu J, Kong X, Tian H, Duan J-L et al (2020) Differentially charged nanoplastics demonstrate distinct accumulation in Arabidopsis thaliana. Nat Nanotechnol 15:755–760

Sytar O, Kumari P, Yadav S, Brestic M, Rastogi A (2019) Phytohormone priming: regulator for heavy metal stress in plants. J Plant Growth Regul 38(2):739–752

Syu YY, Hung JH, Chen JC, Chuang HW (2014) Impacts of size and shape of silver nanoparticles on Arabidopsis plant growth and gene expression. Plant Physiol Biochem 83:57–64

Thuesombat P, Hannongbua S, Akasit S, Chadchawan S (2014) Effect of silver nanoparticles on rice (*Oryza sativa* L. cv. KDML 105) seed germination and seedling growth. Ecotoxicol Environ Saf 104:302–309

Torres MA, Dangl JL (2005) Functions of the respiratory burst oxidase in biotic interactions, abiotic stress and development. Curr Opin Plant Biol 8(4):397–403

Trinh NN, Huang TL, Chi WC, Fu SF, Chen CC, Huang HJ (2014) Chromium stress response effect on signal transduction and expression of signaling genes in rice. Physiol Plant 150(2):205–224

Tripathi S, Sonkar SK, Sarkar S (2011) Growth stimulation of gram (*Cicer arietinum*) plant by water soluble carbon nanotubes. Nanoscale 3(3):1176–1181

Tripathi D, Rai KK, Pandey-Rai S (2021) Impact of green synthesized WcAgNPs on *in-vitro* plant regeneration and withanolides production by inducing key biosynthetic genes in *Withania coagulans*. Plant Cell Rep 40(2):283–299

Vankova R, Landa P, Podlipna R et al (2017) ZnO nanoparticle effects on hormonal pools in Arabidopsis thaliana. Sci Total Environ 593:535–542

Vinkovic T, Novak O, Strnad M, Goessler W, Jurasin DD, Parađikovic N, Vrcek IV (2017) Cytokinin response in pepper plants (*Capsicum annuum* L.) exposed to silver nanoparticles. Environ Res 156:10–18

Vishwakarma K, Upadhyay N, Kumar N et al (2017) Abscisic acid signaling and abiotic stress tolerance in plants: a review on current knowledge and future prospects. Front Plant Sci 8:161

Wang A, Zheng Y, Peng F (2014) Thickness-controllable silica coating of CdTe QDs by reverse microemulsion method for the application in the growth of rice. J Spectrosc 2014:169245

Wang X, Yang X, Chen S, Li Q, Wang W, Hou C, Gao X, Wang L, Wang S (2016) Corrigendum on zinc oxide nanoparticles affect biomass accumulation and photosynthesis in arabidopsis. Front Plant Sci 7:559

Wang P, Lombi E, Sun S et al (2017) Characterizing the uptake, accumulation and toxicity of silver sulfide nanoparticles in plants. Environ Sci Nano 4(2):448–460

Wang Y, Zhang P, Li M, Guo Z, Ullah S, Rui Y, Lynch I (2020a) Alleviation of nitrogen stress in rice (*Oryza sativa*) by ceria nanoparticles. Environ Sci Nano 7(10):2930–2940

Wang J, Song L, Gong X, Xu J, Li M (2020b) Functions of jasmonic acid in plant regulation and response to abiotic stress. Int J Mol Sci 21(4):1446

Wei J, Zou Y, Li P, Yuan X (2020) Titanium dioxide nanoparticles promote root growth by interfering with auxin pathways in *Arabidopsis thaliana*. Phyton 89(4):883–891

Xia XJ, Zhou YH, Shi K, Zhou J, Foyer CH, Yu JQ (2015) Interplay between reactive oxygen species and hormones in the control of plant development and stress tolerance. J Exp Bot 66(10):2839–2856

Xiao Z, Yue L, Wang C et al (2021) Downregulation of the photosynthetic machinery and carbon storage signaling pathways mediate La2O3 nanoparticle toxicity on radish taproot formation. J Hazard Mater 411:124971

Xie L, Chen F, Du H, Zhang X, Wang X, Yao G, Xu B (2020) Graphene oxide and indole-3-acetic acid cotreatment regulates the root growth of *Brassica napus* L. *via* multiple phytohormone pathways. BMC Plant Biol 20(1):1–12

Xiong T, Zhang S, Kang Z, Zhang T, Li S (2021a) Dose-dependent physiological and transcriptomic responses of lettuce (*Lactuca sativa* L.) to copper oxide nanoparticles—insights into the phytotoxicity mechanisms. Int J Mol Sci 22(7):3688

Xiong T, Zhang T, Xian Y, Kang Z, Zhang S, Dumat C, Shahid M, Li S (2021b) Foliar uptake, biotransformation, and impact of CuO nanoparticles in Lactuca sativa L. var. ramosa. Hortic Environ Geochem Health 43:423–439

Yang Y, Wang Y, Westerhoff P, Hristovski K, Jin VL, Johnson MVV, Arnold JG (2014) Metal and nanoparticle occurrence in biosolid-amended soils. Sci Total Environ 485–486:441–449

Yang J, Cao W, Rui Y (2017) Interactions between nanoparticles and plants: phytotoxicity and defense mechanisms. J Plant Interact 12(1):158–169

Yousefzadeh S, Sabaghnia N (2016) Nano-iron fertilizer effects on some plant traits of dragonhead (*Dracocephalum moldavica* L.) under different sowing densities. Acta Agric Slov 107:429–437

Youssef MS, Elamawi RM (2020) Evaluation of phytotoxicity, cytotoxicity, and genotoxicity of ZnO nanoparticles in *Vicia Faba*. Environ Sci Pollut Res 27:18972–18984

Yue L, Chen F, Yu K, Xiao Z, Yu X, Wang Z, Xing B (2019) Early development of apoplastic barriers and molecular mechanisms in juvenile maize roots in response to La2O3 nanoparticles. Sci Total Environ 653:675–683

Zahedi SM, Abdelrahman M, Hosseini MS, Hoveizeh NF, Tran LSP (2019) Alleviation of the effect of salinity on growth and yield of strawberry by foliar spray of selenium-nanoparticles. Environ Pollut 253:246–258

Zhao LJ, Peralta-Videa JR, Ren M, Varela-Ramirez A, Li C, Hernandez-Viezcas JA, Aguilerad RJ, Gardea-Torresdeya JL (2012) Transport of Zn in a sandy loam soil treated with ZnO NPs and uptake by corn plants: electron microprobe and confocal microscopy studies. Chem Eng J 184:1–8

Zhao L, Peralta-Videa JR, Rico CM, Hernandez-Viezcas JA, Sun Y, Niu G, Servin A, Nunez JE, Duarte-Gardea M, Gardea-Torresdey JL (2014) CeO2 and ZnO nanoparticles change the nutritional qualities of cucumber (*Cucumis sativus*). J Agric Food Chem 62:2752–2759

Chapter 5
Organic Nanoparticles as Delivery Tools for Bio-Based Antimicrobials

Paulo R. Oliveira-Pinto, Juliana Oliveira-Fernandes, Leandro Pereira-Dias, Rose M. O. F. Sousa, and Conceição Santos

Abstract The application of antibiotics and agrochemicals has been increasingly restricted due to health and environmental concerns and the development of resistance by phytopathogens. There is an urge to develop eco-friendly and sustainable strategies for phytopathogen management. Recently, a myriad of natural antimicrobials such as plant-based products (e.g., essential oils, plant extracts, and secondary metabolites), antimicrobial peptides (e.g., peptaibiotics), and other less-studied molecules, like enzymes, have been proposed as green solutions. However, there are some important challenges to the application of these natural antimicrobials in the field, mainly related to their rapid degradation and/or volatility, as well as quick lixiviation from the plant. Several strategies to overcome these handicaps have been proposed, with the most promising being the encapsulation of these natural antimicrobials in stable, organic-based nanoparticles (e.g., lignin, chitosan, and zein). Recent works have shown that nanoparticle encapsulation increased the antimicrobial activity of natural antimicrobials against multiresistant plant pathogens, as well

---

P. R. Oliveira-Pinto · J. Oliveira-Fernandes · C. Santos
Biology Department, Faculty of Sciences, University of Porto (FCUP), Porto, Portugal

LAQV-REQUIMTE, iB2Lab, Faculty of Sciences, University Porto (FCUP), Porto, Portugal

L. Pereira-Dias
Biology Department, Faculty of Sciences, University of Porto (FCUP), Porto, Portugal

LAQV-REQUIMTE, iB2Lab, Faculty of Sciences, University Porto (FCUP), Porto, Portugal

Instituto de Conservación y Mejora de la Agrodiversidad Valenciana (COMAV), Universitat Politècnica de València (UPV), Valencia, Spain

Rose M. O. F. Sousa (✉)
Biology Department, Faculty of Sciences, University of Porto (FCUP), Porto, Portugal

GreenUPorto/Inov4Agro, Faculty of Sciences, University of Porto (FCUP), Porto, Portugal

CITAB/Inov4Agro, Universidade de Trás-os-Montes e Alto Douro, Vila Real, Portugal
e-mail: rose.sousa@fc.up.pt

M. Khan, J.-T. Chen (eds.), *Nanoparticles in Plant Biotic Stress Management*,
https://doi.org/10.1007/978-981-97-0851-2_5

as their solubility and stability, while being nontoxic for the plant. Furthermore, recent research points to the possibility that chitosan and zein-based nanoparticles might present antimicrobial activity against some important plant pathogens, and thus could pose an interesting alternative to the more common Ag or Zn-based nanoparticles, currently employed for phytopathogen management. These were recently found to affect root morphology and impair the growth of maize as well as to be potentially harmful to the central nervous system of mammals. Moreover, nature-based nanoparticles might display a potential elicitation effect on the plant, enhancing its ability to respond to pathogen infection. Due to the insufficient knowledge in this field, nanoparticles should be the focus of future research regarding their antimicrobial properties against plant pathogens. Furthermore, their elicitor potential should be further investigated. This review provides the first insight into the current strategies employed for antimicrobials' encapsulation and the effectiveness of current nanoformulations to control plant pathogens.

Keywords Antibiotics · Agrochemicals · Nanoformulations · Phytopathogen management · Plant extracts

5.1 Introduction

Phytopathogens are responsible for crop production losses costing over 290 billion dollars to the global economy each year, jeopardizing food security and safety worldwide (IPPC Secretariat 2021). Current agricultural systems heavily rely on synthetic agrochemicals to control phytopathogens (Brauer et al. 2019; Sundin and Wang 2018). Agrochemicals, although effective, have numerous health and environmental drawbacks associated with their application (Brauer et al. 2019). Plus, conventional agrochemical formulas are hugely inefficient. It is estimated that 90% of applied agrochemicals are lost during or after application through lixiviation and degradation, and less than 10% will indeed be effective against the target organisms (Menossi et al. 2021; Nuruzzaman et al. 2016). Moreover, these formulations lack sustained-release function, which leads to the need for periodical and excessive applications to maintain an effective level of protection (Nuruzzaman et al. 2016; Zhao et al. 2018). As a result, there is a pressing need for cost-efficient, high-performing, bio-based formulas that are less harmful to the environment while providing a good level of protection to crops (Zhao et al. 2018).

In addition to the damage caused during cultivation, phytopathogens can have a tremendous impact during the postharvest stages (Garcia et al. 2022; Mohammadi et al. 2015a; Upadhyay et al. 2021). Moreover, along with lipid peroxidation, these are the main agents causing food quality deterioration (i.e., development of off-flavor and or off-odor and/or loss of essential dietary constituents) and mycotoxin contamination (Ali et al. 2021; Das et al. 2023). Normally, refrigeration is the method of choice for slowing down ripening and fungal decay; however, its efficacy is limited. Hence, it is frequently coupled with other methods to avoid produce spoilage. Among those, biocides and chemical preservatives' application is particularly common to extend the shelf life of food products (Chaudhari et al. 2022; Mondéjar-López et al. 2022). However, these also carry health and environmental drawbacks (Ali et al. 2021; Menossi et al. 2021). Thus, the agrifood sector is in urgent need of bio-based solutions to minimize postharvest losses without increasing the toll on the environment and the consumer's health.

Thus, during the last decade, a myriad of eco-friendly and sustainable strategies have been proposed as viable alternatives to conventional agrochemicals for the mitigation of phytopathogens (Menossi et al. 2021; Oliveira-Fernandes et al. 2023). Among these, the application of microbial organisms acting as pathogen antagonists (Rahman et al. 2018), plant host symbionts (White et al. 2018), plant growth promoters (Fusco et al. 2022), natural antimicrobial compounds (Menossi et al. 2021), enzymes (Narendrakumar et al. 2018), and antimicrobial peptides (Pereira-Dias et al. 2023), which directly act on the phytopathogen's cells and/or stimulate plant defenses, have been in the spotlight. The use of bio-based antimicrobial products carries major advantages, such as the absence of bioaccumulation phenomena or the development of phytopathogenic resistance (Chandra et al. 2017; Pereira-Dias et al. 2023; Yu et al. 2020).

However, these also carry several handicaps that hinder their application. For instance, plant-based products like essential oils, (EOs) plant extracts, phytohormones, phenolics, etc. are extremely volatile and degradable and have poor solubility in water-based solutions (Menossi et al. 2021; Wang et al. 2022). Plus, their direct application in foods is not always ideal due to the negative impact on food organoleptic properties (Mahanta et al. 2021; Martínez-Hernández et al. 2017; Shavisi et al. 2017). Likewise, the application of biological control agents (BCAs) and plant growth promoters (PGPs) to control phytopathogens could disrupt beneficial endemic communities (Huang et al. 2021). In addition, most BCAs and PGPs exhibit a certain degree of specificity and dependency on the crop and the environmental conditions, which impairs broad-range application (Köhl et al. 2019). Likewise, fermented microbial molecules and compounds' direct application can be hindered by inhibitors produced by some pathogens (York et al. 2004) and environmental degradation (Atienza et al. 2021). Finally, the production costs of some of these are still prohibitive to the majority of producers (Pereira-Dias et al. 2023). Therefore, in recent years, a trend has emerged to make up for these limitations: the incorporation of antimicrobials in stable matrices (Stoleru and Brebu 2021).

Nanotechnology has been gaining prominence in material and life sciences. This rapidly growing interdisciplinary field has provided the prospect for the development and functional understanding of several types of nanoparticles. The application of this multifaceted technology in agriculture has recently received attention in crop protection, where is already being explored to develop effective nanoformulations against phytopathogens (Kalagatur et al. 2018; Pan et al. 2023; Parisi et al. 2015). Likewise, increasing efforts have been devoted to changing modern agriculture's reliance on nonrenewable, synthetic resources by reusing agricultural, food, and other biomass waste to produce high-value bio-based products, like biopolymers, with many applications across industries (Tolisano and Del Buono 2023).

Thus, biomass biorefining has opened sustainable alternatives to the disposal of the agrifood sector waste. Among those alternatives is the development of organic nanoparticles obtained through biogenic synthesis of waste and repurposing of its richness in bioactive compounds, to be used in agricultural applications (Ali et al. 2021; Tolisano and Del Buono 2023).

Up until now, nanoparticles most frequently used in crop protection include metalloids, metallic oxides, and carbon (e.g., silver, copper, zinc oxide, gold, titanium oxide, iron). However, because these were found to produce deleterious effects on plant development and can be harmful to the central nervous system of mammals, bio-based nanoparticles have been gaining traction as a safer and more effective alternative (Ali et al. 2021). Environmentally friendly nanoparticles have shown the potential to act either alone, as protectants, plant defense boosters, or as nanocarriers for bioactive compounds (Ingle et al. 2022; Maximiano et al. 2022; Poznanski et al. 2023; Tolisano and Del Buono 2023). These compounds can be incorporated into the nanoparticles through chemical or physical methods for a precise and targeted delivery, increasing their stability and efficacy at low dosages (Kučuk et al. 2023; Pateiro et al. 2021; Wang et al. 2022) and culminating in the mitigation of plant pathogens, and potentially the enhancement of crop yields (Worrall et al. 2018). Other beneficial biological properties of nanoparticles, synthesized from natural polymers, generally include biodegradability, biocompatibility, non-immunogenicity, increased antimicrobial activity, lower administration frequency and dosage, decreased toxicity, and therefore a diminished impact on the environment and human health compared to their conventional synthetic counterparts (Worrall et al. 2018; Yetisgin et al. 2020).

Herein we provide a comprehensive analysis of the developments in the application of organic nanoparticles in the control of phytopathogens, with particular focus on their employment as nanocarriers of bio-based antimicrobials in crop protection. For that, the different types of nanoparticles, the main green-based antimicrobials used in nanoformulations and their synthesis methodologies, along with the factors affecting their bioactivity and the controlled release of antimicrobials, are discussed. Moreover, we compiled the works assessing the nanoparticles' antimicrobial activity against bacterial, fungal, and viral phytopathogens, highlighting their main conclusions regarding effectivity, application, and mode of action, to make that information readily available to researchers. The implications of the application of nanoparticles in plant defenses and their biotic interactions are also discussed.

5.2 Types of Organic Nanoparticles and Carriers

Typically, nanoparticles can be divided into organic, inorganic, ceramic, and bio-nanoparticles (Fig. 5.1). This classification is based on specific attributes such as size, shape, composition, and morphology (Ijaz et al. 2020). From this panoply of different nanoparticles, herein, emphasis will be placed on organic nanoparticles (also referred to as carbon-based nanoparticles), which have been studied for their

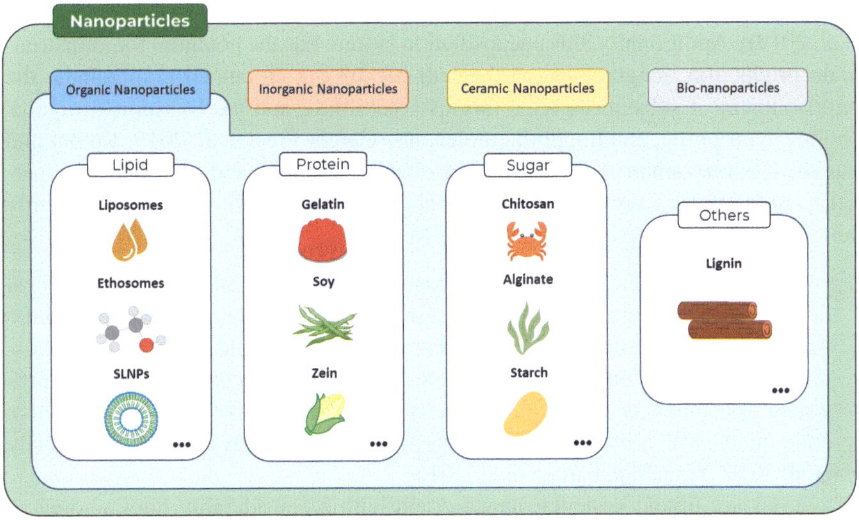

Fig. 5.1 Schematical representation of the several types of nanoparticles, with an emphasis on the most promising system for biocontrol of phytopathogens, organic nanoparticles, their different subclasses (lipid-, protein- and sugar-based), and their respective sources

potential use in plant disease management. Organic-based delivery systems, including lipid, protein, and polysaccharide-based nanoparticles (Fig. 5.1) and carriers have been demonstrated to exhibit plant protective properties (Campos et al. 2015; Sathiyabama and Parthasarathy 2016).

5.2.1 Lipid-Based Nanoparticles

Lipid-based nanoparticles are one of the most encouraging nanoencapsulation strategies in nanobiotechnology (Mashaghi et al. 2013). Despite having earned considerable attention in the biomedical (Sheoran et al. 2022) and food (Yousefi et al. 2019) science fields, mainly due to their biocompatible properties as nanocarriers, the relevance of these nanosystems in plant disease control has also been explored, albeit to a lesser extent (Jin et al. 2019). In a recent work, Xu et al. (2022) showed that solid liposomes loaded with tea tree (*Melaleuca alternifolia*) EO were able to prevent the growth of *Monilinia fructicola*, the causal agent of brown rot, in postharvest peach fruit, and postpone fruit senescence, successfully overcoming the limitations imposed by the chemical nature of the EO.

These colloidal carrier systems are spherical vesicles usually composed of phospholipids, although other lipids may also be used (Kumar and Lal 2014; Singh et al. 2019). Since they are made of functional and/or biodegradable materials, these delivery systems are referred to as "nano-safe" carriers, some of which are also classified as Generally Regarded as Safe (GRAS) by the US Food and Drug

Administration (FDA) [(e.g., solid lipid nanoparticles) (Pardeike et al. 2009; Samimi et al. 2019)]. Additionally, this encapsulation system has the potential for industrial-scale production, lengthening of the half-life of the entrapped compounds, the imprisonment of substances with various solubilities, and the transport of hydrophobic, hydrophilic, and lipophilic molecules (García-Pinel et al. 2019; Kumar and Lal 2014; Nuruzzaman et al. 2016). Liposomes, ethosomes, and solid lipid nanoparticles (SLNPs) are a few examples of lipid-based nanostructures that are currently being studied for bioactive compound delivery.

Liposomes, the first generation of lipid-based nanoparticles and one of the earliest delivery nanosystems produced, are an organized structure made of amphiphilic lipid molecules arranged in a membrane, consisting of lipid bilayers (e.g., phospholipids). These bilayers self-assemble into a spherical colloidal structure providing liposomes the distinctive property to load and deliver biologically active compounds with varying solubility properties (Guimarães et al. 2021; Sebaaly et al. 2016; Yingchoncharoen et al. 2016).

Ethosomes are lipidic vesicular nanocarriers with a considerably high amount of ethanol in their composition (20–45%). Ethanol is a widely recognized permeation-enhancing solution and confers ethosomes distinctive properties, such as strong elasticity and deformability (Fu et al. 2019; Paiva-Santos et al. 2021). However, due to their composition, ethosomes are not very suited for in planta applications.

Solid lipid nanoparticles (SLNPs) are the most widely recognized nanosystems of lipid-based nanoparticles (Samimi et al. 2019). These colloidal delivery systems are composed of a lipidic layer surrounding a solid lipid core, at room temperature, where the compounds of interest are ingrained, preventing particle coalescence (Mirchandani et al. 2021). Attributable to their many advantages, SLNPs have arisen to overcome the drawbacks of the aforementioned lipid-based carrier systems, namely, their poor efficiency of entrapment (% EE), complex manufacturing process, and challenging scalability (Duan et al. 2020). Suitability of production methods and scalability, enhanced stability, controlled compound release, high loading capacity, exclusion of organic solvents, and recognition of all formulation components as safe ("GRAS") make SLNPs a more desirable choice among lipid-based nanoparticles (Doktorovova et al. 2014; Mirchandani et al. 2021).

5.2.2 *Protein-Based Nanoparticles*

Protein-based nanoparticles predominantly consist of proteins. These natural biopolymers possess an inherent ability to undergo self-assembly, orchestrating themselves into highly organized structures. Their exceptional ability to fold into precise three-dimensional configurations bestows upon them a range of distinctive functional properties, including the formation of gels and emulsions (Kumar and Lal

2014; Solomonov and Shimanovich 2020). These characteristics make protein-based nanoparticles an intriguing material for the advancement of nanodelivery systems (Kaltbeitzel and Wich 2023). These types of nanoparticles have been initially and widely explored for biomedical purposes (e.g., cancer/tumor imaging) (An et al. 2020). Notably, the increasing appeal of harnessing protein-based nanoparticles' unique features has sparked a rising interest in their applicability in mitigating plant diseases. They have been shown to be simultaneously able to deliver bioactive compounds with antimicrobial properties against phytopathogenic agents (e.g., *Xanthomonas euvesicatoria*), as well as to elicit plant defense responses (Oliveira-Pinto et al. 2021). This emerging area of research shows potential for creating cutting-edge, long-lasting, and sustainable approaches to ensure crop health and enhance agricultural practices.

Overall, the proteins employed for producing protein-based nanoparticles can be categorized into animal proteins (e.g., gelatin) or plant proteins (e.g., zein), the latter being preferred as they are considered safer for many applications (Kumar and Lal 2014; Tarhini et al. 2017). Due to their inherent biocompatibility and biodegradability features, with some being regarded as "GRAS" (Hassan et al. 2022a; Lohcharoenkal et al. 2014), protein-based materials offer a compelling replacement for the inorganic/synthetic polymers frequently used in nanoparticle production. Among the several benefits of protein-based nanoparticles are the abundance of natural sources, as well as the ability for surface modification of the particles by conjugation of various protein and carbohydrate ligands, enabling tailored distribution, which further mitigates difficulties associated with toxicity and cost-effectiveness (Hong et al. 2020; Jain et al. 2018; Kianfar 2021). Examples of organic protein-based nanoparticles include, among others, gelatin, soy, and zein nanoparticles.

Gelatin nanoparticles are nanodelivery systems whose main component is gelatin, a denatured protein obtained from the partial hydrolysis of collagen found in connective tissues, bones, and skin of animals, predominantly sourced from bovine and porcine origins (Dille et al. 2021; Hathout and Metwally 2019). This protein possesses the amino acid combination arginine—lysine—glycine in its primary structure, which distinguishes it from other natural or synthetic polymers lacking these recognition and binding sites (Kianfar 2021). Moreover, gelatin-active groups allow a variety of chemical modifications directly or through different linkers. This attribute allows gelatin to act as an efficient drug deliver, able to bind and target-deliver several types of bioactive substances (Hussain et al. 2020).

Soy nanoparticles are derived from soy proteins, which are among the most abundant and extensively used plant proteins for active compound encapsulation. Soy protein isolate (SPI) is an enriched form of soy protein with a particularly high protein content (85–90%), commonly used in food applications (Tang 2019). Predominantly consisting of glycinin and β-conglycinin, SPI presents a well-balanced blend of polar, nonpolar, and essential amino acids, rendering SPI-based nanoparticles highly versatile for a wide range of biotechnological applications (Kianfar 2021).

Zein nanoparticles, derived from zein, the major protein found in maize comprising around 50% of the total protein content in corn kernels, have garnered remarkable attention as a carrier and release system for diverse active substances. Moreover, due to their high coating ability, biocompatibility, and biodegradability, the natural biopolymer zein was approved as "GRAS" by the FDA and has been employed in several industries, particularly in the biomedical, pharmaceutical, and food activities (Garavand et al. 2022; Pascoli et al. 2018). This prolamin is rich in hydrophobic amino acids such as proline, leucine, and alanine, which contribute to zein capacity to easily self-assemble into zein nanoparticles in presence of polar solvents like water (Bidyarani et al. 2020) and accommodate and protect high amounts of less water-soluble compounds, making zein nanoparticles an optimum carrier for phytochemicals (Hassan et al. 2022b). Recent findings revealed that zein nanoparticles are able to enhance the translocation of agrochemicals within soybean plants, from the roots to the leaves (Hanna et al. 2022), which inspires the use of this versatile system for nanodelivery of bio-based antimicrobial.

5.2.3 Polysaccharide-Based Nanoparticles

Polysaccharide-based nanoparticles primarily consist of simple repeating sugar monomers linked by glycosidic bonds that form the polysaccharide chain. These biopolymers are the most abundant biomolecules and can be readily retrieved from a diverse range of natural resources, including plants (e.g., starch), algae (e.g., alginate), animals (e.g., chitosan), or even from microbial origin (e.g., dextran) (De Anda-Flores et al. 2021). The inherent natural origin of these ubiquitous biopolymers endows polysaccharide-based nanoparticles with exceptional properties, such as stability, safety, low toxicity, hydrophilicity, and biodegradability, making them attractive, cheap, and promising biomaterials (Debele et al. 2016; Huh et al. 2018). Moreover, the versatility of polysaccharides stems from their ease of undergoing chemical and biochemical modifications, owing to the presence of various reactive hydrophilic functional groups (such as hydroxyl, carboxyl, and amino groups) in their backbone. These offer an array of polysaccharide derivatives, contributing to their remarkable functional and structural diversity (Debele et al. 2016; Saravanakumar et al. 2012). For instance, the hydrophilic groups facilitate the establishment of non-covalent bonds with biological tissues, enabling bioadhesion. This, in turn, extends the residency period of the polysaccharide-based carriers, enhancing the absorption of encapsulated bioactive compounds and presenting promising potential for a vast array of compound delivery applications (Liu et al. 2008). As an illustration, chitosan displays robust adhesion to leaves and stems epidermis, extending contact time and improving the uptake of bioactive compounds (Malerba and Cerana 2016).

In recent years, the development and innovation of polysaccharide-based nanoparticles have gained widespread interest and accomplished significant advancements across multiple fields. These nanocarriers have captured considerable attention in biomedicine for applications such as gene therapy and tissue engineering, as well as in the pharmaceutical industry. Furthermore, the "GRAS" status of polysaccharides attributed by the FDA generalized their broad usage in the food industry (Miao et al. 2018). In the context of plant disease management, the exploration of this type of organic nanoparticles has also been fruitful (Kumaraswamy et al. 2019). In addition to chitosan, alginates and starch have also been studied for the development of polymeric polysaccharide-based nanoparticles.

Chitosan nanoparticles are comprised primarily of chitosan, a natural, cationic, and linear polysaccharide derived from the alkaline deacetylation of chitin, through demineralization and deproteinization processes (Chandrasekaran et al. 2020; Pirzada et al. 2020). Chitosan has attracted considerable interest from different research areas, not only owing to its "GRAS" status but also due to its notable biological properties, namely, biocompatibility, biodegradability, low toxicity, and good miscibility with other polymers (Pirzada et al. 2020). Additionally, chitosan is non-allergenic, has a broad-spectrum antimicrobial activity, and potential effects on plant growth and immunity (Divya et al. 2020; Maluin and Hussein 2020). Hence, this material shows enormous potential both as cost-effective a nanoparticle alone or as a nanocarrier within the biotechnological and agricultural sectors (Hu and Luo 2021; Chandrasekaran et al. 2020).

Alginate nanoparticles' main component is alginate, a naturally occurring polysaccharide derived from the cell walls of brown algae (*Phaeophyceae*), which possesses the exceptional ability to bind diverse cations leading to the formation of stable hydrogels (Cao et al. 2020). In addition to its remarkable set of physicochemical properties, which allow chemical modifications to enable site-specific targeting, and its biological attributes such as biocompatibility, biodegradation, and mucoadhesiveness, alginate has garnered FDA approval, making alginate nanoparticles a compelling option across many industries, including the agrifood sector (Dodero et al. 2021; Severino et al. 2019).

Starch nanoparticles exhibit distinctive physicochemical and biological characteristics in comparison to native starch, a plentiful, cost-effective, and versatile plant-derived polysaccharide. These variations encompass smaller particle sizes, with a predominantly spherical shape, increased surface area per unit mass, minimal diffusion restrictions, enhanced absorption capacity, accelerated biological penetration rate, and improved solubility compared to natural starch granules (Campelo et al. 2020; Torres and De-la-Torre 2022).

5.2.4 Lignin-based nanoparticles

Lignin nanoparticles cannot be categorized into the previously mentioned three groups as lignin is a highly cross-linked aromatic biopolymer (Tang et al. 2020). Lignin nanoparticles, thus, belong to a distinct class of organic-based systems.

Furthermore, lignin is considered a multifunctional biodegradable material owing to its renewable nature, biocompatibility, inherent biocidal activity, and antioxidant properties associated with its polyphenolic nature (Ullah et al. 2022) and holds significant promise as an eco-friendly option for a sustainable future in agriculture. Derived from byproducts of pulp and paper industries and modern biorefinery processes, lignin (a major component of plant cell walls and the second most abundant natural polymer) offers a readily available and cost-effective resource with diverse potential applications across various industries, namely, in agriculture. Notably, lignin has already been assessed as a nanocarrier for biocides and fertilizers, showcasing their potential in agricultural applications (Gigli et al. 2022; do Espirito Santo Pereira et al. 2022). Moreover, the design of new protection strategies relying on lignin nanoparticles is particularly attractive because it is aligned with the principles of the circular economy, by using lignocellulosic biomass for the development of value-added products.

5.3 Bio-Based Antimicrobial Ingredients Used in Organic Nanoformulations

5.3.1 Plant-Based Bioactive Products/Compounds

Plant-derived antimicrobials are still a poorly explored subject in agriculture. These natural products have a long tradition of being used as food preservatives, flavor enhancers, and dietary supplements; however, they have long been employed to treat bacterial and fungal diseases (Ahmed and Mikail 2017). The use of plant-based antimicrobial products to treat plant diseases carries major advantages, mainly through the absence of bioaccumulation phenomena, like those observed for conventional, chemical pesticides. Furthermore, to this date, there are no reports of developed resistance to these products (Chandra et al. 2017). It should also be noted that there is an increasing consumer demand for organic and/or sustainably produced food, and plant-based antimicrobials could be employed as organic alternatives to conventional pesticides.

One of the major drawbacks of using plant-based products as antimicrobials is their high volatility and degradability. Furthermore, the majority of these products, e.g., EOs, plant extracts, phytohormones, flavonoids, etc., possess poor water solubility (Fig. 5.2). Therefore, in recent years, a myriad of solutions have emerged to make these bioactive ingredients more stable and improve their solubility. Among the various solutions proposed, the incorporation in stable matrices has been by far the most exploited (Stoleru and Brebu 2021). A myriad of matrices has been exploited so far, namely, biopolymers, films, and nanoparticles (Acosta et al. 2016; De Moraes et al. 2020; Oliveira-Pinto et al. 2021). However, nanoparticles, due to their long-term stability and controlled-release properties, seem to be the preferred encapsulation system for these compounds (Nasseri et al. 2016). With the major

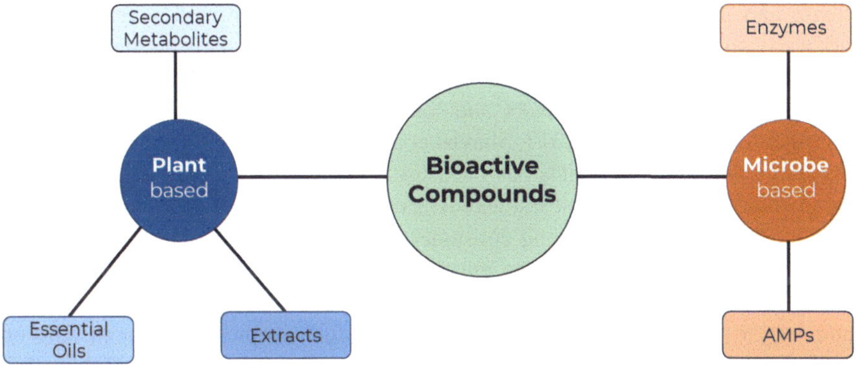

Fig. 5.2 Different antimicrobial natural compounds of plant and microbial origin, are frequently loaded in organic nanoparticles

advances nanotechnology has experienced in recent years, several organic compounds have been successfully used for the encapsulation of plant-based products.

Essential Oils (EOs)

EOs are the most widely employed plant-based antimicrobials for incorporation in nanoparticles. They are generally recognized as biodegradable and have been regarded as safe (GRAS) by the FDA (Menossi et al. 2021). Furthermore, EOs have been in the spotlight due to their potent antimicrobial and antioxidant properties (Oliveira-Fernandes et al. 2023; Valdivieso-Ugarte et al. 2019). EOs, like those obtained from *Cananga odorata*, *Carum copticum*, *Cinnamomum zeylanicum*, *Coriandrum sativum*, *Cymbopogon martinii*, *Eugenia caryophyllata*, *Allium sativum*, *Origanum vulgare*, *Pimenta dioica*, *Valeriana officinalis*, *Zataria multiflora*, *Thymus vulgaris*, *Cuminum cyminum*, *Mentha piperita*, *Hyssopus officinalis*, etc., are complex mixtures of secondary metabolites (e.g., terpenes, terpenoids, phenylpropenes, aldehydes, ketones, ethers, and phenolics) obtained by hydrodistillation or cold pressing (*Citrus* spp.) (Ferhat et al. 2007; Katekar et al. 2022). The composition of EOs can vary widely within the same species, according to the bioclimatic conditions, development stage of the source material, and genetics (Ćavar et al. 2008; Guedri Mkaddem et al. 2022). The most relevant bioactive components found within EOs are carvacrol, carvone, cinnamaldehyde, citral, *p*-cymene, eugenol, limonene, menthol, linalool, vanillin, and thymol (Huang et al. 2019). Fruit of the complexity of compounds and, consequently, different modes of action against pathogens, the likelihood of resistance development is extremely low (Yu et al. 2020).

However, EOs are extremely reactive, volatile, and hydrophobic, limiting their application and hindering their efficacy. Plus, their direct application in foods is not always ideal due to the strong aroma and negative impact on food organoleptic properties (Mahanta et al. 2021). Nanoparticle encapsulation can protect EOs from heat and UV degradation while improving their water solubility and controlled release and in some cases even boost the antimicrobial properties against plant pathogens since they enhance EOs' diffusion capacity through biological membranes by increasing their surface-to-volume ratio (Nair et al. 2022). Likewise, they could prevent undesirable flavors and minimize interaction with food matrices (Martínez-Hernández et al. 2017; Shavisi et al. 2017).

A large variety of materials has been used to successfully produce EO-loaded nanoparticles, and while most studies focus on metal-based nanoparticles (e.g., Ag, Au, Fe, Cu), with the increasing consumer awareness for organic agriculture, the search for nontoxic alternatives to metal-based nanoparticles has been increasing (Menossi et al. 2021). As a result, there have been reports of lignin, zein, and chitosan nanoparticles loaded with EOs, being successfully employed, either in vitro or in planta against a myriad of fungal and bacterial phytopathogens (Tables 5.2 and 5.3).

Plant Extracts

Like EOs, plant extracts can be defined as a complex mixture of secondary metabolites with biological activities (Venditti et al. 2016). This richness in secondary metabolites is also heavily linked with the antimicrobial activity of these products. However, plant extracts can be obtained from a far wider group of medicinal and aromatic plants (MAPs). To obtain plant extracts, several organic solvents can be employed, with the most common being methanol, ethanol, hexane, chloroform, and diethyl ether (Guerrero-Rodríguez et al. 2007). Moreover, it should be noted that the use of these solvents to obtain the extracts implies that they cannot be used in organic farming systems (Castillo et al. 2010). To counteract this handicap, there has been increasing research for novel polar solvents that comply with organic farming practices. However, the fact that these plant products are biodegradable and have a very low likelihood of triggering pathogen resistance development (Satish et al. 1999) should not be overlooked when considering their application as viable substitutes for conventional (Cu-based) pesticides. Furthermore, some works have developed water-based plant extracts, which do not have the hindrances of polar solvent-based extracts in terms of their application in the context of organic farming (Reeve et al. 2016).

In fact, studies have reported that plant extracts obtained from a plethora of MAPs, using a wide variety of extraction methods, can be used as suitable antimicrobials to control a myriad of phytopathogens in vitro, like *Xanthomonas campestris*, *Alternaria solani*, *A. zinnia*, *Fusarium oxysporum*, *Rhizoctonia solani*, and *Curvularia lunata* (Satish et al. 1999; Tapwal et al. 2011). Besides the type of solvent employed in their extraction, overall, plant extracts suffer from the same impairments as EOs regarding their large-scale use in phytopathogen management, namely, high volatility and thermal and oxidative degradation. These handicaps can be overcome by several strategies, among which, encapsulation in nanoparticles, where chitosan or lignin seem to be particularly promising (Armendáriz-Barragán et al. 2016). In fact, some recent works have pointed out that the encapsulation of plant extracts in organic nanoparticles not only significantly increases their stability but also triggers major increases in the antimicrobial activity of the plant extracts against phytopathogens (Sánchez-Hernández et al. 2022).

Besides protecting plant extracts from environmental degradation, organic nanoparticles can increase the availability of these products by enhancing surface-to-volume ratio and promoting the controlled release of the compounds contained in the plant extracts (Beig et al. 2022). Moreover, some studies have already pointed to a synergistic effect between the plant extract and the nanoparticles' antimicrobial effect (Muzzalupo et al. 2020; Sánchez-Hernández et al. 2022), which could reduce the dosage required for an effective pathogen control and thus shrink the costs for the producers. Furthermore, since there is a multitude of plant extracts with antimicrobial activity, the plethora of cocktails that could be developed would decrease the possibility of resistance development even further.

Despite the abovementioned advantages of plant extract-loaded nanoparticles for organic farming, studies that focus on this solution have been sparse. To our knowledge, only three works have explored the use of plant extracts encapsulated in organic nanoparticles against plant pathogens (Table 5.3).

Plant Secondary Metabolites

Secondary metabolites are often regarded as small molecules that are not fundamental for the survival of the organism that produces them (Yang et al. 2018b). However, it is known that these compounds, despite not intervening in the primary metabolism, play key roles in the development and adaptation of plants to abiotic and biotic stress (Dong and Lin 2021; Pott et al. 2019). Among the most common secondary metabolites are terpenoids, polyphenols, and polyacetylenes.

Plant secondary metabolites exhibit a wide range of biological activities, among which is antimicrobial activity, which opens the door for these compounds to be used in phytopathogen management. In fact, a plethora of secondary metabolites has already been studied regarding their in vitro antimicrobial activity against notorious plant pathogens (e.g., *Xanthomonas perforans*, *Pseudomonas syringae*, *Aspergillus* spp., *Alternaria* spp., etc.) (Abbaszadeh et al. 2014; Ni et al. 2020; Qiao et al. 2020). Carvacrol, thymol, linalool, and menthol are among the most studied plant secondary metabolites against phytopathogens (Abbaszadeh et al. 2014). However, it should be noted that most of these compounds can also be found in EOs and plant extracts. Hence, secondary metabolites also suffer from a series of handicaps to their application in the field, mostly associated with thermal and UV stability and oxidative degradation. Moreover, as it is with plant extracts and EOs, the encapsulation of secondary metabolites in organic-based nanoparticles has been suggested as a viable way to solve these issues. Some works seem to point out that organic nanoparticles (e.g., chitosan nanoparticles) can provide a controlled release of the secondary metabolites, improve their antimicrobial properties, and mitigate the off- flavors these products may leave on food products (Martínez-Hernández et al. 2017).

Despite all the advantages these strategies appear to possess, studies on the actual potential of secondary metabolite-loaded organic nanoparticles to control plant pathogens are sparse (Tables 5.2 and 5.3), compared to the clinical works pointing this strategy as a promising way for disease management (Compean and Ynalvez 2014).

5.3.2 *Microbial-Based Bioactive Products/Compounds*

Microorganisms are regarded as important sources of compounds with antibacterial, antifungal, or cytotoxic properties (Karpiński 2019). Moreover, the discovery of new microorganisms with the potential to produce antimicrobial compounds was boosted by the development of new genomics and bioinformatics tools in the last decades (Maghembe et al. 2020). Bacteria and fungi are, by far, the most widely scoped groups of microorganisms in terms of the production of antimicrobial compounds, mainly due to their capacity to produce enzymes and peptides, respectively (Amaning Danquah et al. 2022). The application of microbial-based compounds to treat plant diseases does not lead to bioaccumulation or lixiviation phenomena, like Cu-based antimicrobials; furthermore, it is also aligned with the need for the reduction in conventional antimicrobials usage in agriculture.

As it occurs with plant-based antimicrobials, microbe-based ones also suffer from drawbacks associated with degradability and poor water solubility (Pereira-Dias et al. 2023; Rodrigo et al. 2022). However, the search for ways to minimize these drawbacks is still in its early stages, and works focusing on new encapsulation/controlled delivery methods are sparse. However, some recent works point out that organic nanoparticles might be a viable solution to this problem (Narendrakumar and Namasivayam 2021). Nonetheless, these works are quite preliminary, and there is an undoubted need for a deeper approach to this very promising topic.

Enzymes

Enzymes are proteins with the ability to act as biocatalysts, speeding biochemical reactions in living organisms (Robinson 2015). Moreover, it has been known that some types of bacteria, namely, plant growth-promoting rhizobacteria (PGPR), as well as fungi, are able to produce and secrete several hydrolytic enzymes, like chitinase, glucanase, protease, and cellulase, which degrade the cell wall of other microorganisms, namely, phytopathogens (Jadhav et al. 2017; Thambugala et al. 2020).

In recent years there has been some focus on the use of PGPRs and other hydrolytic enzyme-producing bacteria as BCAs against phytopathogens (Shaikh and Sayyed 2014). However, the application of BCAs in the field still raises some uncertainties. Firstly, the introduction of non-native microbes might disrupt endemic communities (Huang et al. 2021). Furthermore, most BCAs exhibit a certain degree of pathogen specificity, which impairs their large-scale application, as well as a strong dependency on crop physiology and environmental conditions, which are also major handicaps for their use as a broad-range method for phytopathogen control (Köhl et al. 2019). Recently a new trend has emerged that highlights the major potential of purified bacterial enzymes to be employed directly as a control tool against plant pathogens (Hernández-Melchor et al. 2023; Hjort et al. 2014; Khairah et al. 2023) (Fig. 5.2). Despite the promising results some works have shown, it should be noted that this strategy presents handicaps for field applications. It is widely known that some pathogens can produce enzyme inhibitors that could impair the activity of these antimicrobials (York et al. 2004). Furthermore, environmental degradation should also be taken into account when considering the long-term viability of using enzymes as phytopathogen control tools. In fact, some studies have shown that chitinase suffers a significant loss in activity after just 10 min at 50 °C (Xie et al. 2021).

To overcome these limitations, the encapsulation of enzymes into organic matrixes, such as nanoparticles, has been suggested (Narendrakumar and Namasivayam 2021). In fact, nanoparticles may protect the enzymes from the action of pathogen-produced inhibitors and environmental degradation. Furthermore, there is the possibility of controlled release of enzymes into the environment, effectively reducing the dosage required for successful pathogen control. Another interesting perspective is the combination of enzyme-loaded nanoparticles with other bioactive products (e.g., plant extracts/EOs), which could not only potentiate the activity of the formulation but also diversify its mode of action, significantly diminishing the possibility of antimicrobial resistance development (Narendrakumar and Namasivayam 2021).

Notwithstanding these promising indications, it should be noted that the research on the topic of enzyme-loaded organic nanoparticles is particularly scarce. To our knowledge, only a few studies have explored this path.

Antimicrobial Peptides (AMPs)

AMPs are small (under 50 amino acids) cationic molecules capable of interacting with negatively charged membranes of pathogens, leading to destabilization of its surface, through pore formation, and subsequent leakage of cytoplasmatic content, ultimately inducing cell apoptosis (Huan et al. 2020). Thousands of these molecules have been identified as part of the innate immune system of amphibians, insects, mammals, microorganisms, and plants and have been shown to exhibit antibacterial, antifungal, and antiviral properties, being able to induce plant resistance (Huan et al. 2020; Pereira-Dias et al. 2023). Accordingly, these have been prospected for their antimicrobial activity to ultimately develop low-cost, safe, bio-based antibiotics (Rosa et al. 2022; Zhang et al. 2023).

Despite presenting several advantages compared to chemical pesticides, including easy metabolization, specificity, and effectiveness at low dosages, AMPs' full-scale application in agriculture has been hindered by high production costs, toxicity effects, low stability, proneness to hydrolysis by proteases, and poor solubility (Huan et al. 2020; Pereira-Dias et al. 2023). In that sense, nanotechnology has been used to associate nanoscale delivery systems with AMPs to stabilize these molecules and mitigate some of their limitations (Maximiano et al. 2022). Encapsulation of AMPs in bio-based nanoformulations and their application as delivery systems could improve bioactivity by protecting them from protease and or environmental elements' degradation, besides improving bioavailability at the desired target site (Lin et al. 2020; Maximiano et al. 2022). Plus, nanoparticles can be tailored to gradually release AMPs, lingering their effects on the pathogen's cells and increasing their half-life (Lin et al. 2020; Maximiano et al. 2022). Because nanoparticles increase bioavailability and the contact surface, lower dosages would be needed, which in turn would reduce potential toxicity effects on crops and frequency of application (Lin et al. 2020; Maximiano et al. 2022). Moreover, a cocktail of different AMPs could be simultaneously administered with different targets (e.g., by combining antimicrobial peptides with others capable of boosting plant defenses), including synthetic analogs with improved function. Finally, the application of AMPs in nanoformulations could have synergistic effects that go beyond the effectiveness of AMPs or nanoparticles when applied separately (Garcia et al. 2022; Maximiano et al. 2022).

Despite the aforementioned advantages and potential applicability in agriculture, the exploitation of nanoparticles as delivery tools for bioactive AMPs to mitigate phytopathogens has been underwhelming (Maximiano et al. 2022), compared to clinical research (Martin-Serrano et al. 2019). To this day, only two works have tried this type of nanoformulation against phytopathogens using organic matrices (Table 5.3).

5.4 Organic Nanoparticle Synthesis and Factors Influencing the Antimicrobial Activity and Controlled Release

The application of bio-based antimicrobials (Fig. 5.2) is hampered by their instability and degradation under environmental conditions (Atienza et al. 2021). Certain types of organic materials have been used as nanocarriers to improve the stability, water solubility, and bioavailability of lipophilic compounds (Oh et al. 2019; Vettraino et al. 2022). A variety of organic nanoparticles has been synthesized and optimized to retain and regulate the release of antimicrobials upon certain stimuli, showing promising application in phytopathogen management (Vettraino et al. 2022; Wang et al. 2022) or preservation of the agricultural commodities from mycotoxigenic fungi during postharvest storage (Das et al. 2020; Kalagatur et al. 2018; Kaliamurthi et al. 2019). Examples of different organic nanoparticles developed with this perspective are listed in Table 5.1. Simple or functionalized chitosan, lignin, and zein-nanoparticles currently rise as promising carriers of nature-based antimicrobials with potential application in agriculture (Table 5.1). Several

Table 5.1 Characteristics of organic antimicrobial nanoparticles (NPs) with application in phytopathogens control and their use for delivery of bio-based antimicrobials

Organic molecules/ polymers used for nanoparticles fabrication	Synthesis method	Loaded bio-based antimicrobial/elicitor agent	Carrier:antimicrobial ratio	Particle size (diameter)	Entrapment efficiency (%)	Reference
Bovine serum albumin	Coacervation modified method	Bacterial chitinase	ND	110–120 nm[a]	ND	Narendrakumar et al. (2018)
Chitosan	Ionotropic gelation	–	–	20–70 nm[b]	–	Divya et al. (2017)
	Ionotropic gelation	–	–	≈100–1000 nm[c]	–	Oh et al. (2019)
	Ionotropic gelation	–	–	78.4–300.1 nm[c]	–	Boruah and Dutta (2021)
	Ionotropic gelation	Salicylic acid (SA)	0.1:0.4 (w/w), SA:chitosan ratio	368.7 nm[c]	80%	Kumaraswamy et al. (2019)
	Ionotropic gelation	Harpin$_{Pss}$	0.1:1 (w/w), Harpin$_{Pss}$:chitosan ratio	133.7 nm[d]	89.7%	Nadendla et al. (2018)
	Emulsification-ionotropic gelation	*Allium sativum* EO	1:0, 1:0.25, 1:0.5, 1:0.75, and 1:1 (w/w), chitosan:EO ratio	200–400 nm[a]	23.8–32.8%	Mondéjar-López et al. (2022)
	Emulsification-ionotropic gelation and freeze-drying	*Illicium verum* EO	1:0.8 (w/v), chitosan:EO ratio	<200 nm[a]	20.9%	Dwivedy et al. (2018)
	Emulsification-ionotropic gelation	*Peganum harmala* extract (PhE) and *Carum copticum* EO (CcEO)	1:1 (v/v), chitosan emulsion:PhE ratio for PhE-loaded NPs	80 nm for CcEO-loaded NPs and 110 nm for PhE-loaded NPs[a]	80% for CcEO-loaded NPs and 87% for PhE-loaded NPs	Izadi et al. (2021)
	Emulsification-ionotropic gelation	*Myristica fragrans* EO	1:0.8 (w/v), chitosan:EO ratio	28.2–39.0 nm[d]	18.1–82.5%	Das et al. (2020)
	Emulsification-ionotropic gelation	*Syzygium aromaticum* EO	1:1 (w/w), chitosan:EO ratio	100 nm[d]	31–45.8%	Hasheminejad et al. (2019)
	Emulsification followed by self-aggregation	*Cymbopogon martinii* EO	ND	445–480 nm[c]	ND	Kalagatur et al. (2018)
	Electrospraying deposition	*Origanum vulgare* EO	0.0625:1, 0.125:1, 0.25:1, and 0.5:1 (v/w), EO:chitosan ratio	290 and 483 nm[c]	70.1% and 79.6%	Yilmaz et al. (2019)
Chitosan-oleoyl	O/W emulsification method followed by self-aggregation	–	–	296.9 nm[c]	ND	Xing et al. (2016)

Table 5.1 (continued)

Chitosan-cinnamic acid	Coupling reaction mediated by EDC followed by self-aggregation	–	–	15.8–30.6 nm[a]	–	Kujur et al. (2019)
	Coupling reaction mediated by EDC followed by self-aggregation and ultrasonication	Methyl salicylate (MeSA)	1:0.5 (w/w), chitosan-cinnamic acid nanogel:MS ratio	35.2–78.7 nm[a]	32–34%	Kujur et al. (2019)
	Coupling reaction mediated by EDC followed self-aggregation and ultrasonication	*Mentha piperita* EO	ND	≤100 nm[a,b]	ND	Beyki et al. (2014)
Chitosan glycol-β-cyclodextrin	Kneading method (KM) followed by emulsification-ionotropic gelation	Carvacrol and linalool	ND	175.2 nm for carvacrol-loaded NP and 245.8 nm for linalool-loaded NP[c]	95.3% for carvacrol-loaded NPs and 93.1% for linalool-loaded NPs	Campos et al. (2018)
Chitosan-isoleucine-polyethylene glycol methyl ether	Ultrasonic self-assembly	Spinosad	ND	454.9 nm[a]	40.4–82.2%	Dong et al. (2019)
Chitosan-g-poly (acrylic acid)	Graft polymerization	–	–	50.8 nm[c]	–	Sahab et al. (2015)
Chitosan-lecithin	Self-aggregation	Kaempferol	ND	273.9 nm[c]	93.80%	Ilk et al. (2017)
β-Cyclodextrin (β-CD)	Kneading method (KM)	Carvacrol and linalool	1:1, β-CD:compound molar ratio	ND	86.2% for carvacrol-loaded NPs and 74.2% for linalool-loaded NPs	Campos et al. (2018)
	Kneading method (KM); freeze-drying (FD)	Carvacrol	1:1, β-CD:compound molecular ratio	441 nm (KM method) and 899 nm (FD method)[c]	83.8% (KM method) and 91.3% (FD method)	Santos et al. (2015)
Glyceryl monostearate-GMS and Precirol® ATO 5	High shear homogenization and ultrasonication	*Zataria multiflora* EO	ND	255.5 nm[a]	84%	Nasseri et al. (2016)
Lecithin-cholesterol	Thin-film ultrasonic method	Quercetin (Q)	10:2:1, lecithin:cholesterol:quercetin mass ratio	273 nm for Q-NPs, 318.67 nm for TQ-NPs and 117.50 nm for H-TQ-NPs[c]	82.3% for TQ-NPs and 88.0% for H-TQ-NPs	Wang et al. (2022)

Lignin	Hydrochloric acidolysis	–	–	32.8–120.0 nm[c]	–	Yang et al. (2018a)
	Hydrochloric acidolisis followed by incorporation into nanocomposite films[e]	–	–	–	–	Yang et al. (2016a)
	Hydrochloric acidolysis followed by solvent casting (film formation)[f]	–	–	–	–	Yang et al. (2016b)
	Hydrochloric acidolysis and polydispersion	*Thymus serpyllum* EO	1:3 (w/w), lignin:EO ratio	≈100 nm[a]	61.90%	Vettraino et al. (2022)
	Fast anti-solvent method	*Cinnamomum zeylanicum* (Cz), *Thymus serpyllum* (Ts), *Thymus vulgaris* (Tv) EOs	ND	346.0, 369.0, 244.0 nm for Cz, Ts and Tv EO-loaded LNPs, respectively[c]	35.5%, 60.7%, and 70.0% for Cz, Ts, and Tv EO-loaded LNPs, respectively	Zikeli et al. (2020)
	Nanoprecipitation	–	–	215 nm[d]	–	Amat et al. (2023)
Lignin-tannic acid	Enzyme-assisted functionalization and sonication	–	–	217 nm[b]	–	Morena et al. (2022)
PCDA-DMPC (mixture of polymerizable lipids, 8:2 molar ratio)	Thin-film rehydration method followed by sonication and UV irradiation	PGPB ethanolic extract from *Lysinibacillus fusiformis*	ND	110 nm[c]	ND	Atienza et al. (2021)
Zein	Nano precipitation	Thymol and carvacrol	2:1 (w/w) zein:compound ratio	109.2 nm for thymol-loaded NPs and 111.2 nm for carvacrol-loaded NPs[c]	91.9% for thymol-loaded NPs and 88.5% for carvacrol-loaded NPs	da Rosa et al. (2015)
	Anti-solvent precipitation method	*Satureja montana* EO	ND	159.1 nm[c]	20.58%	Oliveira-Pinto et al. (2021)
Zein-sodium caseinate	Anti-solvent precipitation method	Rotenone	ND	471.3 nm[a]	95.80%	Bidyarani and Kumar (2019)
Zein-rhamnolipid (1:0.5, 1:1, and 1:2 ratio w/w)	Anti-solvent precipitation method	Carvacrol	ND	276 nm[c]	95.50%	Bidyarani et al. (2020)

DMPC 1,2-dimyristoyl-*sn*-glycero-3-phosphocholine, *EDC* *N*-(3-dimethylaminopropyl)-*N*-ethyl carbodiimide hydrochloride, *EO* essential oil, *ND* not described, *PCDA* 10,12-pentacosadiynoic acid, *PGPB* plant growth-promoting bacteria, *TQ-NPs* Tween 80® quercetin nanoliposomes, *H-TQ-NPs* HACC Tween 80® quercetin nanoliposomes

[a] Size measured through scanning electron microscopy

[b] Size measured through transmission electron microscopy

[c] Hydrodynamic diameter measured through dynamic light scattering

[d] Size measured through field-emission scanning electron microscopy

[e] Lignin NPs combined with poly (lactic acid) film

[f] Lignin combined with polyvinyl alcohol-chitosan film

physicochemical, biological, and hybrid methods have been explored for the synthesis of organic nanoparticles. Physicochemical methods, namely, coacervation, anti-solvent precipitation, nanoprecipitation, ionic gelation, self-assembly, and sonochemical approaches, have been the most used techniques (Table 5.1). However, many of these commonly implicate the use of toxic chemical components, organic solvents, and/or cross-linking agents for covalent bonding. The fabrication process should focus on green chemistry avoiding the use of synthetic chemicals to design more biocompatible nanomaterials with reduced hazardous impacts (Sipponen et al. 2019). According to the latest developments, green synthesis making use of plant cell-free extracts (acting as both a reducing and stabilizing agent of nanoparticles) or enzymes from living microorganisms (used as nanofactories for the extracellular formation of nanoparticles) has progressively been acknowledged by researchers as a more environmentally friendly process of nanofabrication (Choudhary et al. 2022; Hong et al. 2021).

Favorable storage conditions and long-term stability are key aspects that should be first investigated when creating a controlled delivery system, as particle aggregation and leakage of encapsulated bioactive molecules are major inconveniences leading to changes in size and burst release, respectively (Reimhult 2015). As regards the application to control the target species, it has been suggested that multiple factors, including light, temperature, vapor pressure, dispersibility, application methods, leaf affinity, physiological conditions in plants, bioavailability, and ultimately the interaction between the compounds and the target species, could greatly influence the bioactivity of bio-based antimicrobials (Campos et al. 2018; Zhao et al. 2018). Moreover, after being released from the nanoparticles, the bioactive agent may be subjected to such variable environmental conditions that the overall practical efficacy of encapsulated substances is affected. Besides, transformation products with unexpected side effects on human health and the environment could be formed (Das et al. 2020).

In the following subsections, we will explore different types of organic nanoparticles described in the literature and technical aspects related to their synthesis and stability study. In addition, optimization approaches to improve organic nanoparticles bioactivity and stability, as well as factors triggering the release of bio-based antimicrobials, will be discussed.

5.4.1 Lipid-Based Nanoparticles

Lipids have been used as nanocarriers for the delivery of bioactive compounds, namely, in the form of nanoemulsions, solid lipid nanoparticles, and liposomes. These nanosystems find numerous applications in the food industry, as natural food preservatives to prevent food deterioration (Yousefi et al. 2019), as well as controlling agents of food-borne pathogens and pests affecting stored products (grains and flours) (Kaliamurthi et al. 2019).

Lipid-based nanoparticle application in agriculture for loading of crop protectants is much less explored, compared to other nanocarriers. This might be related to disinterest in developing lipid-based nanosystems for application at a large scale owing to the standing notion that fabrication processes are more costly and time-consuming and frequently involve the use of organic solvents (dichloromethane, trichloromethane, etc.), which must be removed by rotary evaporation. Furthermore, liposomes may present a major inconvenient regarding instability, rapid drug release, and poor loading efficacy (Hasheminejad et al. 2019). However, their mechanical and chemical stability is fundamental to ensure their purpose and release behavior under harsh environmental conditions such as in soil or high ionic conditions (Atienza et al. 2021). If we contemplate exclusively their use as controlled delivery of nature-based antimicrobial agents, the number of publications is minimal. Therefore, this section will briefly analyze some studies and the possible conceptual and technical approaches that might improve or provide stability to lipid-based nanoparticles, in particular nanoliposomes.

A mixture of polymerizable lipids PCDA and DMPC (8:2 molar ratio) was used for the synthesis of a nanoliposome with antifungal activity against *Fusarium* spp. infecting bananas, tomatoes, and cucumbers (Atienza et al. 2021). The nanobiofungicide was synthesized using a thin-film rehydration method, and then sonication and UV-mediated polymerization were used to encapsulate PGPBs' ethanolic extracts (*Lysinibacillus fusiformis*) within nanoliposomes. According to the authors, storage at 4 °C overnight is important to facilitate the self-assembly of diacetylene monomers before the photopolymerization, bringing high mechanical stability as compared to unpolymerized nanoliposomes. Although the delivery kinetic under storage was not assessed, the authors studied the physicochemical and morphological characteristics of the nanosystem and demonstrated, through different in vitro growth inhibition assays, that nanoliposomes enhanced the antifungal activity of the PGPB extract (Atienza et al. 2021).

In another study, lecithin was used as an amphiphilic compound in combination with cholesterol to obtain nanoliposomes able to deliver quercetin for the control of the tobacco mosaic virus (TMV) (Wang et al. 2022). Initial embedding of quercetin was achieved through thin-film sonication, and its successful encapsulation was verified by FTIR (Fourier-transform infrared spectroscopy) analysis. The formulation was optimized with the addition of a surfactant (Tween 80®) and hydroxypropyl trimethyl ammonium chloride chitosan (HACC) to improve physicochemical stability, loading efficiency, quercetin retention, and release in tobacco plants. The addition of HACC and/or Tween 80® significantly reduced the particle's average size, and both contributed to the interface stability since no significant change of average particle sizes was found after 20 days of storage, whereas the non-optimized quercetin liposome had its average particle size increased from 273 nm, for newly prepared, to approximately 500 nm on the 20th day. FTIR analysis also demonstrated that chitosan had successfully attached to the surface of the liposomes increasing the electrostatic repulsion between the nanoparticles. On the other hand, Tween 80-HACC-quercetin nanoliposomes (H-TQ-NPs) had good stability in the pH range of 5.0–7.0. H-TQ-NPs prepared under the conditions of 0.3 g/L of HACC and pH 6.0 had the best performance in retaining quercetin (>70%) after 20 days of storage and good stability at low temperatures and dark conditions. According to Wang et al. (2022), the optimized nanoformulation (H-TQ-NPs) significantly improved the solubility of quercetin, its transport, and penetration into plants cells, thus inhibiting the expression of hsp70 protein and ultimately improving the prevention of TMV infection. Considering the production cost of the carrier used for the synthesis (about US$0.5/g), the effectiveness in pot and field experiments and performance in symptom alleviation with no visible phytotoxic side effects (Wang et al. 2022), this innovative antiviral nanobiopesticide is an excellent example that demonstrates the great potential of lipid-based nanoparticles when combined with stabilizers.

5.4.2 Protein-Based Nanoparticles

Proteins from botanical, microbial, or animal sources have been utilized either as nanocarriers (da Rosa et al. 2015; Narendrakumar et al. 2018), as antimicrobial ingredients (Narendrakumar et al. 2018), or as stabilizing agents (Bidyarani and Kumar 2019) for the synthesis of biodegradable nanoformulations to control plant pathogens. In addition to their biodegradable characteristics, nanomaterials derived from proteins are amenable to surface modification and covalent attachment (Narendrakumar et al. 2018), which is useful for adding coating materials and the synthesizing of hybrid nanosystems, respectively.

Zein-based nanoparticles have been synthesized through the anti-solvent precipitation method at room temperature, pH-controlled nanoprecipitation, microfluidization or high-pressure homogenization, ultrasonic dialysis procedure, the electrospraying method, or the solution-enhanced dispersion by supercritical fluids (SEDS) technique. While the first two methods generally use minimal amounts of organic solvent, being specifically useful for encapsulating phytochemicals, the latter two have the advantage of producing zein nanoparticles in a dry form (Hassan et al. 2022b). The possible factors affecting zein nanoparticle formation, such as protein concentration, pH, operating conditions, ethanol concentration, and percentage of anti-solvent, were discussed in the recent review of Hassan et al. (2022b).

Some researchers have explored zein nanoparticles in a simple form or complexed with other organic molecules/polymers (zein-based nanoparticles), to carry bio-based antimicrobial agents aiming at preventing plant diseases. The bioactive components summarized in Table 5.1 are derived from plants and have a hydrophobic nature, which makes these good candidates for loading into zein-based nanoparticles.

In a recent study, Oliveira-Pinto et al. (2021) reported the application of zein nanoparticles and EO-loaded zein nanoparticles for the control of tomato bacterial spot. *S. montana* EO-loaded zein nanoparticles were monodisperse with a surface charge of approximately +30 mV, indicating that this nanosystem has uniform size and stable character. Indeed, the nanoparticles remained stable through the 90 days of storage at 4 °C with sizes ranging between 122 and 160 nm. It is worth noting that a small increase in nanoparticle size was noticed when comparing EO-loaded nanoparticles with empty ones. A possible relaxation of the matrix could have occurred over time as a consequence of differentiated interactions with the several constituents of the EO (hydrocarbons and oxygen-containing terpenoids).

The nanodelivery performance of zein nanoparticles through time for individual EO compounds, thymol and carvacrol, was evaluated by da Rosa et al. (2015). No burst effects were observed when tested on a buffer solution of citric acid-sodium phosphate, indicating strong interaction between the terpenoids and the zein matrix, with a diffusion release kinetic progressing at moderate speed according to the Korsmeyer-Peppas model (approximately 50% of carvacrol and 60% of thymol were released in 72 h). Besides, the authors verified that thymol and carvacrol-loaded zein nanoparticles were physically and chemically stable at 6 and 20 °C during 90 days of monitoring, when considering the encapsulation efficiency parameter, which remained very high and close to the initial value, as well as antioxidant activity, average nanoparticle size (Z-average), and polydispersity index. Thus, both nanosystems warranted excellent storage with no decrease in the encapsulated compound during the 90 days (da Rosa et al. 2015).

Other alternative approaches for the design of zein nanopaticles may include the addition of another starting material during nanoparticle formation (other proteins, lipids, or polysaccharides) to create zein-based nanocomplexes with core-shell architecture or hybrid matrix (Hassan et al. 2022b). For example, zein nanoparticles stabilized with sodium caseinate (Bidyarani and Kumar 2019) and zein-rhamnolipid complex (Bidyarani et al. 2020) were produced by anti-solvent precipitation for effective entrapment (≈96%) of rotenone and carvacrol, respectively, and tested for their antibacterial effect against *Pseudomonas syringae* (bacterial canker) and antifungal activity on *Fusarium oxysporum* (Fusarium wilt). Nevertheless, none of these studies investigated the release characteristics of the zein-based complexes in the medium, or existing experimental conditions, to better support the exhibited bioactivity. However, the interaction of carvacrol with the zein-rhamnolipid complex was confirmed through FTIR, H^1NMR, and molecular docking studies. The latter revealed that the rhamnolipid has stronger interactions and binding affinity with zein than carvacrol (Bidyarani et al. 2020).

On the other hand, the animal-derived protein bovine serum albumin (BSA) was proposed as an effective material for the nanodelivery of a hydrolytic enzyme with biocidal properties. Small average-size BSA nanoparticles were prepared through a modified method of simple coacervation for entrapment of extracellular chitinase, an important antimicrobial that has been explored for the control of phytopathogenic fungi (Narendrakumar et al. 2018). The method consisted of suspending lyophilized chitinase enzyme extracted from *Serratia marcescens* in an aqueous solution of BSA, followed by mixing with ethanol. The method also employs the addition of glutaraldehyde for cross-linking and Tween 20® (0.01% v/v) to stabilize the chitinase-loaded BSA nanoparticle prior to further processing (dispersion, concentration, and freeze-drying). Results from FTIR analysis suggested that BSA nanoparticles were surface-modified after loading with chitinase. Narendrakumar et al. (2018) studied several factors including time-dependent release and enzyme kinetic properties under variable pH, temperature, and subtract concentration and assessed the biocontrol efficacy of chitinase-loaded BSA nanoparticle against *Alternaria alternata*. They verified that the total release of the nanoformulated chitinase was sustained for 96 h, which they attributed to the solubilizing ability of the BSA carrier. Besides, the authors found that chitinase-loaded BSA nanoparticles greatly enhanced the enzyme activity and thermostability, by recording higher activity at the optimum temperature of 70 °C (61.2 U), which was much superior to the free enzyme activity at optimum temperature (25.2 U at 40 °C). In addition, the nanoformulated chitinase performed much better in alkaline conditions (25 U for the optimal pH of 9.5), and a pH-dependent increase of activity was reported, in contrast to the free enzyme (6 U at optimal pH of 8.0). The authors evidenced that chitinase-loaded BSA nanoparticles have enhanced performance and could reveal an effective alternative against a wide range of economically important phytopathogenic fungi (Narendrakumar et al. 2018).

5.4.3 Polysaccharide-Based Nanoparticles

Several polysaccharides have been exploited as bulk materials, or in combination with other molecules, for the fabrication of organic nanoparticles with enhanced properties. Cyclodextrins (CDs) are cyclic oligosaccharides produced from the enzymatic conversion of starch with an interesting diversity of applications. Their distinctive structure characterized by a hydrophilic outer surface and a hydrophobic inner cavity allows the formation of inclusion of complexes with compounds exhibiting low aqueous solubility (Pirzada et al. 2020). Several methods have been used to prepare cyclodextrin inclusion complexes such as coprecipitation, neutralization, spray-drying, coevaporation, kneading, and freeze-drying (Liu et al. 2010). Among them, the kneading method, also known as slurry complexation, gives a very good yield of inclusion with a small amount of solvent in the preparation (Santos et al. 2015). β-CD inclusion complexes were used for the entrapment of two volatile terpenoids, carvacrol and linalool, both antimicrobial compounds of botanical origin showing well-described antimicrobial properties over several pathogens (Campos et al. 2018; Santos et al. 2015). The complexation of β-CD with carvacrol and linalool was efficient (EE% ˃70), yet superior for carvacrol (86%) (Campos et al. 2018). The reported encapsulation efficiency for carvacrol was in accordance with the former study (EE% = 84%) (Santos et al. 2015). In their investigation, Campos et al. (2018) also obtained carvacrol-loaded β-CD nanoparticles through freeze-drying, which revealed an effective method, as well (EE% = 91%).

It has been suggested that chitosan nanoparticles and their derivatives are one of the best candidates for nanomaterials (Chandrasekaran et al. 2020). In general, chitosan nanoparticles have been prepared by the ionic gelation technique (Table 5.1), in which ionic interactions between the positively charged amino groups of chitosan and the negative charges of polyanions, such as pentasodium tripolyphosphate

(TPP), result in the formation of inter- and intramolecular bonds (Campos et al. 2018) and spontaneous formation of nanoparticles. Ionic gelation is a mild, simple, and organic solvent-free approach for the formation of stable nanogel particles (Hasheminejad et al. 2019). Chitosan nanoparticle synthesis through ionic gelation requires controlled pH conditions (pH 3.5–4) of the chitosan solution, which potentiate inter- and intramolecular cross-linkages (Gooneh-Farahani et al. 2020). Also, to facilitate the loading of poorly soluble bioactive compounds, a previous step of emulsification with the chitosan solution (typically 1% aqueous acetic acid) is performed making use of emulsifiers, high shear homogenization, or ultrasonication before the cross-linking. Yilmaz et al. (2019) described an alternative approach for the fabrication of chitosan nanoparticles loaded with different concentrations of *Origanum vulgare* EO, using the electrospraying deposition method. The implementation of a high voltage to the EO/chitosan solution forces the solution to emerge from a nozzle and flow into an electric field creating highly charged droplets, which after solvent evaporation formed EO-loaded chitosan particles at the nano-size level. This technique was efficient for EO encapsulation (about 70–80%) and delivered EO-loaded chitosan nanoparticles with antifungal activity against *Alternaria alternata*.

Chitosan nanoparticles' antimicrobial activities depend evidently on the target and growth stage. Nevertheless, these may be triggered by several factors intrinsic to the nanosystems, their fabrication process, and physicochemical characteristics (viz., zeta potential, surface properties, concentration, pH, molecular weight, and degree of acetylation) (Chandrasekaran et al. 2020), in conjunction with the release mechanism (i.e., pore diffusion, surface desorption, capsule swelling, and degradation) (Maluin and Hussein 2020). A positive correlation between chitosan nanoparticle size and molecular weight (MW) and degree of acetylation have been described in several reports, with smaller nanoparticles exhibiting greater inhibition of bacterial growth (Chandrasekaran et al. 2020; Yien et al. 2012). For example, low MW chitosan nanoparticles had improved inhibitory effects against *Erwinia carotovora* compared to medium MW chitosan nanoparticles (Mohammadi et al. 2016).

A factor limiting the use of chitosan in sustained-release applications is its hydrophilicity, resulting in a poor capacity to carry hydrophobic molecules (Pirzada et al. 2020). Some investigations demonstrated that modifications of its hydrophobic groups could be achieved by grafting, using hydrophobic or amphipathic compounds to form an amphiphilic complex (Beyki et al. 2014; Dong et al. 2019; Sahab et al. 2015; Xing et al. 2016). When exposed to water, amphiphilic-modified chitosan can self-assemble into micelles by intra- and intermolecular forces, thereby encapsulating hydrophobic molecules to improve their water solubility. Xing et al. (2016) and Ilk et al. (2017) coupled a fatty acid and a phospholipid (lecithin), respectively, to chitosan to obtain nanosystems intended to be used as natural antifungals. The lecithin-chitosan nanocarrier system was used for the sustained release of kaempferol, which presented significantly higher antifungal activity (67%) compared to the pure compounds (no inhibition) after being stored for 60 days (Ilk et al. 2017). A different hybrid nanosystem consisting of chitosan with cinnamic acid was used for the preservation and controlled release of *Mentha piperita* EO and MeSA, respectively (Beyki et al. 2014; Kujur et al. 2019). On the other hand, chitosan glycol, a commercially available chitosan derivative, has been selected for the development of functionalized polymers, to form chitosan glycol-β-cyclodextrin (Campos et al. 2018). The complex was studied as a release system for two volatile oxygen-containing terpenoids, carvacrol and linalool, which presented good colloidal characteristics (size, polydispersity, and zeta potential), higher complexation efficiency values, and very high encapsulation efficiencies for both compounds (95% and 93%, respectively) (Campos et al. 2018).

Chitosan-based nanoparticles, variable in size and complexity, have been largely investigated for their ability to carry natural bioactive products from plants, such as kaempferol (Ilk et al. 2017), SA (Kumaraswamy et al. 2019), MeSA (Kujur et al. 2019), and a variety of EOs, including some of their volatile constituents (Table 5.1). A few works also successfully explored the nanodelivery properties of chitosan-based nanoparticles to load antimicrobials of bacterial origins, such as harpin peptide (Nadendla et al. 2018) and spinosad (Dong et al. 2019). Overall, most researchers complemented the morphological, molecular, and functional studies with bioassays to find evidence of chitosan-based nanoparticles' enhanced antimicrobial activity against several phytopathogens. Nevertheless, very few presented more comprehensive investigations as regards long-term stability and controlled-release assays (Das et al. 2020; Ilk et al. 2017; Kalagatur et al. 2018; Kumaraswamy et al. 2019).

The state of the art provided new insights into methodological approach and the bioactive potential of functionalized chitosan, reinforcing the understanding that chitosan-based nanoparticles are versatile and effective alternatives to safeguard different types of bio-based and biodegradable antimicrobials and improve their lifetimes when exposed to the environment.

5.4.4 Polyphenol-Based Nanoparticles

There are different types of lignin derived from varied natural sources (Ullah et al. 2022), and their structure strongly depends on the botanical origin and the isolation process (Sipponen et al. 2019). Such variability of its chemical structure makes difficult its application in many fields. Innovative lignin nanoparticles could exhibit a strong capacity as nanocarriers besides delivering a sustained release of bio-active compounds. Nonetheless, the efficiency of the process will depend not only on the amount but also on the structure of the lignin (Vettraino et al. 2022) since its structure is pointed as a key aspect to be controlled in lignin nanoparticles synthesis. Moreover, the antibacterial activity of lignin is specifically related to the presence of phenolic compounds and different functional groups containing oxygen in its structure (Yang et al. 2016a). Thus, more research should be conducted to correlate structural moieties and functional groups with antimicrobial properties (Ullah et al. 2022) to understand the structural factors contributing to the observed activities.

Sipponen et al. (2019) and Chauhan (2020) previously discussed the state of the art concerning the several chemical, physical, and mechanical processes that have been described for the synthesis of lignin nanoparticles, loading and releasing of active substances and their potential applications in different fields. According to Chauhan (2020), varied approaches have been used to synthesize lignin nanoparticles, such as polymerization, acid precipitation, solvent exchange, ultrasonication, self-assembly, interfacial crosslinking and emulsion, anti-solvent precipitation, microbial and enzyme-mediated synthesis, and many more. Depending on the method used, the loading of bioactive antimicrobials can occur during or after the formation of lignin nanoparticles through entrapment (by infusion, solvent displacement, ion exchange, or coating), encapsulation by forming emulsions, or adsorption (Morena and Tzanov 2022). More recently, Morena and Tzanov (2022) presented the latest developments of nanoformulated lignin including the antibacterial potential of the dispersed formulation or embedded in nanocomposites, as well as their use as carriers of other antibacterial agents. Herein, we focused our efforts on discussing the methods used to develop lignin nanoparticles with the specific intend to control phytopathogens.

Despite being described as a promising and versatile material for green nanofabrication, only a few studies have disclosed the possible uses of lignin nanoparticles in crop protection as a bio-based nanosystem for delivery of antimicrobial agents (Vettraino et al. 2022) or as an antimicrobial per se (Yang et al. 2018a). Lignin nanoparticles could be a versatile tool for plant pathogens control, yet the possibility of using unmodified lignin nanoparticles in plants should be investigated, specifically penetration, transport, accumulation, and absorption mechanisms (Yang et al. 2018a).

Yang et al. (2018a) prepared stable spherical lignin nanoparticles with low zeta potential values (<-30.0 mV) and average diameters varying from 33 to 120 nm, through the dissolution of pristine alkali lignin into ethylene glycol, followed by acid hydrolysis under different conditions (HCl at pH 2.5 and 4.6, H_2SO_4 at pH 2.9 and 4.7, and H_3PO_4 at pH 2.6 and 3.3), to investigate how different acids and pH could influence thermal, morphological, and surface properties as well as the size of precipitated lignin nanoparticles. The authors verified that the average diameter of the nanoparticles was pH dependent and the optimum conditions to attain the highest yield of lignin nanoparticles (87.9%) required the use of HCl at pH 2.5. Their findings suggest that the formation of more methoxy groups in the lignin nanoparticles' structure under a condition of pH < 3.0 would be a major factor responsible for antibacterial performance. The same authors (Yang et al. 2018a) also draw attention to the fact that the side chain structure and nature of the functional groups of the phenolic compounds have an important influence on antimicrobial behavior.

Chemical modification or combination might be required to increase its antimicrobial functionality (Yang et al. 2018a) and optimize its application in crop protection. For example, functionalization of lignin through grafting of tannic acid was obtained using laccase oxidation and high-intensity ultrasound to prepare antibacterial phenolated lignin nanoparticles with mean diameters of 200 nm. This bio-based material, which combines both phenolic compounds, exhibited enhanced antimicrobial activity when compared to nonfunctionalized lignin nanoparticles and phenolated lignin in the bulk form (Morena et al. 2022). This sonoenzymatic method offers a more environmentally friendly approach than nanofabrication involving HCl treatment for lignin hydrolysis.

Recently, lignin-based nanocarriers have been used for the delivery of plant EOs (Vettraino et al. 2022; Zikeli et al. 2020). These studies described different approaches and used different nanolignin/EO ratios, to synthesize EO-loaded lignin nanoparticles of distinct sizes. Zikeli et al. (2020) used the fast anti-solvent method to successfully entrap EOs from three different plant species, *Cinnamomum zeylanicum*, *Thymus vulgaris*, and *T. serpyllum*, into lignin nanoparticles, but reported large differences in the encapsulation efficiency. Py-GCMS and FTIR spectroscopy analysis supported that molecular interactions of *T. vulgaris* and *T. serpyllum* EO compounds with lignin nanoparticles were substantial, which contributed to significantly delayed EO release, whereas *C. zeylanicum* EO, enriched in cinnamaldehyde, was found less compatible for entrapment in lignin nanoparticles. Nevertheless, both studies revealed that lignin nanoparticles are a suitable nanodelivery system that provides a controlled release for *T. serpyllum* (wild thyme) EO and satisfactory encapsulation efficiency (about 60%). On the other hand, *T. serpyllum* EO-loaded lignin nanoparticles were prepared via simultaneous entrapment, which resulted in highly loaded nanoparticles, despite the low quantity of lignin used compared to the loaded EOs (Vettraino et al. 2022). The authors observed that the EO-lignin nanoparticles had irregular rounded shapes; nonetheless, the system provided stability to the EO by retarding its release to the aqueous medium in a 48 h in vitro experiment.

On the other hand, some authors have explored the application potential of lignin nanoparticles into innovative nanocomposite films to control bacterial plant/fruit pathogens. Yang et al. (2016a) incorporated a lignin nanoparticle suspension, prepared from pristine lignin, into ternary poly(lactic acid) (PLA)-based nanocomposite films to investigate the biocidal activity against tomato bacterial plant pathogen *Pseudomonas syringae* pv. *tomato* (Pst). Likewise, Yang et al. (2016b) used the solvent casting method to develop binary and ternary polymeric films containing 1 and 3 wt% of lignin nanoparticles in combination with polyvinyl alcohol (PVA) and chitosan polymers and assessed their antimicrobial activity over time against Gram-negative bacteria *Erwinia carotovora* subsp. *carotovora* and *Xanthomonas arboricola* pv. *pruni*. The findings showed that lignin nanoparticles improved the thermal stability of the binary and ternary nanocomposite systems showing enhanced antimicrobial effect as well, which underlines another interesting application of nanolignin in green food packaging.

5.5 Bioassays and Assessment of Nanoparticles' Antimicrobial Activity Against Phytopathogens

Organic nanoparticles have been shown to possess a wide spectrum of biological activities and great potential to control phytopathogens (Tables 5.2 and 5.3). Despite that, their application in agriculture has been underwhelming. Fortunately, this paradigm seems to be about to change. The advent of agricultural nanotechnology is providing researchers with a new generation of biocides capable of combining the advantages of organic nanoparticles with those of bio-based antimicrobial compounds without the harmful consequences to the environment associated with conventional pesticides (Ayyaril et al. 2023; Hussain et al. 2023).

Traditionally, nanoparticle-derived protection occurs via what the particles' properties can offer to plants directly (Poznanski et al. 2023). For example, chitosan nanoparticles were reported to improve the innate immune response of plants by activating a cascade of defense enzymes and by upregulating defense-related genes against *Rhizoctonia solani* (Divya et al. 2020). Likewise, lignin and iodo-lignin nanoparticles were able to directly increase the lipophilicity of fungi cell membranes due to the releasing of free iodine leading to the coagulation of cytoplasmic constituents and leading to cell death in several phytopathogenic fungal isolates (Amat et al. 2023).

In addition to these mechanisms, bio-based nanoparticles can act as carriers for a plethora of green active substances, such as EOs, plant extracts, phytohormones, secondary metabolites, proteins and AMPs, that can be applied by spraying or drenching/soaking seeds, foliar tissue, and or roots, offering a whole new level of biological activities and efficacy (Ayyaril et al. 2023; Ingle et al. 2022; Maximiano et al. 2022). These nanoformulations offer several advantages compared to the application of empty nanoparticles or bio-based antimicrobial compounds alone, such as enhanced solubility, stability, bioavailability, and controlled release (Hussain et al. 2023). Plus, direct application of high concentrations of some antimicrobial compounds could lead to toxicity problems. Thus, nanoencapsulation effectively reduces the practical dose needed for pathogen control due to increased surface area-to-volume ratios, which enables targeted site-specific delivery and efficient absorption through cells, leading to boosted antimicrobial activity (Hussain et al. 2023). Likewise, the controlled release of the antimicrobial significantly increases its half-life and the exposure period, thus increasing its bioactivity (Mohammadi et al. 2015a; Nadendla et al. 2018; Zhaveh et al. 2015). Furthermore, the synergism of the nanoformulation reduces the probability of resistance development among phytopathogens, which is common to single synthetic compound treatments (Bidyarani et al. 2020; Elsharkawy et al. 2022; Muzzalupo et al. 2020).

Organic nanoparticles are usually made from polymers such as bovine serum albumin, chitosan, lignin, lipids, and zein. Chitosan is by far the most studied polymer regarding phytopathogenic control, probably due to its abundancy and bioactive properties. Works involving bio-based nanoparticles have been performed in every stage of plant development, from seeds to postharvest, which provides a wealth of putative solutions to reduce production losses caused by bacterial, fungal, and viral infections (Tables 5.2 and 5.3). Plus, the availability of data regarding the nanoparticles' behavior on different plant tissues provides a better picture of their applicability, efficiency, and bioavailability.

Table 5.2 Application of organic nanoparticles loaded or not with bio-based antimicrobials, for phytopathogenic bacteria management

Nanoparticle (organic molecules/polymers)	Compound type	Antimicrobial compound (natural-based)	Phytopathogens assessed	Approach description	Reference
Chitosan	–	–	*Pectobacterium carotovorum* subsp. *carotovorum*, *Xanthomonas arboricola* pv. *vesicatoria* (0.5–5.0 mg/mL)	In vitro assay to evaluate the antibacterial activity against pathogen growth	Oh et al. (2019)
Chitosan	–	–	*Pectobacterium carotovorum* (2 mg/mL)	In vitro assay to evaluate the antibacterial activity against pathogen growth	El-Naggar et al. (2022)
Chitosan	EO	Thyme and lime EO	*Pectobacterium carotovorum*	In vitro assay to evaluate the antibacterial activity against pathogen growth	Sotelo-Boyás et al. (2015, 2016)
Chitosan	EO	*Athyrium sinense* EO	*Pectobacterium carotovorum* subsp. *carotovorum*	In vitro assay to evaluate the antibacterial activity against pathogen growth	Cai et al. (2023)
Chitosan	EO	*Cinnamomum zeylanicum* EO	*Pectobacterium carotovorum* (375 ppm)	In vitro assay to evaluate the antibacterial activity against pathogen growth	Mohammadi et al. (2020)
Chitosan	EO	Thyme EO	*Pectobacterium carotovorum*	In vitro assay to evaluate the antibacterial activity against pathogen cellular alterations and viability	Sotelo-Boyás et al. (2019)
Chitosan	–	–	*Pectobacterium carotovorum*, *Pseudomonas fluorescens* (0.009–0.15%)	In vitro assay to evaluate the antibacterial activity against pathogen growth	Mohammadi et al. (2016)
Chitosan	–	–	*Ralstonia solanacearum* (50–200 µg/mL)	In vivo assay to evaluate disease severity	Khairy et al. (2022)

(continued)

Table 5.2 (continued)

Chitosan	–	–	*Ralstonia solanacearum*	In vivo assay to evaluate disease impact on the plant metabolism	Narasimhamurthy et al. (2022)
Chitosan	Monoterpenoid; phenylpropanoid	Thymol; eugenol	*Ralstonia solanacearum* (22.5–45 μg/mL)	In vitro assay to evaluate the antibacterial activity against pathogen growth	Oluoch et al. (2021)
Chitosan	Monoterpenoid; phenylpropanoid	Thymol; eugenol	*Ralstonia solanacearum* (45–90 μg/mL)	In vitro and in vivo assays to evaluate the antibacterial activity against pathogen growth and disease severity	Oluoch et al. (2022)
Chitosan	–	–	*Xanthomonas campestris* (1 mg/mL)	In vivo assay to evaluate disease severity	Esyanti et al. (2020)
Chitosan	Monoterpenoid	Thymol	*Xanthomonas arboricola* pv. *arboricola* (100–600 μg/mL)	In vitro assay to evaluate the antibacterial activity against pathogen cellular alterations and viability	Sreelatha et al. (2022a)
Chitosan	Monoterpenoid	Thymol	*Xanthomonas campestris* pv. *campestris*	In vitro assay to evaluate the cytotoxic activity against the pathogen	Sreelatha et al. (2022b)
Chitosan	Biosurfactant	Rhamnolipid	*Xanthomonas campestris* (4–128 μg/mL)	In vitro assay to evaluate the antibacterial activity against pathogen cellular alterations and viability	Karamchandani et al. (2022)
Chitosan	–	–	*Xanthomonas oryzae* pv. *oryzae* (4–16 μg/mL)	In vitro assay to evaluate the antibacterial activity against pathogen cellular alterations and viability	Abdallah et al. (2020)
Chitosan	–	–	*Xanthomonas oryzae* (10 μL)	In vitro assay to evaluate the antibacterial activity against pathogen growth	Chitra et al. (2022)

(continued)

Table 5.2 (continued)

Nanoparticle (organic molecules/ polymers)	Compound type	Antimicrobial compound (natural-based)	Phytopathogens assessed	Approach description	Reference
Chitosan	–	–	*Xanthomonas oryzae* pv. *oryzae* (6 mg/100 mL)	In vitro assay to evaluate the antibacterial activity against pathogen growth	Praveen et al. (2022)
Chitosan guar	–	–	*Xanthomonas oryzae* (0.01%)	In vitro assay to evaluate the antibacterial activity against pathogen growth	Sathiyabama and Muthukunar (2020)
Chitosan and lignin	–	–	*Pectobacterium carotovorum* subsp. *odoriferum*, *Xanthomonas arboricola* pv. *pruni* (1 and 3 wt%)	In vitro assay to evaluate the antibacterial activity against pathogen growth	Yang et al. (2016b)
Lignin	–	–	*Pseudomonas syringae* pv. *tomato*, *Xanthomonas axonopodis* pv. *arboricola*, *X. arboricola* pv. *pruni*	In vitro assay to evaluate the antibacterial activity against pathogen growth	Yang et al. (2016a)
Lignin	–	–	*Pseudomonas syringae* pv. *tomato*	In vitro assay to evaluate the antibacterial activity against pathogen growth	Yang et al. (2016c)
Lignin	–	–	*Pseudomonas syringae* pv. *tomato*, *Xanthomonas axonopodis* pv. *arboricola*, *X. arboricola* pv. *pruni* (5–8 wt%)	In vitro assay to evaluate the antibacterial activity against pathogen growth	Yang et al. (2018a)
Zein	EO	*Satureja montana* EO	*Xanthomonas euvesicatoria*	In vivo assay to evaluate disease impact on the plant metabolism	Oliveira-Pinto et al. (2021)
Zein	Isoflavone	Rotenone	*Pseudomonas syringae* (135 μg/mL)	In vitro assay to evaluate the antibacterial activity against pathogen growth	Bidyarani and Kumar (2019)
Zein-rhamnolipid	Monoterpenoid	Carvacrol	*Pseudomonas syringae* (16 μg/mL)	In vitro assay to evaluate the antibacterial activity against pathogen growth	Bidyarani et al. (2020)

Table 5.3 Application of organic nanoparticles loaded or not with bio-based antimicrobials, for phytopathogenic fungi management

Nanoparticles (organic molecules/polymers)	Type of encapsulated product	Antimicrobial compound (natural-based)	Phytopathogens assessed (MIC)	Approach description	Reference
Bovine serum albumin	Protein	Chitinase	*Alternaria alternata* (1 U)	In vitro assay to evaluate the antifungal activity against spore germination and mycelial growth	Narendrakumar et al. (2018)
Chitosan	Bacteria-derived compounds	Spinosad	*Fusarium oxysporum* (8 mg/L)	In vitro assay to evaluate the antifungal activity against mycelial growth	Dong et al. (2019)
Chitosan	–	–	*Rhizoctonia solani* (1 mg/mL)	In vitro, in vivo, and in planta assay to evaluate the antifungal activity against pathogen growth, spore germination and disease development	Divya et al. (2020)
Chitosan	–	–	*Pyricularia grisea*, *Alternaria solani*, *Fusarium oxysporum* (100 μg/mL)	In vitro and in planta assay to evaluate the antifungal activity against pathogen growth, germination, and shoot development	Sathiyabama and Parthasarathy (2016)
Chitosan	EO	*Cananga odorata* essential oil	*Aspergillus flavus* (0.75–1 μL/mL)	In vitro and in vivo assay to evaluate the antifungal activity against pathogen growth, aflatoxin B_1 production, and storage quality	Upadhyay et al. (2021)
Chitosan	EO/plant extract	*Carum copticum* EO/*Peganum harmala* extract	*Alternaria alternata* (200 ppm)	In vitro and in planta assay to evaluate the antifungal activity against fungal growth	Izadi et al. (2021)

(continued)

Table 5.3 (continued)

Nanoparticles (organic molecules/polymers)	Type of encapsulated product	Antimicrobial compound (natural-based)	Phytopathogens assessed (MIC)	Approach description	Reference
Chitosan	EO	*Cinnamomum zeylanicum* EO	*Phytophthora drechsleri* (0.75 g/L)	In vitro and in vivo assay to evaluate the antifungal activity against pathogen growth, disease development, and storage quality	Mohammadi et al. (2015b)
Chitosan	EO	*Coriandrum sativum* EO	*Aspergillus niger*, *Aspergillus fumigatus*, *Aspergillus sydowii*, *Aspergillus repens*, *Aspergillus versicolor*, *Aspergillus luchuensis*, *Alternaria alternata*, *Penicillium italicum*, *Penicillium chrysogenum*, *Penicillium spinulosum*, *Mycelia sterilia*, *Cladosporium herbarum*, *Fusarium poae*, *Fusarium oxysporum* (0.5 μL/mL)	In vitro and in vivo assay to evaluate the antifungal activity against pathogen growth, aflatoxin B_1 production	Das et al. (2019)
Chitosan	EO	*Cymbopogon toloni* EO	*Fusarium graminearum* (421.7 ppm)	In vitro assay to evaluate the antifungal activity against mycelial growth and mycotoxin production	Kalagatur et al. (2018)
Chitosan	EO	*Eugenia caryophyllata* EO	*Aspergillus niger* (1.5 mg/mL)	In vitro assay to evaluate the antifungal activity against mycelial growth	Hasheminejad et al. (2019)
Chitosan	EO	*Allium sativum* EO	*Aspergillus versicolor* (3.33–5.0 mg/mL), *Aspergillus niger* (0.37–5.0 mg/mL), *Fusarium oxysporum* (1.11–2.5 mg/mL)	In vitro assay to evaluate antifungal activity against spores and mycelia.	Mondéjar-López et al. (2022)
Chitosan	EO	*Origanum vulgare* EO	*Alternaria alternata* (0.005% w/v)	In vitro assay to evaluate the antifungal activity against mycelial growth	Yilmaz et al. (2019)

Chitosan	EO	*Pimenta dioica* EO	*Aspergillus niger*, *Aspergillus fumigatus*, *Aspergillus sydowii*, *Aspergillus repens*, *Aspergillus versicolor*, *Aspergillus luchuensis*, *Alternaria alternata*, *Penicillium italicum*, *Penicillium chrysogenum*, *Penicillium spinulosum*, *Mycelia sterilia*, *Cladosporium herbarum*, *Fusarium poae*, *Fusarium oxysporum* (1.6 μL/mL)	In vitro and in vivo assay to evaluate the antifungal activity against spore germination, mycelial growth, and aflatoxin B_1 production	Chaudhari et al. (2022)
Chitosan	EO	*Valeriana officinalis* EO	*Aspergillus flavus* (6.5 μL/mL), *Aspergillus niger* (5 μL/mL), *Aspergillus versicolor* (4 μL/mL), *Penicillium italicum* (5.5 μL/mL), *Fusarium oxysporum* (3.5 μL/mL)	In vitro and in vivo assay to evaluate the antifungal activity against pathogen growth, disease development, and storage quality	Das et al. (2023)
Chitosan	EO	*Zataria multiflora* EO	*Botrytis cinerea* (1500 ppm)	In vitro and in vivo assay to evaluate the antifungal activity against pathogen growth and disease development	Mohammadi et al. (2015a)
Chitosan-benzoic acid	EO	*Thymus vulgaris* EO	*Aspergillus flavus* (500 mg/L)	In vitro and in vivo assay to evaluate the antifungal activity against pathogen growth and storage time	Tahereh et al. (2015)
Chitosan-caffeic acid	EO	*Cuminum cyminum* EO	*Aspergillus flavus* (350 ppm)	In vitro assay to evaluate the bioactivity against fungal growth	Zhaveh et al. (2015)

(continued)

Table 5.3 (continued)

Nanoparticles (organic molecules/ polymers)	Type of encapsulated product	Antimicrobial compound (natural-based)	Phytopathogens assessed (MIC)	Approach description	Reference
Chitosan-cinnamic acid	EO	*Mentha piperita* EO	*Aspergillus flavus* (500–1000 ppm)	In vitro and in vivo assay to evaluate the antifungal activity against pathogen growth and storage time	Beyki et al. (2014)
Chitosan-pea protein	EO	*Hyssopus officinalis* EO	*Botrytis cinerea* (2 mg/mL)	In vitro and in vivo assay to evaluate the antifungal activity against pathogen growth and disease development	Hadidi et al. (2021)
Chitosan	Plant extract	*Byrsonima crassifolia* extract	*Colletotrichum gloeosporioides* (0.05% w/v)	In vitro assay to evaluate the bioactivity against fungal growth	Barrera-Necha et al. (2018)
Chitosan	Plant extract	*Mentha longifolia* extract	*Aspergillus niger* (2 mg/mL)	In vitro assay to evaluate the bioactivity against fungal growth	El-Aziz et al. (2018)
Chitosan	Plant extract	*Olea europaea* extract	*Fusarium proliferatum* (1200 mg/L)	In vitro assay to evaluate the potency against spore germination and mycelial growth	Muzzalupo et al. (2020)
Chitosan	Plant hormone	MeSA	*Aspergillus flavus* (1 μL/mL)	In vitro assay to evaluate the potency against spore germination and mycelial growth and aflatoxin B_1 production	Kujur et al. (2019)
Chitosan	Plant hormone	SA	*Alternaria alternata* (400 ppm)	In vitro, in vivo, and in planta assay to evaluate the antifungal activity against pathogen growth, spore germination, and disease development	Hoang et al. (2023)

Chitosan	Plant hormone	SA	*Fusarium verticillioides* (0.16% w/v)	In vitro, in vivo, and in planta assay to evaluate the antifungal activity against pathogen growth, spore germination, and disease development	Kumaraswamy et al. (2019)
Chitosan	Plant hormone	SA	*Puccinia striiformis* f. sp. *tritici* (not specified)	In planta assay to evaluate the antifungal activity against disease development	Hembade et al. (2022)
Chitosan-lecithin	Flavonol	Kaempferol	*Fusarium oxysporum* (300 ppm)	In vitro assay to evaluate the antifungal activity against mycelial growth and hyphal extension	Ilk et al. (2017)
N-acylated chitosan	Monoterpene	Limonene	*Rhizopus tolonifera*, *Botrytis cinerea* (0.02% w/v)	In vivo assay to evaluate the antifungal activity against mold growth and storage quality	Vu et al. (2011)
Chitosan	Protein	*Pseudomonas syringae* Harpin	*Rhizoctonia solani* (20 μg)	In vivo assay to evaluate the antifungal activity against pathogen growth and disease development	Nadendla et al. (2018)
Lignin and iodo-lignin	–	–	*Cunninghamella* sp., *Fusarium equiseti*, *Penicillium chermesinum*, *Aspergillus flavus*, *Aspergillus niger*, *Trichoderma piluliferum* (12 μM)	In vitro assay to evaluate the antifungal activity against mycelial growth	Amat et al. (2023)

(continued)

Table 5.3 (continued)

Nanoparticles (organic molecules/ polymers)	Type of encapsulated product	Antimicrobial compound (natural-based)	Phytopathogens assessed (MIC)	Approach description	Reference
Lignin	Essential oil	*Thymus serpyllum* essential oil	*Phytophthora cactorum* (48 μg/mL)	In vitro and in planta assay to evaluate antifungal activity and ability to control mortality caused by the pathogen in seedlings	Vettraino et al. (2022)
Lignin-tannins	Plant oil	*Azadirachta indica* (neem) oil	*Botrytis cinerea*, *Phaeoacremonium minimum* (0.45% w/v), *Phaeomoniella chlamydospora*	In vitro assay to evaluate the antifungal activity against mycelial growth and spore germination	Falsini et al. (2023)
Lipid	Bacterial extract	*Lysinibacillus fusiformis* extract	*Fusarium oxysporum* f. sp. *cubense*, *Fusarium oxysporum* f. sp. *cucumerinum*, *Fusarium oxysporum* f. sp. *lycopersici* (500 ppm)	In vitro assay to evaluate the antifungal activity against mycelial growth	Atienza et al. (2021)
Lipid	EO	*Zataria multiflora* EO	*Aspergillus niger* (100 ppm), *Aspergillus flavus* (100 ppm), *Aspergillus ochraceus* (100 ppm), *Alternaria solani* (50 ppm), *Rhizoctonia solani* (50 ppm), *Rhizopus stolonifer* (50 ppm)	In vitro assay to evaluate the antifungal activity against mycelial growth	Nasseri et al. (2016)
Lipid	EO	*Melaleuca alternifolia* EO	*Monilinia fructicola* (3 g/L)	In vitro and in vivo assay to evaluate the antifungal activity against mycelia growth and disease development during storage	Xu et al. (2022)

Lipid	Peptide	Cyclic lipopeptide	*Aspergillus niger*, *Aspergillus carbonarius* (1 mg/mL)	In vitro assay to evaluate the antifungal activity against mycelial growth and mycotoxin production	Lin et al. (2020)
Zein	Peptide	ε-Polylysine	*Colletotrichum* spp. (5 mg/mL)	In vivo assay to evaluate the antifungal activity and storage quality	Garcia et al. (2022)
Zein	Diarylheptanoid	Curcumin	*Colletotrichum gloeosporioides* (100 μg/mL)	In vitro and in vivo *a*ssay to evaluate the antifungal activity against mycelia grow	Chuacharoen and Sabliov (2022)
Zein	Isoflavone	Rotenone	*Fusarium oxysporum* (48 mg/mL)	In vitro assay to deliver rotenone and evaluate its potency against plant pathogenic microbes	Bidyarani and Kumar (2019)
Zein-rhamnolipid	Monoterpene	Carvacrol	*Fusarium oxysporum* (270 μg/mL)	In vitro assay to evaluate the potency against plant pathogenic microbes	Bidyarani et al. (2020)

5.5.1 Antibacterial

The management of plant diseases caused by bacterial phytopathogens through the application of organic nanoparticles has recently come into the spotlight as a potential and relatively untapped resource. We have identified 27 articles exploring the antibacterial activity of organic nanoparticles, alone or loaded with antimicrobial compounds. These studies encompassed nine different species, across four different genera. Remarkably, *Xanthomonas* emerges as the most prominently represented genus, with five species, underscoring its notable significance and potential threat to agriculture. Additionally, *Pseudomonas* spp., *Ralstonia solanacearum*, and *Pectobacterium carotovorum* are also notorious targets in these studies. Moreover, within the scope of the compiled studies, chitosan-based nanoparticles were utilized in 20 research works, clearly illustrating their prominence as a choice among organic nanoparticles. Following this, lignin- and zein-based nanoparticles were employed in four and three works, respectively. Overall, the majority of the examined organic nanoparticles were subjected to evaluation for their antibacterial activity. Additionally, ten studies explored their combined efficacy with EOs or aromatic compounds while two studies examined their encapsulation alongside other compounds (Table 5.2).

Zein Nanoparticles' Antibacterial Properties

Zein nanoparticles have also been harnessed as carriers for the delivery of *Satureja montana* EOs aimed at treating *X. euvesicatoria*-infected tomato plants. Although *S. montana* EO entrapment efficiency in zein nanoparticles was low (about 20%), this approach resulted in a significant reduction of the pathogen load. Notably, the use of *S. montana* EO-loaded zein nanoparticles and free zein nanoparticles produced a reduction in the bacterial quantity by 38- and 202-fold, respectively (Oliveira-Pinto et al. 2021). Likewise, a decrease in bacterial load was observed in zein-nanoparticle-treated plants, supporting the hypothesis that zein nanoparticles would exhibit antimicrobial properties, as well (Oliveira-Pinto et al. 2021).

In vitro antibacterial studies of zein nanoparticles loaded with rotenone, a naturally occurring isoflavone, revealed a MIC of 16 μg/mL against *P. syringae* (Bidyarani and Kumar 2019), while rhamnolipid-stabilized carvacrol-loaded zein nanoparticles exhibited a MIC value at 135 μg/mL against the same phytopathogen (Bidyarani et al. 2020).

Chitosan Nanoparticles' Antibacterial Properties

Chitosan nanoparticles have emerged as one of the most promising antibacterial agents, either alone or in combination with other natural bioactive compounds, displaying noteworthy efficacy against diverse phytopathogens, including bacteria

(Cai et al. 2023). In addition to other factors, this is partially attributed to their heightened affinity toward microorganisms' membranes, enabling chitosan nanoparticles to readily penetrate the pathogen's cells (Nguyen et al. 2013).

Pectobacterium carotovorum is the causal agent of soft rot in several economically important crops. Oh et al. (2019) conducted a study on chitosan nanoparticles' antimicrobial potential against *P. carotovorum* subsp. *carotovorum*, previously acknowledged as *Erwinia carotovora* subsp. *carotovora*, and reported, for a selection of three different strains of the pathogen, significant antimicrobial activity of chitosan nanoparticles after 48 h of incubation. Similarly, El-Naggar et al. (2022) reported on the antimicrobial potential of chitosan nanoparticles against *P. carotovorum*. Here, the authors reported a minimal inhibitory concentration (MIC) value of 2 mg/mL and an inhibition zone of 19 mm for these nanoparticles, surpassing the inhibition observed with bulk chitosan (11 mm), suggesting the enhanced efficacy of the nanoparticles. Similarly, Sotelo-Boyás et al. (2015) reported an inhibition zone diameter of 15 mm, further supporting their antibacterial capacity against this pathogen.

As mentioned in preceding topics, chitosan nanoparticles are often loaded with antimicrobial compounds, such as EOs, to further enhance their efficacy. Cai et al. (2023) found that chitosan nanoparticles loaded with *Athyrium sinense* EO produced MIC and minimum bactericidal concentration (MBC) values of 0.16 and 0.63 mg/mL, correspondingly, against *P. carotovorum* subsp. *carotovorum*, significantly enhancing the EOs' antibacterial activity (0.31 and 1.25 mg/mL). Likewise, Mohammadi et al. (2020) also observed this synergistic effect between the encapsulated EO and the chitosan nanoparticles. The study reported that *Cinnamomum zeylanicum* EO-loaded chitosan nanoparticles displayed an improved antibacterial effect against *P. carotovorum*, leading to complete inhibition of bacterial growth (MIC) at 375 ppm and MBC at 1500 ppm. Notably, when the chitosan nanoparticles and the EO were assessed individually, they demonstrated much higher MIC values at 1500 ppm. Thyme and lime EO-loaded chitosan nanoparticles were also subjected to evaluation (Sotelo-Boyás et al. 2015, 2016). The findings indicated that thyme EO-loaded chitosan nanoparticles exhibited the highest inhibitory effect against *P. carotovorum*, producing an inhibition diameter of 24 mm. In contrast, lime EO-loaded chitosan nanoparticles and free chitosan nanoparticles showed smaller inhibition diameters, measuring 13 and 15 mm, respectively. Moreover, for thyme and lime EO-loaded chitosan nanoparticles, the count of colonies forming units (CFU) was found to be 240 and 450, respectively, whereas, for the unloaded chitosan nanoparticles, this number of CFU could not be determined. Furthermore, in a subsequent study by Sotelo-Boyás et al. (2019), the impact of thyme EO-loaded chitosan nanoparticles on the cell viability of *P. carotovorum* was investigated using both transmission electron microscopy (TEM) and two colorimetric assays: 3-(4,5-dimethylthiazol-2-yl)-2,5-diphenyl tetrazolium bromide (MTT) and Alamar blue (AMB). The outcomes of this study revealed a deformation of the bacteria's plasma membrane and cell wall, alongside a loss of cellular content leading to cell lysis. The reduction of cell viability was notable, with a decrease of 80% observed for the MTT assay and a 100% reduction for the AMB assay, after 48 and 6 h of incubation, respectively.

Mohammadi et al. (2016) further demonstrated a concentration-dependent increase in the antibacterial effects of chitosan nanoparticles, with different molecular weights, encompassing a range of concentrations from 0.009% to 0.15% (w/v), not only against *P. carotovorum* but also against another soft rot bacterial pathogen, *Pseudomonas fluorescens*. These concentrations led to a significant decrease in the bacterial optical density. Additionally, across all chitosan nanoparticles evaluated, the MIC and MBC values for both phytopathogens were consistently determined to be 0.037% (w/v) and 0.075% (w/v), respectively. The IC_{50} recorded values were 0.028 (w/v) for *P. fluorescens* and ranged from 0.018% to 0.024% (w/v) for *P. carotovorum*.

Chitosan nanoparticles were also explored as a potential control measure against *Ralstonia solanacearum*, a significant threat causing bacterial wilt in economically important *Solanaceae* crops. Khairy et al. (2022) conducted an extensive evaluation of the antibacterial efficacy of chitosan nanoparticles, ranging from 50 to 200 μg/mL, against three isolates of *R. solanacearum* (RS1, RS3, and RS5). The results obtained in vitro undoubtedly displayed the susceptibility of all bacterial isolates to the chitosan nanoparticles, correlated with the tested nanoparticle concentration. The highest inhibition zone was recorded at 200 μg/mL with the inhibition diameter ranging from 20 to 25.9 mm. Moreover, in vivo assays showed a significant reduction in disease severity following the foliar application of chitosan nanoparticles at concentrations of 100 and 200 μg/mL on *R. solanacearum*-infected potato and tomato plants, respectively. Specifically, for treated potato plants, the results indicated a reduction of disease incidence and severity of 78.93% and 71.85%, respectively, at 200 μg/mL, compared to the untreated and infected controls. Similarly, for tomato plants, the reductions were 81.64% and 77.63%, respectively, at 200 μg/mL, in comparison to the untreated and infected controls. Additionally, through TEM observation, the authors reported that chitosan nanoparticles could induce notable morphological modifications on the external surface of the bacterial cell, including flagella loss, and, ultimately, cell lysis. Furthermore, in another recent study, Narasimhamurthy et al. (2022) showed that tomato seedlings inoculated with *R. solanacearum* and subsequently treated with chitosan nanoparticles displayed a noteworthy reduction in disease incidence of 34%. This value was nearly three times lower compared to the disease incidence observed in *R. solanacearum*-inoculated tomato seedlings that remained untreated (96%). Besides, this study also revealed that these nanoparticles were also capable of eliciting defense mechanisms in tomato seedlings, which, in turn, led to the enhancement of multiple growth parameters, suggesting the dual potential of chitosan nanoparticles by mitigating the impact of the disease and fostering plant growth.

The antibacterial effectiveness of loaded chitosan nanoparticles was also assessed against *R. solanacearum*. Previous studies demonstrated that chitosan nanoparticles loaded with thymol and eugenol displayed enhanced antibacterial activity against this pathogen through in vitro assays. These loaded nanoparticles achieved a notable inhibition of bacterial growth, reaching 92% and 94% respectively, in comparison to an inhibition of 80% observed for chitosan nanoparticles alone. Likewise, prior to encapsulation, thymol and eugenol also exhibited a MIC value of 175 and 275 μg/mL which, after encapsulation, was significantly reduced to 22.5 and 45 μg/mL, correspondingly (Oluoch et al. 2021). In a following study, the same authors conducted an in vivo experiment under greenhouse conditions, where *R. solanacearum*-infected potato plants treated with 90 and 45 μg/mL of eugenol and thymol loaded chitosan nanoparticles, respectively, presented the most substantial decrease in disease severity (Oluoch et al. 2022).

Xanthomonas spp. represent also a concerning group of phytopathogenic bacteria with an extensive host range, leading to severe production losses. The challenges posed by these phytopathogens have prompted several studies highlighting the potential of chitosan nanoparticles as a viable alternative for their control since it was reported that these nanoparticles were able to significantly reduce the optical density of *X. campestris* pv. *vesicatoria*, emphasizing their antimicrobial effect (Oh et al. 2019). Chitosan nanoparticles were also shown to be effective against *X. campestris* by reducing disease severity in three different chili pepper cultivars (Esyanti et al. 2020). Similarly, when loaded with thymol, chitosan nanoparticles showed the ability to disrupt in vitro the viability of *X. campestris* pv. *campestris* across a concentration range spanning from 100 to 600 μg/mL, exhibiting increased cytotoxicity in a dose-dependent fashion. Particularly, at a concentration of 300 μg/mL, thymol-loaded chitosan nanoparticles induced a 50% reduction in bacterial viability. This was substantiated by the inhibition of notable bacterial features such as the production of biofilm, exopolysaccharides, and xanthomonadin. Moreover, the authors confirmed through ultrastructure analysis the resultant membrane damage and release of intracellular contents (Sreelatha et al. 2022a). In a subsequent study, the same authors revealed that membrane damage and oxidative stress were the primary distresses triggered by thymol-loaded chitosan nanoparticles. Employing time-kill kinetic assays, they demonstrated that exposure to sublethal concentrations of thymol-loaded chitosan nanoparticles (at 500 μg/mL) led to the gradual mortality of *X. campestris* pv. *campestris* cells in a time-dependent manner. Within 1 h of exposure, a reduction in the number of bacterial colonies was noted, and after 24 h, an absolute absence of bacterial growth was confirmed in cells subjected to the loaded nanoparticles (Sreelatha et al. 2022b). Likewise, Karamchandani et al. (2022) evaluated fungal chitosan-derived nanoparticles against *X. campestris* and reported a MIC value greater than 1.024 μg/mL. Nonetheless, when these nanoparticles were combined with a commercial rhamnolipid biosurfactant, a synergistic effect was displayed. This synergy resulted in a significant reduction in the MIC values down to 4 μg/mL for the chitosan nanoparticles and 128 μg/mL for the biosurfactant. Additionally, scanning electron microscopy (SEM) showed that in the presence of chitosan nanoparticles (MIC >1024 μg/mL), the bacterial cells suffered discernible morphological modifications, including a ruptured appearance. These morphological deformations were intensified following treatment with the synergistic combination of rhamnolipid biosurfactant (at 128 μg/mL) and chitosan nanoparticles (at 4 μg/mL), culminating in a complete distortion of the bacterial cells.

Abdallah et al. (2020) conducted an extensive study of the antibacterial efficacy of chitosan nanoparticles against *X. oryzae* pv. *oryzae*. The authors reported that at concentrations of 4, 8, and 16 μg/mL, these nanoparticles displayed considerable in vitro inhibition halos of 21, 23, and 24 mm, respectively, highlighting an enhancement of the antibacterial activity compared to chitosan alone, which showed inhibition halos of 9, 12, and 13 mm at the same tested concentrations. Furthermore, these nanoparticles induced a notable reduction of bacterial growth, as evidenced by a significant decrease of the optical density at 600 nm, reaching 86.28%, 86.76%, and 86.85% for the respective concentrations of 4, 8, and 16 μg/mL. Besides, a concentration-dependent reduction in biofilm production and motility was reported. At the concentrations of 4, 8, and 16 μg/mL, chitosan nanoparticles caused reductions of 50.44%, 61.91%, and 69.91%, respectively, in biofilm production, and additionally, a corresponding decrease of 10.7%, 19.8%, and 32.1% of the diameter of migration areas was observed. Further, TEM micrographs showed extensive damage and abnormalities in *X. oryzae* pv. *oryzae* cell wall and cytoplasmatic membrane caused by chitosan nanoparticles at 16 μg/mL. This led to the loss of cellular content and ultimately bacterial death. This was further validated by flow cytometry where, upon treatment with chitosan nanoparticles, bacterial cell apoptosis reached 99.51% (Abdallah et al. 2020). Moreover, a MIC value of 10 μL of chitosan (1 mg/mL) nanoparticles against *X. oryzae* was also reported (Chitra et al. 2022), whereas, in another study, these nanoparticles at a concentration of 6 mg/100 mL produced an inhibition zone diameter of 22.9 mm against this phytopathogen (Praveen et al. 2022). Chitosan-guar gum nanoparticles were also assessed against *X. oryzae*, exhibiting a maximum inhibition zone diameter of 18 mm at a concentration of 0.01% (Sathiyabama and Muthukumar 2020).

Lignin Nanoparticles' Antibacterial Properties

Lignin nanoparticles have also garnered considerable interest as a potential tool against numerous phytopathogens. Such attraction stems from the inherent antimicrobial properties of lignin, presenting a promising novel approach for new disease management strategies (Chen et al. 2023).

Lignin nanoparticles have been strategically employed in PVA/chitosan films as fillers. In vitro antimicrobial assays revealed that these nanoparticles were able to decrease the bacterial growth of *P. carotovorum* subsp. *odoriferum* and *X. arboricola* pv. *pruni* (Yang et al. 2016b). Also, Yang et al. (2016a) by blending lignin nanoparticles and cellulose nanocrystals into neat or grafted polylactic acid, revealed that, over time, the binary amalgamation of these nanoparticles with the polylactic acid film, encouraged an effective antibacterial effect against Pst over time. This was discerned through a substantial decrease in bacterial growth, showcasing the enhanced antibacterial activity compared to the polylactic acid control. Subsequently, in a follow-up study, Yang et al. (2016c) assessed the same blend of lignin nanoparticles and cellulose nanocrystals against a broader spectrum of phytopathogenic bacterial strains, including Pst, *X. axonopodis* pv. *vesicatoria*, and *X. arboricola* pv.

pruni. The study's findings unveiled a synergistic effect that led to an overall decrease in the CFU count for all strains, with a particularly pronounced impact on the two *Xanthomonas* spp. strains.

In another study using lignin nanoparticles against the aforementioned phytopathogens, it was reported that an enhanced susceptibility for Pst and *X. axonopodis* pv. *vesicatoria* as concentration increased from 5 to 8 wt%. This heightened susceptibility to lignin nanoparticles was evidenced by inhibition halo diameters of 0.28 and 0.33 mm for Pst at concentrations of 5% and 8%, respectively. Similarly, for *X. axonopodis* pv. *vesicatoria* and *X. arboricola* pv. *pruni*, the corresponding values were 0.26 and 0.33 mm and 0.24 and 0.36 mm. Moreover, across all bacterial strains, there was an overall reduction in bacterial growth compared to the control, with *X. arboricola* pv. *pruni* showcasing the most significant decrease after 24 h. Additionally, the incorporation of lignin nanoparticles led to the complete inhibition of bacterial growth for all tested strains (Yang et al. 2018a).

5.5.2 *Antifungal*

Fungal phytopathogens are a major target for phytopathologists due to their ubiquity, devastating effects, and difficulty to control (Rosa et al. 2022). Naturally, works have been directed at the development of nanoparticles capable of controlling their proliferation. We found 40 works testing the antifungal activity of bio-based nanoparticles, alone or antimicrobial compound-loaded, against 36 different species and across 18 genera (Table 5.3). *Aspergillus* is by far the most represented genus (nine species) due to its implication in the contamination of foods with aflatoxin B_1 during the postharvest stages. Other well-represented phytopathogens are *Fusarium* (six species) and *Penicillium* spp. (four species). Furthermore, chitosan-based nanoparticles were used in 28 of the works compiled herein, demonstrating their predominance compared to other organic polymers. Zein-based nanoformulations follow with four works, while lignin- and lipid-based nanoparticles have only been tested three and four times against fungal phytopathogens, respectively. Finally, albumin-based nanoparticles were tested once (Table 5.3). EOs have been the preferred bio-based antimicrobial agents encapsulated in the nanoparticles (19 different works), followed by plant metabolites (9 works), plant extracts (3 works), bacteria-derived antimicrobials (2 works), and proteins and peptides with 2 works each (Table 5.3). It is also worth mentioning that the vast majority (19) of these works were performed in vitro exclusively, while 14 have tested the nanoparticles' bioactivity in vivo (detached plant organs) alone or in combination with in vitro methods. Inversely, only seven works have performed in planta tests, either exclusively or after in vitro and in vivo trials.

Zein Nanoparticle's Antifungal Properties

Zein has been extensively studied for films and coatings aiming at increasing the postharvest quality and shelf life of foods. Despite having no direct antifungal ability, it can improve delivery efficiency and solubility of the compounds entrapped within its nanoparticles resulting in high levels of protection against phytopathogens (Chuacharoen and Sabliov 2022). Thus, zein nanoparticles have been used as nanocarriers to deliver ε-polylysine, a peptide with a wide range of antimicrobial properties, to protect the surface of avocados from *Colletotrichum* spp. during the postharvest stage while extending the fruits' shelf life without losing quality (Garcia et al. 2022). Treatment of fruits with zein-ε-polylysine nanoparticles (5 mg/mL) helped to mitigate the severity of postharvest fungal disease. On day 6, control avocados displayed 40% fungal incidence, while treated avocados reached that same incidence percent only on day 12. At day 15, the control group had more than 83% incidence, while the treated group had 60%. Plus, fruits treated with this nanoformulation experienced a more discreet weight loss, respired the least, showed a higher firmness, and experienced less color change (Garcia et al. 2022). Researchers pointed out the ε-polylysine cationic composition, targeting anionic fungal cell membranes and causing cytoplasmic leakage, and disruption of fungal growth as the mechanisms behind these results. Also, because zein nanoparticles had similar incidences and results, the hydrophobic properties of zein probably played an important role in hindering gas and water transfer, hampering fungal proliferation and food spoilage (Garcia et al. 2022).

Plant secondary metabolites have also been encapsulated in zein nanoparticles to assess their antifungal activity. Rotenone-loaded zein nanoparticles showed MIC at 48 mg/mL against *F. oxysporum* in vitro. Rotenone is a broad-spectrum insecticide capable of inhibiting oxygen uptake and blocking the electron transport chain, also leading to the cessation of microbial growth (Bidyarani et al. 2020). Also, against *F. oxysporum*, carvacrol-loaded zein-rhamnolipid nanoparticles were able to inhibit its growth in vitro at 270 µg/mL (Bidyarani et al. 2020). The authors hypothesized that the antimicrobial activity could be due to carvacrol interaction with the microbial phospholipid cell membrane increasing permeability. In addition, a synergistic effect of rhamnolipid and zein could promote the increased activity of zein-rhamnolipid nanoparticles compared to the free components. Along with that, rhamnolipid enhanced the access of carvacrol to these regions by increasing its aqueous solubility (Bidyarani et al. 2020). Moreover, curcumin-loaded zein nanoparticles were more effective at inhibiting the mycelial growth of *C. gloeosporioides* in vitro and disease severity in vivo resulting in prolonged shelf life and postharvest quality of mangoes under storage conditions, compared to both the free forms of zein and curcumin at 100 µg/mL (Chuacharoen and Sabliov 2022).

Bovine Serum Albumin Nanoparticles' Antifungal Properties

Protein nanoparticles have been pursued due to their biodegradable nature, surface alteration possibilities, and particularly for their reported ability to encapsulate and deliver other interest compounds against pathogens. BSA nanoparticles carrying chitinase, a lytic enzyme capable of catalyzing the hydrolysis of chitin of fungal phytopathogens produced by *Serratia marcescens*, have been rendered effective at controlling *Alternaria alternata* growth in vitro. At 1 U (units), the nanoformulated chitinase was able to reduce fungal growth by 65%, reaching 100% at 5 U, significantly higher values than those observed for the application of free chitinase (Narendrakumar et al. 2018) (Table 5.3). The authors showed that nanoencapsulated enzymes had more stability, and the in vitro release profile showed prolonged exposure and increased activity. Enzymatic lysis of fungal cell walls through extracellular chitinases is one of many mechanisms of biocontrol. Thus, the combination of the hydrolytic properties of chitinases with nanoparticles could turn into an environmentally safe biocontrol method for phytopathogenic fungi (Narendrakumar et al. 2018).

Chitosan Nanoparticles' Antifungal Properties

Chitosan is highly regarded for its strong antifungal activity while being biocompatible. Because of this, chitosan has been widely used to control fungal phytopathogen proliferation. Chitosans' antifungal activity can be further enhanced by obtaining chitosan nanoparticles, which have been the subject of many works aiming at developing safer, cheaper alternatives to conventional agrochemicals against a plethora of phytopathogenic fungi (Poznanski et al. 2023). In that sense, chitosan nanoparticles (1 mg/mL) are effective against the rice sheath blight causal agent, *Rhizoctonia solani* (Table 5.3). The application of a sprayable solution was found to suppress the disease symptoms by 90% and 75% in detached leaves under greenhouse conditions, respectively (Divya et al. 2020). In vitro mycelial growth was significantly decreased by chitosan nanoparticles' application at a concentration of 100 μg for *P. grisea* (92%), *F. oxysporum* f. sp. *ciceri* (87%), and *A. solani* (72%), whereas bulk chitosan had less impressive inhibition results (Sathiyabama and Parthasarathy 2016). Despite chitosans' molecular weight, degree of deacetylation, acetylation pattern, and dispersity, along with chitosan nanoparticle size, zeta potential, dispersity, and fungal species development stage and cell properties, that determine the antifungal activity (Poznanski et al. 2023), the encapsulation of antimicrobial compounds in chitosan nanoparticles can play a fundamental role in that activity as well.

Spinosad-loaded nanomicelles were evaluated against two isolates of *Fusarium oxysporum* by exposing the pathogen to different concentrations of nanoparticles (Dong et al. 2019). The MIC for spinosad and spinosad-loaded nanoparticles was 11 and 8 mg/L, respectively. In all tests, the nanoformulation performed 5–8% better than the free-form spinosad. Since the empty nanoparticles had no antifungal activity, the improved antifungal action should be due to the extended duration of spinosad action. In addition, authors believe that the spinosad-loaded nanomicelles were able to inhibit fungal growth because a larger surface area of the fungus came in contact with spinosad. Furthermore, spinosad was effectively protected from photodegradation after encapsulation (Dong et al. 2019).

Chitosan nanopartciles encapsulating EOs have been tested in numerous assays for their antifungal abilities, with promising results (Table 5.3). For instance, *Cananga odorata* EO-chitosan nanoemulsion has been proven effective in vitro at controlling *Aspergillus flavus* proliferation and aflatoxin B_1 production at concentrations as low as 1 and 0.75 μL/mL, respectively (Upadhyay et al. 2021). The authors also provided evidence of this ability in vivo by testing the nanoemulsion under storage conditions on groundnut seeds. They reported a 83.52% level of protection at 1 μL/mL (MIC) and 91.76% at the 2MIC, compared to the 67.03% protection offered by the EO alone at 2MIC concentration (Upadhyay et al. 2021). The enhanced efficacy, compared to the free EO, was attributed to the controlled release, to the presence of cationic-charged amine groups on the surface of chitosan molecules, which can alter the fungal membrane fluidity, and to the larger surface area achieved through nanoencapsulation which enables easier passage through the fungal cell membrane. Also, encapsulation might have enhanced the physical stability of EO (Upadhyay et al. 2021).

Likewise, *Carum copticum* EO-loaded chitosan nanoparticles have been shown to inhibit *Alternaria alternata* growth in vitro (at 500 ppm). However, when *C. copticum* EO was encapsulated together with *Peganum harmala* extract in chitosan nanoparticles, this value dropped to 200 ppm (Izadi et al. 2021). Conversely, neither pure nor nanoparticulated chitosan showed complete inhibition while both the EO and plant extract exhibited complete inhibitory effects in vitro at 750 and 500 ppm, respectively. Moreover, the application of *C. copticum-Peganum harmala* EO nanoformulation under greenhouse conditions proved to be as effective as the commercial fungicide chlorothalonil against *A. alternata*. The nanoparticles stabilized the volatile compounds of the EO and provided better infiltration into the fungal tissues, enhancing their antifungal effectiveness (Izadi et al. 2021).

The application of *Cinnamomum zeylanicum* EO encapsulated in chitosan nanoparticles has completely inhibited the growth of *Phytophthora drechsleri* in vitro at 0.75 g/L, while the free chitosan nanoparticles were only able to do it at 3 g/L and the free EO at 1.5 g/L (Mohammadi et al. 2015b). A similar trend was observed in vivo after cucumbers were coated with the nanoformulation. The treatment was not only effective at controlling cucumber decay but also in delaying the onset of symptoms and reducing the disease severity caused by *P. drechsleri* (Mohammadi et al. 2015b). Also, chitosan nanoparticles loaded with *Coriandrum sativum* EO showed complete inhibition of broad-spectrum biodeteriorating storage molds *Aspergillus flavus*, *A. niger*, *A. fumigatus*, *A. sydowii*, *A. repens*, *A. versicolor*, *A. luchuensis*, *A. alternata*, *Penicillium italicum*, *P. chrysogenum*, *P. spinulosum*, *Mycelia sterilia*, *Cladosporium herbarum*, *Fusarium poae*, and *F. oxysporum* at 0.5 μL/mL, while aflatoxin MIC was 0.4 μL/mL (Das et al. 2019). This protective effect was further assayed in vivo using rice seeds. The nanoemulsion showed a level of protection of 84.17% at 2MIC, whereas the EO was only able to provide

62.43% protection. Both completely inhibited aflatoxin contamination at MIC and 2MIC concentrations (Das et al. 2019). The results showed that both the nanoformulation and free EO impaired ergosterol biosynthesis caused irreversible damage to the plasma membrane and leakage of intracellular content and inhibited the production of methylglyoxal, an aflatoxin inducer (Das et al. 2019).

Antifungal activity of *Cymbopogon martinii* EO against postharvest pathogen *Fusarium graminearum* has been shown at 900 ppm, whereas the encapsulation of this EO in chitosan nanoparticles reduced the effective dosage to 700 ppm. In addition to the complete inhibition of fungal growth, the treatment successfully inhibited the production of mycotoxins deoxynivalenol and zearalenone in storage maize grains (Kalagatur et al. 2018). Likewise, *Eugenia caryophyllata* EO-loaded chitosan nanoparticles completely inhibited the growth of *A. niger* in vitro at 1.5 mg/mL, whereas the free EO, even at concentrations as high as 3 mg/mL, was unable to do it (Hasheminejad et al. 2019). *Allium sativum* EO encapsulation in chitosan nanoparticles has been linked to improved stability and antifungal activity (Mondéjar-López et al. 2022). This nanoformulation has been rendered effective against *A. niger*, *A. versicolor*, and *F. oxysporum*. Inhibitory concentrations were assessed against spores and mycelia growth, showing important differences in susceptibility between these structures. For *A. niger* the authors reported MICs of 0.37 and 5.0 mg/mL for mycelia and spores, respectively. For *F. oxysporum* the values were 1.11 and 2.50 mg/mL, and for *A. versicolor* the values were 3.33 and 5.0 mg/mL, respectively (Mondéjar-López et al. 2022). Likewise, *Origanum vulgare* (0.005% w/v) EO-loaded chitosan nanoparticles have been shown to inhibit *A. alternata* growth in vitro in a concentration-dependent manner (Yilmaz et al. 2019). The superior performance of encapsulated forms over free EOs might be due to the controlled release of the preserved volatile EO as well as the inhibitory effect of chitosan nanoparticles, often associated with an increase in permeability of the pathogen's cell membrane as well as with membrane depolarization (Hasheminejad et al. 2019; Mondéjar-López et al. 2022; Yilmaz et al. 2019).

Pimenta dioica EO-loaded chitosan nanoparticles were tested in vitro for their antifungal activity against 15 mold strains, affecting food commodities at the postharvest stage (Chaudhari et al. 2022). The complete suppression of fungal growth and aflatoxin production by *A. flavus* was observed at 1.6 and 1.0 μL/mL, respectively. Furthermore, *Pimenta dioica* EO-loaded chitosan nanoparticles were also reported as able to completely inhibit the growth of *A. niger*, *A. fumigatus*, *A. sydowii*, *A. repens*, *A. versicolor*, *A. luchuensis*, *A. alternata*, *Penicillium italicum*, *P. chrysogenum*, *P. spinulosum*, *Mycelia sterilia*, *Cladosporium herbarum*, *Fusarium poae*, and *F. oxysporum* at 1.6 μL/mL. In vivo, the nanoformulation was able to eliminate aflatoxin B_1 contamination of treated maize grains under storage conditions. This effect was observed up to a year after the experiment (Chaudhari et al. 2022). EO-loaded nanoparticles were shown to inhibit ergosterol biosynthesis, causing a dose-dependent decline in mitochondrial membrane potential and leakage of important cellular constituents. Plus, they interfered with aflatoxin biosynthesis by inhibiting methylglyoxal production and downregulating the *aflR* gene (Chaudhari et al. 2022). Chitosan nanoemulsion carrying *Valeriana officinalis* EO showed antimicrobial activity against *A. flavus*, *A. niger*, *A. versicolor*, *P. italicum*, and *F. oxysporum* at 3.5–6.5 μL/mL concentrations (Das et al. 2023). In addition, the nanoparticles were able to inhibit aflatoxin production at 5.0 μL/mL in vitro. For the in vivo experiment, *Citrus sinensis* fruits were coated with the nanoemulsion at MIC and 2 MIC dosages. These dosages completely inhibited fungal growth and preserved the fruits for a longer period than the control. In addition, aflatoxin accumulation in the treated fruits was completely inhibited at both concentrations (Das et al. 2023). Nanoencapsulation of *Zataria multiflora* EO in chitosan nanoparticles has been put to the test to control *Botrytis cinerea* in vitro (Mohammadi et al. 2015a). Results showed that the encapsulation of the EO completely inhibited fungal growth at 1500 ppm, contrarily to the application of the free EO which struggled to achieve more than 50% growth inhibition. The in vivo experiments yielded similar results, where nanoformulation-treated strawberries (1500 ppm) showed decreased disease severity and incidence of *B. cinerea*, after 2 days of shelf life at 20 °C (Mohammadi et al. 2015a).

Encapsulation of *Thymus vulgaris* EO in chitosan-benzoic acid nanogel was used to enhance its antifungal properties and half-life. Interestingly, the application of the nanogel alone against *A. flavus* was found to completely inhibit fungal growth at 1500 mg/L, whereas the nanoformulation presented a MIC of 300 and 500 mg/L under sealed and non-sealed conditions, respectively (Tahereh et al. 2015). Likewise, the encapsulation of *Mentha piperita* EO in chitosan-cinnamic acid nanoparticles effectively inhibited *A. flavus* growth at 500 and 800 ppm under sealed and non-sealed conditions, values far lower than those of free EOs (2100 and 3000 ppm, respectively) (Beyki et al. 2014). Plus, in the in vivo experiments, both works showed that the encapsulated EOs, at 700 mg/L and 1000 ppm, were capable of preserving tomato fruits during the 1-month storage period (Beyki et al. 2014; Tahereh et al. 2015). Similarly, the encapsulation of *Cuminum cyminum* EO in chitosan-caffeic acid nanogel led to an increase in antimicrobial activity and stability (Zhaveh et al. 2015). Although no *A. flavus* growth was observed for EOs in both their free and encapsulated forms, the concentrations needed differed significantly (650 and 350 ppm, respectively). Plus, as observed by Tahereh et al. (2015) and Beyki et al. (2014), the inhibition ability of free EO in non-sealed flasks was severely affected, whereas the encapsulated form was able to retain that ability at higher concentration (900 ppm) (Zhaveh et al. 2015). Finally, *Hyssopus officinalis* EO was encapsulated in a chitosan-pea protein isolate nanocomplex and tested against fungal phytopathogen *B. cinerea* (Hadidi et al. 2021). The authors reported increased bioactivity after the encapsulation, and 2 mg/mL of EO-loaded chitosan-pea protein nanoformulation inhibited fungal growth by 84.3% in vitro. The treatment was also effective at controlling strawberry spoilage by delaying symptom development and disease severity and reducing the growth of *B. cinerea* during the storage period (Hadidi et al. 2021). In the mentioned works, the superior performance of the encapsulated EOs was due to the slow release of the effective compounds as well as the inhibitory effect of the chitosan (Beyki et al. 2014; Hadidi et al. 2021; Tahereh et al. 2015; Zhaveh et al. 2015).

Other antimicrobial compounds frequently encapsulated in chitosan nanoparticles with antifungal activity are plant extracts and secondary metabolites. In this regard, the incorporation of *Byrsonima crassifolia* methanolic extract into chitosan nanoparticles successfully inhibited mycelia growth and sporulation between 79% and 82%, and spore germination completely at 0.05% w/v concentration (Barrera-Necha et al. 2018). Moreover, chitosan nanoparticles carrying *Mentha longifolia* extract increased the antimicrobial activity of the extract significantly. At 0.5, 1, and 2 mL/mL, it inhibited the mycelial growth of *A. niger* by 54.4%, 66.7%, and 75.6%, respectively (El-Aziz et al. 2018). Authors reported that, contrarily to their counterparts, chitosan nanoparticles were able to introduce themselves intracellularly, leading to organelle and nucleic acid damage (El-Aziz et al. 2018). Likewise, the encapsulation of *Olea europaea* extracts in chitosan nanoparticles showed greater efficacy than the pure extracts and the empty nanoparticles against *F. proliferatum* at 1200 mg/L, likely due to the combined action of the antioxidant and chelating power of the hydroxyl groups of the phenolic compounds found in the extract and the electrostatic interaction between the amine group of chitosan with the negatively charged cell membrane of the pathogen (Muzzalupo et al. 2020).

The nanoencapsulation of the volatile phytohormone methyl salicylate (MeSA) and its application as a control method with antifungal and aflatoxin B_1 inhibitory properties has been tested against *A. flavus*. Both the free MeSA (1.50 μL/mL) and the encapsulated form (1.0 μL/mL) showed significant in vitro efficacy against the mycelial growth and aflatoxin production, although at different dosages (Kujur et al. 2019). The mode of action analysis demonstrated that the nanoformulation led to a significant decrease in ergosterol content, leakage of vital ions, utilization of alternative carbon sources, and interference with *ver1* and *omtA* genes involved in the aflatoxin biosynthesis pathway (Kujur et al. 2019). To assess the effectiveness of chitosan nanoparticles loaded with salicylic acid (SA) on reducing cassava leaf spot and *A. alternata* virulence, Hoang et al. (2023) developed different formulations and used two methods of application. The SA-loaded nanoparticles (400 ppm) were able to control cassava leaf spot disease by 40.77%, reduced the virulence of *A. alternata* through inhibition of conidial germination and mycelia viability, and decreased the rot diameter in detached leaves. Moreover, this nanoformulation proved to be effective as both pre- and postinoculation methods (Hoang et al. 2023). These results may point out a synergy between the direct effect of the chitosan and the SA and their ability to induce plant defenses. The same nanoformulation was tested in a different work against *F. verticillioides* (Kumaraswamy et al. 2019). The authors reported strong fungal growth inhibition in vitro and a significant reduction of disease severity, with a good disease control efficacy index in planta under open-field conditions at 0.16% w/v. The chitosan nanoparticles loaded with SA performed better than the free chitosan and SA. In addition to the antifungal activity reported, the encapsulation enabled the slow release of SA avoiding toxicity effects on the plants despite the high dose used (Kumaraswamy et al. 2019). Likewise, chitosan-SA nanoformulation was able to completely protect rice plants against the causal agent of the yellow rust disease, *Puccinia striiformis* f. sp. *tritici*. The treatment was able to produce symptomless plants by increasing PR gene expression, phenylalanine ammonia-lyase (PAL), TAL, PPO activities, and total soluble sugar content by spraying the plants 24 h before the inoculation (Hembade et al. 2022). In a different work, kaempferol-loaded lecithin-chitosan nanoparticles were developed for the inhibition of *F. oxysporum*. At 300 ppm, hyphal growth was strongly inhibited, compared to free kaempferol and the control. The nanoparticles' treatment caused damage to the hyphal morphology, hampering its growth (Ilk et al. 2017). The higher antifungal activity was linked to the higher contact area with the fungal cell wall, due to the negatively charged membrane of fungus being targeted by the positively charged nanoformulation. Plus, chitosan nanoparticles can penetrate and diffuse into the fungal cells and disrupt the synthesis of DNA and RNA. The sustained release of kaempferol from the lecithin-chitosan nanocarrier increased its antimicrobial activity (Ilk et al. 2017). Likewise, the application of the nanoformulation, consisting of *N*-acylated chitosan encapsulating limonene, at 0.02% w/v, successfully protected strawberries against molds like *Rhizopus stolonifer* and *Botrytis cinerea* under storage conditions (Vu et al. 2011).

Finally, harpin, a protein secreted by the phytopathogen *Pseudomonas syringae* with major implications in triggering non-host-plant defense responses, has been employed as a control method of *Rhizoctonia solani* in tomatoes using chitosan nanoparticles as a carrier (Nadendla et al. 2018). In vivo treatment (20 μg) resulted in sustained induction of peroxidase and PAL while decreasing fungal biomass and infection of detached tomato leaves. The authors reported other important alterations to the tomato's transcriptome for both free harpin and encapsulated forms regarding the oxidative response, host defense, ion transport, and jasmonic acid (JA) signaling. Hence, the sustained release of harpin extended the bioavailability, induced plant defense, and reduced *R. solani* infection.

Lignin Nanoparticle's Antifungal Properties

Due to its abundant, biodegradable, and biocompatible nature, lignin has been functionalized for many agricultural applications, among which we can find antifungal nanoformulations (Gigli et al. 2022; do Espirito Santo Pereira et al. 2022). For instance, lignin and iodo-lignin nanoparticles were tested through the agar diffusion method against *A. flavus*, *A. niger*, *Cunninghamella* sp., *F. equiseti*, *P. chermesinum*, and *T. piluliferum* (Amat et al. 2023). Results showed that the presence of the lignin nanoparticles reduced fungal growth, compared to the control plates. At the maximum concentration tested (12 mM), lignin nanoparticles showed 25.81–65.29% fungal growth inhibition, whereas iodo-lignin nanoparticles showed significantly higher values (83–96%). Because the microbial cell membrane can be damaged by phenolic compounds, lignin nanoparticles were able to reduce fungal growth. However, by iodinating the lignin nanoparticles, this activity was increased due to the added toxicity. iodo-lignin nanoparticles release free iodine which increases the lipophilicity of fungi cell membranes, leading to the coagulation of cytoplasmic constituents, impairing the growth of fungal isolates (Amat et al. 2023).

The bioactivity of lignin nanoparticles against phytopathogenic fungi can be further improved through the encapsulation of antimicrobial compounds, like EOs, in them. Thus, when encapsulated in lignin nanoparticles, *Thymus serpyllum* EO was able to inhibit *Phytophthora cactorum* growth at 48 μg/mL and produced a fungicidal effect at 240 μg/mL, whereas the free EO was only able to produce the same effects at 120 and 480 μg/mL, respectively, under in vitro conditions. When applied as a soil drench on *Pinus* seedlings, the nanoformulation was able to avoid mortality, while 7% of EO-treated seedlings died within the experiment window (Vettraino et al. 2022). *Thymus serpyllum* EO richness in several compounds with proved antimicrobial activity, e.g., carvacrol, linalool, *p*-cymene, and γ-terpinene, capable of causing an increase in permeability as well as affecting biological processes inside the cell, could be behind the results observed. This may include plant defense activation. The delayed release of EOs into the medium allowing a continuous release of the compounds could also have a significant impact effect on *P. cactorum* growth (Vettraino et al. 2022). Likewise, Falsini et al. (2023) showed that lignin-tannin nanoparticles carrying *Azadirachta indica* (neem) EO boosted *P. minimum* growth inhibition with respect to a commercial control. The lignin-based nanoformulation showed the highest delivery efficacy, reducing the effective dose up to 400-fold compared to the control. The authors argued that lignin was able to mask the bioactive compound, facilitating the uptake by fungal mycelium and thus more efficiently exerting its effect (Falsini et al. 2023).

Lipid Nanoparticle's Antifungal Properties

Despite not having antifungal activity on their own, lipid nanoparticles can be used as nanocarriers to deliver antimicrobials to the phytopathogens and disrupt their cells' enzymatic and molecular mechanisms. Thus, nanoliposomes encapsulating *Lysinibacillus fusiformis* bacterial extract (500 ppm) were found effective at controlling fungal growth of *Fusarium oxysporum* f. sp. *cubense* and *F. oxysporum* f. sp. *cucumerinum* in vitro by 66% and 44%, respectively. The nanoformulation was also trialed against *F. oxysporum* f. sp. *lycopersici*, showing only a 19% reduction in mycelial growth. Aversely, both free forms of nanoliposomes and the bacterial extract showed no antifungal activity (Atienza et al. 2021). The authors suggested the mode of action was the result of induced permeability as a consequence of the collision of nanoliposomes with the fungal cells and the release of the extract.

Solid lipid nanoparticles containing *Zataria multiflora* EO have been shown to successfully inhibit the growth of *Aspergillus niger*, *A. flavus*, *A. ochraceus*, *Alternaria solani*, *Rhizoctonia solani*, and *Rhizopus stolonifer* in vitro at concentrations between 50 and 200 ppm (Nasseri et al. 2016). These results were significantly lower than those obtained using the EO by itself (200–300 ppm), which averaged a growth inhibition of 54%, while the nanoformulation averaged 79%. Greater contact surface and better preservation of the EO against degradation were the reasons behind the improved activity (Nasseri et al. 2016). Likewise, solid lipid nanoparticles carrying *Melaleuca alternifolia* EO (3 g/L) are capable of inhibiting the development of brown rot caused by *M. fructicola* and delaying the deterioration of stored fruits, paving the way for safe, green preservation methods in the agrifood sector (Xu et al. 2022).

Lin et al. (2020) investigated the effects of encapsulation on the stability and in vitro antifungal activity of cyclic lipopeptides produced by *Bacillus subtilis*. They reported that nanocrystallization was able to increase the stability of the peptide under storage conditions by inhibiting oxidation. In addition, it also potentiated the antifungal action against *Aspergillus niger* and *A. carbonarius* (1 mg/mL) through its controlled release. The encapsulated form performed better than the free-form peptide in terms of membrane, mycelia, and spore damage, as well as ochratoxin A inhibition. Lin et al. (2020) refer to the strong affinity of nanoparticles with the fungal membrane and induced permeability of the cell membrane to explain the activity. Despite this, the nanoformulation showed poor or moderate dispersibility in liquid solutions, requiring more tests to check its effectivity (Lin et al. 2020).

5.5.3 *Antiviral*

Plant viruses account for nearly half of all plant diseases worldwide, causing serious production and economic losses (Dutta et al. 2022; Wang et al. 2022). Because of that, significant efforts have been put toward developing preventive measures to reduce the number of infections and the nefarious results that derive from them. In vitro production of virus-free propagules, introgression of resistance genes into cultivars, and controlling the vectors' populations have been the most commonly adapted strategies to do so. However, these techniques are not full-proof, and, in case of infection, only antiviral agents can effectively deter the disease progression (Dutta et al. 2022; Wang et al. 2022). In that sense, nanotechnological-based methodologies are gaining momentum to face the threat caused by plant viral diseases by adopting a role as protectants of crops, nanocarriers for antiviral compounds, and biostimulants of plant defenses (Dutta et al. 2022; Wang et al. 2022). In fact, in agriculture, the potential of nanoparticles has been assessed against numerous viral pathogens, namely, the tomato spotted wilt virus, potato virus Y, tobacco mosaic virus (TMV), etc., with results ranging from inhibition of local lesions to reduction of infection and complete reduction of virus replication (Dutta et al. 2022). In these works, metal and metal-oxide nanoparticles have been the main subjects, and works considering organic-based nanoparticles are incredibly sparse. We have found one work, employing a bio-based nanoformulation aiming at controlling viral infection.

Quercetin nanoliposomes were employed, for the first time, for the control of TMV infection of tobacco plants in vivo and in planta experiments (Wang et al. 2022). Results show that the nanoformulation was able to significantly improve the solubility of quercetin, which enabled its efficient transport into the plant cells through penetration. At this point, the nanoparticles were destroyed by lipase in the cell, and quercetin could be released to reach the organelles. Both in vitro and in planta assays suggested that quercetin inhibited *Nbhsp70cp-1* and *Nbhsp70c-A* expression, codign for key proteins involved in virus replication, thus hampering TMV infection. Quercetin nanoliposomes decreased by 40.8% and 26.4% the expression of the *hsp70* genes and the accumulation of the virus' coat protein by 42.1%. Compared with free quercetin, the inhibition effect was increased by 37.4% in the field experiment (Wang et al. 2022).

5.6 Impact of Organic Nanoparticles in Plants and Their Biotic Interactions

The agricultural sector is currently being challenged by a myriad of different threats associated with climate change, namely, abiotic stress and diseases (Fincheira et al. 2021). However, as mentioned in the topics above, recent advances in the agricultural field have uncovered that the application of organic nanoparticles as delivery tools for natural antimicrobials can pose a green alternative to the use of chemical pesticides (Baldassarre et al. 2023). Furthermore, some works have uncovered that the application of organic nanoparticles can have a deeper impact on plants, acting not only as antimicrobial agents but also promoting plant growth and enhancing their defense mechanisms (Sathiyabama et al. 2014). So far, the works focusing on the growth-promoting and/or defense-priming abilities of nanoparticle-compound formulations are sparse; however, the effect of nanoparticles alone in plant tissues has been extensively reported in recent years.

5.6.1 Plant Growth and Defense

Nanoparticles have been described as having a myriad of properties that go beyond the antimicrobial activity these compounds are desired for. The application of nanoparticles to crops can trigger a plethora of biochemical shifts, namely, increases in cytosolic Ca^{2+}, activation of antioxidant enzymes, activation of an oxidative burst, callose apposition, increase in pathogenesis-related (PR) genes' expression, triggering the production of phytoalexins, activating the hypersensitive response (HR), accumulation of hydrogen peroxide, and synthesis of phytohormones.

Several studies have reported that the application of chitosan nanoparticles, at a suitable concentration, can lead to increases in the growth and yield of several important monocots, namely, maize and rice, and some dicot crops like mung bean, safflower, stevia, sunflower, and tomato (Balusamy et al. 2022). In maize, foliar application of 1 mg/mL of chitosan nanoparticles led to significant increases in key parameters of plant productivity, namely, root length and plant height, which increased by 13.7% and 10.7%, respectively, when compared with salt-stressed plants (Turk 2019). The same study also reported that chitosan nanoparticles acted at the molecular level, inducing shifts in antioxidant enzyme activities. Moreover, some works have uncovered that chitosan nanoparticles can prevent pigment degradation by oxidation, namely, chlorophylls and carotenoids, and also promote shoot growth, countering salt stress in milk thistle (*Silybum marianum*) (Helaly et al. 2018).

de Oliveira et al. (2018) assessed the effect of zein nanoparticles loaded with geraniol and citronellal on pea shoot and root length and uncovered that they were able to protect the seedlings from phytotoxic effects observed for pure geraniol (5 mg/mL). Another study demonstrated that the application of SA-loaded chitosan nanoparticles was able to promote several parameters related to shoot growth and vigor, namely, shoot/root length, fresh/dry weight, germination percentage, root number, and seedling vigor index. The authors found a correlation between increases in the activity of α-amylase and proteinase and the promotion of maize seedlings' growth and vigor (Kumaraswamy et al. 2018). The same study also reported that the application of SA-chitosan nanoparticles to potted plants led to increases in the activity of several antioxidant enzymes (SOD, POD, CAT, PPO) as well as PAL. However, when high concentrations of nanoformulation were used, there was a concomitant increase in the hydrogen peroxide content of maize plants. Similarly, Muthukrishnan et al. (2019) reported that *Cicer arietinum* seedlings showed increased germination rate and vigor when treated with thiamine-loaded chitosan nanoparticles, in comparison to a mock control. The authors further hypothesize that this might be due to the formulation enhancing the uptake of water and nutrients by the seedlings. Moreover, the treatment with thiamine-chitosan nanoformulation led to an increase in indole-3-acetic acid (IAA) levels as well as root/shoot length, number of leaves, fresh/dry weight, and number of lateral roots as well as a reduction in cell death of seedlings. Furthermore, increments in the activity of antioxidant enzymes [polyphenol oxidase (PPO), peroxidase (POD), β-1,3-glucanase, chitinase, chitosanase, and protease) in leaves and roots of chickpea seedlings once again corroborate the findings of Kumaraswamy et al. (2018) for monocots (maize). In fact, chitosan has been in the spotlight for its effects on the enzymatic machinery of several plants. One of the most complete works on this topic showcases that chitosan was able to boost the activity of a myriad of defense-related enzymes of *Camellia sinensis*, namely, POD, PPO, PAL and β-1,3-glucanase. The authors observed a similar trend for the antioxidant enzymes catalase (CAT) and superoxide dismutase (SOD), which exhibited a >33% increase in activity when compared to the control. It should be noted that chitosan also increased the non-enzymatic defense mechanisms of tea plants, namely, by boosting the flavonoid and phenol concentrations and increasing the expression of genes in antioxidant enzyme and polyphenol biosynthesis. It is also worthy of note that, in this study, the authors reported that chitosan nanoparticles were significantly more effective at regulating the plants' defense mechanisms when compared to plain chitosan, even when applied at a ten times lower concentration (Chandra et al. 2015).

In addition to the growth-promoting properties showcased above for organic nanoparticles in crop systems, the application of this kind of bioactive compound to plants is known to have important antimicrobial abilities. More recently it has been uncovered that organic nanoparticles, namely, chitosan, are also capable of inducing plants' defense responses against pathogens, namely, by triggering the production of defense-related compounds, like pathogenesis-related (PR) proteins, phytoalexins, chitinases, glucanases, lignin, and reactive oxygen species. On this note, the potential of nanoparticles to elicit plant defenses has been screened on several pathosystems.

Hence, chitosan has been shown to increase the tolerance of *Curcuma longa* to *Pythium aphanidermatum*, when applied as an amendment in a hydroponic nutrient solution, by stimulating the activity of several enzymes, namely, chitinase, chitosanase, and β-1,3-glucanase in both leaves and roots, which in turn points to a systemic effect of this nanoparticle (El Ghaouth et al. 1992). The application of chitosan nanoparticles to pea seedlings challenged by *Fusarium solani* f. sp. *phaseoli* and f. sp. *pisi* was able to induce the typical defense response of pea, namely, a boost in enzyme activity, activation of PR genes, and accumulation of pisatin. Furthermore, Kumaraswamy et al. (2018) also reported that, in the field, besides reducing the incidence of *F. verticillioides*, SA-chitosan nanoparticles also increased several physiological parameters of maize plants, namely, leaf area, shoot/root length, number of roots, stem diameter, and total chlorophyll content. Likewise, SA-chitosan nanoparticles were reported to induce systemic resistance through early upregulation of pathogenesis-related genes and increased activity of defense enzymes in rice against *P. striiformis* f. sp. *tritici* (Hembade et al. 2022).

Despite recent evidence pointing to the potential of organic nanoparticles to act as growth promoters of crops either by serving as controlled-release matrixes for bioactive organic compounds (i.e., plant hormones, EOs, etc.) or by directly influencing their growth through stimulating the uptake of water and nutrients and ultimately leading to a reduction in the need for the usage of water/chemical fertilizers, detailed studies on the mode of action of nanoparticles in plant systems are still lacking. These assessments should be the focus of novel studies in order to accelerate the regulation of these molecules for agricultural applications around the world.

5.6.2 Endosymbiotic Interactions

Besides their potential as inducers of plant growth and defense, organic nanoparticles have recently been reported to play a more indirect role in boosting crop water-use efficiency and nutrient uptake, by exerting a positive effect on beneficial microorganisms, such as arbuscular mycorrhizal fungi (AMF) and rhizobacteria. On this topic, it should be highlighted that the effects of organic nanoparticles on soil-beneficial microorganisms are still poorly understood. A preliminary study uncovered that, in the short term (7 days), chitosan nanoparticles triggered an increase in beneficial microbes in the soil, which was reversed to levels similar to the control after 30 days (Maruyama et al. 2016). Furthermore, the application of chitosan nanoparticles combined with *Pseudomonas taiwanensis* and *Pantoea agglomerans* increased the total CFU count (27.4×10^5) of soil microbes as well as K and P solubilizers specifically. Furthermore, metagenomic data uncovered that, overall, the diversity of microbes in the soil was greater in the chitosan nanoparticles-treated samples than the control (Agri et al. 2022). Conversely, Agbodjato et al. (2016) reported that the positive effects of the application of beneficial microorganisms, such as *Pseudomonas fluorescens*, *P. putida*, and *Azospirillum lipoferum* on maize plant's growth parameters were greatly enhanced by the joint application of these bacteria with chitosan. The authors hypothesized that besides its positive action, chitosan was able to enhance the beneficial properties of the rhizobacteria to stimulate plant growth. Furthermore, in the same study, the authors assessed the uptake of macronutrients, namely, nitrogen, phosphorus, and potassium, and concluded that chitosan-*A. lipoferum*-treated maize plants showed higher levels of nitrogen (4.08% of the plant's dry matter) than the control. On the same note, the application of chitosan nanoparticles, combined with *Bacillus licheniformis* in an alginate matrix, led to an increase in *Capsicum annuum* plant growth. Furthermore, the authors highlighted that the plants treated with this formulation showed a lower southern blight disease incidence (52%) than the control (90%) (Panichikkal et al. 2021). On the same note, it has been evidenced that the application of chitosan nanoparticles in conjugation with *Bacillus* spp. in the soil of maize fields led to an increase in macronutrient concentrations as well as an improvement of beneficial microorganisms' populations, namely, *Firmicutes*, *Bacteroidetes*, *Chloroflexi*, and *Cyanobacteria* (Chaudhary et al. 2021).

Regarding AMF, several studies have uncovered the potential of organic nanoparticles to potentiate the colonization of plant roots when applied in combination with AMF. Goudarzian et al. (2020) highlighted that, when applied to *Mentha piperita*, chitosan combined with a formulation of AMF of the *Glomus* genus (namely, *Glomus mosseae*, *G. intraradices*, *G. hoi*, *G. etunicatum*, *G. clarideum*, *G. versiforme*, *G. fasciculatum*, *G. acaulospora longula*, and *G. margarita*) triggered an increase in plants' growth parameters, namely, fresh and dry weight, while also boosting EO content and increasing the percentage of menthol. It should also be noted that chitosan alone was able to increase the concentrations of limonene, menthone, (+)-4-carene, and β-caryophyllene, while AMF by its own did not have any significant impacts on the relative abundance of these compounds in *Mentha* EO. Further works have pointed out that this impact of organic nanoparticles in AMF-host interactions can be primarily attributed to an action on the host's metabolism, which in consequence, affects the symbiont. Thus, it could be hypothesized that nanoparticles could modulate the plant's secondary metabolism pathways and thus promote AMF colonization (Fig. 5.3). Recently Amer and Shoala (2020) uncovered that chitosan nanoparticles led to the activation of MAPK cascades, which in turn resulted in a stimulation of phytohormone production of oregano. This phenomenon was also observed in cucumber and *Arabidopsis*, where IAA levels were also increased, via a stimulation of the tryptophan-dependent pathway. When applied in combination with *Rhizophagus irregularis*, chitosan nanoparticles promoted an increase in IAA and strigolactone levels. It should also be noted that chitosan nanoparticles also exerted a positive role in the exudation of flavonoids into the rhizosphere, which, together with phytohormones, plays a major role in mycorrhizal establishment. In consequence, the authors conclude that the influence of chitosan nanoparticles in these physiological parameters results in an increase in

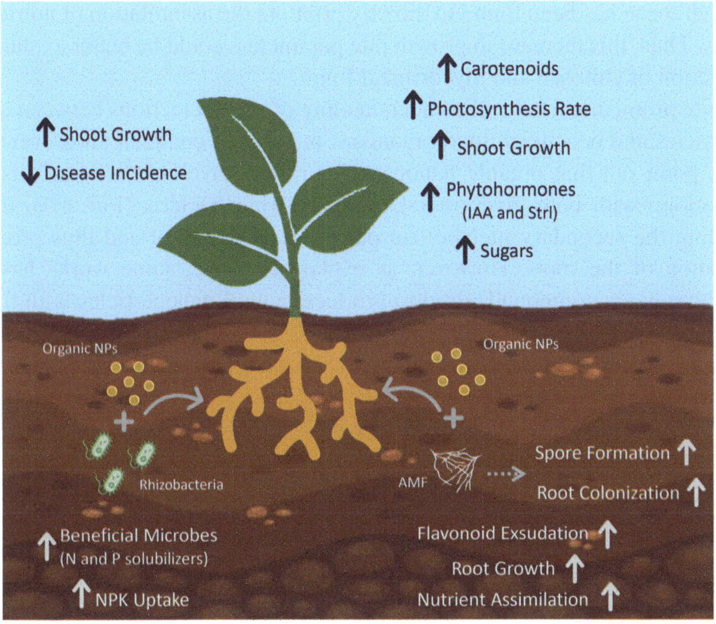

Fig. 5.3 Schematic summary of the beneficial role that formulations of organic nanoparticles (NPs) and plant growth promoters (PGPs) can have in crop systems. Rhizobacteria in combination with organic NPs can improve beneficial microbes and nutrient uptake, resulting in increased shoot growth and reduced disease incidence. Arbuscular mycorrhizal fungi (AMF), combined with organic NPs trigger the recruitment of beneficial microbes and plant growth, through a stimulation of the secondary metabolism

AMF colonization, as well as hyphal length and number of arbuscules. This boost in mycorrhization was accompanied by an increase in photosynthesis rate, associated with stomatal conductance, carotenoids, and sugar concentrations. Furthermore, the activity of key enzymes in sugar metabolism was also boosted by the joint application of *R. irregularis* and chitosan nanoparticles (Saleh et al. 2023).

This positive role of chitosan in promoting root colonization when applied in combination with mycorrhizal fungi might be particularly important with the rise of climate change-related issues, as, in some parts of the world, soils become dryer and poorer in nutrients. Some studies have highlighted that, in low-nutrient soils, chitosan showed a remarkable ability to promote AMF colonization of tomatoes, mainly by *Glomus* sp., *Sclerocystis* sp., and *Acaulospora* sp. after 8 weeks of application (up to 97% colonization). In contrast, significant AMF colonization of the roots was only observed after 12 weeks of culture for nutrient-rich soils (El Amerany et al. 2020). Some studies suggest that, besides influencing plant's secondary metabolic pathways (Fig. 5.3), chitosan also increases the extra-radical mycelium length and promotes the formation of fungal spores (Velásquez and Pirela 2016). As described earlier, this effect on root colonization in turn produced increases in leaf area, root length, and fresh and dry mass of both roots and shoots. It should also be pointed out that chitosan has been found to directly promote the assimilation of nutrients by the roots. Thus, this increase in growth rate parameters could be better explained by a joint action of chitosan and mycorrhizal fungi.

Despite promising results, the understanding of the interactions between organic nanoparticles and beneficial microorganisms is far from complete; however, recent findings point out that organic nanoparticles might have beneficial effects when applied along with both mycorrhizal fungi and rhizobacteria (Fig. 5.3), through modulating the secondary metabolism pathways of the plant and thus promoting colonization of the roots. However, as explained above, some works have also linked the organic compounds used to produce natural nanoparticles with the promotion of fungal growth and sporulation (Fig. 5.3). Thus, the combined action of these materials in the plants' and microbes' metabolism might better explain the remarkable effects regarding plant growth. Nevertheless, further studies are needed to better comprehend which nanoparticles have a more relevant role in promoting the interactions between plants and beneficial microbes.

5.7 Concluding Remarks

Organic nanoparticles have a huge potential to be used as effective stabilizers and to promote the controlled release of natural antimicrobial compounds. But despite recent, promising findings in this field, there is still room for improvement in several fronts, including field application, with only a small number of articles assessing in planta effects. Another major handicap for the integration of organic nanoparticles in organic farming is the synthesis process, which still involves hazardous solvents. Hence, future strategies of nanoparticle synthesis should aim at replacing these solvents with nontoxic alternatives, namely, water and ionic liquids. It should also be noted that, as seen here, most of those articles focus on a few genera of plant pathogens, namely, *Fusarium*, *Aspergillus*, *Pseudomonas*, and *Xanthomonas*, with other important phytopathogens, namely, *Ustilago* spp. and *Xylella* spp. being absent. Moreover, the off-target effects of the application of organic nanoparticles as carriers of natural antimicrobials in the field are also extremely understudied, mainly regarding their impact on the native soil/phyllosphere microbiome. Hence, research toward new formulations should account for their impact on plant pathogens and on the native microbiomes as well as plant-microbe interacions. As this topic matures with in-depth long-term research, organic nanoparticles can become the prime matrix for the controlled release of natural antimicrobials uplifting their usage in agriculture and increasing crop protection against a wide range of phytopathogens while reducing the toll on the environment.

References

Abbaszadeh S, Sharifzadeh A, Shokri H, Khosravi AR, Abbaszadeh A (2014) Antifungal efficacy of thymol, carvacrol, eugenol and menthol as alternative agents to control the growth of food-relevant fungi. J Mycol Med 24(2):e51–e56

Abdallah Y, Liu M, Ogunyemi SO, Ahmed T, Fouad H, Abdelazez A, Yan C, Yang Y, Chen J, Li B (2020) Bioinspired green synthesis of chitosan and zinc oxide nanoparticles with strong antibacterial activity against rice pathogen *Xanthomonas oryzae* pv. *oryzae*. Molecules 25(20):4795

Acosta S, Chiralt A, Santamarina P, Rosello J, González-Martínez C, Cháfer M (2016) Antifungal films based on starch-gelatin blend, containing essential oils. Food Hydrocoll 61:233–240

Agbodjato NA, Noumavo PA, Adjanohoun A, Agbessi L, Baba-Moussa L (2016) Synergistic effects of plant growth promoting rhizobacteria and chitosan on *in vitro* seeds germination, greenhouse growth, and nutrient uptake of maize (*Zea mays* L.). Biotechnol Res Int 2016:7830182

Agri U, Chaudhary P, Sharma A, Kukreti B (2022) Physiological response of maize plants and its rhizospheric microbiome under the influence of potential bioinoculants and nanochitosan. Plant Soil 474(1–2):451–468

Ahmed IA, Mikail MA (2017) Paradigm shift: focusing on plant-based natural antimicrobials. J Microbiol Exp 5(2):00145

Ali SS, Al-Tohamy R, Koutra E, Moawad MS, Kornaros M, Mustafa AM, Mahmoud YAG, Badr A, Osman MEH, Elsamahy T, Jiao H, Sun J (2021) Nanobiotechnological advancements in agriculture and food industry: applications, nanotoxicity, and future perspectives. Sci Total Environ 792:148359

Amaning Danquah C, Minkah PAB, Osei Duah Junior I, Amankwah KB, Somuah SO (2022) Antimicrobial compounds from microorganisms. Antibiotics 11(3):285

Amat M, Chin SF, Wasli ME, Ngieng NS, Husaini AASA (2023) Eco-friendly synthesis of lignin-based nanoparticles as antifungal agent. Mater Lett 340:134166

Amer A, Shoala T (2020) Physiological and phenotypic characters of sweet marjoram in response to pre-harvest application of hydrogen peroxide or chitosan nanoparticles. Sci Hortic 268:109374

An F, Chen N, Conlon WJ, Hachey JS, Xin J, Aras O, Rodriguez EA, Ting R (2020) Small ultra-red fluorescent protein nanoparticles as exogenous probes for noninvasive tumor imaging *in vivo*. Int J Biol Macromol 153:100–106. https://doi.org/10.1016/j.ijbiomac.2020.02.253

Armendáriz-Barragán B, Zafar N, Badri W, Galindo-Rodríguez SA, Kabbaj D, Fessi H, Elaissari A (2016) Plant extracts: from encapsulation to application. Expert Opin Drug Deliv 13(8):1165–1175

Atienza MTJA, Magpantay MDA, Santos KLT, Mora NB, Balaraman RP, Gemeinhardt ME, Dela Cueva FM, Paterno ES, Fernando LM, Kohli P (2021) Encapsulation of plant growth-promoting bacterial crude extract in nanoliposome and its antifungal property against *Fusarium oxysporum*. ACS Agric Sci Technol 1(6):691–701

Ayyaril SS, Shanableh A, Bhattacharjee S, Rawas-Qalaji M, Cagliani R, Shabib AG, Imran khan, M. (2023) Recent progress in micro and nano-encapsulation techniques for environmental applications: a review. Results Eng 18:10

Baldassarre F, Schiavi D, Ciarroni S, Tagliavento V, De Stradis A, Vergaro V, Suranna GP, Balestra GM, Ciccarella G (2023) Thymol-nanoparticles as effective biocides against the quarantine pathogen Xylella fastidiosa. Nanomaterials 13(7):1285

Balusamy SR, Rahimi S, Sukweenadhi J, Sunderraj S, Shanmugam R, Thangavelu L, Mijakovic I, Perumalsamy H (2022) Chitosan, chitosan nanoparticles and modified chitosan biomaterials, a potential tool to combat salinity stress in plants. Carbohydr Polym 284:119189

Barrera-Necha LL, Correa-Pacheco ZN, Bautista-Baños S, Hernández-López M, Jiménez JEM, Mejía AFM (2018) Synthesis and characterization of chitosan nanoparticles loaded botanical extracts with antifungal activity on *Colletotrichum gloeosporioides* and *Alternaria* species. Adv Microbiol 8(4):286–296

Beig B, Niazi MBK, Sher F, Jahan Z, Malik US, Khan MD, Américo-Pinheiro JHP, Vo D-VN (2022) Nanotechnology-based controlled release of sustainable fertilizers. A review. Environ Chem Lett 20(4):2709–2726

Beyki M, Zhaveh S, Khalili ST, Rahmani-Cherati T, Abollahi A, Bayat M, Tabatabaei M, Mohsenifar A (2014) Encapsulation of *Mentha piperita* essential oils in chitosan-cinnamic acid nanogel with enhanced antimicrobial activity against *Aspergillus flavus*. Ind Crop Prod 54:310–319

Bidyarani N, Kumar U (2019) Synthesis of rotenone loaded zein nano-formulation for plant protection against pathogenic microbes. RSC Adv 9(70):40819–40826

Bidyarani N, Srivastav AK, Gupta SK, Kumar U (2020) Synthesis and physicochemical characterization of rhamnolipid-stabilized carvacrol-loaded zein nanoparticles for antimicrobial application supported by molecular docking. J Nanopart Res 22(10):307

Boruah S, Dutta P (2021) Fungus mediated biogenic synthesis and characterization of chitosan nanoparticles and its combine effect with *Trichoderma asperellum* against *Fusarium oxysporum*, *Sclerotium rolfsii* and *Rhizoctonia solani*. Indian Phytopathol 74(1):81–93

Brauer VS, Rezende CP, Moreira Pessoni A, De Paula RG, Rangappa KS, Chandra Nayaka S, Kumar Gupta V, Almeida F (2019) Antifungal agents in agriculture: friends and foes of public health. Biomolecules 9:521. https://doi.org/10.3390/biom9100521

Cai J, Yang D, Wang Q (2023) Preparation and characterization of chitosan nanoparticles loaded with *Athyrium sinense* essential oil with antibacterial properties against *Pectobacterium carotovorum* subsp. *carotovorum*. Ind Crop Prod 195:116382

Campelo PH, Sant'Ana AS, Clerici MTPS (2020) Starch nanoparticles: production methods, structure, and properties for food applications. Curr Opin Food Sci 33:136–140

Campos EVR, de Oliveira JL, da Silva CMG, Pascoli M, Pasquoto T, Lima R, Abhilash PC, Fraceto LF (2015) Polymeric and solid lipid nanoparticles for sustained release of carbendazim and Tebuconazole in agricultural applications. Sci Rep 5(1):13809. https://doi.org/10.1038/srep13809

Campos EVR, Proença PLF, Oliveira JL, Melville CC, Vechia JFD, De Andrade DJ, Fraceto LF (2018) Chitosan nanoparticles functionalized with β-cyclodextrin: a promising carrier for botanical pesticides. Sci Rep 8(1):2067

Cao L, Lu W, Mata A, Nishinari K, Fang Y (2020) Egg-box model-based gelation of alginate and pectin: a review. Carbohydr Polym 242:116389

Castillo F, Hernández D, Gallegos G, Mendez M, Rodríguez R, Reyes A, Aguilar CN (2010) *In vitro* antifungal activity of plant extracts obtained with alternative organic solvents against *Rhizoctonia solani* Kühn. Ind Crop Prod 32(3):324–328

Ćavar S, Maksimović M, Šolić ME, Jerković-Mujkić A, Bešta R (2008) Chemical composition and antioxidant and antimicrobial activity of two *Satureja* essential oils. Food Chem 111(3):648–653

Chandra S, Chakraborty N, Dasgupta A, Sarkar J, Panda K, Acharya K (2015) Chitosan nanoparticles: a positive modulator of innate immune responses in plants. Sci Rep 5(1):15195

Chandra H, Bishnoi P, Yadav A, Patni B, Mishra AP, Nautiyal AR (2017) Antimicrobial resistance and the alternative resources with special emphasis on plant-based antimicrobials—a review. Plants 6(2):16

Chandrasekaran M, Kim KD, Chun SC (2020) Antibacterial activity of chitosan nanoparticles: a review. Processes 8(9):1173

Chaudhari AK, Singh VK, Das S, Deepika, Dubey NK (2022) Fabrication, characterization, and bioactivity assessment of chitosan nanoemulsion containing allspice essential oil to mitigate *Aspergillus flavus* contamination and aflatoxin B1 production in maize. Food Chem 372:131221

Chaudhary P, Sharma A, Chaudhary A, Khati P, Gangola S, Maithani D (2021) Illumina based high throughput analysis of microbial diversity of maize rhizosphere treated with nanocompounds and *bacillus* sp. Appl Soil Ecol 159:103836

Chauhan PS (2020) Lignin nanoparticles: eco-friendly and versatile tool for new era. Bioresour Technol Rep 9:100374

Chen M, Li Y, Liu H, Zhang D, Shi Q-S, Zhong X-Q, Guo Y, Xie X-B (2023) High value valorization of lignin as environmental benign antimicrobial. Mater Today Bio 18:100520

Chitra K, Rani J, Sherief SH (2022) The fabrication of chitosan nanoparticles and its *in-vitro* antibacterial activity against plant pathogen. In: Recent trends of innovation in chemical and biological sciences, vol 7. Bhumi Publishing

Choudhary M, Jones JB, Paret ML (2022) Natural or green synthesis nanomaterials and impact on plant pathogens. In: Nanotechnology-based sustainable alternatives for the management of plant diseases. Elsevier, pp 5–29

Chuacharoen T, Sabliov CM (2022) Development of coating material by incorporating curcumin-loaded zein nanoparticles to maintain the quality of mango (*Mangifera indica* L. cv. Nam Dokmai). J Agric Food Res 10:100444

Compean KL, Ynalvez RA (2014) Antimicrobial activity of plant secondary metabolites: a review. Res J Med Plant 8(5):204

da Rosa CG, de Oliveira Brisola Maciel MV, de Carvalho SM, de Melo APZ, Jummes B, da Silva T, Martelli SM, Villetti MA, Bertoldi FC, Barreto PLM (2015) Characterization and evaluation of physicochemical and antimicrobial properties of zein nanoparticles loaded with phenolics monoterpenes. Colloids Surf A Physicochem Eng Asp 481:337–344

Das S, Singh VK, Dwivedy AK, Chaudhari AK, Upadhyay N, Singh P, Sharma S, Dubey NK (2019) Encapsulation in chitosan-based nanomatrix as an efficient green technology to boost the antimicrobial, antioxidant and *in situ* efficacy of *Coriandrum sativum* essential oil. Int J Biol Macromol 133:294–305

Das S, Singh VK, Dwivedy AK, Chaudhari AK, Upadhyay N, Singh A, Deepika, Dubey NK (2020) Fabrication, characterization and practical efficacy of *Myristica fragrans* essential oil nanoemulsion delivery system against postharvest biodeterioration. Ecotoxicol Environ Saf 189:110000

Das S, Kumar A, Kumar V, Kumar A, Kishore N (2023) Chitosan based encapsulation of *Valeriana officinalis* essential oil as edible coating for inhibition of fungi and aflatoxin B1 contamination, nutritional quality improvement, and shelf life extension of *Citrus sinensis* fruits. Int J Biol Macromol 233:123565

De Anda-Flores Y, Carvajal-Millan E, Campa-Mada A, Lizardi-Mendoza J, Rascon-Chu A, Tanori-Cordova J, Martínez-López AL (2021) Polysaccharide-based nanoparticles for colon-targeted drug delivery systems. Polysaccharides 2(3):626–647

De Moraes MA, Da Silva CF, Vieira RS (2020) Biopolymer membranes and films: health, food, environment, and energy applications. Elsevier

de Oliveira JL, Campos EVR, Pereira AES, Pasquoto T, Lima R, Grillo R, de Andrade DJ, dos Santos FA, Fraceto LF (2018) Zein nanoparticles as eco-friendly carrier systems for botanical repellents aiming sustainable agriculture. J Agric Food Chem 66(6):1330–1340

Debele TA, Mekuria SL, Tsai H-C (2016) Polysaccharide based nanogels in the drug delivery system: application as the carrier of pharmaceutical agents. Mater Sci Eng C 68:964–981

Dille MJ, Haug IJ, Draget KI (2021) Chapter 34: Gelatin and collagen. In: Phillips GO, Williams PA (eds) Handbook of hydrocolloids, 3rd edn. Woodhead Publishing, pp 1073–1097

Divya K, Vijayan S, George TK, Jisha MS (2017) Antimicrobial properties of chitosan nanoparticles: mode of action and factors affecting activity. Fibers Polymers 18:221–230

Divya K, Thampi M, Vijayan S, Varghese S, Jisha MS (2020) Induction of defence response in *Oryza sativa* L. against *Rhizoctonia solani* (Kuhn) by chitosan nanoparticles. Microb Pathog 149:104525

do Espirito Santo Pereira A, Luiz de Oliveira J, Maira Savassa S, Barbara Rogério C, Araujo de Medeiros G, Fraceto LF (2022) Lignin nanoparticles: new insights for a sustainable agriculture. J Clean Prod 345:131145

Dodero A, Alberti S, Gaggero G, Ferretti M, Botter R, Vicini S, Castellano M (2021) An up-to-date review on alginate nanoparticles and nanofibers for biomedical and pharmaceutical applications. Adv Mater Interfaces 8(22):2100809

Doktorovova S, Souto EB, Silva AM (2014) Nanotoxicology applied to solid lipid nanoparticles and nanostructured lipid carriers—a systematic review of *in vitro* data. Eur J Pharm Biopharm 87(1):1–18

Dong NQ, Lin HX (2021) Contribution of phenylpropanoid metabolism to plant development and plant–environment interactions. J Integr Plant Biol 63(1):180–209

Dong E, Yang Z, Zhou C, Wang C, Li S, Ouyang Q, Kong L, He Z, Xie J, Li P, Yang P (2019) pH-responsive ultrasonic self-assembly spinosad-loaded nanomicelles and their antifungal activity to *Fusarium oxysporum*. React Funct Polym 141:123–132

Duan Y, Dhar A, Patel C, Khimani M, Neogi S, Sharma P, Kumar NS, Vekariya RL (2020) A brief review on solid lipid nanoparticles: part and parcel of contemporary drug delivery systems. RSC Adv 10(45):26777–26791

Dutta P, Kumari A, Mahanta M, Biswas KK, Dudkiewicz A, Thakuria D, Abdelrhim AS, Singh SB, Muthukrishnan G, Sabarinathan KG (2022) Advances in nanotechnology as a potential alternative for plant viral disease management. Front Microbiol 13:935193

Dwivedy AK, Singh VK, Prakash B, Dubey NK (2018) Nanoencapsulated *Illicium verum* Hook. f. essential oil as an effective novel plant-based preservative against aflatoxin B1 production and free radical generation. Food Chem Toxicol 111:102–113

El Amerany F, Rhazi M, Wahbi S, Taourirte M, Meddich A (2020) The effect of chitosan, arbuscular mycorrhizal fungi, and compost applied individually or in combination on growth, nutrient uptake, and stem anatomy of tomato. Sci Hortic 261:109015

El Ghaouth A, Ponnampalam R, Castaigne F, Arul J (1992) Chitosan coating to extend the storage life of tomatoes. HortScience 27(9):1016–1018

El-Aziz AR, Al-Othman MR, Mahmoud MA, Shehata SM, Abdelazim NS (2018) Chitosan nanoparticles as a carrier for *Mentha longifolia* extract: synthesis, characterization and antifungal activity. Curr Sci 114(10):2116–2122

El-Naggar NE-A, Bashir SI, Rabei NH, Saber WIA (2022) Innovative biosynthesis, artificial intelligence-based optimization, and characterization of chitosan nanoparticles by Streptomyces microflavus and their inhibitory potential against *Pectobacterium carotovorum*. Sci Rep 12(1):21851

Elsharkawy MM, Omara RI, Mostafa YS, Alamri SA, Hashem M, Alrumman SA, Ahmad AA (2022) Mechanism of wheat leaf rust control using chitosan nanoparticles and salicylic acid. J Fungi 8(3):304

Esyanti RR, Farah N, Bajra BD, Nofitasari D, Martien R, Sunardi S, Safitri R (2020) Comparative study of nano-chitosan and synthetic bactericide application on chili pepper (Capsicum annuum L.) infected by Xanthomonas campestris. AGRIVITA J Agric Sci 42(1):13–23

Falsini S, Nieri T, Paolini A, Schiff S, Papini A, Mugnai L, Gonnelli C, Ristori S (2023) Tannins-lignin mixed nanoformulations for improving the potential of neem oil as fungicide agent. Environ Sci Pollut Res 30(13):39131–39141

Ferhat MA, Meklati BY, Chemat F (2007) Comparison of different isolation methods of essential oil from citrus fruits: cold pressing, hydrodistillation and microwave 'dry' distillation. Flavour Fragr J 22(6):494–504

Fincheira P, Quiroz A, Tortella G, Diez MC, Rubilar O (2021) Current advances in plant-microbe communication via volatile organic compounds as an innovative strategy to improve plant growth. Microbiol Res 247:126726

Fu X, Shi Y, Wang H, Zhao X, Sun Q, Huang Y, Qi T, Lin G (2019) Ethosomal gel for improving transdermal delivery of thymosin β-4. Int J Nanomedicine 14:9275–9284

Fusco GM, Nicastro R, Rouphael Y, Carillo P (2022) The effects of the microbial biostimulants approved by EU regulation 2019/1009 on yield and quality of vegetable crops. Foods 11(17):2656

Garavand F, Khodaei D, Mahmud N, Islam J, Khan I, Jafarzadeh S, Tahergorabi R, Cacciotti I (2022) Recent progress in using zein nanoparticles-loaded nanocomposites for food packaging applications. Crit Rev Food Sci Nutr:1–21. https://doi.org/10.1080/10408398.2022.2133080

Garcia F, Lin WJ, Mellano V, Davidov-Pardo G (2022) Effect of biopolymer coatings made of zein nanoparticles and ε-polylysine as postharvest treatments on the shelf-life of avocados (*Persea americana* Mill. cv. Hass). J Agric Food Res 7:100260

García-Pinel B, Porras-Alcalá C, Ortega-Rodríguez A, Sarabia F, Prados J, Melguizo C, López-Romero JM (2019) Lipid-based nanoparticles: application and recent advances in cancer treatment. Nanomaterials 9(4). https://doi.org/10.3390/nano9040638

Gigli M, Fellet G, Pilotto L, Sgarzi M, Marchiol L, Crestini C (2022) Lignin-based nano-enabled agriculture: a mini-review. Front Plant Sci 13:976410

Gooneh-Farahani S, Naghib SM, Naimi-Jamal MR (2020) A novel and inexpensive method based on modified ionic gelation for pH-responsive controlled drug release of homogeneously distributed chitosan nanoparticles with a high encapsulation efficiency. Fibers Polymers 21:1917–1926

Goudarzian A, Pirbalouti AG, Hossaynzadeh M (2020) Menthol, balance of menthol/menthone, and essential oil contents of *Mentha* × *piperita* L. under foliar-applied chitosan and inoculation of arbuscular mycorrhizal fungi. J Essent Oil Bearing Plants 23(5):1012–1021

Guedri Mkaddem M, Zrig A, Ben Abdallah M, Romdhane M, Okla MK, Al-Hashimi A, Alwase YA, Hegab MY, Madany MMY, Hassan AHA (2022) Variation of the chemical composition of essential oils and total phenols content in natural populations of *Marrubium vulgare* L. Plants 11(5):612

Guerrero-Rodríguez E, Solís-Gaona S, Hernández-Castillo FD, Flores-Olivas A, Sandoval-López V, Jasso-Cantú D (2007) Actividad biológica in vitro de extractos de *Flourensia cernua* DC en patógenos de postcosecha: *Alternaria alternata* (Fr.: Fr.) Keissl., *Colletotrichum gloeosporioides* (Penz.) Penz. y Sacc. y *Penicillium digitatum* (Pers.: Fr.) Sacc. Revista Mexicana de Fitopatología 25(1):48–53

Guimarães D, Cavaco-Paulo A, Nogueira E (2021) Design of liposomes as drug delivery system for therapeutic applications. Int J Pharm 601:120571

Hadidi M, Motamedzadegan A, Jelyani AZ, Khashadeh S (2021) Nanoencapsulation of hyssop essential oil in chitosan-pea protein isolate nano-complex. LWT 144:111254

Hanna EA, Mendez Lopez OE, Salinas F, Astete CE, Tamez C, Wang Y, Wu H, Eitzer BD, Elmer WH, Louie S, White JC, Sabliov CM (2022) Zein Nanoparticles for Enhanced Translocation of Pesticide in Soybean (*Glycine max*). ACS Agric Sci Technol 2(5):1013–1022

Hasheminejad N, Khodaiyan F, Safari M (2019) Improving the antifungal activity of clove essential oil encapsulated by chitosan nanoparticles. Food Chem 275:113–122

Hassan EA, Hathout RM, Gad HA, Sammour OA (2022a) Multi-purpose zein nanoparticles for battling hepatocellular carcinoma: a green approach. Eur Polym J 176:111396

Hassan EA, Hathout RM, Gad HA, Sammour OA (2022b) A holistic review on zein nanoparticles and their use in phytochemicals delivery. J Drug Deliv Sci Technol 73:103460

Hathout RM, Metwally AA (2019) Gelatin nanoparticles. In: Pharmaceutical nanotechnology: basic protocols. Humana Press, pp 71–78

Helaly M, Farouk S, Arafa S, Amhimmid N (2018) Inducing salinity tolerance of rosemary (*Rosmarinus officinalis* L.) plants by chitosan or zeolite application. Asian J Adv Agric Res 5(4):1–20

Hembade VL, Yashveer S, Taunk J, Sangwan S, Tokas J, Singh V, Redhu NS, Grewal S, Malhotra S, Kumar M (2022) Chitosan-salicylic acid and zinc sulphate nano-formulations defend against yellow rust in wheat by activating pathogenesis-related genes and enzymes. Plant Physiol Biochem 192:129–140

Hernández-Melchor DJ, Guerrero-Chávez AC, Ferrera-Rodríguez MR, Ferrera-Cerrato R, Larsen J, Alarcón A (2023) Cellulase and chitinase activities and antagonism against *Fusarium oxysporum* f. sp. *cubense* race 1 of six Trichoderma strains isolated from Mexican maize cropping. Biotechnol Lett 45(3):387–400

Hjort K, Presti I, Elväng A, Marinelli F, Sjöling S (2014) Bacterial chitinase with phytopathogen control capacity from suppressive soil revealed by functional metagenomics. Appl Microbiol Biotechnol 98:2819–2828

Hoang NH, Le Thanh T, Sangpueak R, Thepbandit W, Saengchan C, Papathoti NK, Treekoon J, Kamkaew A, Phansak P, Buensanteai K (2023) The effect of chitosan nanoparticle formulations for control of leaf spot disease on cassava. Phytoparasitica 51(3):621–636

Hong S, Choi DW, Kim HN, Park CG, Lee W, Park HH (2020) Protein-based nanoparticles as drug delivery systems. Pharmaceutics 12(7):604

Hong J, Wang C, Wagner DC, Gardea-Torresdey JL, He F, Rico CM (2021) Foliar application of nanoparticles: mechanisms of absorption, transfer, and multiple impacts. Environ Sci Nano 8(5):1196–1210

Hu Q, Luo Y (2021) Chitosan-based nanocarriers for encapsulation and delivery of curcumin: a review. Int J Biol Macromol 179:125–135

Huan Y, Kong Q, Mou H, Yi H (2020) Antimicrobial peptides: classification, design, application and research progress in multiple fields. Front Microbiol 11:582779

Huang H-T, Lin C-C, Kuo T-C, Chen S-J, Huang R-N (2019) Phytochemical composition and larvicidal activity of essential oils from herbal plants. Planta 250:59–68

Huang Z, Liu B, Yin Y, Liang F, Xie D, Han T, Liu Y, Yan B, Li Q, Huang Y (2021) Impact of biocontrol microbes on soil microbial diversity in ginger (*Zingiber officinale* Roscoe). Pest Manag Sci 77(12):5537–5546

Huh MS, Lee EJ, Koo H, Yhee JY, Oh KS, Son S, Lee S, Kim SH, Kwon IC, Kim K (2018) Polysaccharide-based nanoparticles for gene delivery. In: Cheng Y (ed) Polymeric gene delivery systems. Springer International Publishing, pp 65–83. https://doi.org/10.1007/978-3-319-77866-2_3

Hussain A, Hasan A, Babadaei MMN, Bloukh SH, Edis Z, Rasti B, Sharifi M, Falahati M (2020) Application of gelatin nanoconjugates as potential internal stimuli-responsive platforms for cancer drug delivery. J Mol Liquids 318:114053

Hussain M, Zahra N, Lang T, Zain M, Raza M, Shakoor N, Adeel M, Zhou H (2023) Integrating nanotechnology with plant microbiome for next-generation crop health. Plant Physiol Biochem 196:703–711

Ijaz I, Gilani E, Nazir A, Bukhari A (2020) Detail review on chemical, physical and green synthesis, classification, characterizations and applications of nanoparticles. Green Chem Lett Rev 13(3):223–245

Ilk S, Saglam N, Özgen M (2017) Kaempferol loaded lecithin/chitosan nanoparticles: preparation, characterization, and their potential applications as a sustainable antifungal agent. Artif Cells Nanomed Biotechnol 45(5):907–916

Ingle PU, Shende SS, Shingote PR, Mishra SS, Sarda V, Wasule DL, Rajput VD, Minkina T, Rai M, Sushkova S, Mandzhieva S, Gade A (2022) Chitosan nanoparticles (ChNPs): a versatile growth promoter in modern agricultural production. Heliyon 8(11):e11893

IPPC Secretariat (2021) Scientific review of the impact of climate change on plant pests—a global challenge to prevent and mitigate plant pest risks in agriculture, forestry and ecosystems. FAO on behalf of the IPPC Secretariat

Izadi M, Moosawi Jorf SA, Nikkhah M, Moradi S (2021) Antifungal activity of hydrocolloid nano encapsulated *Carum copticum* essential oil and *Peganum harmala* extract on the pathogenic fungi *Alternaria alternata*. Physiol Mol Plant Pathol 116:101714

Jadhav HP, Shaikh SS, Sayyed RZ (2017) Role of hydrolytic enzymes of rhizoflora in biocontrol of fungal phytopathogens: an overview. In: Rhizotrophs: plant growth promotion to bioremediation. Springer, pp 183–203

Jain A, Singh SK, Arya SK, Kundu SC, Kapoor S (2018) Protein nanoparticles: promising platforms for drug delivery applications. ACS Biomater Sci Eng 4(12):3939–3961

Jin P, Yao R, Qin D, Chen Q, Du Q (2019) Enhancement in antibacterial activities of eugenol-entrapped ethosome nanoparticles via strengthening its permeability and sustained release. J Agric Food Chem 67(5):1371–1380

Kalagatur NK, Nirmal Ghosh OS, Sundararaj N, Mudili V (2018) Antifungal activity of chitosan nanoparticles encapsulated with *Cymbopogon martinii* essential oil on plant pathogenic fungi *Fusarium graminearum*. Front Pharmacol 9:610

Kaliamurthi S, Selvaraj G, Hou L, Li Z, Wei Y, Gu K, Wei D (2019) Synergism of essential oils with lipid based nanocarriers: emerging trends in preservation of grains and related food products. Grain Oil Sci Technol 2(1):21–26
Kaltbeitzel J, Wich PR (2023) Protein-based nanoparticles: from drug delivery to imaging, nanocatalysis and protein therapy. Angew Chem Int Ed:e202216097. https://doi.org/10.1002/ange.202216097
Karamchandani BM, Maurya PA, Dalvi SG, Waghmode S, Sharma D, Rahman PKSM, Ghormade V, Satpute SK (2022) Synergistic activity of rhamnolipid biosurfactant and nanoparticles synthesized using fungal origin chitosan against phytopathogens. Front Bioeng Biotechnol 10:917105
Karpiński TM (2019) Marine macrolides with antibacterial and/or antifungal activity. Mar Drugs 17(4):241
Katekar VP, Rao AB, Sardeshpande VR (2022) Review of the rose essential oil extraction by hydrodistillation: an investigation for the optimum operating condition for maximum yield. Sustain Chem Pharm 29:100783
Khairah M, Mubarik NR, Manaf LA (2023) Bacterial selection and characterization of chitinase enzyme from bacteria controlling *Fusarium proliferatum*. Biodivers J Biol Divers 24(3):1926
Khairy AM, Tohamy MRA, Zayed MA, Mahmoud SF, El-Tahan AM, El-Saadony MT, Mesiha PK (2022) Eco-friendly application of nano-chitosan for controlling potato and tomato bacterial wilt. Saudi J Biol Sci 29(4):2199–2209
Kianfar E (2021) Protein nanoparticles in drug delivery: animal protein, plant proteins and protein cages, albumin nanoparticles. J Nanobiotechnol 19(1):159
Köhl J, Kolnaar R, Ravensberg WJ (2019) Mode of action of microbial biological control agents against plant diseases: relevance beyond efficacy. Front Plant Sci 10:845
Kučuk N, Primožič M, Knez Ž, Leitgeb M (2023) Sustainable biodegradable biopolymer-based nanoparticles for healthcare applications. Int J Mol Sci 24(4):3188
Kujur A, Yadav A, Kumar A, Singh PP, Prakash B (2019) Nanoencapsulated methyl salicylate as a biorational alternative of synthetic antifungal and aflatoxin B1 suppressive agents. Environ Sci Pollut Res 26(18):18440–18450
Kumar R, Lal S (2014) Synthesis of organic nanoparticles and their applications in drug delivery and food nanotechnology: a review. J Nanomater Mol Nanotechnol 3:4
Kumaraswamy RV, Kumari S, Choudhary RC, Pal A, Raliya R, Biswas P, Saharan V (2018) Engineered chitosan based nanomaterials: bioactivities, mechanisms and perspectives in plant protection and growth. Int J Biol Macromol 113:494–506
Kumaraswamy RV, Kumari S, Choudhary RC, Sharma SS, Pal A, Raliya R, Biswas P, Saharan V (2019) Salicylic acid functionalized chitosan nanoparticle: a sustainable biostimulant for plant. Int J Biol Macromol 123:59–69
Lin N, Wang C, Ding J, Su L, Xu L, Zhang B, Zhang Y, Fan J (2020) Efficacy of nanoparticle encapsulation on suppressing oxidation and enhancing antifungal activity of cyclic lipopeptides produced by *Bacillus subtilis*. Colloids Surf B Biointerfaces 193:111143
Liu Z, Jiao Y, Wang Y, Zhou C, Zhang Z (2008) Polysaccharides-based nanoparticles as drug delivery systems. Adv Drug Deliv Rev 60(15):1650–1662
Liu J, Setijadi E, Liu Y, Whittaker MR, Boyer C, Davis TP (2010) PEGylated gold nanoparticles functionalized with β-cyclodextrin inclusion complexes: towards metal nanoparticle–polymer–carbohydrate cluster biohybrid materials. Aust J Chem 63(8):1245–1250
Lohcharoenkal W, Wang L, Chen YC, Rojanasakul Y (2014) Protein nanoparticles as drug delivery carriers for cancer therapy. Biomed Res Int 2014:180549
Maghembe R, Damian D, Makaranga A, Nyandoro SS, Lyantagaye SL, Kusari S, Hatti-Kaul R (2020) Omics for bioprospecting and drug discovery from bacteria and microalgae. Antibiotics 9(5):229
Mahanta BP, Bora PK, Kemprai P, Borah G, Lal M, Haldar S (2021) Thermolabile essential oils, aromas and flavours: degradation pathways, effect of thermal processing and alteration of sensory quality. Food Res Int 145:110404

Malerba M, Cerana R (2016) Chitosan effects on plant systems. Int J Mol Sci 17(7). https://doi.org/10.3390/ijms17070996
Maluin FN, Hussein MZ (2020) Chitosan-based agronanochemicals as a sustainable alternative in crop protection. Molecules 25(7):1611
Martínez-Hernández GB, Amodio ML, Colelli G (2017) Carvacrol-loaded chitosan nanoparticles maintain quality of fresh-cut carrots. Innov Food Sci Emerg Technol 41:56–63
Martin-Serrano Á, Gómez R, Ortega P, de la Mata FJ (2019) Nanosystems as vehicles for the delivery of antimicrobial peptides (AMPs). Pharmaceutics 11(9):448
Maruyama CR, Guilger M, Pascoli M, Bileshy-José N, Abhilash PC, Fraceto LF, De Lima R (2016) Nanoparticles based on chitosan as carriers for the combined herbicides imazapic and imazapyr. Sci Rep 6(1):19768
Mashaghi S, Jadidi T, Koenderink G, Mashaghi A (2013) Lipid nanotechnology. Int J Mol Sci 14(2):4242–4282
Maximiano MR, Rios TB, Campos ML, Prado GS, Dias SC, Franco OL (2022) Nanoparticles in association with antimicrobial peptides (NanoAMPs) as a promising combination for agriculture development. Front Mol Biosci 9:890654
Menossi M, Ollier RP, Casalongué CA, Alvarez VA (2021) Essential oil-loaded bio-nanomaterials for sustainable agricultural applications. J Chem Technol Biotechnol 96(8):2109–2122
Miao T, Wang J, Zeng Y, Liu G, Chen X (2018) Polysaccharide-based controlled release systems for therapeutics delivery and tissue engineering: from bench to bedside. Adv Sci 5(4):1700513
Mirchandani Y, Patravale VB, S., B. (2021) Solid lipid nanoparticles for hydrophilic drugs. J Control Release 335:457–464
Mohammadi A, Hashemi M, Hosseini SM (2015a) Nanoencapsulation of *Zataria multiflora* essential oil preparation and characterization with enhanced antifungal activity for controlling *Botrytis cinerea*, the causal agent of gray mould disease. Innov Food Sci Emerg Technol 28:73–80
Mohammadi A, Hashemi M, Masoud S (2015b) Chitosan nanoparticles loaded with *Cinnamomum zeylanicum* essential oil enhance the shelf life of cucumber during cold storage. Postharvest Biol Technol 110:203–213
Mohammadi A, Hashemi M, Hosseini SM (2016) Effect of chitosan molecular weight as micro and nanoparticles on antibacterial activity against some soft rot pathogenic bacteria. LWT Food Sci Technol 71:347–355
Mohammadi A, Hosseini SM, Hashemi M (2020) Emerging chitosan nanoparticles loading-system boosted the antibacterial activity of *Cinnamomum zeylanicum* essential oil. Ind Crop Prod 155:112824
Mondéjar-López M, Rubio-Moraga A, López-Jimenez AJ, García Martínez JC, Ahrazem O, Gómez-Gómez L, Niza E (2022) Chitosan nanoparticles loaded with garlic essential oil: a new alternative to tebuconazole as seed dressing agent. Carbohydr Polym 277:118815
Morena AG, Tzanov T (2022) Antibacterial lignin-based nanoparticles and their use in composite materials. Nanoscale Adv 4(21):4447–4469
Morena AG, Bassegoda A, Natan M, Jacobi G, Banin E, Tzanov T (2022) Antibacterial properties and mechanisms of action of sonoenzymatically synthesized lignin-based nanoparticles. ACS Appl Mater Interfaces 14(33):37270–37279
Muthukrishnan S, Murugan I, Selvaraj M (2019) Chitosan nanoparticles loaded with thiamine stimulate growth and enhances protection against wilt disease in chickpea. Carbohydr Polym 212:169–177
Muzzalupo I, Badolati G, Chiappetta A, Picci N, Muzzalupo R (2020) In vitro antifungal activity of olive (*Olea europaea*) leaf extracts loaded in chitosan nanoparticles. Front Bioeng Biotechnol 8:151
Nadendla SR, Rani TS, Vaikuntapu PR, Maddu RR, Podile AR (2018) HarpinPss encapsulation in chitosan nanoparticles for improved bioavailability and disease resistance in tomato. Carbohydr Polym 199:11–19

Nair A, Mallya R, Suvarna V, Khan TA, Momin M, Omri A (2022) Nanoparticles—attractive carriers of antimicrobial essential oils. Antibiotics 11(1):108

Narasimhamurthy K, Udayashankar AC, De Britto S, Lavanya SN, Abdelrahman M, Soumya K, Shetty HS, Srinivas C, Jogaiah S (2022) Chitosan and chitosan-derived nanoparticles modulate enhanced immune response in tomato against bacterial wilt disease. Int J Biol Macromol 220:223–237

Narendrakumar G, Namasivayam SKR (2021) Surface-modified nanosilica–chitinase (SiNp-Chs)-doped nano enzyme conjugate and its synergistic pesticidal activity with plant extracts against armyworm *Spodoptera litura* (fab.) (Lepidoptera: Noctuidae). IET Nanobiotechnol 15(1):117–134

Narendrakumar G, Namasivayam SKR, Manikanta M, Saha M, Dasgupta T, Divyasri N, Anusha C, Arunkumar B, Preethi TV (2018) Enhancement of biocontrol potential of biocompatible bovine serum albumin (BSA) based protein nanoparticles loaded bacterial chitinase against major plant pathogenic fungi *Alternaria alternata*. Biocatal Agric Biotechnol 15:219–228

Nasseri M, Golmohammadzadeh S, Arouiee H, Jaafari MR, Neamati H (2016) Antifungal activity of *Zataria multiflora* essential oil-loaded solid lipid nanoparticles *in vitro* condition. Iran J Basic Med Sci 19(11):1231

Nguyen VC, Nguyen VB, Hsieh M-F (2013) Curcumin-loaded chitosan/gelatin composite sponge for wound healing application. Int J Polymer Sci 2013:106570

Ni P, Wang L, Deng B, Jiu S, Ma C, Zhang C, Almeida A, Wang D, Xu W, Wang S (2020) Combined application of bacteriophages and carvacrol in the control of *Pseudomonas syringae* pv. *actinidiae* planktonic and biofilm forms. Microorganisms 8(6):837

Nuruzzaman M, Rahman MM, Liu Y, Naidu R (2016) Nanoencapsulation, nano-guard for pesticides: a new window for safe application. J Agric Food Chem 64(7):1447–1483

Oh JW, Chun SC, Chandrasekaran M (2019) Preparation and in vitro characterization of chitosan nanoparticles and their broad-spectrum antifungal action compared to antibacterial activities against phytopathogens of tomato. Agronomy 9(1):21

Oliveira-Fernandes J, Oliveira-Pinto PR, Mariz-Ponte N, Sousa RMOF, Santos C (2023) *Satureja montana* and *Mentha pulegium* essential oils' antimicrobial properties against *Pseudomonas syringae* pv. *actinidiae* and elicitor potential through the modulation of kiwifruit hormonal defenses. Microbiol Res 277:127490

Oliveira-Pinto PR, Mariz-Ponte N, Sousa RMOF, Torres A, Tavares F, Ribeiro A, Cavaco-Paulo A, Fernandes-Ferreira M, Santos C (2021) *Satureja montana* essential oil, zein nanoparticles and their combination as a biocontrol strategy to reduce bacterial spot disease on tomato plants. Horticulturae 7(12):584

Oluoch G, Matiru V, Mamati EG, Nyongesa M (2021) Nanoencapsulation of thymol and eugenol with chitosan nanoparticles and the effect against *Ralstonia solanacearum*. Adv Microbiol 11(12):723–739

Oluoch G, Nyongesa M, Mamati EG, Matiru V (2022) Thymol and eugenol nanoparticles elicit expression of Ralstonia solanacearum virulence and potato defense genes and are potential bactericides against potato bacterial wilt. Arch Phytopathol Plant Protect 55(9):1136–1157

Paiva-Santos AC, Silva AL, Guerra C, Peixoto D, Pereira-Silva M, Zeinali M, Mascarenhas-Melo F, Castro R, Veiga F (2021) Ethosomes as nanocarriers for the development of skin delivery formulations. Pharm Res 38(6):947–970

Pan X, Guo X, Zhai T, Zhang D, Rao W, Cao F, Guan X (2023) Nanobiopesticides in sustainable agriculture: developments, challenges, and perspectives. Environ Sci Nano 10(1):41–61

Panichikkal J, Prathap G, Nair RA, Krishnankutty RE (2021) Evaluation of plant probiotic performance of *Pseudomonas* sp. encapsulated in alginate supplemented with salicylic acid and zinc oxide nanoparticles. Int J Biol Macromol 166:138–143

Pardeike J, Hommoss A, Müller RH (2009) Lipid nanoparticles (SLN, NLC) in cosmetic and pharmaceutical dermal products. Int J Pharm 366(1):170–184

Parisi C, Vigani M, Rodríguez-Cerezo E (2015) Agricultural nanotechnologies: what are the current possibilities? Nano Today 10(2):124–127

Pascoli M, de Lima R, Fraceto LF (2018) Zein nanoparticles and strategies to improve colloidal stability: a mini-review. Front Chem 6:6

Pateiro M, Gómez B, Munekata PES, Barba FJ, Putnik P, Kovačević DB, Lorenzo JM (2021) Nanoencapsulation of promising bioactive compounds to improve their absorption, stability, functionality and the appearance of the final food products. Molecules 26(6):1547

Pereira-Dias L, Oliveira-Pinto PR, Fernandes JO, Regalado L, Mendes R, Teixeira C, Mariz-Ponte N, Gomes P, Santos C (2023) Peptaibiotics: harnessing the potential of microbial secondary metabolites for mitigation of plant pathogens. Biotechnol Adv 68:108223

Pirzada T, de Farias BV, Mathew R, Guenther RH, Byrd MV, Sit TL, Pal L, Opperman CH, Khan SA (2020) Recent advances in biodegradable matrices for active ingredient release in crop protection: towards attaining sustainability in agriculture. Curr Opin Colloid Interface Sci 48:121–136

Pott DM, Osorio S, Vallarino JG (2019) From central to specialized metabolism: an overview of some secondary compounds derived from the primary metabolism for their role in conferring nutritional and organoleptic characteristics to fruit. Front Plant Sci 10:835

Poznanski P, Hameed A, Orczyk W (2023) Chitosan and chitosan nanoparticles: parameters enhancing activity. Molecules 28:2996. https://doi.org/10.3390/molecules28072996

Praveen N, GowdaA SC, Murthy KN, Srinivas C (2022) Efficacy of chitosan nanoparticles against bacterial leaf blight of rice caused by *Xanthomonas oryzae* pv. *oryzae*. Mysore J Agric Sci 56(3):211

Qiao K, Liu Q, Huang Y, Xia Y, Zhang S (2020) Management of bacterial spot of tomato caused by copper-resistant *Xanthomonas perforans* using a small molecule compound carvacrol. Crop Prot 132:105114

Rahman SFS, Singh E, Pieterse CMJ, Schenk PM (2018) Emerging microbial biocontrol strategies for plant pathogens. Plant Sci 267:102–111

Reeve JR, Hoagland LA, Villalba JJ, Carr PM, Atucha A, Cambardella C, Davis DR, Delate K (2016) Organic farming, soil health, and food quality: considering possible links. Adv Agron 137:319–367

Reimhult E (2015) Nanoparticle-triggered release from lipid membrane vesicles. New Biotechnol 32(6):665–672

Robinson PK (2015) Enzymes: principles and biotechnological applications. Essays Biochem 59:1

Rodrigo S, García-Latorre C, Santamaria O, Suárez A (2022) Metabolites produced by fungi against fungal phytopathogens: review, implementation and perspectives. Plants 11:81

Rosa S, Pesaresi P, Mizzotti C, Bulone V, Mezzetti B, Baraldi E, Masiero S (2022) Game-changing alternatives to conventional fungicides: small RNAs and short peptides. Trends Biotechnol 40(3):320–337

Sahab AF, Waly AI, Sabbour MM, Nawar LS (2015) Synthesis, antifungal and insecticidal potential of chitosan (CS)-g-poly (acrylic acid) (PAA) nanoparticles against some seed borne fungi and insects of soybean. Int J ChemTech Res 8(2):589–598

Saleh AM, El-Soud WMA, Alotaibi MO, Beemster GTS, Mohammed AE, AbdElgawad H (2023) Chitosan nanoparticles support the impact of arbuscular mycorrhizae fungi on growth and sugar metabolism of wheat crop. Int J Biol Macromol 235:123806

Samimi S, Maghsoudnia N, Eftekhari RB, Dorkoosh F (2019) Chapter 3: Lipid-based nanoparticles for drug delivery systems. In: Mohapatra SS, Ranjan S, Dasgupta N, Mishra RK, Thomas S (eds) Characterization and biology of nanomaterials for drug delivery. Elsevier, pp 47–76

Sánchez-Hernández E, Langa-Lomba N, González-García V, Casanova-Gascón J, Martín-Gil J, Santiago-Aliste A, Torres-Sánchez S, Martín-Ramos P (2022) Lignin–chitosan nanocarriers for the delivery of bioactive natural products against wood-decay phytopathogens. Agronomy 12(2):461

Santos EH, Kamimura JA, Hill LE, Gomes CL (2015) Characterization of carvacrol beta-cyclodextrin inclusion complexes as delivery systems for antibacterial and antioxidant applications. LWT Food Sci Technol 60(1):583–592

Saravanakumar G, Jo D-G, Park H, J. (2012) Polysaccharide-based nanoparticles: a versatile platform for drug delivery and biomedical imaging. Curr Med Chem 19(19):3212–3229

Sathiyabama M, Muthukumar S (2020) Chitosan guar nanoparticle preparation and its *in vitro* antimicrobial activity towards phytopathogens of rice. Int J Biol Macromol 153:297–304

Sathiyabama M, Parthasarathy R (2016) Biological preparation of chitosan nanoparticles and its in vitro antifungal efficacy against some phytopathogenic fungi. Carbohydr Polym 151:321–325

Sathiyabama M, Akila G, Charles RE (2014) Chitosan-induced defence responses in tomato plants against early blight disease caused by *Alternaria solani* (Ellis and Martin) Sorauer. Arch Phytopathol Plant Protect 47(16):1963–1973

Satish S, Raveesha KA, Janardhana GR (1999) Antibacterial activity of plant extracts on phytopathogenic *Xanthomonas campestris* pathovars. Lett Appl Microbiol 28(2):145–147

Sebaaly C, Greige-Gerges H, Agusti G, Fessi H, Charcosset C (2016) Large-scale preparation of clove essential oil and eugenol-loaded liposomes using a membrane contactor and a pilot plant. J Liposome Res 26(2):126–138

Severino P, da Silva CF, Andrade LN, de Lima Oliveira D, Campos J, Souto EB (2019) Alginate nanoparticles for drug delivery and targeting. Curr Pharm Des 25(11):1312–1334

Shaikh SS, Sayyed RZ (2014) Role of plant growth-promoting rhizobacteria and their formulation in biocontrol of plant diseases. In: Plant microbes symbiosis: applied facets. Springer, pp 337–351

Shavisi N, Khanjari A, Basti AA, Misaghi A, Shahbazi Y (2017) Effect of PLA films containing propolis ethanolic extract, cellulose nanoparticle and *Ziziphora clinopodioides* essential oil on chemical, microbial and sensory properties of minced beef. Meat Sci 124:95–104

Sheoran S, Arora S, Samsonraj R, Govindaiah P, vuree, S. (2022) Lipid-based nanoparticles for treatment of cancer. Heliyon 8(5):e09403

Singh T, Singh A, Wang W, Yadav D, Kumar A, Singh PK (2019) Biosynthesized nanoparticles and its implications in agriculture. In: Biological synthesis of nanoparticles and their applications. CRC Press, pp 257–274

Sipponen MH, Lange H, Crestini C, Henn A, Österberg M (2019) Lignin for nano-and microscaled carrier systems: applications, trends, and challenges. ChemSusChem 12(10):2039–2054

Solomonov A, Shimanovich U (2020) Self-assembly in protein-based bionanomaterials. Israel J Chem 60(12):1152–1170

Sotelo-Boyás ME, Valverde-Aguilar G, Plascencia-Jatomea M, Correa-Pacheco ZN, Jiménez-Aparicio A, Solorza-Feria J, Barrera-Necha L, Bautista-Baños S (2015) Characterization of chitosan nanoparticles added with essential oils: in vitro effect on *Pectobacterium carotovorum*. Revista Mexicana de Ingeniería Química 14(3):589–599

Sotelo-Boyás ME, Bautista-Baños S, Correa-Pacheco ZN, Jiménez-Aparicio A, Sivakumar D (2016) Chapter 13: Biological activity of chitosan nanoparticles against pathogenic fungi and bacteria. In: Bautista-Baños S, Romanazzi G, Jiménez-Aparicio A (eds) Chitosan in the preservation of agricultural commodities. Academic Press, pp 339–349

Sotelo-Boyás ME, Correa-Pacheco Z, Corona-Rangel ML, Villanueva-Arce R, Bautista-Baños S (2019) Cellular alterations in *Pectobacterium carotovorum* treated with nanostructured formulations during the incubation time. Arch Microbiol 201(5):615–622

Sreelatha S, Kumar N, Yin TS, Rajani S (2022a) Evaluating the antibacterial activity and mode of action of thymol-loaded chitosan nanoparticles against plant bacterial pathogen *Xanthomonas campestris* pv. *campestris*. Front Microbiol 12:792737

Sreelatha S, Kumar N, Rajani S (2022b) Biological effects of thymol loaded chitosan nanoparticles (TCNPs) on bacterial plant pathogen *Xanthomonas campestris* pv. *campestris*. Front Microbiol 13:1085113

Stoleru E, Brebu M (2021) Stabilization techniques of essential oils by incorporation into biodegradable polymeric materials for food packaging. Molecules 26(20):6307

Sundin GW, Wang N (2018) Antibiotic resistance in plant-pathogenic bacteria. Annu Rev Phytopathol 56:161–180

Tahereh S, Mohsenifar A, Beyki M, Zhaveh S, Rahmani-Cherati T, Abdollahi A, Bayat M, Tabatabaei M (2015) Encapsulation of thyme essential oils in chitosan-benzoic acid nanogel with enhanced antimicrobial activity against *Aspergillus flavus*. LWT Food Sci Technol 60(1):502–508

Tang C (2019) 10—Nanostructures of soy proteins for encapsulation of food bioactive ingredients. In: Jafari SM (ed) Biopolymer nanostructures for food encapsulation purposes. Academic Press, pp 247–285

Tang Q, Qian Y, Yang D, Qiu X, Qin Y, Zhou M (2020) Lignin-based nanoparticles: a review on their preparations and applications. Polymers 12(11):2471

Tapwal A, Garg S, Gautam N, Kumar R (2011) *In vitro* antifungal potency of plant extracts against five phytopathogens. Braz Arch Biol Technol 54:1093–1098

Tarhini M, Greige-Gerges H, Elaissari A (2017) Protein-based nanoparticles: from preparation to encapsulation of active molecules. Int J Pharm 522(1):172–197

Thambugala KM, Daranagama DA, Phillips AJL, Kannangara SD, Promputtha I (2020) Fungi vs. fungi in biocontrol: an overview of fungal antagonists applied against fungal plant pathogens. Front Cell Infect Microbiol 10:604923. https://doi.org/10.3389/fcimb.2020.604923

Tolisano C, Del Buono D (2023) Biobased: biostimulants and biogenic nanoparticles enter the scene. Sci Total Environ 885:163912

Torres FG, De-la-Torre GE (2022) Synthesis, characteristics, and applications of modified starch nanoparticles: a review. Int J Biol Macromol 194:289–305

Turk H (2019) Chitosan-induced enhanced expression and activation of alternative oxidase confer tolerance to salt stress in maize seedlings. Plant Physiol Biochem 141:415–422

Ullah I, Chen Z, Xie Y, Khan SS, Singh S, Yu C, Cheng G (2022) Recent advances in biological activities of lignin and emerging biomedical applications: a short review. Int J Biol Macromol 208:819–832

Upadhyay N, Singh VK, Dwivedy AK, Chaudhari AK, Dubey NK (2021) Assessment of nano-encapsulated *Cananga odorata* essential oil in chitosan nanopolymer as a green approach to boost the antifungal, antioxidant and in situ efficacy. Int J Biol Macromol 171:480–490

Valdivieso-Ugarte M, Gomez-Llorente C, Plaza-Díaz J, Gil A (2019) Antimicrobial Antioxidant and Immunomodulatory Properties of Essential Oils: A Systematic Review. Nutrients 11(11): 2786

Velásquez CL, Pirela MR (2016) Biochemical aspects of the chitin fungicidal activity in agricultural uses. In: Chitosan in the preservation of agricultural commodities. Academic Press, pp 279–298

Venditti A, Bianco A, Frezza C, Serafini M, Giacomello G, Giuliani C, Bramucci M, Quassinti L, Lupidi G, Lucarini D (2016) Secondary metabolites, glandular trichomes and biological activity of *Sideritis montana* L. subsp. *montana* from Central Italy. Chem Biodivers 13(10):1380–1390

Vettraino AM, Zikeli F, Scarascia Mugnozza G, Vinciguerra V, Tabet D, Romagnoli M (2022) Lignin nanoparticles containing essential oils for controlling *Phytophthora cactorum* diseases. For Pathol 52(2):e12739

Vu KD, Hollingsworth RG, Leroux E, Salmieri S, Lacroix M (2011) Development of edible bioactive coating based on modified chitosan for increasing the shelf life of strawberries. Food Res Int 44(1):198–203

Wang J, Hao K, Yu F, Shen L, Wang F, Yang J, Su C (2022) Field application of nanoliposomes delivered quercetin by inhibiting specific hsp70 gene expression against plant virus disease. J Nanobiotechnol 20:1–16

White JF, Kingsley KL, Verma SK, Kowalski KP (2018) Rhizophagy cycle: an oxidative process in plants for nutrient extraction from symbiotic microbes. Microorganisms 6(3):95

Worrall EA, Hamid A, Mody KT, Mitter N, Pappu HR (2018) Nanotechnology for plant disease management. Agronomy 8(12):285

Xie X-H, Fu X, Yan X-Y, Peng W-F, Kang L-X (2021) A broad-specificity chitinase from *Penicillium oxalicum* k10 exhibits antifungal activity and biodegradation properties of chitin. Mar Drugs 19(7):356

Xing K, Shen X, Zhu X, Ju X, Miao X, Tian J, Feng Z, Peng X, Jiang J, Qin S (2016) Synthesis and in vitro antifungal efficacy of oleoyl-chitosan nanoparticles against plant pathogenic fungi. Int J Biol Macromol 82:830–836. https://doi.org/10.1016/j.ijbiomac.2015.09.074

Xu Y, Wei Y, Jiang S, Xu F, Wang H, Shao X (2022) Preparation and characterization of tea tree oil solid liposomes to control brown rot and improve quality in peach fruit. LWT 162:113442. https://doi.org/10.1016/j.lwt.2022.113442

Yang W, Fortunati E, Dominici F, Giovanale G, Mazzaglia A, Balestra GM, Kenny JM, Puglia D (2016a) Effect of cellulose and lignin on disintegration, antimicrobial and antioxidant properties of PLA active films. Int J Biol Macromol 89:360–368

Yang W, Fortunati E, Dominici F, Giovanale G, Mazzaglia A, Balestra GM, Kenny JM, Puglia D (2016b) Synergic effect of cellulose and lignin nanostructures in PLA based systems for food antibacterial packaging. Eur Polym J 79:1–12

Yang W, Owczarek JS, Fortunati E, Kozanecki M, Mazzaglia A, Balestra GM, Kenny JM, Torre L, Puglia D (2016c) Antioxidant and antibacterial lignin nanoparticles in polyvinyl alcohol/chitosan films for active packaging. Ind Crop Prod 94:800–811

Yang W, Fortunati E, Gao D, Balestra GM, Giovanale G, He X, Torre L, Kenny JM, Puglia D (2018a) Valorization of acid isolated high yield lignin nanoparticles as innovative antioxidant/antimicrobial organic materials. ACS Sustain Chem Eng 6(3):3502–3514

Yang L, Wen K-S, Ruan X, Zhao Y-X, Wei F, Wang Q (2018b) Response of plant secondary metabolites to environmental factors. Molecules 23(4):762

Yang W, Fortunati E, Gao D, Balestra GM, Giovanale G, He X, Torre L, Kenny JM, Puglia D (2018) Valorization of Acid isolated high yield lignin nanoparticles as innovative antioxidant/antimicrobial organic materials. ACS Sustain Chem Eng 6(3):3502–3514

Yetisgin AA, Cetinel S, Zuvin M, Kosar A, Kutlu O (2020) Therapeutic nanoparticles and their targeted delivery applications. Molecules 25(9):2193

Yien L, Zin NM, Sarwar A, Katas H (2012) Antifungal activity of chitosan nanoparticles and correlation with their physical properties. Int J Biomater 2012:632698

Yilmaz MT, Yilmaz A, Akman PK, Bozkurt F, Dertli E, Basahel A, Al-Sasi B, Taylan O, Sagdic O (2019) Electrospraying method for fabrication of essential oil loaded-chitosan nanoparticle delivery systems characterized by molecular, thermal, morphological and antifungal properties. Innov Food Sci Emerg Technol 52:166–178

Yingchoncharoen P, Kalinowski DS, Richardson DR (2016) Lipid-based drug delivery systems in cancer therapy: what is available and what is yet to come. Pharmacol Rev 68(3):701–787

York WS, Qin Q, Rose JKC (2004) Proteinaceous inhibitors of endo-β-glucanases. Biochim Biophys Acta 1696(2):223–233

Yousefi M, Ehsani A, Jafari SM (2019) Lipid-based nano delivery of antimicrobials to control food-borne bacteria. Adv Colloid Interf Sci 270:263–277

Yu Z, Tang J, Khare T, Kumar V (2020) The alarming antimicrobial resistance in ESKAPEE pathogens: can essential oils come to the rescue? Fitoterapia 140:104433

Zhang Y-M, Ye D-X, Liu Y, Zhang X-Y, Zhou Y-L, Zhang L, Yang X-L (2023) Peptides, new tools for plant protection in eco-agriculture. Adv Agrochem 2(1):58–78

Zhao X, Cui H, Wang Y, Sun C, Cui B, Zeng Z (2018) Development strategies and prospects of nano-based smart pesticide formulation. J Agric Food Chem 66(26):6504–6512

Zhaveh S, Mohsenifar A, Beiki M, Khalili ST, Abdollahi A, Rahmani-Cherati T, Tabatabaei M (2015) Encapsulation of *Cuminum cyminum* essential oils in chitosan-caffeic acid nanogel with enhanced antimicrobial activity against *Aspergillus flavus*. Ind Crop Prod 69:251–256

Zikeli F, Vinciguerra V, Sennato S, Scarascia Mugnozza G, Romagnoli M (2020) Preparation of lignin nanoparticles with entrapped essential oil as a bio-based biocide delivery system. ACS Omega 5(1):358–368

Chapter 6
Nanoparticles in Phytovirus Management

Mohsen Mohamed Elsharkawy

Abstract Plant viral diseases have a significant negative economic impact and endanger sustainable agriculture. Organic and inorganic nanoparticle-based technologies have been developed in recent years, opening up exciting new research opportunities. Unique nanoparticles have different physicochemical features that enable them to interact with host plants, viruses, and their vectors in a variety of beneficial and novel ways. Due to their special association with the production of antivirals, these substances were widely employed to combat phytoviruses. It is essential to understand all parameters, such as the time of application, the concentration of nanoparticles, application intervals, and their mode of action, to effectively control plant viruses and investigate the antiviral function of nanoparticles. Considering the antiviral function and the complex and multidimensional viewpoint of nanotechnology in the treatment, detection, and management of plant viral diseases, this chapter focuses on many features of nanoparticles, such as their interaction with plant viruses and their mechanism of action and the toxicological hazards.

Keywords Nano-encapsulation · Nanotechnology · Plant viral diseases · Toxicological hazards

M. M. Elsharkawy (✉)
Agricultural Botany Department, Faculty of Agriculture, Kafrelsheikh University, Kafr El-Sheikh, Egypt
e-mail: mohsen.abdelrahman@agr.kfs.edu.eg

M. Khan, J.-T. Chen (eds.), *Nanoparticles in Plant Biotic Stress Management*,
https://doi.org/10.1007/978-981-97-0851-2_6

6.1 Introduction

Agriculture is an underlying science concerned with the widespread cultivation of food crops to alleviate global hunger. Increases in agricultural yield are of paramount importance because of the growing global population, which is predicted to reach 10 billion by 2050 (Gill and Garg 2014). The primary goal of food production is to increase both the quality and quantity of the food produced. Losses in global food production are substantial, however, due to biotic and abiotic variables. About 20% to 40% of the world's food output is lost every year because of pests, according to the Food and Agriculture Organization (FAO). The prevention of plant diseases requires the careful management of pathogens in agricultural production. Bacteria, viruses, and fungi are the principal pathogens in the development of various plant diseases, and there is now no effective method for their avoidance. Additionally, new virulent pest types are always developing owing to the considerable pest populations and their quick life cycles. Viruses are a special kind of intracellular parasite that can only replicate inside a living cell, where they may use the host's resources to grow and eventually infect and harm the host. The widespread impact of viral infection on plants, humans, and animals poses serious challenges to global health. The SARS-CoV-2, which recently developed COVID-19, is only one example of the many human viral infections that have caused massive deaths across the globe. The virus has the largest genetic diversity, making control efforts more challenging (Mehetre et al. 2021). Viruses in plants are the major cause of many devastating plant diseases that are difficult to control and, in some circumstances, incurable. Viruses are responsible for almost half of all plant diseases and 20% of all yield loss globally (Vargas-Hernandez et al. 2020). There are around 900 different types of pathogenic viruses that may infect plants (Chauhan et al. 2019). Food insecurity is a common result of pandemics and epidemics of plant viral diseases that cause massive crop devastation. Viruses produce a wide variety of devastating plant diseases, including tobacco mosaic virus, cucumber mosaic virus, tomato yellow leaf curl virus, maize dwarf mosaic virus, papaya leaf curl virus, potato leaf roll virus, okra yellow vein mosaic virus, cauliflower mosaic virus, and numerous others. Stunting, distortion, chlorosis, and mottling of fruits and leaves of diseased plants are some of the many symptoms caused by phytoviruses. These symptoms, which are caused by changes at the molecular, physiological, and biochemical levels of the host plant, lead to a loss in growth and yield, which in turn leads to a loss in profit. Further, viral pathogen infection may have a significant impact on blooming and fruiting conditions. Tobacco streak virus (TSV) necrosis is the most common and devastating viral disease of sunflowers in different places where they are grown (Prasada Rao et al. 2000; Kumar et al. 2006; Jain et al. 2008). Cotton, legumes, tomatoes, okra, chili peppers, and other commercially significant crops are all vulnerable to the destruction of geminiviruses (Reddy and Zehr 2004; Biswas et al. 2005; Biswas et al. 2015). Viral infections in plants may be spread from one plant to another by the process of vegetative reproduction or with the use of biotic vectors such as nematodes, fungi, insects, and others (Boualem et al. 2016). Therefore, preventing the spread of viruses via their insect vectors is crucial for the effective management of viral diseases (Khurana and Marwal 2016; Marwal and Gaur 2019). Embargoes against economically significant crops and items like grains were implemented by several nations, successfully halting the introduction and spread of disease and numerous insect vectors (Almeida 2018).

Controlling plant diseases is essential to meeting the growing global demand for plant-based foods, fibers, and other goods. The most common method used for controlling plant diseases is increasingly focusing on the resistance of the host plant. Traditional plant breeding methods include raising and evaluating a large population of crops over many generations in order to find the best candidates for admission in the breeding population (Vincelli 2016; Dong and Ronald 2019). Transgenics is the process of manipulating an organism's genes to improve its traits. It is possible to use transgenics as an alternative to the constraints of traditional breeding (Jain et al. 2008). Long-term resistance to one or more infections is shown by transgenic plants due to the introduction of gene(s) from distantly or unrelated species, which also decreases the risk of the pathogen attaining resistance (Wally and Punja 2010). The advantages of transgenic technology have not been completely realized, however, because of consumer worries and legislation from different governments concerning the safety of transgenic crops (van Esse et al. 2019). The goal of using chemical, biological, and integrated pest control strategies is to ensure food safety while improving profitability. Unfortunately, several insect pests and a wide range of bacterial, fungal, and viral pathogens continue to destroy agricultural production despite these efforts.

Controlling virus vectors may significantly minimize the spread of plant viruses. However, the widespread use of chemical pesticides has resulted in a devastating effect on ecosystems worldwide. Plant viruses may be treated with a variety of antiviral medications, including ribavirin, benzamide riboside, tylophorine B, tiazofurin, etc. (Park et al. 2006). Although chemotherapy is efficient in preventing the production of new viral particles, it has no impact on preexisting viruses. RNA interference (RNAi), a kind of posttranscriptional gene silencing, is another strategy for combating plant viruses (Lamsal et al. 2011). An important focus of research is the creation of new antiviral medications and the improvement of existing antiviral substances. Antipathogenic compounds may be developed on a wide variety of platforms made possible by nanotechnology, including those for the conversion of metallic, nonmetallic, and biocompounds into their nanoforms. Potential applications for nanoparticles (NPs) are quickly emerging, and nanotechnology has been one of the fastest-evolving scientific fields in recent years. Due to the cheap cost of manufacturing in agriculture, nanoparticles are anticipated to increase crop yields because of their many useful uses in fields such as disease diagnostics, high-sensitivity biomolecular detection, and therapeutic and antimicrobial chemicals (Dutta 2012; Dutta et al. 2021). Nanoparticles have several potential uses in plant and human diseases because of their small size, high reactivity, and large surface area (Jeevanandam et al. 2018). It has also been shown that nanoparticles may strengthen plant resistance to disease and boost plant health. The use of nanotechnology in the diagnosis and treatment of disease is expected to become standard practice in the near future (Mahmood et al. 2017). Promising findings from recent investigations point to nanoparticles as a possible antiviral treatment and plant defense activator. In this chapter, we will summarize the dispersed literature on nanotechnology's role in managing viral diseases and concentrate on its several applications.

6.2 Nano-encapsulation

Effective agricultural strategies have been developed using nanoparticles, which typically function as carriers to absorb, encapsulate, or attach active chemicals. Nanocarriers for fungicides, herbicides, insecticides, and RNA-inducing compounds include chitosan, silica, solid lipid, and layered double hydroxide nanoparticles (Worrall et al. 2018). In the contest against plant pathogens, these nanoparticle carriers have shown to be very successful. The benefits of nanoparticles in efficient pesticide administration stem from their low toxicity, substantial surface area, and rapid mass transfer (Ghormade et al. 2011). Agrochemicals could have micronic or submicronic particles added to them by absorption, encapsulation, surface ionic, or weak bond attachments (Shang et al. 2019). Among the most interesting future research is the nano-encapsulation of pesticides in controlled-release matrices. It has been shown that using nano-encapsulation may make pesticides more effective, less volatile, and less hazardous to the environment and crops (González-Melendi et al. 2008; Sopeña et al. 2009). Additionally, it has been shown that using it reduces the pesticides needed to manage diseases and reduces their unfavorable impact on both people and the environment. The small size, kinetic stability, large surface area, low viscosity, and optical transparency of nano-emulsions provide a superior pesticide delivery mechanism (Xu et al. 2010). Therefore, agri-nanotechnology offers the potential for regulated pesticide release and site-specific delivery of different macromolecules. The applied chemicals and macromolecules may contribute to increased plant growth and improved nutrient uptake and disease resistance.

6.3 Application of Nanoparticles Against Plant Pathogens

Resistance to synthetic compounds used against plant diseases is a common problem, especially as the usage of agrochemicals and bulk metals (like copper) has increased. Thus, researchers have prioritized the creation of nanoformulations since their very small size facilitates the penetration of bioactive chemicals in plant cells. Materials having a nanosize of 1–100 nm are referred to as nanoparticles (NPs). It is possible to categorize NPs depending on their form or size and, most crucially, their composition. Their huge surface area in relation to their mass and new therapeutic interactions with biological systems all contribute to their distinctive physiochemical characteristics. The distinctive features and traits have attracted interest from a wide variety of industries and disciplines, including commercial and household use, medicine (Loureiro et al. 2016; Nikalje 2015), agriculture (Srivastava et al. 2021), environment (Ripp and Henry 2011; Golobič et al. 2012), and energy research (Avasare et al. 2015; Ning et al. 2016). Encapsulating bioactive compounds in biodegradable polymer matrices allows for the controlled release of these compounds. Additionally, it was shown that by using pesticides in nanoform, a smaller amount of pesticide is needed to produce the same or better results than by using bulk pesticide formulations. Sustainable agriculture using nanoparticles was considered (Srivastava and Singh 2021; Sinha et al. 2017; Balaure et al. 2017). Various nanoparticles, including nanopesticides and nanofertilizers (Tarafder et al. 2020; Ahmadian et al. 2021), are employed in the development of biosensors for plant disease detection. Nanoparticles might be created by either biological, physical, or chemical process. Nanoparticles may be synthesized in a wide variety of ways using plants and microbes in a biosynthetic approach. The inorganic nanoparticles are safe, biocompatible with water, and very stable. Calcium phosphate, iron oxide, gold, silver, zinc oxide, and other metal oxides and ions are all examples of inorganic nanoparticles (Paul and Sharma 2010; Kurtjak et al. 2017). Ions found in inorganic nanoparticles act as antimicrobial compounds. However, few researchers are actively investigating the mechanism through which NPs inhibit microbial growth. Even in low quantities, gold and silver are very poisonous to microorganisms. Nanoparticles are used to cause damage to the cell membranes of harmful pathogens. Reactive oxygen species (ROS), metal ions, and free radicals are all produced after application of nanoparticles. Plant biomolecules may be impacted by ROS production caused by nanoparticles. ROS generated by the zinc and copper then suffocate and kill the enveloped pathogen. Protein and membrane rupturing, ROS release, cytoplasmic and nuclear leakage, and a damaged cell wall were all reported by Hajipour et al. (Hajipour et al. 2012; Lemire et al. 2013; Zhang et al. 2018; Alghuthaymi et al. 2015; Masum et al. 2019). Metal ion attraction by proteins causes cellular accumulation and ultimate cell death (Mittapally et al. 2018). Plants respond to threats by inducing the transcription of genes involved in systemic resistance, which includes salicylic acid, ethylene, jasmonic acid, and phenylalanine ammonia-lyase synthesis (Derbalah and Elsharkawy 2019). Several plant pathogens may be countered simultaneously by activating these mechanisms. Several studies have shown that metallic oxides, metalloids, and nonmetals have all been considered for their potential uses as bactericides, fungicides, and nanofertilizers in the battle against plant diseases (Ahmadian et al. 2021; Ni et al. 2017; Iravani 2011; Mittal et al. 2013; Alghuthaymi et al. 2015). Seeds, leaves, and roots may all benefit from a direct application of nanoparticles to control plant pathogens. The actions of nanoparticles may be similar to those of chemical pesticides. Pheromones, compounds that induce systemic acquired resistance, active components of pesticides may all be encapsulated through nanomaterials for controlled release. There is evidence that nanoparticles may inhibit the growth of infectious viruses, fungi, and bacteria. Antifungal properties of silver nanoparticles were shown against many different species of fungi, including *Candida albicans*, *Candida tropicalis*, *Candida glabrata*, *Candida krusei*, and *Aspergillus brasiliensis* (Bryaskova et al. 2011). The fungal pathogens *Alternaria alternate*, *Botrytis cinerea*, *Macrophomina phaseolina*, *Sclerotinia sclerotiorum*, *Rhizoctonia solani*, and *Curvularia lunata* were all suppressed by the presence of silver nanoparticles (Krisnaraj et al. 2012). ZnO nanoparticles showed fungicidal activity against *Penicillium expansum* and *B. cinerea*. The nanoparticles were shown to impede the growth and development of fungal hyphae as well as the formation of conidia and conidiophores. Additionally, these chitosan nanoparticles have shown to have fungistatic properties both in vivo and in vitro against a wide range of fungi, including *Aspergillus niger*, *Alternaria alternate*, *Aspergillus parasiticus*, *Colletotrichum gloeosporioides*, *Botrytis cinerea*, *Puccinia triticina*, and *Rhizopus stolonifer* (El Ghaouth et al. 1994; Bautista-Baños et al. 2006; Martínez-Camacho et al. 2010; Cota-Arriola et al. 2011; Elsharkawy et al. 2022). The antibacterial experiment demonstrated that MgO NPs effectively inhibited the development of *Ralstonia solanacearum*, which causes tomato bacterial wilt disease (Imada et al. 2016). MgO NPs were shown to be more effective against soilborne fungi than macroscale MgO in a comparison study between the two types of MgO (Chen et al. 2020). The colonization of *Staphylococcus aureus*, *Escherichia coli*, *Pseudomonas aeruginosa*, and *Klebsiella pneumonia* is inhibited by silver nanoparticles produced by *Ocimum tenuiflorum* and *Solanum trilobatum*. *P. aeruginosa* is sensitive to the antibacterial effects of zinc nanoparticles (Jayaseelan et al. 2012). Additionally, CuO nanoparticles have been proven to possess antibacterial action against a variety of bacterial species (Azam et al. 2012). Several isolates of bacteria and viruses are sensitive to chitosan nanoparticles. Gram-negative bacteria, including *Pseudomonas aeruginosa*, *Escherichia coli*, *Pseudomonas fluorescens*, *Salmonella typhimurium*, and *Vibrio parahaemolyticus* were reported to be sensitive to the antimicrobial actions of chitosan nanoparticles (No et al. 2002; Rabea et al. 2003; Sun et al. 2007; Palma-Guerrero et al. 2010).

6.4 Antiviral Activities of Nanoparticles

A wide range of plant diseases may be traced back to phytoviruses, which are notoriously devastating plant pathogens. They cause significant qualitative and/or quantitative yield losses and are characterized by genetic diversity and quick and dynamic development. Several viruses and vectors were introduced to new ecosystems in the twenty-first century, and many plant viruses are transmitted from generation to generation via seeds. Abiotic and biotic stressors may also have an impact on the infection cycle. Controlling a perfect viral environment and preventing coincidental spread across crops, vectors, and viruses require robust disease-prevention measures.

As a result of their size and shape, nanoparticles are able to inhibit viral replication and spread. The antiviral properties of silver nanoparticles have been well-documented. These nanoparticles were tested to see whether they might be used as an antiviral agent to trigger systemic acquired resistance against *Tomato mosaic virus* and *Potato virus Y*. Under a transmission electron microscope, silver nanoparticles were observed attaching to the protein of the viral particle in the sap of infected tomato plants. Additionally, chemical evidence supported the generation of systemic acquired resistance to *Potato virus Y* and *Tomato mosaic virus*. Leaves of cluster beans infected with Sunn-hemp rosette virus were completely protected after being sprayed with an aqueous solution of silver nanoparticles at a concentration of 50 ppm. The application of therapeutically effective amounts of polydispersed gold nanoparticles into the plant has induced resistance to *Barley yellow mosaic virus*. Mechanical abrasion was utilized to apply the gold nanoparticles. Nanoparticles ranged in size from roughly 0.5 to 200 nm in terms of their effective diameter. The virus particles were dissolved and melted by the gold nanoparticles, which then conferred resistance on the plant. Antiviral effects have been demonstrated for chitosan nanoparticles (Alkubaisi et al. 2015). The plant tissues were protected against the *Snuff mosaic virus*, *Alfalfa mosaic virus*, *Potato mosaic virus*, *Peanut mosaic virus*, and *Cucumber mosaic virus* by the chitosan, which induces resistance in the plant (Pospieszny et al. 1991; Kochkina et al. 1994; Chirkov 2002). Zinc oxide (ZnO) nanoparticles and silica nanoparticles were effective against *Tobacco mosaic virus*. Significant aggregation and rupture of virus particles occur after in vitro pretreatment with these nanoparticles for 2 h. Tobacco plants were also infected with a variety of nanoparticle mixtures to study their effects under in vivo conditions. Virus proliferation was found to be much lower than in the control group 2 days after infection (Cai et al. 2019). ZnO nanoparticles synthesized biologically have been examined for their antiviral efficacy against TMV. The disease severity and viral accumulation were reduced by 90% after repeated foliar application of the synthesized nanoparticles, 24 hours before and after TMV inoculation (Abdelkhalek and Al-Askar 2020).

6.5 Management of Plant Viruses Using Nanoparticles

6.5.1 Silica Nanoparticles

Systemic acquired resistance was triggered by silica NPs in *Arabidopsis thaliana*, a plant defense mechanism (El-Shetehy et al. 2021). Tomato yellow leaf curl virus infection was delayed by one day in plants prepared with horsemint (*Mentha longifolia*) and ginger (*Zingiber officinale*) plant extracts and silica nanoparticles. The findings showed that the levels of peroxidase and polyphenol in tomato leaves were increased when treated with plant extracts of ginger and horsemint, as well as nanoparticles (El-Sawy et al. 2017). Cucumber (*Cucumis sativus*) plants infected with *Papaya ringspot virus* benefit from SiO_2 NPs and culture filtrate of *Penicillium simplicissimum* GP17-2. *Phenylalanine ammonia-lyase*, *peroxidase*, and *PR-1* gene expression were all highly expressed in NPs 1 day after inoculation treatment and remained highly expressed 6 days after inoculation (Elsharkawy and Mousa 2015). SiO_2 NPs activated systemic resistance in tobacco plants infected with the cucumber mosaic virus (Elsharkawy et al. 2020).

6.5.2 Chitosan Nanoparticles

The deacetylation of chitin results in chitosan, a biocompound with effective antimicrobial characteristics. The severity and accumulation *Alfalfa mosaic virus* in mechanically infected tobacco plants were both reduced in an antiviral experiment using chitosan/dextran NPs (CDNPs) as a model system. Total carbohydrates and phenolic levels were risen tremendously by CDNPs, and the genes *PAL*, *peroxidase*, and *PR-1* were triggered (Abdelkhalek et al. 2021).

6.5.3 Plant Extract Nanoparticles

The rhizome of the turmeric (*Curcuma longa*) plant is where we get curcumin, a biocompound that is both powerful and safe to use. Curcumin-milk-protein nanoparticles inhibited the spread of *Potato virus Y* (PVY) in infected potato plants by stimulating the production of polyphenol oxidase and peroxidase (Taha et al. 2019). Tobacco mosaic virus was inhibited in the datura plants by the use of both pomegranate peel extract and nanoparticles (Dawi et al. 2021). Pomegranate peel nanoparticles were superior to pomegranate peel extract alone in reducing the number of local lesions and the concentration of the virus in infected plants.

6.5.4 Carbon-Based Nanomaterial

Incredibly potent antimicrobial capabilities were observed in nanocarbon. Carbon-based nanomaterials include graphene oxide (GO), carbon nanotubes (CNTs), and fullerene nanoparticles. Metallic NPs are harmful to soil microorganisms, but fullerenes and their derivatives are less hazardous. Three weeks before inoculation with *Tobacco mosaic virus*, tobacco plants were treated with carbon nanotubes, TiO_2 NPs, C60 fullerenes, and AgNPs (Dinesh et al. 2012). The findings demonstrated that the Ag and TiO_2 NPs did not affect the symptoms of the diseases. The phytohormones salicylic acid and abscisic acid are involved in plant defense against viral infection, and they were activated in CNT and C60-treated plants, resulting in total resistance to viral infection (Adeel et al. 2021).

6.5.5 Zinc Nanoparticles

Numerous advantages, such as low toxicity, strong photocatalytic activity, and simplicity of production, have led to zinc nanoparticles' widespread use. Evidence of ZnO NPs' virucidal activity against a wide range of plant and animal viruses has been collected via a number of different lines of inquiry. Antiviral activity against *Tomato mosaic virus* was observed in tomato leaves pretreated with green-produced ZnO nanoparticles using *Mentha spicata* leaf extract (Abdelkhalek and Al-Askar 2020). TMV was effectively suppressed by both ZnO NPs and SiO_2 NPs. Nanoparticles interacting with the glycoprotein envelope of TMV lead to shell damage and TMV aggregation, resulting in the inactivation of the virus. Additionally, nanoparticles elevate peroxidase activity and activate tobacco resistance genes in diseased plants (Cai et al. 2019). To combat the tomato mosaic virus, tomato plants treated with ZnO NPs demonstrated enhanced photosynthetic indices, growth rate, and enzymatic and nonenzymatic antioxidants. The results suggest that ZnO NPs boost tomato plants' resistance to tomato mosaic virus (Sofy et al. 2021).

6.5.6 Gold Nanoparticles

Due to their exceptional physicochemical features, gold NPs have great antimicrobial effects. The infectivity and severity of *Barley yellow dwarf virus PAV* were reduced after incubation with gold nanoparticles for varying periods. The production of viral capsid proteins and the structure of viral coat proteins were both influenced by AuNPs (Alkubaisi and Aref 2017). For virus identification using a visual detection approach, Au NPs are also utilized (Lavanya and Arun 2021).

6.5.7 Silver Nanoparticles

The most potent nanoparticles with antiviral, antibacterial, antifungal, and nematocidal properties are thought to be silver nanoparticles (Ag NPs). Complete resistance to *Sun hemp rosette virus* was observed in infected cluster bean plants after being treated with biosynthesized silver nanoparticles (Jain and Kothari 2014). The faba bean plants were protected against BYMV due to the AgNPs. AgNPs treated 48 h after inoculation had more virucidal activity than those applied 24 h before inoculation (El Gamal et al. 2022). AgNPs were applied topically to tomato plants that were already infected with the *Tomato mosaic virus* and the *Potato virus Y*, and the plants developed systemic resistance (Noha et al. 2018). *Pseudomonas fluorescens*-based silver nanoparticle synthesis reduces local lesions in tobacco plants infected with the tobacco mosaic virus (Ahsan 2020). However, simultaneous NPs treatment and virus inoculation resulted in the maximum inhibition of the virus. A reduction was noticed in virus accumulation and disease incidence of *Potato virus Y* in plants that had been sprayed with AgNPs, salicylic acid alone, or a combination of the two for 3 days, 7 days before inoculation, and 24 h after inoculation, respectively (El-Shazly et al. 2017). The spraying of AgNPs on plants 24 h after inoculation resulted in either a strong inhibition or a weak infection, whereas the spraying of AgNPs with salicylic acid 3 days and 7 days before inoculation resulted in a significant inhibition. A similar effect was shown in tomato spotted wilt virus-infected potato plants, where the Ag NPs sprayed 24 h after inoculation at a concentration of 200 ppm resulting in enhanced growth and decreased infectivity (Shafie et al. 2018). Plants treated with nanoparticles after a simulated viral inoculation show moderate inhibition, whereas plants treated before inoculation show a reduced inhibitory potential.

6.5.8 Iron Nanoparticles

Iron is one of the most reactive metals. Treatment of tobacco leaves with iron nanoparticles resulted in antiviral activity against *Tobacco mosaic virus*. The results demonstrated that salicylic acid-induced expression of PR genes and production of antioxidants activated the plant's resistance response against *Tobacco mosaic virus*. Iron nanoparticles penetrated plant leaf cells, accumulated throughout the plant, and improved both dry and fresh weight (Cai et al. 2020).

6.5.9 Nickel Nanoparticles

Nickel oxide nanoparticles inhibited the spread of cucumber mosaic virus in cucumber plants. Defense genes involved in the jasmonic acid, salicylic acid, and ethylene signaling pathways were upregulated, decreasing disease severity and viral accumulation in infected cucumber plants. The number of leaves, both fresh and dry weight, increased in cucumber plants treated with NiONPs (Derbalah and Elsharkawy 2019).

6.5.10 Titanium Nanoparticles

The ability of titanium dioxide nanoparticles to oxidize biomolecules gives them powerful antiviral action. Treatment of *Broad bean stain virus*-infected faba bean plants with titanium dioxide nanoparticles significantly reduces disease severity relative to untreated plants (Elsharkawy and Derbalah 2019). The expression of the gene involved in pathogenesis (*PR-1*) was found to be elevated. Twenty-one days of daily pretreatment of TiO_2 and two carbon-based nanomaterials resulted in tobacco plants that were resistant to *Turnip mosaic virus*. Metallic NPs accumulated in chloroplasts, whereas carbon-based NMs boosted shoot biomass without affecting cellular integrity (Hao et al. 2018). Treatment of faba bean plants with nano-carboxymethyl chitosan-titanium resulted in resistance to *Bean yellow mosaic virus*. Plants that were given treatments had higher concentrations of total soluble proteins, as well as enzymatic and nonenzymatic antioxidants (Sofy et al. 2020).

6.6 Induction of Defense Response Against Plant Viruses by Nanoparticles

6.6.1 *Antioxidant Activities*

The release of defense proteins and other local and systemic defense responses are examples of how plants respond defensively to stress (Table 6.1). ROS are produced in response to these stresses. Several biotic and abiotic factors trigger ROS production in plants. The production of oxidative products and the disruption of the balance between antioxidants and ROS occur when ROS levels exceed a certain threshold. Plants have an antioxidant mechanism to combat the effects of oxidants. Antioxidant enzymes include guaiacol peroxidase (GPX), catalase (CAT), superoxide dismutase (SOD), and ascorbate peroxidase (APX) (Tan et al. 2018). Nanoparticles (NPs) may either increase oxidative stress in cells or decrease its incidence, hence altering the cellular redox homeostasis (Soares et al. 2018). Depending on the needs of the host plant, NPs may either promote ROS generation or suppress the oxidative burst. In the instance of virus infection, a quick burst of ROS is released and impacts the host cells; in this situation, NPs stimulate an antioxidant mechanism to regulate the oxidative burst. For example, many different types of crops may benefit from the presence of antioxidant-inducing compounds, including CuO, CeO_2, TiO_2, ZnO, Fe_3O_4, Ag, $CoFe_2O_4$, and NiO (Soares et al. 2018). Antioxidant enzymes, including POD and CAT were also stimulated by Fe_3O_4, ZnO, and SiO_2 NPs. In addition to these enzymes, the production of other enzymes was increased in the presence of the virus. As an example, after 24 h of

Table 6.1 The involved mechanisms of plant viral diseases management by nanoparticles

Virus	Nanoparticles	Mechanisms of resistance	References
Barley yellow dwarf virus	Au NPs	Synthesis of viral capsid protein and configuration of viral coat protein	Alkubaisi and Aref (2017)
Broad bean stain virus	TiO_2 NPs	Expression of salicylic acid-responsive defense-related gene	Lavanya and Arun (2021)
Cucumber mosaic virus	NiO NPs	Expression of defense-related genes associated with SA, JA, and ethylene	Derbalah and Elsharkawy (2019)
Bean yellow mosaic virus	Carboxymethyl chitosan-Ti nanocomposites	Total soluble proteins, enzymatic, and nonenzymatic antioxidants	Sofy et al. (2020)
Tomato yellow leaf curl virus	SiO_2 NPs	Increased activity of polyphenols and peroxidase	El-Sawy et al. (2017)
Tomato bushy stunt virus	AgNPs	Reduction in disease severity and virus concentration	El-Shazly et al. (2017)
Tobacco mosaic virus	Fe NPs	Activation of SA-responsive PR genes	Cai et al. (2020)

Papaya ringspot virus inoculation, cucumber plants treated with SiO_2NP expressed pal and pox genes (Elsharkawy and Mousa 2015). NiONP treatment of cucumber plants after 4 days of CMV infection also resulted in elevated pod gene expression. Meanwhile, NPs may prime plants to create ROS as a defensive mechanism before virus particles ever enter the cells. Additionally, the activation of reactive oxygen species in foliar-sprayed tobacco leaves treated with Fe_3O_4NPs was confirmed (Cai et al. 2020). Tobacco leaves may develop resistance after being exposed to a stimulus that triggers the release of reactive oxygen species. Hydrogen peroxide enhances in response to ZnO and SiO_2 NP application, whether or not a virus is present.

6.6.2 Plant Hormones

Plants are resistant mostly due to their complex network of defense mechanisms, which are regulated by signaling molecules such as salicylic acid (SA), jasmonic acid (JA), and ethylene (ET) to generate the correct defensive reactions (Fig. 6.1). Additionally, gibberellin, abscisic acid (ABA), cytokinin, brassinosteroids (BRs), auxin, and strigolactone all can influence defensive responses as well as other growth-promoting activities in plants. Multiple plant hormones communicate with each other to maintain the delicate balance between defense and growth. There is evidence that NPs may affect the hormonal equilibrium in plants (Rastogi et al. 2017). *Arabidopsis thaliana* treated with CuONPs exhibited increased concentrations of SA and JA, as well as increased transcription of genes involved in defensive signaling. The combination of polyvinyl alcohol hydrogels with CuNPs

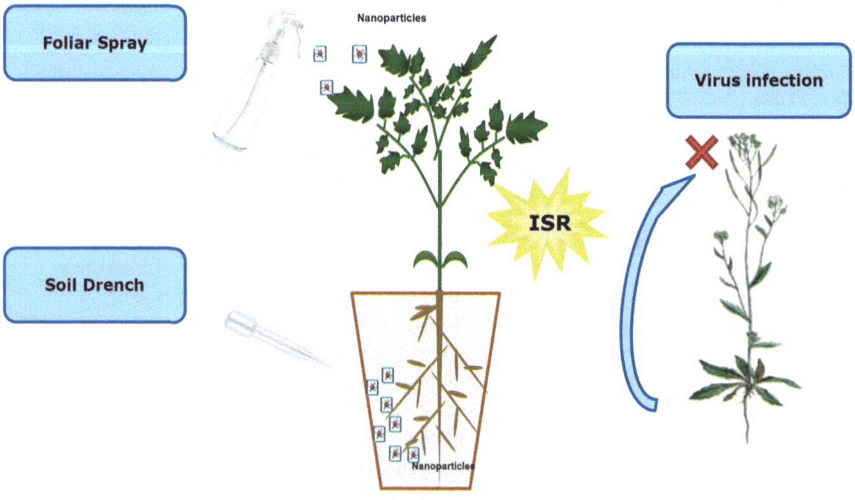

Fig. 6.1 Nanoparticles directly reduce virus infection through foliar spray treatment or by induced systemic resistance (ISR) through the soil drench application method

under salt stress-induced upregulation of the JA gene in tomato plants (Hernández-Hernández et al. 2018). Elevated SA and ABA levels were reported in *N. benthamiana*, whereas JA levels remained unchanged (Cai et al. 2020). Thus, the aforementioned findings show that the production of a specific plant hormone relies on the unique interaction between NPs and plants, application time, and dosage. *Capsicum annuum* treated with AgNPs exhibited a rise in the production of other hormones, including cytokinin (Vinkovic et al. 2017). ZnONPs produced an increase in cytokinin biosynthesis at low dosages and moderate and high concentrations; they may also operate as a stressor molecule that stimulates plant defenses and stress hormones, including SA and ABA (Landa et al. 2017).

Multiple methods are used by plants to modify plant development and regulate defense responses against plant diseases, such as viruses. The levels of phytohormones such as zeatin riboside, ABA, and brassinolide are altered in TMV-infected tobacco plants exposed to TiO_2NPs and Fe_2O_3 (Hao et al. 2018). Furthermore, the addition of AgNPs led to a rise in proline levels in tomato plants (El-Dougdoug and El-Dougdoug 2018). PR proteins, which serve as diagnostic molecular markers of plant defense signaling pathways, are an additional significant component of a plant's immune system. Upregulation of the SA signaling pathway is shown by the expression of *PR1*, *PR2*, and *PR5* (Ali et al. 2018). Upregulation of SA-inducible PR genes was detected in noninfected tobacco plants (*N. benthamiana*) after treatment with SiO_2 NPs, ZnONPs, or Fe_3O_4NPs (Cai et al. 2020). Therefore, the activation of the antiviral plant defense system may be seen in the increased production of phytohormones and PR proteins in NP-treated plants.

6.6.3 Synthesis of Secondary Metabolites

Several types of research have been undertaken to investigate the impact of NPs on secondary metabolite synthesis in plants, but the underlying mechanism remains unclear (Marslin and Sheeba 2017). The phenylalanine ammonia-lyase enzyme is regarded as a connection between the primary and secondary metabolisms. It is released in response to biotic and abiotic stress and stimulates SA/JA production. Trans-cinnamate is produced from phenylalanine through a non-oxidative deamination reaction catalyzed by PAL. Both NiONP and SiO_2NP treatments increased the amount of JA-induced PAL in plants infected with CMV or PRSV, respectively (Derbalah and Elsharkawy 2019; Elsharkawy and Mousa 2015). The impact of CuONPs and the induction of the Pal gene and increased phenolic and glucosinolate phytochemicals in *Brassica rapa* were reported (Chung et al. 2019). Similar outcomes were seen in tobacco plants that had PAL enzyme activity and accumulated flavonoid and phenolic compounds after being exposed to ZnONPs (Mazaheri-Tirani et al. 2019). Additionally, ZnONPs increased the level of total phenolics and anthocyanins in potato plants (Raigond et al. 2017).

6.7 Induction of Gene Silencing

Nanotechnology and RNAi are the two captivating sciences of the twenty-first century; when combined, they have the ability to completely improve plant disease management. RNAi is a method that alters mRNA integrity and protein synthesis to confer resistance to endogenous parasites or exogenous harmful nucleic acids through dsRNA molecules (Hannon 2002). To develop plants resistant to the target virus, NPs may be carried with dsRNA or viral coat protein. NPs enable RNAi treatments to be delivered to particular targets with a higher likelihood of uptake by the target plant cells or tissues. A noteworthy investigation was conducted to confirm resistance to *Cucumber mosaic virus* and *Pepper mild mottle virus*. This method included spraying the affected plants with layered double hydroxides that had been loaded with dsRNA known as BioClay. The plants showed 20 days of resistance to the aforementioned viruses, which was much longer than the control group (Elbeshehy et al. 2015). Topical treatment of dsRNA to trigger RNA interference was tested and found to be successful in a variety of plants (Worrall et al. 2018). They discovered that dsRNA applied topically was effective against *Bean common mosaic virus* infection by aphids and mechanical inoculation. *N. benthamiana* and *Vigna unguiculata* plants were resistant to mechanical inoculation with BCMV after receiving a treatment of dsRNA targeting either the potyviral nuclear inclusion b (Nib) coding region or the coat protein-coding region. Specifically, layered double hydroxide nanoparticles were loaded with *Bean common mosaic virus* coat protein to create *Bean common mosaic virus*-coat protein-coding region-BioClay. Additional research showed that BioClay-applied dsRNA prevented aphid-mediated *Bean common mosaic virus* transmission, opening the door to adopting this method in practice to monitor crops. The nanoliposomes now being employed to transport the antiviral biological chemical quercetin are only one example of how nanomaterials may be exploited as carriers for the delivery of antiviral treatments. As Hsp70 is quercetin's target gene during abiotic and biotic stress, the inhibitory impact of nanoliposomes was increased by 33.6% at the gene level and by 42.0% at the protein level. In addition, the effectiveness of the control was improved in field experiments (38% more than with regular quercetin formulations). According to the results of the research, incorporating nanomaterials into disease management strategies has the potential to both increase the effectiveness of treatments and decrease the number of active substances needed to achieve that success (Wang et al. 2022).

6.8 Detection of Plant Viruses Using Nanoparticles

6.8.1 Nano-biosensors

Biosensors are devices that can detect the presence of any biological structure, including biomolecules and microorganisms. A portion that detects the analyte and generates a signal, a signal transducer, and a reading device are the major components of the device (Cash and Plaxco 2021). When combined with aptamer-based detection techniques, various nanomaterials, basic metallic nanoparticles (such as carbon and gold nanoparticles), and nanospheres increase the sensitivity of the test.

6.8.2 Antibodies

Several papers in recent years have shown the ability to identify plant viruses using antibodies (Perdikaris et al. 2011; James 2013; Lin et al. 2014). When compared to conventional ELISA techniques, biosensors for the detection of the *Cowpea mosaic virus*, *Lettuce mosaic virus*, and *Tobacco mosaic virus* had a twofold improvement in sensitivity of detection (James 2013; Lin et al. 2014). Indirect ELISA to detect the cucumber mosaic virus was used. The three phases in this ELISA procedure are as follows: fixing the virus's antigen onto the surface, using certain antibodies to identify the virus, and incubating an enzyme with a secondary antibody that has been HRP-labeled. The mercury electrode was used to keep the reaction. The sensitivity of an electrochemical enzyme-linked immunoassay sensor-based technique for detecting CMV was four times that of the conventional spectrophotometric ELISA. These results were also used with other plant viruses such as *Potato virus Y*, *Tomato aspermy virus*, *Tobacco mosaic virus*, and *Southern bean mosaic virus*. For the diagnosis of plant pathogens, immunosensors made use of self-assembled monolayers. For the purposes of this technique, gold electrodes are the standard for detecting *Plum pox virus* (Jarocka et al. 2011). Jarocka et al. (Jarocka et al. 2013) used the same technique for diagnosing *Prunus necrotic ringspot virus* (PNRSV), and they came to the same conclusion: the biosensor is superior to ELISA (Jarocka et al. 2013). Another biosensor-based method of detecting plant viruses was described (Huang et al. 2014). A quartz crystal microbalance immunological sensor based on self-assembled monolayers was employed for the diagnosis of *Maize chlorotic mottle virus*. The biosensor's detection limit of 250 ng/mL is comparable to that of ELISA, and it demonstrated great sensitivity with closely related viruses like *Wheat streak mosaic virus* (Huang et al. 2014). *Tobacco mosaic virus* was detected using a lateral flow immunoassay, a form of optical immunosensor. Several additional viruses, including *Citrus tristeza virus*, *Potato virus X*, *Potato virus Y*, *Potato virus M*, and *Potato virus A*, were later diagnosed using this technique with a stated sensitivity of 2 ng/mL (Salomone et al. 2004; Danks and Barker 2000; Drygin et al. 2012). An immunoassay using fluorophore antibodies and fluorescence-loaded magnetic microspheres was introduced for the detection of different substances, including plant pathogens and biomarkers (Kim et al. 2010; Mushaben et al. 2013). Antibodies against three plant viruses—*Watermelon silver mottle virus*, *Chili vein-banding mottle virus*, and *Melon yellow spot virus*—have been used in an investigation. Even though the methods have shown both excellent detection sensitivity and the ability to perform numerous detections within a single test, the complexity of the assays and fluorescence readers has prevented them from gaining widespread use. Reports have discussed the application of SPR-based label-free biosensors for the detection of *Tobacco mosaic virus*, *Cucumber mosaic virus*, *Orchid viruses*, *Cymbidium mosaic virus*, and *Lettuce mosaic virus* (Lin et al. 2014; Boltovets et al. 2002; Torrance et al. 2006; Skottrup et al. 2007a, b).

6.8.3 *Quantum Dots (QD)*

Small semiconductor nanocrystals called quantum dots have gained use in biosensor development (Frasco and Chaniotakis 2009). As it possesses a special optical feature useful in fluorescence resonance energy transfer, it has been put to use in disease diagnosis (Algar and Krull 2008). *Candidatus Phytoplasma aurantifolia* (witches' broom disease of lime) was detected using this method (Rad et al. 2012). Consistently, this method yielded a 100% specific and sensitive result for just 5 *Candidatus Phytoplasma aurantifolia*/μL. The *Rhizoctonia* fungus, the disease's vector *Beet necrotic yellow vein virus*, was detected using this method (Safarpour et al. 2012).

6.9 Plant Virus-Loaded Nanoparticles

To improve the bioavailability of abamectin (Abm) against *Meloidogyne hapla* root-knot nematodes, encapsulation of Abm with the *Red clover necrotic mosaic virus* and then preparation of Abm-loaded plant virus NPs were shown to be an effective strategy. Tomato plants treated with Abm-loaded plant virus NPs showed increased root development and less root galling, demonstrating the viability of this delivery strategy for enhancing Abm's capacity to control nematodes (Cao et al. 2015). *Melaleuca alternifolia* oil has in vitro and in vivo nematicidal action against *Meloidogyne* sp., regardless of its formulation (Madalosso et al. 2017). The outcome was comparable to conventional treatments with abamectin, a biological pesticide exhibiting remarkable effectiveness against various plant parasitic nematodes.

6.10 Bio-stimulation of Infected Plants Using Nanoparticles

The term "plant biostimulants" refers to any material or substance, outside of conventional pesticides and plant nutrients, that can influence the physiology and biochemistry of plants in a manner that promotes photosynthesis, growth, development, and resistance to pathogens (du Jardin 2015). An increase in plant growth was observed in healthy tobacco plants treated with SiO_2, ZnO, and Fe_2O_3 NPs (Cai et al. 2019; Cai et al. 2020). NPs can counterbalance the negative effects that they may have; therefore they may have attained higher effects on both healthy and stressed plants (e.g., infected plants with a virus). NiONPs were sprayed on diseased cucumber plants and resulted in an increase in the number of leaves and fresh and dry weight (Derbalah and Elsharkawy 2019). Tobacco plants infected with the *Turnip mosaic virus*, which were treated with FeO_3 NPs and TiO_2 (50 mg/L) had a substantial increase in fresh and dry weight (Hao et al. 2018). PVY-infected plants treated with Ag NPs produced higher-quality potato tubers compared to untreated plants (El-Shazly et al. 2017). One potential explanation is that the use of NPs induces resistance or impedes viral entry. Furthermore, it may be utilized to protect seeds against viruses. Viruses may contaminate seeds by the gametes of infected plants, through the contaminated seed coat, or by directly invading the embryo of the seed before fertilization (Cobos et al. 2019). Due to their size and antimicrobial activities, NPs have the potential to be utilized as seed antivirals. This is accomplished by reducing the accumulation of virus particles in the seed. Additionally, NPs may be employed to avoid seedling contamination; however, in vitro culture has shown to be the most effective method for eradicating the virus in seedlings (Wang et al. 2016). Additional research showed that the antiviral action of NPs is more obvious before viral infection in the cultures (Cai et al. 2019; Cai et al. 2020). Hence, NPs are offered as protective control measures that may be employed for early-stage seed or seedling treatments to safeguard against fungi, viruses, or bacteria (Sánchez-López et al. 2020).

6.11 Negative Effects of Nanoparticles on Plants

Nanoparticles have many potential benefits, but there are also major risks associated with their continual presence in the environment. Both beneficial and detrimental impacts of NPs on plant development and growth were noted. The impacts of NPs on plants vary from plant species to plant species. Additionally, the concentration, size, and chemical and physical characteristics should be considered (Yadav and Yadav 2018). Direct contact, solubilization, and molecules' cotransport all contribute to the interactions of nanoparticles with plants (Yadav and Yadav 2018). Plant viability, seed germination, root development, photosynthesis, and overall plant growth are all negatively impacted by NPs (Rajput et al. 2021). The increase of NPs in plants and soil led to their incorporation into the food chain and the subsequent health concerns it produced for humans and other animals (Rajput et al. 2020). Carbon-based NPs sprayed onto plant leaves can enter the plant's cells through the stomata and concentrate in the thylakoids and stroma, increasing photosynthesis (Giraldo et al. 2014). However, maize plants' chlorophyll content and photosynthetic rate were lowered by a larger concentration of iron nanoparticles (Racuciu and Creanga 2007). Treatment of broccoli with ZnO-embedded fungal NPs resulted in increased NP bioaccumulation (Singhal et al. 2017). Silver nanoparticles affected the cell wall and vacuoles of rice root cells (Mazumdar and Ahmed 2011). In the first generation of treated maize plants, iron NPs boosted photosynthetic rate and hydrogen peroxide scavenging ability, but in the second generation, chlorophyll content, biomass, and hydrogen peroxide scavenging capacity were decreased

(Jalali et al. 2017). Soil nanoparticle accumulation alters soil microbiology, which in turn affects plant growth and reproduction (Javed et al. 2019). The accumulation and phytotoxicity of ZnO NPs were studied in nine different agricultural plants (García-Gómez et al. 2018). The composition of nitrogen fixers (biofilm-forming actinobacteria and proteobacteria) is influenced by silver and ZnO NPs. These are the bacteria that promote plant development, and their concentration in the soil has a direct effect on plant growth. The results indicated that the abundance of these three phyla was altered by 25–45% due to the presence of AgNPs in the soil, but the presence of ZnO NPs had no discernible impact.

6.12 Conclusion

Crop yields have significantly declined as a consequence of the rising occurrence of plant diseases throughout the globe. Furthermore, scenarios that take into account predicted climate change foresee an even more accelerated spread of phytopathogens as a result of significant changes in temperature, rainfall, and the frequency of extreme events. As a consequence, it is difficult to exclude the invasion of new pathogens or a rapid expansion of existing pathogens. Therefore, the necessity for efficient plant disease control and protection measures is important. While addressing global environmental concerns, we should restrict the use of agrochemicals and seek superior alternatives. The use of nanotechnology in the field of plant disease control is a relatively new field of study. Pesticides and direct interaction between plants and pathogens are just a few of how nanoparticles are utilized in plant disease control. However, in order to employ nanoparticles in plant disease management without endangering the environment or human health, it is important to create specific approaches or methodologies. Plant, animal, and human viral diseases have all grown increasingly severe and pervasive in recent years. Hence, there is an undeniable need to discover an effective means of treating viral infection, making the study of viruses and the treatment of viral disease more crucial and difficult. Even while the findings are consistent and suggest that NPs have powerful antiviral activities, it is important to note that they undergo a number of changes in the environment after being released. Therefore, we need to learn more about the potential dangers of excessively using nanotechnology in the future. According to recent research, nanoparticles may be found accumulated in soil and plant tissues. It should be underlined that nanoformulations for practical use must be stable for a long time, prevent NP aggregation, and, when administered by spraying, provide excellent adherence of the nanoformulation to the leaves by the selection of the appropriate excipients. The deposition of NPs in consumable plant parts may have downstream consequences on human health. Therefore, scientists need to investigate nanoparticles' (NPs') possible uses in agriculture while also finding ways to decrease their negative effects.

References

Abdelkhalek A, Al-Askar AA (2020) Green synthesized ZnO nanoparticles mediated by Mentha spicata extract induce plant systemic resistance against tobacco mosaic virus. Appl Sci 10:5054. https://doi.org/10.3390/app10155054

Abdelkhalek A, Qari SH, Abu-Saied MAAR, Khalil AM, Younes HA, Nehela Y, Behiry SI (2021) Chitosan nanoparticles inactivate alfalfa mosaic virus replication and boost innate immunity in *Nicotiana glutinosa* plants. Plants 10(12):2701. https://doi.org/10.3390/plants10122701

Adeel M, Farooq T, White JC, Hao Y, He Z, Rui Y (2021) Carbon-based nanomaterials suppress tobacco mosaic virus (TMV) infection and induce resistance in *Nicotiana benthamiana*. J Hazard Mater 404:124167. https://doi.org/10.1016/j.jhazmat.2020.124167

Ahmadian K, Jalilian J, Pirzad A (2021) Nano-fertilizers improved drought tolerance in wheat under deficit irrigation. Agric Water Manag 244:106544

Ahsan T (2020) Biofabrication of silver nanoparticles from *Pseudomonas fluorescens* to control tobacco mosaic virus. Egyp J Biol Pest Control 30:1–4

Algar WR, Krull UJ (2008) Quantum dots as donors in fluorescence resonance energy transfer for the bioanalysis of nucleic acids, proteins, and other biological molecules. Anal Bioanal Chem 391(5):1609–1618

Alghuthaymi MA, Almoammar H, Rai M, Said-Galiev E, Abd-Elsalam KA (2015) Myconanoparticles: synthesis and their role in phytopathogens management. Biotechnol Biotechnol Equip 29(2):221–236. https://doi.org/10.1080/13102818.2015.1008194

Ali S, Ganai BA, Kamili AN, Bhat AA, Mir ZA, Bhat JA, Rawat S et al (2018) Pathogenesis-related proteins and peptides as promising tools for engineering plants with multiple stress tolerance. Microbiol Res 212:29–37. https://doi.org/10.1016/j.micres.2018.04.008

Alkubaisi NA, Aref NM (2017) Dispersed gold nanoparticles potentially ruin gold barley yellow dwarf virus and eliminate virus infectivity hazards. Appl Nanosci 7(1–2):31–40. https://doi.org/10.1080/13102818.2015.1008194

Alkubaisi NAO, Aref NMMA, Hendi AA (2015) Method of inhibiting plant virus using gold nanoparticles. US Patent No. 9198434B1

Almeida RPP (2018) Emerging plant disease epidemics: biological research is key but not enough. PLoS Biol 16:e2007020. https://doi.org/10.1371/journal.pbio.2007020

Avasare V, Zhang Z, Avasare D, Khan I, Qurashi A (2015) Room-temperature synthesis of TiO_2 nanospheres and their solar driven photoelectrochemical hydrogen production. Int J Energy Res 39(12):1714–1719

Azam A, Ahmed AS, Oves M, Khan M, Memic A (2012) Size-dependent antimicrobial properties of CuO nanoparticles against gram-positive and -negative bacterial strains. Int J Nanomedicine 7:3527–3535

Balaure PC, Gudovan D, Gudovan I (2017) Nanopesticides: a new paradigm in crop protection. In: New pesticides and soil sensors. Academic, pp 129–192

Bautista-Baños S, Hernandez-Lauzardo AN, Velazquez-Del Valle MG, Hernández-López M, Barka EA, BosquezMolina E, Wilson C (2006) Chitosan as a potential natural compound to control pre and postharvest diseases of horticultural commodities. Crop Prot 25:108–118

Biswas KK, Pun KB, Pant RP, Ahlawat YS (2005) Mosaic disease in chilli (*Capsicum annuum*) cv Kalimpong local in Darjeeling Hill of West Bengal and its management. Indian Phytopathol 58:456–461

Biswas KK, Biswas K, Tarafdar A (2015) Multiple and mixed infections with yellow mosaic, leaf crinkle and bud necrosis disease complex in mungbean: a threat to cultivation of mungbean in India. Legume Res 38:382–388. https://doi.org/10.5958/0976-0571.2015.00122.8

Boltovets PM, Boyko VR, Kostikov IY, Dyachenko NS, Snopok BA, Shirshov YM (2002) Simple method for plant virus detection: effect of antibody immobilization technique. J Virol Methods 105(1):141–146

Boualem A, Dogimont C, Bendahmane A (2016) The battle for survival between viruses and their plants. Curr Opin Virol 17:32–38. https://doi.org/10.1016/j.coviro.2015.12.001

Bryaskova R, Pencheva D, Nikolov S, Kantardjiev T (2011) Synthesis and comparative study on the antimicrobial activity of hybrid materials based on silver nanoparticles (AgNps) stabilized by polyvinylpyrrolidone (PVP). J Chem Biol 4:185–191

Cai L, Liu C, Fan G, Liu C, Sun X (2019) Preventing viral disease by ZnONPs through directly deactivating TMV and activating plant immunity in *Nicothiana benthamiana*. Environ Sci Nano 6:3653–3669

Cai L, Cai L, Jia H, Liu C, Wang D, Sun X (2020) Foliar exposure of Fe_3O_4 nanoparticles on *Nicotiana benthamiana*: evidence for nanoparticles uptake, plant growth promoter and defense response elicitor against plant virus. J Hazard Mater 393:122415. https://doi.org/10.1016/j.jhazmat.2020.122415122415

Cao J, Guenther RH, Sit TL, Lommel SA, Opperman CH, Willoughby JA (2015) Development of abamectin loaded plant virus nanoparticles for efficacious plant parasitic nematode control. ACS Appl Mater Interfaces 7(18):9546–9553

Cash KJ, Plaxco KW (2021) Signal transduction with a swing. Nat Chem 13(5):392–393

Chauhan P, Singla K, Rajbhar M, Singh A, Das N, Kumar K (2019) A systematic review of conventional and advanced approaches for the control of plant viruses. J Appl Biol Biotechnol 7(4):89–98

Chen J, Wu L, Lu M, Lu S, Li Z, Ding W (2020) Comparative study on the fungicidal activity of metallic MgO nanoparticles and macroscale MgO against soilborne fungal phytopathogens. Front Microbiol 11:365. https://doi.org/10.3389/fmicb.2020.00365

Chirkov S (2002) The antiviral activity of chitosan (review). Appl Biochem Microbiol 38:1–8

Chung I-M, Rekha K, Venkidasamy B, Thiruvengadam M (2019) Effect of copper oxide nanoparticles on the physiology, bioactive molecules, and transcriptional changes in *Brassica rapa* seedlings. Water Air Soil Pollut 230:48. https://doi.org/10.1007/s11270-019-4084-2

Cobos A, Montes N, López-Herranz M, Gil-Valle M, Pagán I (2019) Within-host multiplication and speed of colonization as infection traits associated with plant virus vertical transmission. J Virol 93:e01078–e01119. https://doi.org/10.1128/JVI.01078-19

Cota-Arriola O, Cortez-Rocha MO, Rosas-Burgos EC, Burgos-Hernández A, López-Franco YL, Plascencia-Jatomea M (2011) Antifungal effect of chitosan on the growth of *Aspergillus parasiticus* and production of aflatoxin B1. Polym Int 60:937–944

Danks C, Barker I (2000) On-site detection of plant pathogens using lateral-flow devices∗. EPPO Bull 30:421–426

Dawi F, El-Beltagi HS, Abdel-Mobdy YE, Salah SM, Ghaly IS, AbdelRahim EA, Soliman AM (2021) Synergistic impact of the pomegranate peels and its nanoparticles against the infection of tobacco mosaic virus (TMV). Fresenius Environ Bull 30(1):731–746

Derbalah ASH, Elsharkawy MM (2019) A new strategy to control cucumber mosaic virus using fabricated NiO-nanostructures. J Biotechnol 306:134–141. https://doi.org/10.1016/j.jbiotec.2019.10.003

Dinesh R, Anandaraj M, Srinivasan V, Hamza S (2012) Engineered nanoparticles in the soil and their potential implications to microbial activity. Geoderma 173:19–27. https://doi.org/10.1016/j.geoderma.2011.12.018

Dong OX, Ronald CP (2019) Genetic engineering for disease resistance in plants: recent progress and future perspectives. Plant Physiol 180:26–38

Drygin YF, Blintsov AN, Grigorenko VG et al (2012) Highly sensitive field test lateral flow immunodiagnostics of PVX infection. Appl Microbiol Biotechnol 93(1):179–189

du Jardin P (2015) Plant biostimulants: definition, concept, main categories and regulation. Sci Hortic 196:3–14

Dutta A (2012) How helpful is nanotechnology in agriculture? Adv Nat Sci Nanosci Nanotechnol 3:033002. https://doi.org/10.1088/2043-6262/3/3/033002

Dutta P, Das G, Boruah S, Kumari A, Mahanta M, Yasin A et al (2021) Nanoparticles as nanopriming agent for antifungal and antibacterial activity against plant pathogens. Biol Forum Int J 13:476–482

El Gamal AY, Tohamy MR, Abou-Zaid MI, Atia MM, El Sayed T, Farroh KY (2022) Silver nanoparticles as a viricidal agent inhibit plant-infecting viruses and disrupt their acquisition and transmission by their aphid vector. Arch Virol 167(1):85–97. https://doi.org/10.21203/rs.3.rs-422790/v1

El Ghaouth A, Arul J, Wilson C, Benhamou N (1994) Ultrastructural and cytochemical aspects of the effect of chitosan on decay of bell pepper fruit. Physiol Mol Plant Pathol 44:417–432

Elbeshehy EK, Elazzazy AM, Aggelis G (2015) Silver nanoparticles synthesis mediated by new isolates of *Bacillus* spp., nanoparticle characterization and their activity against *Bean Yellow Mosaic Virus* and human pathogens. Front Microbiol 6:453. https://doi.org/10.3389/fmicb.2015.00453

El-Dougdoug NKAM, El-Dougdoug KA (2018) Evaluation of silver nanoparticles as antiviral agent against ToMV and PVY in tomato plants. Middle East J Appl Sci 8:100–111

El-Sawy MM, Elsharkawy MM, Abass JM, Hagag ES (2017) Inhibition of tomato yellow leaf curl virus by *Zingiber officinale* and *Mentha longifolia*. Int J Antiviral Antiretrovirol 1(1):1–6

Elsharkawy MM, Derbalah A (2019) Antiviral activity of titanium dioxide nanostructures as a control strategy for broad bean strain virus in faba bean. Pest Manag Sci 75(3):828–834. https://doi.org/10.1002/ps.5185

Elsharkawy MM, Mousa KM (2015) Induction of systemic resistance against papaya ring spot virus (PRSV) and its vector *Myzus persicae* by *Penicillium simplicissimum* GP17-2 and silica (SiO_2) nano powder. Int J Pest Manag 61(4):353–358. https://doi.org/10.1080/09670874.2015.1070930

Elsharkawy MM, Suga H, Shimizu M (2020) Systemic resistance induced by *Phoma* sp. GS8-3 and nanosilica against cucumber mosaic virus. Environ Sci Pollut Res 27(16):19029–19037. https://doi.org/10.1007/s11356-018-3321-3

Elsharkawy MM, Omara RI, Mostafa YS, Alamri SA, Hashem M, Alrumman SA, Ahmad AA (2022) Mechanism of wheat leaf rust control using chitosan nanoparticles and salicylic acid. J Fungi 8:304. https://doi.org/10.3390/jof8030304

El-Shazly MA, Attia YA, Kabil FF, Anis E, Hazman M (2017) Inhibitory effects of salicylic acid and silver nanoparticles on potato virus Y-infected potato plants in Egypt. Middle East J Agric Res 6(3):835–848

El-Shetehy M, Moradi A, Maceroni M, Reinhardt D, Petri-Fink A, Rothen Rutishauser B, Mauch F, Schwab F (2021) Silica nanoparticles enhance disease resistance in *Arabidopsis* plants. Nat Nanotechnol 16(3):344–353

Frasco MF, Chaniotakis N (2009) Semiconductor quantum dots in chemical sensors and biosensors. Sensor 9(9):7266–7286

García-Gómez C, Obrador A, González D, Babín M, Fernández MD (2018) Comparative study of the phytotoxicity of ZnO nanoparticles and Zn accumulation in nine crops grown in a calcareous soil and an acidic soil. Sci Total Environ 644:770–780. https://doi.org/10.1016/j.scitotenv.2018.06.356

Ghormade V, Deshpande MV, Paknikar KM (2011) Perspectives for nanobiotechnology enabled protection and nutrition of plants. Biotechnol Adv 29:792–803

Gill HK, Garg H (2014) Pesticide: environmental impacts and management strategies. Pesticides Toxic Aspects 8:187. https://doi.org/10.5772/57399

Giraldo JP, Landry MP, Faltermeier SM, McNicholas TP, Iverson NM, Boghossian AA, Reuel NF, Hilmer AJ, Sen F, Brew JA, Strano MS (2014) Plant nanobionics approach to augment photosynthesis and biochemical sensing. Nat Mater 13(4):400–408

Golobič M, Jemec A, Drobne D, Romih T, Kasemets K, Kahru A (2012) Upon exposure to Cu nanoparticles, accumulation of copper in the isopod Porcellio scaber is due to the dissolved Cu ions inside the digestive tract. Environ Sci Technol 46(21):12112–12119

González-Melendi P, Fernández-Pacheco R, Coronado MJ, Corredor E, Testillano P, Risueño MC, Marquina C, Ibarra MR, Rubiales D, Pérez-de-Luque A (2008) Nanoparticles as smart treatment-delivery systems in plants: assessment of different techniques of microscopy for their visualization in plant tissues. Ann Bot 101:187–195

Hajipour MJ, Fromm KM, Akbar Ashkarran A, Jimenez de Aberasturi D, de Larramendi IR, Rojo T, Serpooshan V, Parak WJ, Mahmoudi M (2012) Antibacterial properties of nanoparticles. Trends Biotechnol 30(10):499–511
Hannon G (2002) RNA interference. Nature 418:244–251. https://doi.org/10.1038/418244a
Hao YI, Yuan W, Ma C, White JC, Zhang Z, Adeel M, Zhou T, Rui Y, Xing B (2018) Engineered nanomaterials suppress *Turnip mosaic virus* infection in tobacco (*Nicotiana benthamiana*). Environ Sci Nano 5(7):1685–1693
Hernández-Hernández H, Juárez Maldonado A, Benavides-Mendoza A, Ortiz H, Cadenas-Pliego G, Aspeytia D, González-Morales S (2018) Chitosan-PVA and copper nanoparticles improve growth and overexpress the SOD and JA genes in tomato plants under salt stress. Agronomy 8:175. https://doi.org/10.3390/agronomy8090175
Huang X, Xu J, Ji H-F, Li G, Chen H (2014) Quartz crystal microbalance-based biosensor for rapid and sensitive detection of maize chlorotic mottle virus. Anal Methods 6(13):4530–4536
Imada K, Sakai S, Kajihara H, Tanaka S, Ito S (2016) Magnesium oxide nanoparticles induce systemic resistance in tomato against bacterial wilt disease. Plant Pathol 65(4):551–560. https://doi.org/10.1111/ppa.12443
Iravani S (2011) Green synthesis of metal nanoparticles using plants. Green Chem 13(10):2638–2650
Jain D, Kothari SL (2014) Green synthesis of silver nanoparticles and their application in plant virus inhibition. J Mycol Plant Pathol 44(1):21–24
Jain RK, Vemana K, Bag S (2008) Tobacco streak virus an emerging virus in vegetable crops. In: Rao GP, Kumar PL, Holuguin-Pena RJ (eds) Characterization, diagnosis and management of plant viruses, Vegetable and pulse crops, vol 3. Stadium Press LLC, Irving, pp 203–212
Jalali M, Ghanati F, Modarres-Sanavi AM, Khoshgoftarmanesh AH (2017) Physiological effects of repeated foliar application of magnetite nanoparticles on maize plants. J Agron Crop Sci 203(6):593–602. https://doi.org/10.1111/jac.12208
James C (2013) Polypyrrole nanoribbon based chemiresistive immunosensors for viral plant pathogen detection. Anal Methods 5(14):3497–3502
Jarocka U, Wąsowicz M, Radecka H, Malinowski T, Michalczuk L, Radecki J (2011) Impedimetric immunosensor for detection of *plum pox virus* in plant extracts. Electroanalysis 23(9):2197–2204
Jarocka U, Radecka H, Malinowski T, Michalczuk L, Radecki J (2013) Detection of *Prunus necrotic ringspot virus* in plant extracts with impedimetric immunosensor based on glassy carbon electrode. Electroanalysis 25(2):433–438
Javed Z, Dashora K, Mishra M, Fasake VD, Srivastva A (2019) Effect of accumulation of nanoparticles in soil health-a concern on future. Front Nanosci Nanotechnol 5:1–9
Jayaseelan C, Rahuman AA, Kirthi AV, Marimuthu S, Santhoshkumar T, Bagavan A, Gaurav K, Karthik L, Rao KB (2012) Novel microbial route to synthesize ZnO nanoparticles using *Aeromonas hydrophila* and their activity against pathogenic bacteria and fungi. Spectrochim Acta A Mol Biomol Spectrosc 90:78–84
Jeevanandam J, Barhoum A, Chan YS, Dufresne A, Danquah MK (2018) Review on nanoparticles and nanostructured materials: history, sources, toxicity and regulations. Beilstein J Nanotechnol 9:1050–1074. https://doi.org/10.3762/bjnano.9.98
Khurana SMP, Marwal A (2016) Recent developments towards detection and diagnosis for management of plant viruses. Indian Phytopathol 69:30–34
Kim JH, Kim SK, Wang KC (2010) Ischemia/angiogenesis-related molecules and cells. In: Cho BK, Tominaga T (eds) Moyamoya disease update. Springer, Tokyo
Kochkina Z, Pospeshny G, Chirkov S (1994) Inhibition by chitosan of productive infection of T-series bacteriophages in the *Escherichia coli* culture. Microbiology 64:211–215
Krisnaraj C, Ramachandran R, Mohan K, Kalaichelvan P (2012) Optimization for rapid synthesis of silver nanoparticles and its effect on phytopathogenic fungi. Spectrochim Acta A Mol Biomol Spectrosc 93:95–99
Kumar AN, Lakshminarasa M, Zehr UB, Ravi KS (2006) Natural occurrence and distribution of tobacco streak virus in South India. Indian J Plant Protect 34:54–58

Kurtjak M, Anicic N, Vukomanovicc M (2017) Inorganic nanoparticles: innovative tools for antimicrobial agents. In: Kumawat RN (ed) Antibacterial agents. InTech, Rijeka, pp 39–60

Lamsal K, Kim SW, Jung JH, Kim YS, Kim KS, Lee YS (2011) Application of silver nanoparticles for the control of *Colletotrichum* species in vitro and pepper anthracnose disease in field. Mycobiology 39(3):194–199. https://doi.org/10.5941/MYCO.2011.39.3.194

Landa P, Dytrych P, Prerostova S, Petrova S, Vankova R (2017) Transcriptomic response of *Arabidopsis thaliana* exposed to CuO nanoparticles, bulk material, and ionic copper. Environ Sci Technol 51:10814–10824. https://doi.org/10.1021/acs.est.7b02265

Lavanya R, Arun V (2021) Detection of Begomovirus in chilli and tomato plants using functionalized gold nanoparticles. Sci Rep 11(1):1–13

Lemire JA, Harrison JJ, Turner RJ (2013) Antimicrobial activity of metals: mechanisms, molecular targets and applications. Nat Rev Microbiol 11(6):371–384. https://doi.org/10.1038/nrmicro3028

Lin HY, Huang CH, Lu SH, Kuo IT, Chau LK (2014) Direct detection of orchid viruses using nanorod-based fiber optic particle plasmon resonance immunosensor. Biosens Bioelectron 51:371–378

Loureiro A, Azoia NG, Gomes AC, CavacoPaulo A (2016) Albumin-based nanodevices as drug carriers. Curr Pharm Des 22(10):1371–1390

Madalosso RG, Verdi CM, Souza ME, Cunha JA, Santos PS, Balardin RS et al (2017) In vitro and in vivo effects of *Melaleuca alternifolia* nanoparticles in Meloidogyne sp. J Asia Pac Entomol 20(4):1344–1349

Mahmood S, Lakra N, Marwal A, Sudheep NM, Anwar K (2017) Crop genetic engineering: an approach to improve fungal resistance in plant system. In: Singh DP (ed) Plant-microbe interactions in agro-ecological perspectives. Springer, Singapore, pp 581–591. https://doi.org/10.1007/978-981-10-6593-4_23

Marslin G, Sheeba CJ (2017) Nanoparticles alter secondary metabolism in plants via ROS burst. Front Plant Sci 8:832. https://doi.org/10.3389/fpls.2017.00832

Martínez-Camacho A, Cortez-Rocha M, Ezquerra-Brauer J, Graciano-Verdugo A, Rodriguez-Félix F, Castillo-Ortega M, Yépiz-Gómez M, Plascencia-Jatomea M (2010) Chitosan composite films: thermal, structural, mechanical and antifungal properties. Carbohydr Polym 82:305–315

Marwal A, Gaur R (2019) Nanophytovirology: an emerging field for disease management. In: Plant diseases-current threats and management trends. IntechOpen, London. https://doi.org/10.5772/intechopen.86653

Masum MMI, Siddiqa MM, Ali KA, Zhang Y, Abdallah Y, Ibrahim E, Qiu W, Yan C, Li B (2019) Biogenic synthesis of silver nanoparticles using *Phyllanthus emblica* fruit extract and its inhibitory action against the pathogen *Acidovorax oryzae* strain RS-2 of rice bacterial brown stripe. Front Microbiol 10:820. https://doi.org/10.3389/fmicb.2019.00820

Mazaheri-Tirani M, Haghjou MM, Ismaili A (2019) Hydroponic grown tobacco plants respond to zinc oxide nanoparticles and bulk exposures by morphological, physiological and anatomical adjustments. Funct Plant Biol 46:360. https://doi.org/10.1071/FP18076

Mazumdar H, Ahmed GU (2011) Phytotoxicity effect of silver nanoparticles on *Oryza sativa*. Int J ChemTech Res 3(3):1494–1500

Mehetre GT, Leo VV, Singh G, Sorokan A, Maksimov I, Yadav MK, Upadhyaya K, Hashem A, Alsaleh AN, Dawoud TM, Almaary KS, Singh BP (2021) Current developments and challenges in plant viral diagnostics: a systematic review. Viruses 13(3):412

Mittal AK, Chisti Y, Banerjee UC (2013) Synthesis of metallic nanoparticles using plant extracts. Biotechnol Adv 31(2):346–356

Mittapally S, Taranum R, Parveen S (2018) Metal ions as antibacterial agents. Drug Deliv Therap 8:411–419

Mushaben EM, Brandt EB, Khurana Hershey GK, Le Cras TD (2013) Differential effects of rapamycin and dexamethasone in mouse models of established allergic asthma. PLoS One 8(1):e54426

Ni D, Bu W, Ehlerding EB, Cai W, Shi J (2017) Engineering of inorganic nanoparticles as magnetic resonance imaging contrast agents. Chem Soc Rev 46(23):7438–7468

Nikalje AP (2015) Nanotechnology and its applications in medicine. Med Chem 5(2):1–9

Ning F, Shao M, Xu S et al (2016) TiO2/graphene/NiFe-layered double hydroxide nanorod array photoanodes for efficient photoelectrochemical water splitting. Energy Environ Sci 9(8):2633–2643

No HK, Park NY, Lee SH, Meyers SP (2002) Antibacterial activity of chitosans and chitosan oligomers with different molecular weights. Int J Food Microbiol 74:65–72

Noha K, Bondok AM, El-Dougdoug KA (2018) Evaluation of silver nanoparticles as an antiviral agent against ToMV and PVY in tomato plants. Sciences 8(1):100–111

Palma-Guerrero J, Lopez-Jimenez J, Pérez-Berná A, Huang IC, Jansson HB, Salinas J, Villalaín J, Read N, Lopez-Llorca L (2010) Membrane fluidity determines sensitivity of filamentous fungi to chitosan. Mol Microbiol 75:1021–1032

Park HJ, Kim SH, Kim HJ, Choi SH (2006) A new composition of nanosized silica silver for control of various plant diseases. Plant Pathol J 22(3):295–302. https://doi.org/10.5423/PPJ.2006.22.3.295

Paul W, Sharma CP (2010) Inorganic nanoparticles for targeted drug delivery. In: Sharma CP (ed) Biointegration of medical implant materials: science and design. CRC Press, Boca Raton, pp 204–235

Perdikaris A, Vassilakos N, Yiakoumettis I, Kektsidou O, Kintzios S (2011) Development of a portable, high throughput biosensor system for rapid plant virus detection. J Virol Methods 177(1):94–99

Pospieszny H, Chirkov S, Atabekov J (1991) Induction of antiviral resistance in plants by chitosan. Plant Sci 79:63–68

Prasada Rao RDVJ, Reddy AS, Chander Rao AS, Varaprasad KS, Thirumala Devi K, Nagaraju, Muniyappa V, Reddy DVR (2000) Tobacco streak ilarvirus as a causal agent of sunflower necrosis disease in India. J Oilseeds Res 17:400–401

Rabea EI, Badawy MET, Stevens CV, Smagghe G, Steurbaut W (2003) Chitosan as antimicrobial agent: applications and mode of action. Biomacromolecules 4:1457–1465

Racuciu M, Creanga DE (2007) TMA-OH coated magnetic nanoparticles internalized in vegetal tissue. Roman J Phys 52(3/4):395

Rad F, Mohsenifar A, Tabatabaei M et al (2012) Detection of *Candidatus Phytoplasma aurantifolia* with a quantum dots fret-based biosensor. J Plant Pathol 94:525–534

Raigond P, Raigond B, Kaundal B, Singh B, Joshi A, Dutt S (2017) Effect of zinc nanoparticles on antioxidative system of potato plants. J Environ Biol 38:435–439. https://doi.org/10.22438/jeb/38/3/MS-209

Rajput V, Minkina T, Mazarji M, Shende S, Sushkova S, Mandzhieva S, Burachevskaya M, Chaplygin V, Singh A, Jatav H (2020) Accumulation of nanoparticles in the soil-plant systems and their effects on human health. Ann Agric Sci 65(2):137–143

Rajput V, Chaplygin V, Gorovtsov A, Fedorenko A, Azarov A, Chernikova N, Barakhov A, Minkina T, Maksimov A, Mandzhieva S, Sushkova S (2021) Assessing the toxicity and accumulation of bulk-and nano-CuO in *Hordeum sativum* L. Environ Geochem Health 43(6):2443–2454

Rastogi A, Zivcak M, Sytar O, Kalaji H, Xiaolan H, Mbarki S et al (2017) Impact of metal and metal oxide nanoparticles on plant: a critical review. Front Chem 5:78. https://doi.org/10.3389/fchem.2017.00078

Reddy KVS, Zehr UB (2004) Novel strategies for overcoming pests and diseases in India. New directions for a diverse planet. In: Proceedings of 4th international crop science, Brisbane, Australia, pp 26

Ripp S, Henry TB (2011) Biotechnology and nanotechnology risk assessment: minding and managing the potential treats around US. ACS symposium series

Safarpour H, Safarnejad MR, Tabatabaei M et al (2012) Development of a quantum dots FRET-based biosensor for efficient detection of *Polymyxa betae*. Can J Plant Pathol 34(4):507–515

Salomone A, Mongelli M, Roggero P, Boscia D (2004) Reliability of detection of citrus tristeza virus by an immunochromatographic lateral flow assay in comparison with ELISA. J Plant Pathol 86:43–48

Sánchez-López E, Gomes D, Esteruelas G, Bonilla L, Lopez-Machado AL, Galindo R, Cano A (2020) Sci Hortic 196:3–14. https://doi.org/10.3390/nano10020292

Shafie RM, Salama AM, Farroh KY (2018) Silver nanoparticles activity against tomato spotted wilt virus. Middle East J Appl Sci 7:1251–1267

Shang Y, Hasan MK, Ahammed GJ, Li M, Yin H, Zhou J (2019) Applications of nanotechnology in plant growth and crop protection: a review. Molecules 24:2558. https://doi.org/10.3390/molecules24142558

Singhal U, Khanuja M, Prasad R, Varma A (2017) Impact of synergistic association of ZnO-nanorods and symbiotic fungus *Piriformospora indica* DSM 11827 on *Brassica oleracea* var. botrytis (Broccoli). Front Microbiol 8:1909. https://doi.org/10.3389/fmicb.2017.01909

Sinha K, Ghosh J, Sil PC (2017) New pesticides: a cutting-edge view of contributions from nanotechnology for the development of sustainable agricultural pest control. In: Grumezescu AM (ed) New pesticides and soil sensors. Academic, Cambridge, pp 47–79

Skottrup P, Hearty S, Frøkiær H et al (2007a) Detection of fungal spores using a generic surface plasmon resonance immunoassay. Biosens Bioelectron 22(11):2724–2729

Skottrup P, Nicolaisen M, Justesen AF (2007b) Rapid determination of *Phytophthora infestans* sporangia using a surface plasmon resonance immunosensor. J Microbiol Methods 68(3):507–515

Soares C, Pereira R, Fidalgo F (2018) Metal-based nanomaterials and oxidative stress in plants: current aspects and overview. In: Faisal M, Saquib Q, Alatar A, Al-Khedhairy A (eds) Phytotoxicity of nanoparticles. Springer International Publishing, Cham, pp 197–227

Sofy AR, Hmed AA, Alnaggar A-A, Dawoud RA, Elshaarawy RFM, Sofy MR (2020) Mitigating effects of *Bean yellow mosaic virus* infection in faba bean using new carboxymethyl chitosan-titania nanobiocomposites. Int J Biol Macromol 163:1261–1275

Sofy AR, Sofy MR, Hmed AA, Dawoud RA, Alnaggar AEAM, Soliman AM, El-Dougdoug NK (2021) Ameliorating the adverse effects of tomato mosaic tobamovirus infecting tomato plants in Egypt by boosting immunity in tomato plants using zinc oxide nanoparticles. Molecules 26(5):1337. https://doi.org/10.3390/molecules26051337

Sopeña F, Maqueda C, Morillo E (2009) Controlled release formulations of herbicides based on micro-encapsulation. Cienc Inv Agr 36:27–42

Srivastava A, Singh R (2021) Nanoparticles for sustainable agriculture and their effect on plants. Curr Nanosci 17(1):58–69

Srivastava A, Chuhan SP, Singh R (2021) Effect of silver nanoparticles on the growth and development of Indian brassica and *Cicer arietinum*. Res J Biotechnol 16:94–102

Sun T, Zhou D, Xie J, Mao F (2007) Preparation of chitosan oligomers and their antioxidant activity. Eur Food Res Technol 225:451–456

Taha SH, El-Sherbiny IM, Salem AS, Abdel-Hamid M, Hamed AH, Ahmed GA (2019) Antiviral activity of curcumin loaded milk proteins nanoparticles on potato virus Y. Pak J Biol Sci 22(12):614–622. https://doi.org/10.3923/pjbs.2019.614.622

Tan BL, Norhaizan ME, Liew WPP, Sulaiman Rahman H (2018) Antioxidant and oxidative stress: a mutual interplay in age-related diseases. Front Pharmacol 9:1162. https://doi.org/10.3389/fphar.2018.01162

Tarafder C, Daizy M, Alam MM et al (2020) Formulation of a hybrid nanofertilizer for slow and sustainable release of micronutrients. ACS Omega 5(37):23960–23966

Torrance L, Ziegler A, Pittman H, Paterson M, Toth X, Eggleston I (2006) Oriented immobilisation of engineered single chain antibodies to develop biosensors for virus detection. J Virol Methods 134(1–2):164–170

van Esse HP, Reuber TL, van der Does D (2019) Genetic modification to improve disease resistance in crops. New Phytol 225:70–86

Vargas-Hernandez M, Macias-Bobadilla I, Guevara-Gonzalez RG, RicoGarcia E, Ocampo-Velazquez RV, Avila-Juarez L, Torres-Pacheco I (2020) Nanoparticles as potential antivirals in agriculture. Agriculture 10(10):444. https://doi.org/10.3390/agriculture10100444

Vincelli PC (2016) Genetially engineered crops: emerging opportunities. Agriculture and Natural Resources Publications, pp 122

Vinkovic T, Novák O, Strnad M, Goessler W, Jurašin DD, Paradikovic N (2017) Cytokinin response in pepper plants (*Capsicum annuum* L.) exposed to silver nanoparticles. Environ Res 156:10–18. envres.03, 015. https://doi.org/10.1016/j.envres.2017.03.015

Wally O, Punja KZ (2010) Genetic engineering for increasing fungal and bacterial disease resistance in crop plants. GM Crops 1:199–206

Wang Y, Sun C, Xu C, Wang Z, Zhao M, Wang C et al (2016) Preliminary experiments on nano-silver against tobacco mosaic virus and its mechanism. Tob Sci Technol 49:22–30. https://doi.org/10.16135/j.issn1002-0861.20160104

Wang J, Hao K, Yu F (2022) Field application of nanoliposomes delivered quercetin by inhibiting specific hsp70 gene expression against plant virus disease. J Nanobiotechnol 20:16. https://doi.org/10.1186/s12951-021-01223-6

Worrall EA, Hamid A, Mody KT, Mitter N, Hanu HR (2018) Nanotechnology for plant disease management. Agronomy 8:285. https://doi.org/10.3390/agronomy8120285

Xu L, Liu Y, Bai R, Chen C (2010) Applications and toxicological issues surrounding nanotechnology in the food industry. Pure Appl Chem 82:349–372

Yadav A, Yadav K (2018) Nanoparticle-based plant disease management: tools for sustainable agriculture. In: Nanobiotechnology applications in plant protection. Springer, Cham, pp 29–61

Zhang LI, Wu L, Si Y, Shu K, Mishra YK (2018) Size-dependent cytotoxicity of silver nanoparticles to *Azotobacter vinelandii*: growth inhibition, cell injury, oxidative stress and internalization. PLoS One 13(12):e0209020

Chapter 7
Antifungal Nanoparticles and Their Impacts on Plant Immunity

Muhammad Afzal, Niaz Muhammad, Mehmood Jan, Sidra Kaleem, Abdul Rehman, Sajjid Muhammad, Mo Jingzhi, Abid Ali Abbas, Asim Shahzad, Mona H. Soliman, Mubbashir Hussain, Chen Xiaoyuan, and Zhiyuan Tan

Abstract This paper looks at how nanoparticles (NPs) can be used as antifungal treatments and how they interact with plant defense systems. It discusses numerous antifungal nanoparticles and their usage against various plant-pathogenic fungi. Furthermore, the combination of nanoparticles with plant essential oils is addressed as a potential synergistic method. The methods, by which nanoparticles demonstrate antifungal action, as well as their influence on improving plant defense systems, are investigated. The interaction of nanoparticles with various plant species, particularly their impact on the natural flora of plant roots, is also investigated. The goal of this review is to offer a thorough knowledge of nanoparticles' potential in increasing plant defense against fungal infections. Future research should focus on developing novel nanoparticle formulations, conducting mechanistic studies to understand the underlying antifungal mechanisms, evaluating long-term effects, improving nanoparticle delivery systems, integrating nanoparticles with other disease management techniques, and taking regulatory and safety considerations into account. Researchers can enhance the development of novel and long-lasting methods for managing fungal disease in plants by tackling these future objectives.

Keywords Nanoparticles · Antimicrobial agents · Antifungal agents · Plant source

M. Afzal (✉)
College of Agriculture, South China Agricultural University, Guangzhou City, People's Republic of China

Guangdong Provincial Key Laboratory of Utilization and Conservation of Food and Medicinal Resources in Northern Region, Shaoguan University, Shaoguan, People's Republic of China

N. Muhammad · A. Rehman · M. Hussain
Department of Microbiology, Kohat University of Science and Technology (KUST), Kohat, Pakistan

M. Khan, J.-T. Chen (eds.), *Nanoparticles in Plant Biotic Stress Management*, https://doi.org/10.1007/978-981-97-0851-2_7

M. Jan · M. Jingzhi · Z. Tan (✉)
College of Agriculture, South China Agricultural University,
Guangzhou City, People's Republic of China
e-mail: zytan@scau.edu.cn

S. Kaleem
Riphah Institute of Pharmaceutical Sciences, Islamabad, Pakistan

S. Muhammad · A. A. Abbas
College of Agriculture and Biotechnology, Zhejiang University,
Hangzhou, People's Republic of China

A. Shahzad
Department of Microbiology, Hazara University, Garden Campus, Mansehra, Pakistan

M. H. Soliman
Biology Department, Faculty of Science, Taibah University, Yanbu, Saudi Arabia

C. Xiaoyuan
Guangdong Provincial Key Laboratory of Utilization and Conservation of Food and
Medicinal Resources in Northern Region, Shaoguan University,
Shaoguan, People's Republic of China

7.1 Introduction

Nanotechnology is quickly expanding in industries such as medicine, electronics, energy, materials, agriculture, and livestock (Prasad et al. 2014). The 1959 speech "Plenty of Room at the Bottom" by Feynman is regarded as a watershed point in nanotechnology (Leong 2022). The nanotechnology lecture by Feynman emphasized the manipulation of matter at the molecular and atomic scale, resulting in substantial improvements led by materials science, nanofabrication techniques, and characterization methodologies (Baig et al. 2021). Nanomaterials with specific mechanical, electrical, thermal, and chemical characteristics may now be designed, synthesized, and manipulated by researchers. This opens up new possibilities for creativity and innovation (Liang et al. 2014).

Engineered nanoparticles (ENPs), on the other hand, are rapidly being used in a variety of sectors, including electronics, paints, medicinal goods, sunscreens, clothes, and vehicles. As a result, their influence on agriculture, including soil, plants, and the whole ecosystem, has piqued the curiosity of many. Understanding how nanoparticles affect agricultural soil, particularly soil bacteria and plant development, is critical for agricultural production and the economy (Phogat et al. 2016).

The use of *Melia azedarach* leaves extract in the green production of nanoparticles, especially selenium nanoparticles (SeNPs), improved physiological and biochemical parameters while suppressing *Fusarium mangifera* development and lowering biotic stress in *Mangifera indica* (Shahbaz et al. 2023). Saffron extract-derived silicon dioxide nanoparticles (SiO_2 NPs) protected wheat seedlings against *R. solani*. *R. solani* growth was suppressed by SiO_2 NPs, but wheat seedling germination and growth were improved, as were photosynthetic pigments and defense-related chemicals. SiO_2 nanoparticles stimulated both enzymatic and non-enzymatic antioxidant defense systems, inhibiting the pathogen and lowering oxidative stress (Abdelrhim et al. 2021).

Similar to gold nanoparticles (AgNPs), silver nanoparticles (AgNPs) have antifungal action against a variety of fungi, including *Aspergillus* spp., *Trichophyton rubrum*, *Candida albicans*, and *Penicillium* spp. This action is ascribed to AgNPs producing reactive oxygen species and free radicals, which cause cellular damage and death. AgNPs might be used in agricultural antifungal treatments (Mansoor et al. 2021).

Biosynthesized AgNPs have shown efficacy against bacterial, fungal, and viral diseases, making nanotechnology a feasible alternative for plant disease control. These nanoparticles cause pathogen membrane damage, induce oxidative stress, and alter host plant defense systems (Paul and Roychoudhury 2021).

The expression of MsDef1 and MsSN1 against *Fusarium oxysporum* was elevated in *M. sativa* roots by foliar application of SA, CaO NPs, and their mixtures. The small increase in PPO/PAL enzyme activity and antioxidant enzyme activity suggests improved plant defense. The research emphasizes the significance of antimicrobial peptides in the first line of defense of plants (Turgut and Bezirganoğlu 2022). Another study found that silver nanoparticles (AgNPs) have the potential to replace chemical fungicides by totally inhibiting *Alternaria solani* development at low doses. As antioxidants, selenium nanoparticles (SeNPs) can improve plant immunity (Ismail et al. 2016).

Ferric oxide nanoparticles were also applied to tomato plants to prevent *Fusarium* wilt and stimulate plant development. Fe_2O_3 nanoparticles showed antifungal properties and decreased illness indices. Additionally, they enhanced the activity of antioxidant enzymes, osmolytes, and photosynthetic pigments. Antimicrobial delivery, pathogen suppression, plant defense activation, vector control, and plant growth promotion are only a few of the many uses for nanotechnology (Elbasuney et al. 2022). These findings demonstrate the crucial roles that nanoparticles play in preventing fungal plant infections and enhancing plant immunity.

This chapter aims to briefly review the function of nanoparticles as antifungal agents and their connection to plant immunity. Additionally, to understand how plant immune responses might be improved by nanoparticles to successfully treat fungal infections in plants. We have cited different attempts by which nanoparticles exert their antifungal effects by investigating the relationship between nanoparticles and the plant immune system. Furthermore, this chapter covers how nanoparticles can replace conventional antifungal methods in the future.

7.1.1 Types of Antifungal Nanoparticles

The creation of antifungal nanoparticles has applications in both agriculture and medicine. They effectively and safely fight fungi-related disorders. Various mycoses in people, animals, and plant diseases are combated by inorganic nanoparticles and nanocarriers. These fungi affect oxygen delivery systems and food supply (Asghari et al. 2016). AgNPs, AuNPs, ZnONPs, biogenic selenium, and tellurium nanoparticles have all undergone substantial research and development to achieve their antifungal capabilities.

Silver nanoparticles (AgNPs): Silver has a long history of usage in cutlery and jewelry. It has curative and preserving qualities. For the management of plant diseases, AgNPs exhibit fungicidal activities and provide a safer alternative to conventional fungicides (Table 7.1) (Logeswari et al. 2015; Al-Surhanee et al. 2021). In vitro and in vivo tests showed that silver ions and nanoparticles were

Table 7.1 Different nanoparticles elevated against different phytopathogen

Types of NPs	Plants used in research	Used against pathogen	With or without plant essential oil	References
Silver nanoparticles (AgNPs)	*Solanum lycopersicum*, *Cyamopsis tetragonoloba*, *Vicia faba*, *Solanum tuberosum*, *Lactuca sativa* *Solanum tuberosum* L. *cv. Spunta*, autotrophic plants, *S. tuberosum*, *N. tabacum*	*Bipolaris sorokiniana* and *Magnaporthe grisea*, *A. alternata*, *Rhizoctonia solani*, *Penicillium* sp., *Aspergillus* sp., *Fusarium* sp., *Ralstonia solanacearum*, and *plant viruses and bacteria*	Both	Haroon et al. (2019); Tariq et al. (2022)
Gold nanoparticles (AuNPs)	Tomato, papaya, lime, orchid leaves, *Solanaceae* plant, potato, wheat, rose, banana plants, potatoes, and tomatoes	*Candida*, *Fusarium*, *Ganoderma*, *Rhizoctonia*, *Melampsora*, *Erysiphe*, and *Armillari*, *Tomato yellow leaf curl virus*, *Cucumber mosaic virus* (CMV), and *Papaya ringspot virus* (PRSV), witches' broom disease (*Candidatus Phytoplasma aurantifolia*), *Odontoglossum ringspot virus* (ORSV), bacterial spot disease by *Xanthomonas axonopodis*, *Ralstonia solanacearum* (potato brown rot), Karnal bunt disease, powdery mildew, *Pseudocercospora fijiensis* Black Sigatoka (leave streak disease), late blight in potatoes and tomatoes (caused by *Phytophthora infestans*), *Acidovorax avenae* subsp. citrulli	Both	Kulabhusan et al. (2022)

Table 7.1 (continued)

Types of NPs	Plants used in research	Used against pathogen	With or without plant essential oil	References
Zinc nanoparticles (ZnONPs)	*Nicotiana tabacum*, rice, Capsicum annuum L	*B. cinerea* and *P. expansum*, *Peronospora tabacina*, *Pythium ultimum*, *P. aphanidermatum*, *A. alternata*, *A. flavus*/*A. fumigatus*, *A. niger*, *Botrytis cinerea*, *Colletotrichum gloeosporioides*, *F. graminearum*, *F. moniliforme*, *F. oxysporum*, *P. expansum*, *Erythricium salmonicolo*, *Peronospora tabacina*	Both	Kalia et al. (2020)
Selenium and tellurium NPs	*Nothofagus pumilio* (lenga), fruits, pearl millet	*Candida albicans*, *F. oxysporum*, *Colletotrichum gloeosporioides*, *Oligoporus pelliculosus*, *Sclerospora graminicola*	Both	Nandini et al. (2017), Lazcano-Ramírez et al. (2023)

efficient against the plant-pathogenic fungus *Bipolaris sorokiniana* and *Magnaporthe grisea* (Jo et al. 2009). *P. longifolia* leaf extract-derived PL-AgNPs successfully converted silver nitrate to nanoparticles. They have strong antifungal efficacy against *A. alternata*, which reduced its influence on plant production. Further studies can be conducted to investigate their potential for increasing crop development (Dashora et al. 2022).

Gold nanoparticles (AuNPs): Gold, an eminent metallic material, is known for its medicinal qualities over a variety of ailments. In the past, gold colloidal solutions were employed to cure infections (Dreaden et al. 2012). AuNPs with biological compatibility, strong reactivity, small dimensions, variable form, and oxidation resistance have distinct features (Nejati et al. 2021). While AuNPs are physiologically inert by nature, they can be engineered to have photochemical and chemical activity (Prasson and Chittaranjan 2012). $SnCl_2$ and $NaBH_4$ were used to create gold nanoparticles with large surface areas and crystalline, spherical shapes (15 nm and 7 nm). UV-vis experiments found optical excitations in the visible spectrum. When compared to the 15 nm nanoparticles, the 7 nm nanoparticles displayed size-dependent antifungal efficacy and better biocidal effects against *Candida* isolates by reducing H^+ efflux (Table 7.1) (Ahmad et al. 2013).

Zinc nanoparticles (ZnONPs): During production, ZnO nanoparticles (NPs) frequently form agglomerates (Yehia and Ahmed 2013). For improved fluorescence qualities, zinc nanoparticles such as ZnS, ZnSe, and CdSe/ZnS quantum dots can be changed. These nanoparticles show promise in protein identification, immunofluorescence analysis, immunohistochemical detection, and 3D confocal investigation of membrane proteins, all of which have potential medical applications (Zhu et al. 2004). By disrupting membranes, hindering development, lowering spore germination, and triggering oxidative stress for cell death, zinc oxide nanoparticles inhibit fungal infections in plants (Abdal Dayem et al. 2017; Singh et al. 2020). Zinc nanoparticles also promote the development of enzymes, defense chemicals, and signaling pathways in plants, which helps them fight off fungi (Raha and Ahmaruzzaman 2022). ZnO nanoparticles (ZnO NPs) were evaluated against *B. cinerea* and *P. expansum* in research. ZnO NPs distort *B. cinerea* hyphae and impede *P. expansum* development, resulting in fungal mortality. In agriculture, ZnO NPs can be efficient fungicides (He et al. 2011).

Selenium and tellurium nanoparticles: Similar to sulfur, selenium and tellurium are metallic compounds with various states of oxidation. Due to their photoconductivity and thermal conductivity, these nanoparticles have garnered a lot of interest (Mirjani et al. 2015; Wu et al. 2012). Tellurium nanoparticles that were produced biologically were investigated for their ability to fight *Candida albicans*. The squalene monooxygenase enzyme was blocked by the nanoparticles, which caused squalene to build up and enhanced ERG1 gene expression (Zare et al. 2014). Fungi create selenium and tellurium nanoparticles in element-rich media. Elemental selenium and tellurium, as well as selenium oxide and tellurium oxide, precipitate on fungal surfaces. The fungal hyphal matrix contributes to the formation of nanoparticles (Liang et al. 2019).

7.1.2 Nanoparticles Coupled with Plant Essential Oils

Essential oils (EOs) have received a lot of attention in recent years because of their antibacterial, antiviral, antifungal, and insecticidal capabilities. However, because of their high volatility, delivering EOs properly to produce desired effects is problematic. As a result, investigating nanotechnology-based ways for developing effective nano-sized carriers for their distribution may give viable answers (Mihai et al. 2020). In a vaporized condition, EOs comprise around 300 volatile chemical components (Jugreet et al. 2020). Because of their low density in comparison to water, they are lipophilic, hydrophobic, and highly soluble in organic solvents (Podgoreanu et al. 2019).

Common plant groups that produce EO include *Lamiaceae*, *Lauraceae*, *Poaceae*, *Asteraceae*, *Verbenaceae*, *Myrtaceae*, *Zingiberaceae*, *Rutaceae*, *Piperaceae*, *Cupressaceae*, and *Umbelliferae* (Ogunsona et al. 2020). Among the techniques used to extract EOs from various plant components include cold pressing, solvent extraction, enfleurage, and distillation (Bava et al. 2023). Extraction yields are affected by variables such as species, segment type, and solvent characteristics (Ali et al. 2019). To retain the composition of the EO, physical means are utilized. Traditional approaches have drawbacks such as high energy usage and carbon dioxide emissions (SeidiDamyeh et al. 2016).

Encapsulation of nanomaterials (polymeric NPs, liposomes, etc.) improves EO effects. While active chemicals are contained in nanocapsules via a core membrane structure (Rai et al. 2017). Similarly, nanospheres are solid colloidal pieces, in which the bioactive components of EOs are distributed, caught, encapsulated, or adsorbed (Ferreira and Nunes 2019). EOs can be coupled with metallic NPs or encapsulated to fight drug-resistant bacteria. *Torreya grandis* aril EO combined with chitosan NPs showed more potent antibacterial properties than chitosan NPs alone (Wu et al. 2018). *Fusarium graminearum* is fought off by chitosan NPs and *Cymbopogon martinii* EO. Silver nanoparticles with eucalyptus oil prevent the growth of *E. coli*, MRSA, *S. enterica*, and *B. subtilis* (Basavegowda et al. 2020). Similarly, *S. aureus* growth and biofilm are controlled by gold nanoparticles linked to *Nigella sativa* EO. While rosemary and oregano essential oils combined with silver/ZnO nanoparticles in pullulan films inhibit the growth of *S. aureus*, *L. monocytogenes*, *E. coli*, and *S. typhimurium* (Morsy et al. 2014).

Only a small amount of research has examined the antibacterial properties of EOs and antibiotics in combination with metal/metal oxide nanoparticles (monometallic NPs) (Theophel et al. 2014). Silver nanoparticles mixed with *Origanum vulgare* essential oil fight *E. coli*, *A. baumannii*, and *S. aureus* (Scandorieiro et al. 2016).

According to Badea et al. (2019), hydroxyapatite NPs combined with peppermint EO increase efficacy against a variety of diseases (Table 7.1). Similar to this, *Siparuna guianensis* EO-coupled chitosan NPs and olive EO-combined lipid nanoparticles both boost effectiveness against certain microbes. Similarly, cinnamon EO and chitosan NPs work together to effectively battle *S. aureus* and *E. coli* (Vahedikia et al. 2019). Additionally, *Z. multiflora* oil and extract coupled with silver nanoparticles demonstrated additive benefits against *S. epidermidis* and *S. aureus* but no discernible difference against MRSA and *P. aeruginosa*. The combination had mixed results, except *S. epidermidis*. For the control of antimicrobials, it may be useful (Sheikholeslami et al. 2016).

7.1.3 Nanoparticles Used Against Different Plant Pathogens Fungi

Various types of plants, including those produced for agriculture, horticulture, decorative uses, and in forest ecosystems, are susceptible to a wide range of diseases brought on by pathogenic fungus. These diseases can have serious negative effects on the environment and the economy, resulting in crop losses, decreased plant vitality, and general declines in plant health (Malcolm et al. 2013; Konno et al. 2011).

Worldwide threats to crops and forests from fungi like *Fusarium solani*, *F. oxysporum*, and *Neofusicoccum* sp. result in negative economic and ecological effects. They enter vascular tissues of plants, causing vessel blockage and wilting (Yan et al. 2018; Roncero et al. 2003). Additionally, *Neofusicoccum* sp. has been linked to dieback and canker in trees and fruit crops (Espinoza et al. 2009; Iturritxa et al. 2011).

Cu-NPs are fungicidal against a variety of pathogenic fungi, including *Fusarium* sp., *Phoma destructiva*, *Curvularia lunata*, *Alternaria alternata*, *Penicillium italicum*, *Penicillium digitatum*, and *Rhizoctonia solani* (Kanhed et al. 2014).

Similarly, silver compounds also exhibit encouraging outcomes when used to combat plant diseases. In vitro, three AgNPs—WA-CV-WA13B, WA-AT-WB13R, and WA-PR-WB13R—successfully inhibit a variety of fungus. Stronger inhibition is shown at higher doses and with WA-CV-WA13B (Kim et al. 2012). Likewise, MgO and ZnO nanoparticles also dramatically reduced the germination of fungal spores. In vitro, MgO nanoparticles inhibited more than zinc nanoparticles at all concentrations (Wani and Shah 2012). In another study, AgNPs were evaluated in two fungal mediums at varying doses (25–100 ppm) against plant-pathogenic fungus (*Alternaria solani*, *Corynespora cassiicola*, and *Fusarium* spp.). At 100 ppm of potato dextrose agar (PDA), *Alternaria solani* and *Fusarium* spp. were completely inhibited, but *Corynespora cassiicola* was only 85% inhibited (Tyagi et al. 2020).

Furthermore, endophytic fungi isolated from *Solanum nigrum* were used to create mycosilver nanoparticles that were evaluated against plant-pathogenic fungus (*F. graminearum*, *F. udum*, *R. solani*, and *Aspergillus niger*). They displayed broad-spectrum antifungal action, suppressing phytopathogen radial growth (Akther and Hemalatha 2019).

The antifungal dispersion system based on O-chitosan nanoparticles demonstrated antifungal effectiveness against plant-pathogenic fungi. Chitosan inhibited chitosan-sensitive fungi (*Nigrospora sphaerica*, *Botryosphaeria dothidea*, *Nigrospora oryzae*, and *Alternaria tenuissima*), but not chitosan-resistant fungi (*Gibberella zeae* and *Fusarium culmorum*). For chitosan-sensitive fungi, greater nanoparticle concentrations enhanced the antifungal index (Xing et al. 2016).

Additionally, different silver nanoparticle concentrations were investigated against plant fungi. Silver nanoparticles displayed a strong inhibitory effect against all pathogens tested at a dosage of 15 mg. This implies that they might be used to manage fungal plant diseases (Krishnaraj et al. 2012).

7.1.4 Mechanism of NP as an Antifungal Agent

Silver nanoparticles (AgNPs) generated from plants have shown great promise as antifungal drugs and have the potential to replace common antibiotics like amphotericin and fluconazole. Ergosterol is present in the membranes of fungi, and these cells maintain different gradients across the cytoplasmic membrane that are essential for sustaining membrane potential. AgNPs disrupt these critical gradients and weaken membrane conformance, resulting in cell death (Mubeen et al. 2021).

The inhibition of budding in yeast has also been linked to gradient breakdown and loss of membrane conformance. The targeting and eradication of fungal growth in spore-producing fungi is another area where AgNPs excel (Kuppusamy et al. 2016).

Other metallic nanoparticles (MNPs), such as gold nanoparticles (AuNPs) and zinc oxide nanoparticles (ZnO-NPs), have also been thoroughly investigated for their antifungal characteristics in addition to AgNPs. These MNPs have proven to have strong antifungal properties against a variety of pathogenic strains, such as

A. solani, *P. graministritci*, *A. flavus*, *A. niger*, *C. albicans*, *T. harzianum*, *R. stolonifer*, and others (Gunalan et al. 2012).

MnO_2 NPs, or manganese oxide nanoparticles were synthesized and tested as biocatalytic inhibitors of fungi development. Application of 200g of the nanoparticles' antifungal efficacy against *Penicillium expansum*, *Monilinia fructigena*, *Penicillium chrysogenum*, and *Aspergillus oryzae* produced measurements of 14.2 86 mm, 8.9 6.0 mm, 17.7 1.26 mm, and 20.7 4.38 mm, respectively (Faisal et al. 2022).

The antifungal properties of garlic's allicin and allicin-loaded PBCA nanoparticles were investigated. In vitro experiments against *Candida albicans*, *Cryptococcus neoformans*, *Trichophyton rubrum*, *Microsporum gypseum*, *M. canis*, and *Epidermophyton floccosum* revealed decreased MIC values for PBCA-allicin NPs. *C. albicans*, *T. rubrum*, *E. floccosum*, and *M. canis* all had much lower MFCs. In vitro antifungal efficacy of pure allicin was higher than that of alliinase and alliin (Luo et al. 2009).

Similarly, the antifungal activity of GKPtNP against *A. parasiticus* and *A. flavus* was investigated. For *A. parasiticus* and *A. flavus*, it had a minimum inhibitory concentration of 10 g mL^{-1} and IC50 values of 37.2 and 29.4 g mL^{-1}, respectively. Destabilizing the fungal outer membrane, producing reactive oxygen species (ROS), and eliciting oxidative stress were all part of the procedure. Pt nanocomposite materials performed as a fungicidal agent having biological uses in mind (Godugu and Beedu 2018).

Another study examined the effectiveness of biogenic PtNPs produced by *Streptomyces* sp. against *A. flavus*, *A. niger*, *Penicillium* sp., *Candida albicans*, and *Candida tropicalis*. According to Sharma (2017), the MIC values for *A. flavus* and *A. niger* were 20 g mL^{-1} and 40 g mL^{-1}, respectively. In contrast to *A. niger* (57% and 63% inhibition), ZnO NPs significantly inhibited *A. flavus* (75% at 1000 ppm) (Al-Dhabi and Valan Arasu 2018).

7.1.5 Improvement in Plant Defense System Due to NP Application

Nanoparticles (NPs) have been investigated for their potential in a novel therapeutic strategy aiming at improving salt tolerance in plants (Fig. 7.1). NPs have emerged as viable carriers for delivering medications and chemicals. In the face of salt stress, the application of nanoparticles is a unique method to improve plant development and performance (Duhan et al. 2017). Compared to traditional bulk fertilization approaches, the application of NPs also improves plant responses to drought stress (Saxena et al. 2016).

In this situation, bitter melon plants were subjected to less salt stress because of the usage of chitosan-selenium nanoparticles (Cs-Se NPs). They enhanced plant development, yield, and production of essential oils. While decreasing damaging

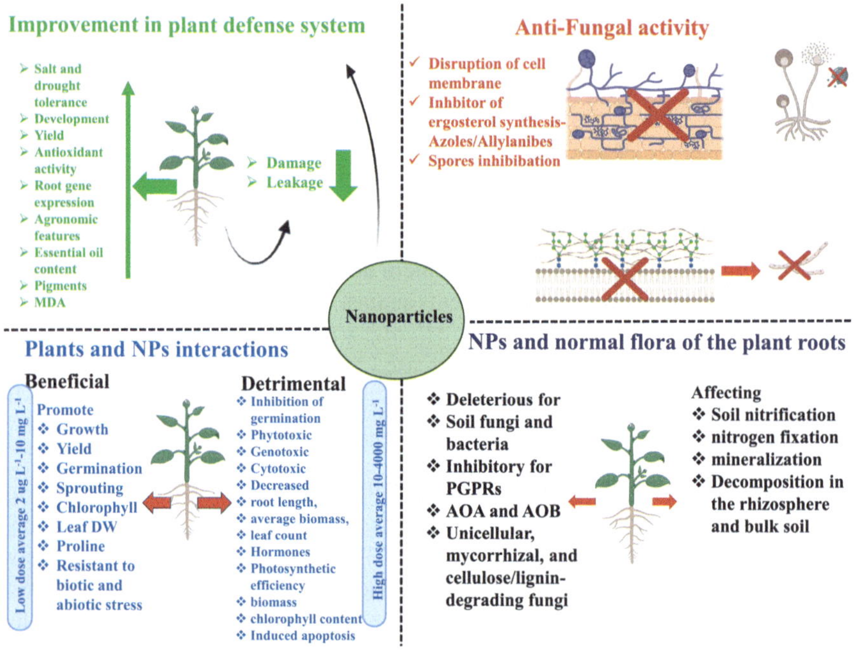

Fig. 7.1 NPs and plant-associated physiology, biochemistry, and its effects on soil microbial community

oxidants and salt buildup, Cs-Se NPs enhanced antioxidant activity, proline concentration, water content, and potassium levels. The highest successful concentration was 20 mg L^{-1}, proving that salt tolerance in bitter melon plants has been successfully increased (Sheikhalipour et al. 2021).

The impact of iron, copper, cobalt, and zinc oxide nanoparticles (NPs) on drought-tolerant soybean plants was examined in the research. The NPs enhanced physiological characteristics and genes involved in drought response. The most notable effect was produced by iron NPs, which improved leaf characteristics and increased root gene expression. In all NP treatments, the GmERD1 gene was induced in the roots and the shoots. These results imply that NPs can increase drought tolerance in soybean plants by activating certain genes (Linh et al. 2020).

In a greenhouse test, titanium dioxide nanoparticles (TiO_2 NPs) were found to significantly enhance the development of Moldavian balm at various salt levels. They decreased H_2O_2 concentration while enhancing agronomic features, antioxidant enzymes, and essential oil content. Geranial, z-citral, geranyl acetate, and geraniol were the most prevalent essential oil constituents. TiO_2 NPs successfully reduced the effects of salt in Moldavian balm (Gohari et al. 2020).

The effects of salt stress and cerium oxide nanoparticles (CeO_2 NPs) on Moldavian balm were also examined. Salinity increased oxidative stress and decreased characteristics, whereas CeO_2-NPs enhanced growth, pigments, and antioxidants. Under salinity, they lessened damage and leakage. For both situations, the optimum concentration was 50 mg L^{-1}. Moldavian balm plant performance was improved by CeO_2-NPs, which also reduced the impacts of salt stress (Mohammadi et al. 2021).

Additionally, the impact of TiO_2 NPs on chickpeas under cold stress was investigated. Hydrogen peroxide, MDA, and electrolyte leakage were all enhanced under cold stress, with ILC533 suffering more damage. Reduced H_2O_2, MDA, and leakage were seen in TiO_2 NP-treated plants, particularly in Sel11439. Lipoxygenase and antioxidant enzyme activity were improved in Sel11439 and TiO_2 NP-treated plants. Levels of carotenoid and chlorophyll were stabilized by TiO_2 NPs. Overall, TiO_2 NPs improved chickpeas' ability to withstand cold stress, minimizing damage and boosting defense mechanisms (Mohammadi et al. 2014).

7.1.6 Different Plants Species and NP Interaction

The interaction of nanoparticles (NPs) with plants has been thoroughly examined in recent research, including both beneficial and detrimental effects (Fig. 7.1) (Kurczyńska et al. 2021; Tarrahi et al. 2021). Despite our growing understanding of the relationship between plants and nanoparticles (NPs), there are still significant issues that need to be researched (Rajput et al. 2018; Zhou et al. 2020; Sanzari et al. 2019). Understanding the function of the cell wall and symplast in NP absorption as well as the role of plasmodesmata (PD) in NP transport through the plant is all part of this. The cell walls in the apoplast and the role of PD in NP translocation inside the symplast must thus be thoroughly examined in both plant compartments (Schwab et al. 2016).

Plant-NP interactions typically include three key steps: (1) NP deposition on surfaces of the plant, such as the leaves, roots, or stems, (2) NP penetration through the cuticle and epidermis, and (3) NP transport and transformation inside the plant. Given that various plant lineages might have distinct leaf and root surfaces and vascular systems, understanding the last two phases necessitates taking these differences into account (Su et al. 2019).

TiO_2 nanoparticle uptake: The absorption and transport of TiO_2 nanoparticles (NPs) in plants, particularly food crops, have received little attention in the literature (Rico et al. 2011). *A. thaliana* seedlings' particular tissues and cells were shown to take in and distribute ultrasmall TiO_2 complexed with Alizarin red S (Kurepa et al. 2010). Another research indicated that maize roots did not absorb TiO_2 NPs (30 nm), probably as a result of their size relative to the pore diameter in the root cell wall. The lack of studies on TiO_2 absorption by plants may be due to the stability of TiO_2 NPs, which prevents their digestion in plant samples (Asli and Neumann 2009).

ZnO nanoparticle uptake: ZnO nanoparticle (NP) absorption in soybean seedlings was investigated in research. Due to reduced NP aggregation, treatment with lower dosages (500 mg L^{-1}) resulted in increased Zn absorption. Agglomerates formed at higher doses (1000–4000 mg L^{-1}), which prevented absorption. In contrast to ZnO NPs, Zn^{2+} was found in plant tissues by X-ray absorption spectroscopy (XAS) analysis and resembled Zn acetate and nitrate. If Zn^{2+} was produced as a result of ZnO, NP transformation on roots was not made clear by the investigation. On the root surface, root exudates may include ionized ZnO NPs (López-Moreno et al. 2010a, b). On the other hand, ZnO NP adsorption and aggregation on the root surface were verified by scanning electron microscopy. TEM pictures with high magnification revealed NPs in vascular cylinder and endodermal cells. There were no X-ray absorption studies offered (Lin and Xing 2008).

Iron oxide nanoparticle uptake: The uptake of magnetite (Fe_3O_4) NP in pumpkin plants was investigated using a vibrating sample magnetometer. NPs were found in the roots, stems, and leaves; however, absorption varied depending on the growing medium. There was no absorption in the soil and just a minor uptake in the sand. There was species-specific uptake detected, with no uptake observed in lima beans. Large NPs may have difficulty penetrating cell walls and transporting them across membranes. Ions and water move via channels and aquaporins with pore diameters smaller than NPs (Zhu et al. 2008; Wang et al. 2011).

CeO_2 nanoparticle uptake: In a recent study, Ce buildup was seen in alfalfa, corn, and tomato seedlings when CeO_2 NP concentration increased. At 4000 mg CeO_2 L^{-1}, the concentration of Ce differed between species. Differential accumulation is ascribed to NP interactions in the rhizosphere and root microstructures. CeO_2 NP was detected in roots by XANES analysis; however, its ability to bind to root hairs and exudates remains uncertain (López-Moreno et al. 2010a, b). CeO_2 nanoparticles (aerosol or suspension) were recently applied to corn leaves and were absorbed, but not translocated to new leaves. No discernible translocation was seen for NPs in irrigation water. Absorption and transfer of nanoceria vary depending on the species (Birbaum et al. 2010).

Nickel hydroxide nanoparticle uptake: In a study on the uptake and translocation of $Ni(OH)_2$ NPs (8.7 nm) in mesquite (Prosopis sp.), it was discovered that X-ray absorption near edge structure (XANES) spectra provided evidence of the presence of NPs in the roots and shoots of uncoated NP-treated plants. Ni NPs were only visible in roots when coated with citrate. However, neither plant size nor chlorophyll production was negatively impacted (Parsons et al. 2010).

CeO_2-NP interaction and translocation in wheat and pumpkin were investigated in hydroponic environments. With or without the addition of fulvic acid (FA) and gum arabic (GA), CeO_2-NPs (17–100 nm) were exposed. NPs were constant with FA or GA, but plant growth altered pH, aggregation, and size. There was no toxicity or decline in growth. Instead of wheat, CeO_2-NPs are translocated to pumpkin shoots. Root association was impacted by FA and GA but not translocation (pure>FA>GA). TEM and SEM demonstrated that NPs attached to root surfaces (Schwabe et al. 2013). Plant secondary metabolism is impacted by the reactive oxygen species (ROS) induction that NPs cause. Secondary metabolites play a critical role in how a plant operates. It is crucial to comprehend how they react to NPs (Marslin et al. 2017).

7.1.7 Nanoparticles and Normal Flora of the Plant Roots

According to Compant et al. (2019), the plant microbiome is made up of microorganisms on plant surfaces (phyllosphere and rhizoplane), in the soil next to roots (rhizosphere), and inside of plants (endosphere). Plant type, age, health, and habitat are only a few of the variables that affect plant-microbiome interactions' complexity (Gupta et al. 2021). For rhizosphere colonization, plants choose certain microbes (Lugtenberg and Kamilova 2009). Even under the same circumstances, microorganisms in plants can alter (Trivedi et al. 2020). Plant microbiome research can help to enhance sustainable agriculture by lowering pesticide usage and enhancing crop nutritional value and yield (Santos and Olivares 2021).

Engineered nanoparticles (Ag NPs and TiO_2 NPs, for example) can influence plant microbiomes, altering plant development and soil microbial populations (Fig. 7.1). Nano TiO_2 and ZnO, for example, have been discovered to influence soil microbial populations, with ZnO being more harmful than TiO_2 (Moll et al. 2017).

According to Khan et al. (2014), ENMs demonstrate antimicrobial action via a variety of pathways. The rupture of membranes, DNA damage, oxidative stress, and enzyme inhibition are a few of them (Khan et al. 2016). Physical and chemical characteristics affect ENM toxicity, and various species have varying sensitivity levels. Compared to Gram +ve bacteria, certain ENMs are more effective (Ali et al. 2022). As with ZnO NPs, gram-positive bacteria were more negatively impacted by ZnO NPs than gram-negative bacteria, and smaller NPs showed stronger antibacterial effects (Emami-Karvani and Chehrazi 2011). As a result, ENM sensitivity is influenced by bacterial growth rate and polysaccharide synthesis.

The ENM treatment decreased bacteria that fix nitrogen and bacteria that oxidize methane while boosting *Sphingomonadaceae*. TiO_2 altered arbuscular mycorrhiza colonization in wheat while did not influence plant development (Grün et al. 2019). Additionally, alterations in the microbial community may be a sign of contaminated soil with TiO_2 NP. Ag NPs have a considerable influence on the microbial communities in soil (Johansen et al. 2008). After exposure to AgNPs, populations of ammonia oxidizers and *Proteobacteria* significantly dropped, whereas populations of *Acidobacteria*, *Actinobacteria*, and *Bacteroidetes* sharply rose. Additionally, it led to a decline in leucine aminopeptidase activity, soil microbial biomass, and nitrogen fixers.

Using 16S *rRNA* sequencing, PCR, and sensitivity testing, the toxicity of Ag_2S-NPs on soil bacteria was assessed. Soil toxicity thresholds (HCx) for Ag NPs, Ag_2S-NPs, and Ag^+ varied. At HC20 (80% species protection), Ag_2S-NPs are less dangerous than AgNPs and Ag^+ (5.9 mg Ag kg^{-1}, 1.4 mg Ag kg^{-1}, and 1.4 mg Ag kg^{-1}, respectively) (Ali et al. 2020).

Similarly, the microbial populations in soil are also impacted by NPs, particularly in the rhizosphere near plant roots. These microorganisms are essential for the intake and cycling of nutrients. NPs can harm DNA and cells and release harmful heavy metals (Hodge et al. 2001). Silver nanoparticles are poisonous to nitrogen-fixing and ammonifying bacteria, causing denitrification and nitrogen fixation to be disrupted. The effects of NP on soil microorganisms vary according to NP composition, shape, and concentration (Burke et al. 2015).

For example, ZnO did not affect nitrogen fixation in soybean crops, but CeO_2 had a detrimental effect at greater concentrations. Selective toxicity has the potential to modify soil microbial populations, hence influencing plant nutrient intake and soil fertility (Priester et al. 2012). According to Yoon et al. (2014), soybean plants treated with ZnO NPs displayed decreased aboveground and root mass as well as altered soil bacterial populations. Additionally, TiO_2 NPs only become hazardous at concentrations above 250 ppm, whereas Ag NPs exhibit toxic effects on duckweed cells at low doses (1 ppm) (Kim et al. 2011).

7.1.8 Mechanism of Promoting Plant Defense

Plant innate immune defense: Plants use physical barriers such as cuticles, cell walls, and antibacterial compounds to protect themselves from microorganisms. If pathogens pass through these barriers, the plant's immune system prevents them from colonizing. Plants, unlike mammals, lack adaptive immunity and instead rely on innate immunity in each cell via systemic signals and cellular memory (Reimer-Michalski and Conrath 2016).

The zig-zag model depicts two lines of defense. The first line of plant defense recognizes MAMP/PAMP/DAMP via PRRs, resulting in MTI-, PTI-, and DAMP-triggered immunity, known together as pattern-triggered immunity (PTI) (Jones and Dangl 2006; Saijo et al. 2018). The second branch, which starts ETI and sets off complicated signaling cascades to stop pathogen assaults, recognizes microbial effectors through R proteins (Nishad et al. 2020). PAMPs or MAMPs are recognized by the plant membrane receptors (PRRs), which then initiate MTI by causing calcium influx, ROS buildup, MAPK phosphorylation, alterations to the cell wall, and the activation of defense genes (Windram and Denby 2015; De Lorenzo et al. 2018).

Like receptor-like kinases (RLKs) or receptor-like proteins (RLP), PRRs are membrane-bound proteins. Flagellin is recognized by FLS2 (flg22), while bacterial elongation factor-Tu is recognized by EFR (elf18) (Bauer et al. 2001; Zipfel et al. 2006). A PRR that recognizes both bacterial peptidoglycan and fungal chitin is called chitin elicitor receptor kinase 1 (CERK1) (Miya et al. 2007; Willmann et al. 2011).

When MAMP is detected, RBOHD, a NADPH oxidase, causes a ROS burst. BIK1 phosphorylates RBOHD to activate flg22-induced ROS (Li et al. 2014). RBOHD generates membrane-impermeable superoxide (O_2), which is transformed to H_2O_2 by superoxide dismutase (SOD) in the apoplast (Podgórska et al. 2017). H_2O_2 causes an increase in cytosolic Ca^{2+} (Yuan et al. 2017). BIK1 controls the extracellular Ca^{2+} influx that is brought on by MAMP perception. Phospholipase C is activated by Ca^{2+}, causing the release of downstream messengers and further Ca^{2+} release. Calmodulin or Ca^{2+}-dependent protein kinases are sensor molecules that bind to Ca^{2+} and phosphorylate protein kinases, which activate enzymes or cause the production of defense genes. Ion fluxes (H^{+}/K^{+}) and effluxes (Cl, K^{+}, NO_3) are also brought on by Ca^{2+} influx, which changes plasma membrane alkalinization and depolarization (Jeworutzki et al. 2010).

Systemic resistance induction by microbial bioagent elicitors in plants: Effector-triggered immunity (ETI) allows pathogens to circumvent PTI and establish infection (Nguyen et al. 2021). The secondary defense of ETI is carried out via intracellular NBS-LRR receptors. PTI and ETI increase plant tissues' ability to defend themselves against injury by inducing systemic acquired resistance (SAR) (Zhang et al. 2020).

Mycorrhizal association and its role in plant immunity: The interaction of plants with mycorrhizal fungi improves plant immune response to soilborne diseases and pests (Sanchez-Bel et al. 2016). Mycorrhiza-induced resistance (MIR) in plants is induced by arbuscular mycorrhizal fungi (AMF) (Cameron et al. 2013). AMF also aids in the development of induced systemic resistance (ISR) to plant diseases and pests (Jung et al. 2012; Song et al. 2015). AMF attracts root-associated microorganisms and stimulates plant immune responses. Exudates from the roots attract microorganisms, including AMF (Akiyama et al. 2005). Plants recognize MAMPs from AMF via PRRs, eliciting immunological responses and increasing SA accumulation (Zhang and Zhou 2010). AMF, like pathogenic fungi, can reduce plant immunity (De Jonge et al. 2011).

Rhizobial associations and the role of rhizobacteria in plant immunity: Plants see both mycorrhizae and beneficial rhizobial bacteria as potentially harmful until they engage with the plant immune system. Although bacteria inhibit it, the rhizobium-legume mutualism causes a defense response (Zamioudis and Pieterse 2012). According to reports, succinoglycan, an exopolysaccharide that chelates calcium and inhibits defense, maybe the cause of the reduction of plant immunity during mutualism (Aslam et al. 2008). The interaction between host plant flavonoids and NodD, a rhizobial transcription factor that activates the nod gene, is essential for root nodule development. Nodule formation and root hair infection are induced by nodule factors (NFs), which are produced by the nod gene (Kouchi et al. 2010). Rhizobial dissemination and the development of nodules depend on NFs and their receptors. Rhizobia are commonly seen in large numbers in infected root nodules (Oldroyd et al. 2011). NFs also partially decrease MAMP-triggered immunity (Liang et al. 2013),

7.1.9 Future Directions

1. **Development of novel nanoparticle formulations:** The development of novel nanoparticle forms with higher stability, decreased toxicity, and enhanced antifungal capabilities might be the subject of research. To maximize their efficiency, this might entail experimenting with various materials, sizes, forms, and surface alterations.

2. **Mechanistic studies:** To fully comprehend the precise processes by which nanoparticles exert their antifungal action and interact with plant immune responses, more research is required. Studying the effects of nanoparticles on plant defense signaling pathways as well as the cellular and molecular interactions between fungal pathogens and nanoparticles might be involved in this.
3. **Evaluation of nanoparticle efficacy in field conditions:** It is crucial to evaluate how antifungal nanoparticles operate in actual agricultural contexts. Field tests can offer insightful information on the potency, longevity, and environmental impact of nanoparticles as antifungal agents.
4. **Optimization of nanoparticle delivery systems:** It is possible to increase the effectiveness of nanoparticles and lessen side effects by developing efficient and focused delivery mechanisms. This might entail creating novel application techniques for effective nanoparticle distribution to plant tissues or encasing nanoparticles in biocompatible carriers.
5. **Assessment of long-term effects:** To assess the possible ecological and physiological consequences of nanoparticle therapies on plants, beneficial microorganisms, and nontarget creatures, long-term research is required. To ensure the safe and sustainable employment of nanoparticles, it is essential to comprehend their long-term effects.
6. **Integration with other disease management strategies:** More efficient and long-lasting methods for controlling fungal infections in agriculture may result from investigating the synergistic effects of nanoparticles with other disease management strategies, such as conventional fungicides, biological control agents, and plant resistance inducers.
7. **Regulation and safety considerations:** As the use of nanoparticles in agriculture increases, it is crucial to address the legal issues and security issues related to their usage. This entails assessing any dangers to the environment, human health, and nontarget creatures as well as developing policies and standards for their secure application.

By concentrating on these future directions, researchers can improve our comprehension of the use of nanoparticles as antifungal agents and their interactions with plant immunity, which will ultimately result in the creation of creative and long-lasting methods for managing fungal diseases in plants.

7.2 Conclusion

NPs are double-faced sword; they have both beneficial and deleterious effects on plants. However, in a large context, they are more effective than the antifungal currently used in agricultural practices. High doses of NPs in different plants affect seed germination, growth, yield, pigments, etc. In addition, its low concentration promotes the plant's capacity for biotic and abiotic stresses by activating the plant's defense-related machinery. NPs enhance the plant's antioxidant enzymes that make the plant avoid the detrimental effects of reactive oxygen species in different abiotic stresses. Generally, NPs increase the defense-related gene expressions in plants. Plants exposed to NPs strengthen the physical barriers, improving the functions of photosystems and defense-related enzyme activation and enhancing anti-stress compounds and genes for defense. However, in the future, we recommend investigating thoroughly the negative effects of NPs on human health because the accumulation and entry of these NPs have already been reported in plants. Further research is needed to search out the pathways to how they accumulate in the edible parts of the plants and their human-related health hazards.

References

Abdal Dayem A, Hossain MK, Lee SB, Kim K, Saha SK, Yang GM et al (2017) The role of reactive oxygen species (ROS) in the biological activities of metallic nanoparticles. Int J Mol Sci 18(1):120

Abdelrhim AS, Mazrou YS, Nehela Y, Atallah OO, El-Ashmony RM, Dawood MF (2021) Silicon dioxide nanoparticles induce innate immune responses and activate antioxidant machinery in wheat against Rhizoctonia solani. Plants 10(12):2758

Ahmad T, Wani IA, Lone IH, Ganguly A, Manzoor N, Ahmad A et al (2013) Antifungal activity of gold nanoparticles prepared by solvothermal method. Mater Res Bull 48(1):12–20

Akiyama K, Matsuzaki KI, Hayashi H (2005) Plant sesquiterpenes induce hyphal branching in arbuscular mycorrhizal fungi. Nature 435(7043):824–827

Akther T, Hemalatha S (2019) Mycosilver nanoparticles: synthesis, characterization and its efficacy against plant pathogenic fungi. BioNanoScience 9(2):296–301

Al-Dhabi NA, Valan Arasu M (2018) Environmentally-friendly green approach for the production of zinc oxide nanoparticles and their anti-fungal, ovicidal, and larvicidal properties. Nanomaterials (Basel) 8(7):500

Ali A, Chua BL, Chow YH (2019) An insight into the extraction and fractionation technologies of the essential oils and bioactive compounds in Rosmarinus officinalis L.: past, present and future. TrAC Trends Anal Chem 118:338–351

Ali MA, Ahmed T, Wu W, Hossain A, Hafeez R, Islam Masum MM et al (2020) Advancements in plant and microbe-based synthesis of metallic nanoparticles and their antimicrobial activity against plant pathogens. Nanomaterials (Basel) 10(6):1146

Ali G, Khan A, Shahzad A, Alhodaib A, Qasim M, Naz I, Rehman A (2022) Phytogenic-mediated silver nanoparticles using Persicaria hydropiper extracts and its catalytic activity against multidrug resistant bacteria. Arab J Chem 15(9):104053

Al-Surhanee AA, Afzal M, Bouqellah NA, Ouf SA, Muhammad S, Jan M, Kaleem S, Hashem M, Alamri S, Abdel LAAH, Omar MA, Mona HS (2021) The antifungal activity of Ag/CHI NPs against *Rhizoctonia solani* linked with tomato plant health. Plants 10(11):2283

Asghari F, Jahanshiri Z, Imani M, Shams-Ghahfarokhi M, Razzaghi-Abyaneh M (2016) Antifungal nanomaterials: synthesis, properties, and applications. In: Nanobiomaterials in antimicrobial therapy. William Andrew Publishing, Norwich, pp 343–383

Aslam SN, Newman MA, Erbs G, Morrissey KL, Chinchilla D, Boller T et al (2008) Bacterial polysaccharides suppress induced innate immunity by calcium chelation. Curr Biol 18(14):1078–1083

Asli S, Neumann PM (2009) Colloidal suspensions of clay or titanium dioxide nanoparticles can inhibit leaf growth and transpiration via physical effects on root water transport. Plant Cell Environ 32(5):577–584

Badea ML, Iconaru SL, Groza A, Chifiriuc MC, Beuran M, Predoi D (2019) Peppermint essential oil-doped hydroxyapatite nanoparticles with antimicrobial properties. Molecules 24(11):2169

Baig N, Kammakakam I, Falath W (2021) Nanomaterials: a review of synthesis methods, properties, recent progress, and challenges. Mater Adv 2(6):1821–1871

Basavegowda N, Patra JK, Baek KH (2020) Essential oils and mono/bi/tri-metallic nanocomposites as alternative sources of antimicrobial agents to combat multidrug-resistant pathogenic microorganisms: an overview. Molecules 25(5):1058

Bauer Z, Gómez-Gómez L, Boller T, Felix G (2001) Sensitivity of different ecotypes and mutants of Arabidopsis thaliana toward the bacterial elicitor flagellin correlates with the presence of receptor-binding sites. J Biol Chem 276(49):45669–45676

Bava R, Castagna F, Palma E, Marrelli M, Conforti F, Musolino V et al (2023) Essential oils for a sustainable control of honeybee varroosis. Vet Sci 10(5):308

Birbaum K, Brogioli R, Schellenberg M, Martinoia E, Stark WJ, Günther D, Limbach LK (2010) No evidence for cerium dioxide nanoparticle translocation in maize plants. Environ Sci Technol 44(22):8718–8723

Burke DJ, Pietrasiak N, Situ SF, Abenojar EC, Porche M, Kraj P et al (2015) Iron oxide and titanium dioxide nanoparticle effects on plant performance and root associated microbes. Int J Mol Sci 16(10):23630–23650

Cameron DD, Neal AL, van Wees SC, Ton J (2013) Mycorrhiza-induced resistance: more than the sum of its parts? Trends Plant Sci 18(10):539–545

Compant S, Samad A, Faist H, Sessitsch A (2019) A review on the plant microbiome: ecology, functions, and emerging trends in microbial application. J Adv Res 19:29–37

Dashora A, Rathore K, Raj S, Sharma K (2022) Synthesis of silver nanoparticles employing Polyalthia longifolia leaf extract and their in vitro antifungal activity against phytopathogen. Biochem Biophys Rep 31:101320

De Jonge R, Bolton MD, Thomma BP (2011) How filamentous pathogens co-opt plants: the ins and outs of fungal effectors. Curr Opin Plant Biol 14(4):400–406

De Lorenzo G, Ferrari S, Cervone F, Okun E (2018) Extracellular DAMPs in plants and mammals: immunity, tissue damage and repair. Trends Immunol 39(11):937–950

Dreaden EC, Alkilany AM, Huang X, Murphy CJ, El-Sayed MA (2012) The golden age: gold nanoparticles for biomedicine. Chem Soc Rev 41(7):2740–2779

Duhan JS, Kumar R, Kumar N, Kaur P, Nehra K, Duhan S (2017) Nanotechnology: the new perspective in precision agriculture. Biotechnol Rep 15:11–23

Elbasuney S, El-Sayyad GS, Attia MS, Abdelaziz AM (2022) Ferric oxide colloid: towards green nano-fertilizer for tomato plant with enhanced vegetative growth and immune response against fusarium wilt disease. J Inorg Organomet Polym Mater 32(11):4270–4283

Emami-Karvani Z, Chehrazi P (2011) Antibacterial activity of ZnO nanoparticle on gram-positive and gram-negative bacteria. Afr J Microbiol Res 5(12):1368–1373

Espinoza JG, Briceño EX, Chávez ER, Úrbez-Torres JR, Latorre BA (2009) Neofusicoccum spp. associated with stem canker and dieback of blueberry in Chile. Plant Dis 93(11):1187–1194

Faisal S, Khan S, Abdullah, Zafar S, Rizwan M, Ali M, Ullah R, Albadrani GM, Mohamed HRH, Akbar F (2022) Fagoniacretica-mediated synthesis of manganese oxide (MnO2) nanomaterials their characterization and evaluation of their bio-catalytic and enzyme inhibition potential for maintaining flavor and texture in apples. Catalysts 12:558. https://doi.org/10.3390/catal12050558

Ferreira CD, Nunes IL (2019) Oil nanoencapsulation: development, application, and incorporation into the food market. Nanoscale Res Lett 14:1–13

Godugu D, Beedu SR (2018) Biopolymer-mediated synthesis and characterisation of platinum nanocomposite and its anti-fungal activity against A. parasiticus and A. flavus. Micro Nano Lett 13(10):1491–1496

Gohari G, Mohammadi A, Akbari A, Panahirad S, Dadpour MR, Fotopoulos V, Kimura S (2020) Titanium dioxide nanoparticles (TiO2 NPs) promote growth and ameliorate salinity stress effects on essential oil profile and biochemical attributes of Dracocephalum moldavica. Sci Rep 10(1):912

Grün AL, Manz W, Kohl YL, Meier F, Straskraba S, Jost C et al (2019) Impact of silver nanoparticles (AgNP) on soil microbial community depending on functionalization, concentration, exposure time, and soil texture. Environ Sci Eur 31(1):1–22

Gunalan S, Sivaraj R, Rajendran V (2012) Green synthesized ZnO nanoparticles against bacterial and fungal pathogens. Progress Nat Sci Mater Int 22(6):693–700

Gupta R, Anand G, Gaur R, Yadav D (2021) Plant–microbiome interactions for sustainable agriculture: a review. Physiol Mol Biol Plants 27:165–179

Haroon M, Zaidi A, Ahmed B, Rizvi A, Khan MSMS, Musarrat J (2019) Effective inhibition of phytopathogenic microbes by eco-friendly leaf extract mediated silver nanoparticles (AgNPs). Indian J Microbiol 59:273–287

He L, Liu Y, Mustapha A, Lin M (2011) Antifungal activity of zinc oxide nanoparticles against Botrytis cinerea and Penicillium expansum. Microbiol Res 166(3):207–215

Hodge A, Campbell CD, Fitter AH (2001) An arbuscular mycorrhizal fungus accelerates decomposition and acquires nitrogen directly from organic material. Nature 413(6853):297–299

Ismail AW, Sidkey N, Arafa R, Fathy R, El-Batal A (2016) Evaluation of in vitro antifungal activity of silver and selenium nanoparticles against Alternaria solani caused early blight disease on potato. Br Biotechnol J 12(3):1–11

Iturritxa E, Slippers B, Mesanza N, Wingfield MJ (2011) First report of Neofusicoccum parvum causing canker and die-back of eucalyptus in Spain. Aust Plant Dis Notes 6:57–59

Jeworutzki E, Roelfsema MRG, Anschütz U, Krol E, Elzenga JTM, Felix G et al (2010) Early signaling through the Arabidopsis pattern recognition receptors FLS2 and EFR involves Ca2+-associated opening of plasma membrane anion channels. Plant J 62(3):367–378

Jo YK, Kim BH, Jung G (2009) Antifungal activity of silver ions and nanoparticles on phytopathogenic fungi. Plant Dis 93(10):1037–1043

Johansen A, Pedersen AL, Jensen KA, Karlson U, Hansen BM, Scott-Fordsmand JJ, Winding A (2008) Effects of C60 fullerene nanoparticles on soil bacteria and protozoans. Environ Toxicol Chem 27(9):1895–1903

Jones JD, Dangl JL (2006) The plant immune system. Nature 444(7117):323–329

Jugreet BS, Suroowan S, Rengasamy RK, Mahomoodally MF (2020) Chemistry, bioactivities, mode of action and industrial applications of essential oils. Trends Food Sci Technol 101:89–105

Jung SC, Martinez-Medina A, Lopez-Raez JA, Pozo MJ (2012) Mycorrhiza-induced resistance and priming of plant defenses. J Chem Ecol 38:651–664

Kalia A, Abd-Elsalam KA, Kuca K (2020) Zinc-based nanomaterials for diagnosis and management of plant diseases: ecological safety and future prospects. J Fungi (Basel) 6(4):222

Kanhed P, Birla S, Gaikwad S, Gade A, Seabra AB, Rubilar O et al (2014) In vitro antifungal efficacy of copper nanoparticles against selected crop pathogenic fungi. Mater Lett 115:13–17

Khan M, Khan ST, Khan M, Adil SF, Musarrat J, Al-Khedhairy AA et al (2014) Antibacterial properties of silver nanoparticles synthesized using Pulicaria glutinosa plant extract as a green bioreductant. Int J Nanomedicine 9:3551–3565

Khan ST, Musarrat J, Al-Khedhairy AA (2016) Countering drug resistance, infectious diseases, and sepsis using metal and metal oxides nanoparticles: current status. Colloids Surf B Biointerfaces 146:70–83

Kim E, Kim SH, Kim HC, Lee SG, Lee SJ, Jeong SW (2011) Growth inhibition of aquatic plant caused by silver and titanium oxide nanoparticles. Toxicol Environ Heal Sci 3:1–6

Kim SW, Jung JH, Lamsal K, Kim YS, Min JS, Lee YS (2012) Antifungal effects of silver nanoparticles (AgNPs) against various plant pathogenic fungi. Mycobiology 40(1):53–58

Konno M, Iwamoto S, Seiwa K (2011) Specialization of a fungal pathogen on host tree species in a cross-inoculation experiment. J Ecol 99(6):1394–1401

Kouchi H, Imaizumi-Anraku H, Hayashi M, Hakoyama T, Nakagawa T, Umehara Y et al (2010) How many peas in a pod? Legume genes responsible for mutualistic symbioses underground. Plant Cell Physiol 51(9):1381–1397

Krishnaraj C, Ramachandran R, Mohan K, Kalaichelvan PT (2012) Optimization for rapid synthesis of silver nanoparticles and its effect on phytopathogenic fungi. Spectrochim Acta A Mol Biomol Spectrosc 93:95–99

Kulabhusan PK, Tripathi A, Kant K (2022) Gold nanoparticles and plant pathogens: an overview and prospective for biosensing in forestry. Sensors (Basel) 22(3):1259

Kuppusamy P, Yusoff MM, Maniam GP, Govindan N (2016) Biosynthesis of metallic nanoparticles using plant derivatives and their new avenues in pharmacological applications—An updated report. Saudi Pharm J 24(4):473–484

Kurczyńska E, Godel-Jędrychowska K, Sala K, Milewska-Hendel A (2021) Nanoparticles—plant interaction: what we know, where we are? Appl Sci 11(12):5473

Kurepa J, Paunesku T, Vogt S, Arora H, Rabatic BM, Lu J et al (2010) Uptake and distribution of ultrasmall anatase TiO2 Alizarin red S nanoconjugates in Arabidopsis thaliana. Nano Lett 10(7):2296–2302

Lazcano-Ramírez HG, Garza-García JJO, Hernández-Díaz JA, León-Morales JM, Macías-Sandoval AS, García-Morales S (2023) Antifungal activity of selenium nanoparticles obtained by plant-mediated synthesis. Antibiotics (Basel) 12(1):115

Leong D (2022) A hundred tiny hands: slavery, nanotechnology, and the anthropocene in midnight robber. Configurations 30(2):171–201

Li L, Li M, Yu L, Zhou Z, Liang X, Liu Z et al (2014) The FLS2-associated kinase BIK1 directly phosphorylates the NADPH oxidase RbohD to control plant immunity. Cell Host Microbe 15(3):329–338

Liang Y, Cao Y, Tanaka K, Thibivilliers S, Wan J, Choi J et al (2013) Nonlegumes respond to rhizobial Nod factors by suppressing the innate immune response. Science 341(6152):1384–1387

Liang H, Zhang XB, Lv Y, Gong L, Wang R, Zhu X et al (2014) Functional DNA-containing nanomaterials: cellular applications in biosensing, imaging, and targeted therapy. Acc Chem Res 47(6):1891–1901

Liang X, Perez MAMJ, Nwoko KC, Egbers P, Feldmann J, Csetenyi L, Gadd GM (2019) Fungal formation of selenium and tellurium nanoparticles. Appl Microbiol Biotechnol 103:7241–7259

Lin D, Xing B (2008) Root uptake and phytotoxicity of ZnO nanoparticles. Environ Sci Technol 42(15):5580–5585

Linh TM, Mai NC, Hoe PT, Lien LQ, Ban NK, Hien LTT et al (2020) Metal-based nanoparticles enhance drought tolerance in soybean. J Nanomater 2020:4056563

Logeswari P, Silambarasan S, Abraham J (2015) Synthesis of silver nanoparticles using plants extract and analysis of their antimicrobial property. J Saudi Chem Soc 19:311–317

López-Moreno ML, de la Rosa G, Hernández-Viezcas JÁ, Castillo-Michel H, Botez CE, Peralta-Videa JR, Gardea-Torresdey JL (2010a) Evidence of the differential biotransformation and genotoxicity of ZnO and CeO2 nanoparticles on soybean (Glycine max) plants. Environ Sci Technol 44(19):7315–7320

López-Moreno ML, de la Rosa G, Hernández-Viezcas JA, Peralta-Videa JR, Gardea-Torresdey JL (2010b) X-ray absorption spectroscopy (XAS) corroboration of the uptake and storage of CeO2 nanoparticles and assessment of their differential toxicity in four edible plant species. J Agric Food Chem 58(6):3689–3693

Lugtenberg B, Kamilova F (2009) Plant-growth-promoting rhizobacteria. Ann Rev Microbiol 63:541–556

Luo DQ, Guo JH, Wang FJ, Jin ZX, Cheng XL, Zhu JC et al (2009) Anti-fungal efficacy of polybutylcyanoacrylate nanoparticles of allicin and comparison with pure allicin. J Biomater Sci Polym Ed 20(1):21–31

Malcolm GM, Kuldau GA, Gugino BK, Jiménez-Gasco MDM (2013) Hidden host plant associations of soilborne fungal pathogens: an ecological perspective. Phytopathology 103(6):538–544

Mansoor S, Zahoor I, Baba TR, Padder SA, Bhat ZA, Koul AM, Jiang L (2021) Fabrication of silver nanoparticles against fungal pathogens. Front Nanotechnol 3:679358

Marslin G, Sheeba CJ, Franklin G (2017) Nanoparticles alter secondary metabolism in plants via ROS burst. Front Plant Sci 8:832

Mihai AD, Chircov C, Grumezescu AM, Holban AM (2020) Magnetite nanoparticles and essential oils systems for advanced antibacterial therapies. Int J Mol Sci 21(19):7355

Mirjani R, Faramarzi MA, Sharifzadeh M, Setayesh N, Khoshayand MR, Shahverdi AR (2015) Biosynthesis of tellurium nanoparticles by Lactobacillus plantarum and the effect of nanoparticle-enriched probiotics on the lipid profiles of mice. IET Nanobiotechnol 9(5):300–305

Miya A, Albert P, Shinya T, Desaki Y, Ichimura K, Shirasu K et al (2007) CERK1, a LysM receptor kinase, is essential for chitin elicitor signaling in Arabidopsis. Proc Natl Acad Sci U S A 104(49):19613–19618

Mohammadi R, Maali-Amiri R, Mantri NL (2014) Effect of TiO2 nanoparticles on oxidative damage and antioxidant defense systems in chickpea seedlings during cold stress. Russ J Plant Physiol 61:768–775

Mohammadi MHZ, Panahirad S, Navai A, Bahrami MK, Kulak M, Gohari G (2021) Cerium oxide nanoparticles (CeO2-NPs) improve growth parameters and antioxidant defense system in Moldavian Balm (Dracocephalum moldavica L.) under salinity stress. Plant Stress 1:100006

Moll J, Klingenfuss F, Widmer F, Gogos A, Bucheli TD, Hartmann M, van der Heijden MG (2017) Effects of titanium dioxide nanoparticles on soil microbial communities and wheat biomass. Soil Biol Biochem 111:85–93

Morsy MK, Khalaf HH, Sharoba AM, El-Tanahi HH, Cutter CN (2014) Incorporation of essential oils and nanoparticles in pullulan films to control foodborne pathogens on meat and poultry products. J Food Sci 79(4):M675–M684

Mubeen B, Ansar AN, Rasool R, Ullah I, Imam SS, Alshehri S et al (2021) Nanotechnology as a novel approach in combating microbes providing an alternative to antibiotics. Antibiotics (Basel) 10(12):1473

Nandini B, Hariprasad P, Prakash HS (2017) Trichogenic-selenium nanoparticles enhance disease suppressive ability of *Trichoderma* against downy mildew disease caused by *Sclerospora graminicola* in pearl millet. Sci Rep 7:2612

Nejati K, Dadashpour M, Gharibi T, Mellatyar H, Akbarzadeh A (2021) Biomedical applications of functionalized gold nanoparticles: a review. J Clust Sci 33:1–16

Nguyen QM, Iswanto ABB, Son GH, Kim SH (2021) Recent advances in effector-triggered immunity in plants: new pieces in the puzzle create a different paradigm. Int J Mol Sci 22(9):4709

Nishad R, Ahmed T, Rahman VJ, Kareem A (2020) Modulation of plant defense system in response to microbial interactions. Front Microbiol 11:1298

Ogunsona EO, Muthuraj R, Ojogbo E, Valerio O, Mekonnen TH (2020) Engineered nanomaterials for antimicrobial applications: a review. Appl Mater Today 18:100473

Oldroyd GE, Murray JD, Poole PS, Downie JA (2011) The rules of engagement in the legume-rhizobial symbiosis. Annu Rev Genet 45:119–144

Parsons JG, Lopez ML, Gonzalez CM, Peralta-Videa JR, Gardea-Torresdey JL (2010) Toxicity and biotransformation of uncoated and coated nickel hydroxide nanoparticles on mesquite plants. Environ Toxicol Chem 29(5):1146–1154

Paul A, Roychoudhury A (2021) Go green to protect plants: repurposing the antimicrobial activity of biosynthesized silver nanoparticles to combat phytopathogens. Nanotechnol Environ Eng 6(1):10

Phogat N, Ali Khan S, Shankar S, Ansary A, Uddin I (2016) Fate of inorganic nanoparticles in agriculture. Adv Mater Lett 7(1):3–12

Podgoreanu P, Negrea SM, Buia R, Delcaru C, Trusca SB, Lazar V, Chifiriuc MC (2019) Alternative strategies for fighting multidrug resistant bacterial infections. Biointerface Res Appl Chem 9(1):3834–3841

Podgórska A, Burian M, Szal B (2017) Extra-cellular but extra-ordinarily important for cells: apoplastic reactive oxygen species metabolism. Front Plant Sci 8:1353

Prasad R, Kumar V, Prasad KS (2014) Nanotechnology in sustainable agriculture: present concerns and future aspects. Afr J Biotechnol 13(6):705–713

Prasson PS, Chittaranjan B (2012) Green synthesis of gold nanoparticles and silver nanoparticles from leaves and bark of Ficus carica for nanotechnology application. Int J Sci Res Pub 2(5):1–4

Priester JH, Ge Y, Mielke RE, Horst AM, Moritz SC, Espinosa K et al (2012) Soybean susceptibility to manufactured nanomaterials with evidence for food quality and soil fertility interruption. Proc Natl Acad Sci U S A 109(37):E2451–E2456

Raha S, Ahmaruzzaman M (2022) ZnO nanostructured materials and their potential applications: progress, challenges and perspectives. Nanoscale Adv 4(8):1868–1925

Rai M, Paralikar P, Jogee P, Agarkar G, Ingle AP, Derita M, Zacchino S (2017) Synergistic antimicrobial potential of essential oils in combination with nanoparticles: emerging trends and future perspectives. Int J Pharm 519(1–2):67–78

Rajput VD, Minkina T, Suskova S, Mandzhieva S, Tsitsuashvili V, Chapligin V, Fedorenko A (2018) Effects of copper nanoparticles (CuO NPs) on crop plants: a mini review. BioNanoScience 8:36–42

Reimer-Michalski EM, Conrath U (2016) Innate immune memory in plants. In: Seminars in immunology, vol 28(4). Academic, pp 319–327

Rico CM, Majumdar S, Duarte-Gardea M, Peralta-Videa JR, Gardea-Torresdey JL (2011) Interaction of nanoparticles with edible plants and their possible implications in the food chain. J Agric Food Chem 59(8):3485–3498

Roncero MIG, Hera C, Ruiz-Rubio M, Maceira FIG, Madrid MP, Caracuel Z et al (2003) Fusarium as a model for studying virulence in soilborne plant pathogens. Physiol Mol Plant Pathol 62(2):87–98

Saijo Y, Loo EPI, Yasuda S (2018) Pattern recognition receptors and signaling in plant–microbe interactions. Plant J 93(4):592–613

Sanchez-Bel P, Troncho P, Gamir J, Pozo MJ, Camañes G, Cerezo M, Flors V (2016) The nitrogen availability interferes with mycorrhiza-induced resistance against Botrytis cinerea in tomato. Front Microbiol 7:1598

Santos LF, Olivares FL (2021) Plant microbiome structure and benefits for sustainable agriculture. Curr Plant Biol 26:100198

Sanzari I, Leone A, Ambrosone A (2019) Nanotechnology in plant science: to make a long story short. Front Bioeng Biotechnol 7:120

Saxena R, Tomar RS, Kumar M (2016) Exploring nanobiotechnology to mitigate abiotic stress in crop plants. J Pharm Sci Res 8(9):974

Scandorieiro S, De Camargo LC, Lancheros CA, Yamada-Ogatta SF, Nakamura CV, De Oliveira AG et al (2016) Synergistic and additive effect of oregano essential oil and biological silver nanoparticles against multidrug-resistant bacterial strains. Front Microbiol 7:760

Schwab F, Zhai G, Kern M, Turner A, Schnoor JL, Wiesner MR (2016) Barriers, pathways and processes for uptake, translocation and accumulation of nanomaterials in plants—critical review. Nanotoxicology 10(3):257–278

Schwabe F, Schulin R, Limbach LK, Stark W, Bürge D, Nowack B (2013) Influence of two types of organic matter on interaction of CeO2 nanoparticles with plants in hydroponic culture. Chemosphere 91(4):512–520

SeidiDamyeh M, Niakousari M, Golmakani MT, Saharkhiz MJ (2016) Microwave and ohmic heating impact on the in situ hydrodistillation and selective extraction of Satureja macrosiphonia essential oil. J Food Process Preserv 40(4):647–656

Shahbaz M, Akram A, Raja NI, Mukhtar T, Mehak A, Fatima N et al (2023) Antifungal activity of green synthesized selenium nanoparticles and their effect on physiological, biochemical, and antioxidant defense system of mango under mango malformation disease. PLoS One 18(2):e0274679

Sharma KD (2017) Antifungal activity of biogenic platinum nanoparticles: an in vitro study. Int J Curr Microbiol App Sci 6(4):334–340

Sheikhalipour M, Esmaielpour B, Behnamian M, Gohari G, Giglou MT, Vachova P et al (2021) Chitosan–selenium nanoparticle (Cs–Se NP) foliar spray alleviates salt stress in bitter melon. Nanomaterials 11(3):684

Sheikholeslami S, Mousavi SE, Ashtiani HRA, Doust SRH, Rezayat SM (2016) Antibacterial activity of silver nanoparticles and their combination with Zataria multiflora essential oil and methanol extract. Jundishapur J Microbiol 9(10):e36070

Singh R, Cheng S, Singh S (2020) Oxidative stress-mediated genotoxic effect of zinc oxide nanoparticles on deinococcusradiodurans. 3 Biotech 10:66

Song Y, Chen D, Lu K, Sun Z, Zeng R (2015) Enhanced tomato disease resistance primed by arbuscular mycorrhizal fungus. Front Plant Sci 6:786

Su Y, Ashworth V, Kim C, Adeleye AS, Rolshausen P, Roper C et al (2019) Delivery, uptake, fate, and transport of engineered nanoparticles in plants: a critical review and data analysis. Environ Sci Nano 6(8):2311–2331

Tariq M, Mohammad KN, Ahmed B, Siddiqui MA, Lee J (2022) Biological synthesis of silver nanoparticles and prospects in plant disease management. Molecules 27(15):4754

Tarrahi R, Mahjouri S, Khataee A (2021) A review on in vivo and in vitro nanotoxicological studies in plants: a headlight for future targets. Ecotoxicol Environ Saf 208:111697

Theophel K, Schacht VJ, Schlüter M, Schnell S, Stingu CS, Schaumann R, Bunge M (2014) The importance of growth kinetic analysis in determining bacterial susceptibility against antibiotics and silver nanoparticles. Front Microbiol 5:544

Trivedi P, Leach JE, Tringe SG, Sa T, Singh BK (2020) Plant–microbiome interactions: from community assembly to plant health. Nat Rev Microbiol 18(11):607–621

Turgut BA, Bezirganoğlu İ (2022) Foliar application of CaO nanoparticles and salicylic acid on Medicago sativa seedlings enhances tolerance against Fusarium oxysporum. Physiol Mol Plant Pathol 122:101926

Tyagi PK, Mishra R, Khan F, Gupta D, Gola D (2020) Antifungal effects of silver nanoparticles against various plant pathogenic fungi and its safety evaluation on Drosophila melanogaster. Biointerface Res Appl Chem 10:6587–6596

Vahedikia N, Garavand F, Tajeddin B, Cacciotti I, Jafari SM, Omidi T, Zahedi Z (2019) Biodegradable zein film composites reinforced with chitosan nanoparticles and cinnamon essential oil: physical, mechanical, structural and antimicrobial attributes. Colloids Surf B Biointerfaces 177:25–32

Wang H, Kou X, Pei Z, Xiao JQ, Shan X, Xing B (2011) Physiological effects of magnetite (Fe3O4) nanoparticles on perennial ryegrass (Lolium perenne L.) and pumpkin (Cucurbita mixta) plants. Nanotoxicology 5(1):30–42

Wani AH, Shah MA (2012) A unique and profound effect of MgO and ZnO nanoparticles on some plant pathogenic fungi. J Appl Pharm Sci 2(3):40–44

Willmann R, Lajunen HM, Erbs G, Newman MA, Kolb D, Tsuda K et al (2011) Arabidopsis lysin-motif proteins LYM1 LYM3 CERK1 mediate bacterial peptidoglycan sensing and immunity to bacterial infection. Proc Natl Acad Sci U S A 108(49):19824–19829

Windram O, Denby KJ (2015) Modelling signaling networks underlying plant defence. Curr Opin Plant Biol 27:165–171

Wu H, Li X, Liu W, Chen T, Li Y, Zheng W et al (2012) Surface decoration of selenium nanoparticles by mushroom polysaccharides–protein complexes to achieve enhanced cellular uptake and antiproliferative activity. J Mater Chem 22(19):9602–9610

Wu J, Shu Q, Niu Y, Jiao Y, Chen Q (2018) Preparation, characterization, and antibacterial effects of chitosan nanoparticles embedded with essential oils synthesized in an ionic liquid containing system. J Agric Food Chem 66(27):7006–7014

Xing K, Shen X, Zhu X, Ju X, Miao X, Tian J et al (2016) Synthesis and in vitro antifungal efficacy of oleoyl-chitosan nanoparticles against plant pathogenic fungi. Int J Biol Macromol 82:830–836

Yan K, Han G, Ren C, Zhao S, Wu X, Bian T (2018) Fusarium solani infection depressed photosystem performance by inducing foliage wilting in apple seedlings. Front Plant Sci 9:479

Yehia RS, Ahmed OF (2013) In vitro study of the antifungal efficacy of zinc oxide nanoparticles against Fusarium oxysporum and Penicillium expansum. Afr J Microbiol Res 7(19):1917–1923

Yoon SJ, Kwak JI, Lee WM, Holden PA, An YJ (2014) Zinc oxide nanoparticles delay soybean development: a standard soil microcosm study. Ecotoxicol Environ Saf 100:131–137

Yuan P, Jauregui E, Du L, Tanaka K, Poovaiah BW (2017) Calcium signatures and signaling events orchestrate plant–microbe interactions. Curr Opin Plant Biol 38:173–183

Zamioudis C, Pieterse CM (2012) Modulation of host immunity by beneficial microbes. Mol Plant Microbe Interact 25(2):139–150

Zare B, Sepehrizadeh Z, Faramarzi MA, Soltany-Rezaee-Rad M, Rezaie S, Shahverdi AR (2014) Antifungal activity of biogenic tellurium nanoparticles against Candida albicans and its effects on squalene monooxygenase gene expression. Biotechnol Appl Biochem 61(4):395–400

Zhang J, Zhou JM (2010) Plant immunity triggered by microbial molecular signatures. Mol Plant 3(5):783–793

Zhang J, Mazur E, Balla J, Gallei M, Kalousek P, Medveďová Z et al (2020) Strigolactones inhibit auxin feedback on PIN-dependent auxin transport canalization. Nat Commun 11(1):3508

Zhou P, Adeel M, Shakoor N, Guo M, Hao Y, Azeem I et al (2020) Application of nanoparticles alleviates heavy metals stress and promotes plant growth: an overview. Nanomaterials (Basel) 11(1):26

Zhu CQ, Zhao DH, Chen JL, Li YX, Wang LY, Wang L et al (2004) Application of L-cysteine-capped nano-ZnS as a fluorescence probe for the determination of proteins. Anal Bioanal Chem 378:811–815

Zhu H, Han J, Xiao JQ, Jin Y (2008) Uptake, translocation, and accumulation of manufactured iron oxide nanoparticles by pumpkin plants. J Environ Monit 10(6):713–717

Zipfel C, Kunze G, Chinchilla D, Caniard A, Jones JD, Boller T, Felix G (2006) Perception of the bacterial PAMP EF-Tu by the receptor EFR restricts Agrobacterium-mediated transformation. Cell 125(4):749–760

Chapter 8
Fungi-Assisted Biosynthesis of Nanoparticles and Their Effect on Plant Pests and Pathogens

Diana Garibo and Masudulla Khan

Abstract Agriculture deals with crop production to feed human beings. Agricultural issues like food security could be solved with the help of nanotechnology applications. Nanotechnology is a modern technology that could play a key role in fulfilling the growing food demand. Nanoparticles may be helpful in the treatment and monitoring of crop diseases. The use of fungi has many applications in the industry such as in food preservation, control, and pathogen detection. Nanomaterials obtained by other methods are expensive. Myco-based synthesis of nanoparticles is cost-effective. In this chapter, we summarized the role of fungi in nanoparticle synthesis and their use in improving crop yield and reducing plant diseases.

Keywords Fungi · Nanomaterials · Nanotechnology · Pathogens · Pests · Plant diseases

8.1 Introduction

The science of nanotechnology is integrated with multidisciplinary sciences including biology material science, medicine, physics, and chemistry dealing with particles of size from 1 to 100 nm (Uskokovic 2008; Rai et al. 2009; Narayanan and Sakthivel 2010). This science also refers to the skill to design, characterize,

D. Garibo
Centro de Nanociencias y Nanotecnología, Universidad Nacional Autónoma de México, Ensenada, Mexico

Consejo Nacional de Ciencia y Tecnología, Ciudad de México, Mexico

M. Khan (✉)
Botany Section, Women's College, Aligarh Muslim University, Aligarh, UP, India

M. Khan, J.-T. Chen (eds.), *Nanoparticles in Plant Biotic Stress Management*,
https://doi.org/10.1007/978-981-97-0851-2_8

produce, and apply structures at the nanometer scale. Nanotechnology is an important interdisciplinary emerging field of modern science, and it is growing rapidly (Mansoori 2005; Khan et al. 2021a). Nowadays, nanoparticles (NPs) have achieved remarkable attention as novel nanomaterials as they possess a high surface area-to-volume ratio and unique physical and chemical properties. Their unique characteristics as building blocks provide different characteristics due to the high area-to-volume ratio. There are experimental methods of NP synthesis, particularly in the biological fields, with many applications, such as therapeutics and catalysis, sensors, DNA sequencing and pharmaceuticals, microelectronics, biomolecular detection and diagnostics, and nanocomposites (Khan et al. 2021a; Moghaddam et al. 2015).

NPs may be inorganic and/or organic materials. Organic-based NPs consist of carbon elements, while several inorganic NPs include gold, silver, titanium oxide, and zinc oxide. However, the synthesis of NPs, either by physical or chemical methods, requires highly trained personnel and highly expensive equipment. Furthermore, for NP synthesis chemical materials that are toxic to the environment and responsible for different biological hazards are used, thus arises the need to search for developing environmentally friendly procedures to obtain NPs using extra-biological methods and green synthesis. In this chapter, we provide a general overview of the implication of fungi in the biosynthesis of nanoparticles, their uses, and their proposed signaling pathways.

Fungi belonging to the kingdom *Fungi* have been classified as eukaryotic cells. About 1.5 million species of them are known on Earth, of which only about 70,000 species have been identified. More recent research estimates the existence of approximately 5.1 million fungal species (Blackwell 2011). In areas such as nanobiotechnology, fungi are highly used in the synthesis of NPs due to their metal tolerance and accumulation capacity. In addition, its biomass is easy to handle, favoring bacteria and other microorganisms (Sastry et al. 2003).

Taking into account that fungi secrete extracellular enzymes, high production of these biomolecules can be achieved (Castro-Longoria et al. 2012). Besides, several species grow fast, and therefore maintenance and scaling of culture is very simple without the requirement of highly expensive equipment (Castro-Longoria et al. 2011). The synthesis of metal NPs and nanostructures can be through their secondary metabolites via extracellularly or intracellularly ways (Ahmad et al. 2003; Durán et al. 2005).

The study and application of fungal species to be explored with applications in nanotechnology is considered new. Mukherjee and collaborators reported one of the first extracellular syntheses of silver nanoparticles (AgNPs) with the filamentous fungus *Verticillium* sp. (Mukherjee et al. 2001a). In most cases, for the extracellular synthesis of NPs, the biomass generated as fungal byproducts is in contact with the metal ion solution (Ahmad et al. 2002). It is possible to obtain spherical NPs with a size range of 20–50 nm using the fungus *F. oxysporum* as a reducing agent (Durán et al. 2005). Factors such as temperature and synthesis time could be influencing a different morphology and size (Riddin et al. 2006).

8.2 An Eco-friendly and Novel Approach: Biosynthesis of Nanomaterials

As mentioned in the previous text, the possibility of obtaining metallic NPs from different methods such as biological, physical, and chemical has its advantages and disadvantages. In the case of synthesis of metallic NPs with physical and chemical methods, it is carried out through a reduction reaction of a precursor salt with secondary metabolites of the fungi, a stabilizing agent for the nanomaterial can be added or not, chemicals can be used, decomposition in organic solvents, and chemical radiation (Rai et al. 2008; Sharma et al. 2009; Thakkar et al. 2010). However, its disadvantages compared to the previous methods are that they involve the use of toxic compounds for the environment and humans; in addition, they require a large expenditure of energy and high temperatures (Sanghi and Verma 2009). Therefore, in this way, the need arises to use simpler methods that do not cause harm to the environment and the use of nontoxic agents, among others to obtain NP. In recent years, the use of microorganisms as reducing agents (extracellular and intracellular products and byproducts) has been of interest for the formation of NPs due to its ease of handling and its use of chemicals, and its high nanomaterial production rate is high. The biogenic nanoparticles obtained from green synthesis can be applied in bio nanocatalysts, such as antimicrobial agents and biosensors (Khan et al. 2021a; Adil et al. 2015; Chen et al. 2014).

8.3 Fungi as an Eco-efficient Mycosystem

Fungi are a group of non-phototrophic eukaryotic microorganisms that have a rigid cell wall (Bowman and Free 2011). They have very simple nutritional requirements, being chemo-organotrophic organisms (Holan and Volesky 1995). Many of them obtain their food from organic material present in the soil. Some of the fungi grow as parasites and other microorganisms. Most fungi feed by secreting enzymes into their environment extracellularly. Enzymes break food down into small molecules. After that, the cells of the fungus absorb these molecules and it is then digested internally (Madigan and Martinko 2006). The fungal cell wall is a dynamic structure that participates in several important processes; it must provide a cell with mechanical resistance to withstand changes in osmotic pressure. In addition, it must retain its plasticity so that it allows the cell to grow, cell divide, and morphogenesis during the life cycle of the fungus. The chemical composition of the cell wall of the fungus influences its morphological characteristics and biological activity (Bowman and Free 2011; Duran and Nombela 2004).

Glucan and chitin are the main glycoproteins and polysaccharides that contain the cell wall of fungi. In general, the cell wall is chemically composed of 80–90% polysaccharides. Among the polysaccharides, glucan and chitin stand out, in addition to glycoprotein, forming a complex network of the structural base of the cell wall of fungi (Bowman and Free 2011). The cell wall mediates cell adhesion to substrates and serves as a signaling center. Any disruption in the cell wall can affect the morphology and growth of the fungus leading to cell lysis (death) (Bowman and Free 2011). The potential use of fungi has been reported in various technological innovations such as biomineralization, bio corrosion, bioleaching, remediation of toxic metals, and biomining (Klaus-Joerger et al. 2001; Yadav et al. 2015). The bioaccumulation capacity and metal tolerance of fungi make them one of the main microorganisms to be used as reducing agents to obtain metallic NPs. This is due to the great capacity of its cell wall to absorb intracellular metals (Yadav et al. 2015; Hemath et al. 2010; Jain et al. 2011). For example, with the use of the fungus *Verticillium* sp. as a silver ion-reducing agent, AgNPs were synthesized below the surface of fungal cells (Mukherjee et al. 2001b). Their practicality in maintaining, cultivating, and being able to synthesize NPs simply and on a large scale makes them attractive in nanomaterial synthesis (Honary et al. 2013). The mycelium of fungi, compared to plants and other microorganisms, is capable of withstanding inflammation and high flow pressure, among other conditions (Soni and Prakash 2012). The scale-up level of NP production on a large scale is economical due to the small amount of biomass required (Vala et al. 2014). The great performance in the formation of NPs is due to the large number of enzymes that are secreted extracellularly by the fungi and are metal-reducing agents (Alani et al. 2012; Birla et al. 2009).

8.4 Importance of Fungi as Mediator of Nanoparticles Synthesis

A new approach known as myco-nanotechnology arises from combining mycology with nanotechnology (Rai et al. 2009a). Fungi play an essential role in the synthesis of NPs since they act as stabilizers but above all as reducing agents of metal as insoluble complexes (i.e., metal sulfides in the form of colloidal particles) (Mehra and Winge 1991). Compared to bacteria and plants, among other microorganisms, fungi are a potential microorganism for bioreduction. This is derived from the ease of being isolated, its maintenance, cultivation, and scaling without the need for expensive and sophisticated instrumentation. The protein and enzymes that are secreted by the fungi facilitate and enhance the reduction process and therefore the formation of NPs that can be biocompatible. Mainly this bioreduction process is carried out when the mycelium is exposed to a metal ion precursor salt. Furthermore, fungi are capable of hydrolyzing metal ions.

The mycelium has rigid structures that allow it to withstand physicochemical conditions during the bioreduction process (Narayanan and Sakthivel 2010; Sastry et al. 2003). The novel approach in the use of fungi for the formation of different shapes, crystals, and inorganic NPs resides in the biomolecules secreted by fungi in an extracellular way that participate in the catalytic bioreduction process. Enzymes are the most important in the process of reduction and stability of the synthesized nanomaterial. A major mediator of bioreduction development is nitrate-dependent reductase, in conjunction with extracellular processes such as quinine shuttle. Several fungi are capable of reducing the metals silver (Ag+1) and gold (Au+1) to form AgNP and AuNP nanomaterials. NP sizes are 10–20 nm from a chemical reduction of metal sales by fungal secondary metabolites (Mukherjee et al. 2001b).

8.5 Development in Myco-fabrication of Nanoparticles

Nanomaterials can be obtained through intracellular (inside the cell) (Vala et al. 2014) and extracellular (outside the cell) synthesis (Gholami-Shabani et al. 2013). Intracellularly, the fungal biomass is in contact with a metal salt solution and incubated for 25 h in the dark. Regarding the extracellular synthesis, the product of the fungal cultures is exposed to the metallic salt, and the synthesis of the nanomaterial can be observed with a color change (Duran et al. 2005; Duran et al. 2006). The advantage of this synthesis concerning the intracellular one is that the synthesis reaction times are shorter. Obtaining smaller sizes of metallic NPs is done through intracellular methods (Narayanan and Sakthivel 2010).

The nucleation of NPs inside the fungal cell may influence the size of the nanomaterial formed. However, the extraction processing of metal NPs becomes complicated because it takes place inside the cell (Dhillon et al. 2012). Unlike intracellular synthesis, extracellular synthesis does not require any additional processing, being faster, cheaper, and more viable. Thus, due to the aforementioned, extracellular synthesis is one of the most used (Devi and Joshi 2015). Some reports of results obtained from the synthesis of metallic NPs via intracellular and extracellular routes are mentioned below.

There are few reports of metallic NPs by intracellular synthesis. Sastry et al. synthesized AgNPs with a size of 2–25 nm intracellularly in *Verticillium* sp. wherein the metallic particles were observed attached to the surface of the cytoplasmic membrane (Sastry et al. 2003). Similar results for AuNPs synthesized with the same fungus were observed by Mukherjee et al. (Mukherjee et al. 2001a). The intracellular synthesis of AgNPs with the fungus *Phoma* sp. 32,883 was obtained by Chen et al. (Chen et al. 2003). In addition, other fungi have been used for the intracellular synthesis of metallic NPs. Moreover, some other fungi such as *Aspergillus flavus*, *Trichothecium* spp., and *Penicillium chrysogenum* are for gold, *Fusarium oxysporum f.* sp. *lycopersici* for platinum, and *Verticillium luteoalbum* (Yadav et al. 2015).

However, the extracellular synthesis of metallic NPs has been widely explored in different fungi. One of the papers presented showed the green synthesis of AuNPs using *Alternaria* sp. The authors showed the influence of chloraurate concentrations on the size of the NPs. The results by TEM showed heterogeneous morphology with rod, spherical, square, hexagonal, and pentagonal shapes with a 1 mM chloroaurate solution. With gold chloride concentrations of 0.3 and 0.5 mM, spherical and quasi-spherical morphologies of size 7–13 and 15–18 nm, respectively, were observed (Dhanasekar et al. 2015). In addition, there are other studies where the endophytic fungi *Aspergillus tamarii* PFL2, *Aspergillus niger* PFR6, and *Penicillium ochromon* PFR8 were isolated from a plant, to be used for the synthesis of AgNPs. The smallest sizes (3.5 ± 3 nm) of NPs were obtained with the fungus *A. tamarii* pfl2, while those synthesized with the fungi *A. niger* pfr6 and *P. ochlorochloron* pfr8 were reported with an average size of 8, 7 ± 6 and 7.7 ± 4.3 nm, respectively (Devi and Joshi 2015).

Similarly, other fungi such as *Fusarium solani, F. acuminatum, F. oxysporum, F. semitectum, Trichoderma asperellum, Aspergillus flavus, A. terreus, A. niger, A. clavatus, Aspergillus* sp., *Trichoderma viride, Phoma glomerata, Pestalotia* sp., *P. gardenia, Trichophyton mentagrophytes, T. rubrum, Microsporum canis, Helminthosporium tetramera* (Yadav et al. 2015), etc. obtained NPs by extracellular synthesis. Table 8.1 shows the list of different fungi used for the synthesis of various NPs.

Table 8.1 Synthesis of nanoparticles from fungi

Fungi	Nanoparticle	Shape and size	Mode of synthesis	References
Fusarium oxysporum	Pt	10–50 nm; triangle, hexagons, square, rectangles	Extracellular	Riddin et al. (2006)
Fusarium acuminatum	Ag	5–40 nm, spherical	Extracellular	Ingle et al. (2008)
Trichothecium sp.	Au	5–200 nm; triangle, hexagonal	Extracellular	Ahmad et al. (2005)
P. lilacinum	Ag	100 nm; spherical	–	Khan et al. (2022)
F. oxysporum	Zr	3–11 quasispherical	Extracellular	Bansal et al. (2004)
F. oxysporum	Si	5–15 nm; quasispherical	Extracellular	Bansal et al. (2005)
Verticillium sp.	Magnetite	100–400 nm; cubo-octahedral	Extracellular	Bharde et al. (2006)
Fusarium solani	Ag	16–23 nm; spherical	Extracellular	Ingle et al. (2009)
Hormoconis resinae	Ag	20–80 nm; spherical with some nanotriangles	Extracellular	Varshney et al. (2009)
Calocybe indica	Ag	5–50 nm; spherical	Extracellular	Sujatha et al. (2013)
Cladosporium cladosporioides	Ag	10–100 nm; spherical	Extracellular	Balaji et al. (2009)
Fusarium semitectum	Ag	10–60 nm; spherical	Extracellular	Basavaraja et al. (2008)
Aspergillus fumigatus	Ag	5–25 nm; spherical, triangular	Extracellular	Bhainsa and D'Souza (2006)

8.6 Mycosynthesis of Metal NPs: Mechanism

As has been commented throughout the text, among the most feasible microorganisms for the synthesis of NPs are fungi. This is because they tend to expel and/or form substances and secondary metabolites extracellularly as a result of a defense mechanism or stress induced by environmental conditions, temperature, humidity, toxic materials, etc. (Mehra and Winge 1991). Extracellularly, the synthesis of metallic NPs is the product of the mycelium of the fungus which, when exposed to a metallic salt solution, due to its survival effect, generates enzymes and proteins, among other secondary metabolites that serve as reducing agents of the precursor salt. In this sense, the catalytic effect of the enzymes in the medium reduces the metal ions to give rise to the formation of metal NPs (Vahabi et al. 2011).

Various mechanisms have been proposed in the extracellular synthesis of metallic NPs. The first report on the mechanism of extracellular NP synthesis was carried out with the fungus *F. oxysporum* (Ahmad et al. 2003). The authors observed that the presence of the NADH-dependent reductase enzyme was responsible for the reduction of Ag+1 to Ag0 to give rise to the formation and nucleation of AgNPs. The reduction reaction consists of the electron required to replace the deficiency of aqueous silver ions (Ag+1) and convert it into neutral silver (Ag0) that was donated by both quinone and NADPH. Ingle et al. (Ingle et al. 2008) demonstrated a biosynthesis mechanism for AgNPs using the fungus *F. acuminatum*. Another type of mechanism was proposed by Mukherjee et al. (Mukherjee et al. 2008), where it is based on the Michaelis-Menten mechanism consisting of pseudo-zero-order kinetics and then higher-order kinetics. That is, the higher the concentration of silver nitrate, the slower the reaction, and as the nucleation reaction progresses, the content of the silver nitrate precursor salt decreases. The authors attribute this to the proteins in the extract with -SH groups and hypothesize that cysteine is dehydrogenated when reacting with AgNO3 to produce AgNPs. The free proteins behave like the stabilizers of the AgNPs.

Das et al. (Das et al. 2009) was the one who suggested that the bioreduction of metal ions was due to the presence of polypeptides/proteins in the cell extracts. This was reaffirmed by the results shown by the FTIR spectra of the fungal groups in the presence of the precursor salt of chloroauric acid (AuCl4) where typical frequencies were reported for the functional groups of amide I, II, and III without observing carboxyls (-COOH) from the mycelium. Understanding the bonds of phosphates in biosynthesis by the frequencies of peaks shown from 1034 to 1075 cm^{-1}. Thus, they reaffirmed that the proteins present could act as reducing agents and stabilizers of the nanomaterial formed. Also, in the biosynthesis of AgNPs, the presence of amide groups I and II is observed (Sanghi and Verma 2009). Like other reported works, the stability of the nanomaterials is attributed to the presence of proteins.

8.7 Nanoparticles Biosynthesized by Using Fungi and Their Antimicrobial Activities

The compounds produced by fungi have different applications. In the biosynthesis of NPs, approximately 6400 bioactive compounds produced by different species of fungi can participate as reducing and/or stabilizing agents for the nanomaterials. The first biosynthesis of cadmium and selenium NPs (Cd and Se) was with the fungus *Candida albicans* (Guilger-Casagrande and de Lima 2019). Since then, various fungi have been investigated and used to synthesize metallic NPs (Khan et al. 2021b).

In the last two decades, fungi such as *Aspergillus niger*, *Colletotrichum* sp., *Coriolus versicolor*, *Cladosporium cladosporioides*, *Aspergillus fumigatus*, *Fusarium semitectum*, *F. oxysporum*, *Fusarium solani*, *Trichoderma viride*, *Phanerochaete chrysosporium*, *Volvariella volvacea*, *Phoma glomerata*, *Penicillium brevicompactum*, *Trichothecium* sp., *Trichoderma asperellum*, and *Penicillium fellutanum* were investigated for both intracellular and extracellular NP synthesis. The biosynthesis of metal NPs (gold and silver) has been reported by many authors on the cytoplasmic membrane and the surface (Yadav et al. 2015; Chen et al. 2003; Ahmad et al. 2005). The morphology of the particles is from spherical to hexagonal, and the size is less than 10–25 nm. The extracellular synthesis of NPs employing fungi has captured the exclusive attention of researchers because of the easy purification process of the particles as they do not have any cellular attachment and hence are easy to apply in different fields without much effort. The common phyla of fungi that have been successfully employed for the green synthesis of nanoparticles mostly include *Ascomycetes*, *Basidiomycetes*, and *Phycomycetes*. The various species of fungi used to obtain a biocompatible metal NP are given in Table 8.1.

An exceedingly pathogenic nematode known as the root-knot nematode, *M. incognita*, was successfully inhibited by the application of spherical and quasi-spherical AgNPs synthesized from the fungus *P. lilacinum* by Khan et al. (Khan et al. 2022). In response to the application of the AgNPs, the fungus *Aspergillus flavus* growth was reduced by 85%, *M. incognita* egg hatching was reduced by four-fold, and *M. incognita* nematode mortality was increased by ninefold. The nematode egg and juvenile of the root-knot nematode were disturbed by synthesized AgNPs, as shown in Figs. 8.1 and 8.2. We are unaware of any studies that have used *P. lilacinum* to create AgNPs that are harmful to *M. incognita*. There are only a few investigations on *A. flavus* where the effect of AgNPs on its growth was evaluated (Bocate et al. 2019; Baker et al. 2021; Li et al. 2012). Bocate et al. found a 50% reduction in mycelium growth of *A. flavus* at 8 ppm of AgNP treatment (Bocate et al. 2019). In other research work, chemically synthesized AgNPs at 100 ppm showed 39% inhibition of *A. flavus* growth (Abdelghany 2020).

Beauveria bassiana fungus culture was used for the synthesis of silver nanoparticles (AgNPs), and these mycosynthesized nanoparticles were found effective in managing the dengue vector *Aedes aegypti* (Banu and Balasubramanian 2014). The endophytic fungi *Pestalotia* sp., was used for the synthesis of AgNPs. This endophytic fungus was isolated from *Syzygium cumini* leaves, and screening of their antibacterial properties has been done against *Salmonella typhi* and *Staphylococcus aureus* (Raheman et al. 2011).

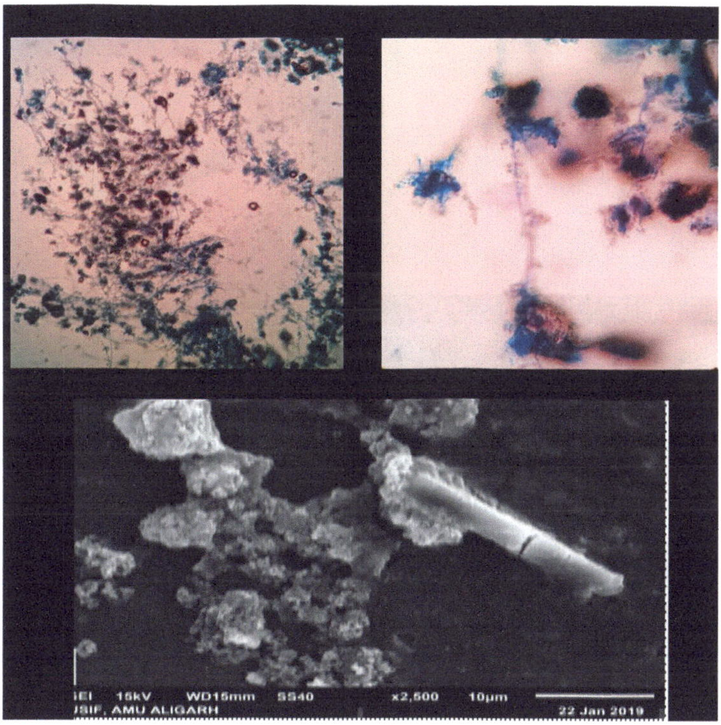

Fig. 8.1 Light microscope images of *A. flavus* fungal mycelium treated with AgNPs; (**c**) SEM image showing disturbed fungal mycelium treated with biosynthesized Ag

Gold nanoparticle (AuNP) formation can occur either in the intracellular or extracellular space (Mukherjee et al. 2001a, b; Yadav et al. 2015). Riddin et al. (Riddin et al. 2006) synthesized triangle, hexagons, square, and rectangle platinum nanoparticles (PtNPs) from *F. oxysporum*. This fungus belongs to the class *Ascomycetes*. The synthesized PtNPs were 10–50 nm in size. Gericke and Pinches (Gericke and Pinches 2006) synthesized spherical gold nanoparticles (AuNPs) from *F. oxysporum* intracellular. This fungus belongs to the class *Ascomycetes*. The synthesized AuNPs were 100 nm in size. Bansal et al. (Bansal et al. 2004) synthesized quasispherical zirconium nanoparticles (ZrNPs) from *F. oxysporum* of size 3–11 nm. Bansal et al. (Bansal et al. 2005) synthesized quasispherical silicon and titanium nanoparticles from *F. oxysporum* of size 5–15 nm. Uddin et al. (Uddin et al. 2008) synthesized quasispherical Bi_2O_3NPs from *F. oxysporum* belonging to the class *Ascomycetes*. The size of synthesized nanoparticles was 5–8 nm. Kumar et al. (Kumar et al. 2007) synthesized spherical CdSe nanoparticles from *F. oxysporum* extracellularly. The size of synthesized nanoparticles was 9–15 nm. Vetchinkina et al. (Vetchinkina et al. 2014) synthesized spherical gold nanoparticles (AuNPs) from *Lentinus edodes* extracellularly. This fungus belongs to the class *Basidiomycetes*. The synthesized AuNPs were in 5–50 nm size. Bharde et al. (Bharde et al. 2006) synthesized magnetite nanoparticles from *F. oxysporum* fungi belonging to the class *Ascomycetes*. The size of synthesized nanoparticles was 20–50 nm and the shape was quasispherical (Fig. 8.3).

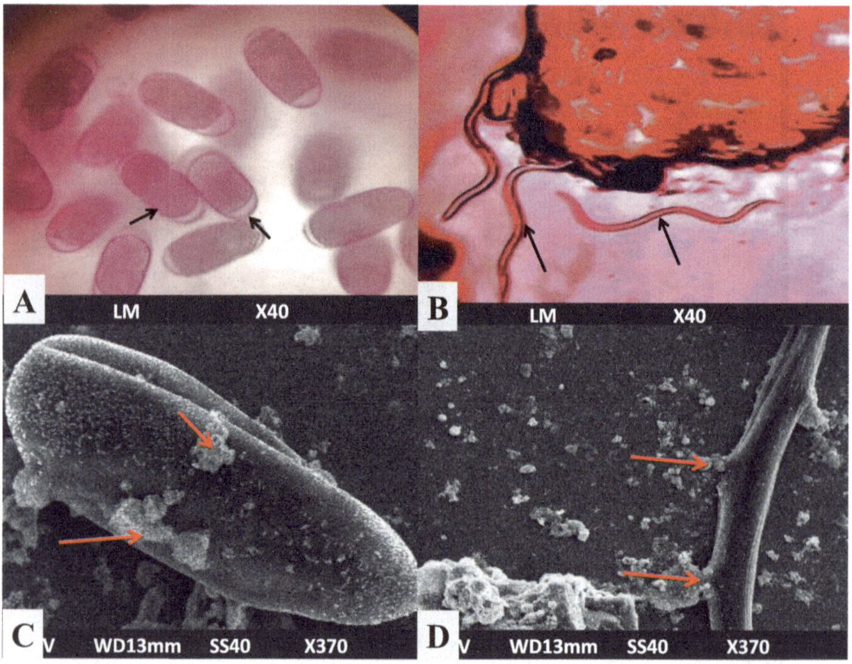

Fig. 8.2 Light microscope images: (**a**) untreated eggs of M. incognita; (**b**) untreated second-stage juveniles of *M. incognita* nematode. SEM images of nematode *M. incognita* treated with AgNPs; (**c**) eggs; (**d**) second-stage juvenile *M. incognita* nematode. Red arrows indicate zone of contacts of the AgNPs with the *M. incognita* eggs and second-stage juvenile nematode

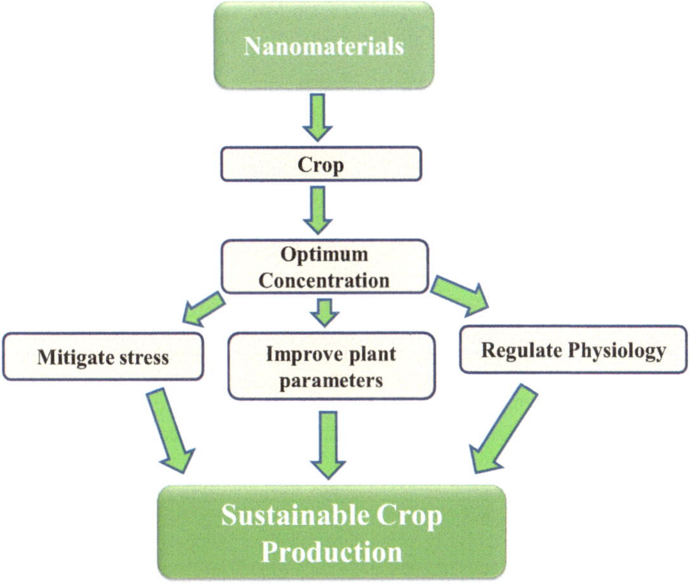

Fig. 8.3 The application of nanomaterials in crop production

8.8 Conclusions

The toxicity of nanomaterials in the environment and humans has intensified the search for new, more eco-friendly, and nontoxic alternatives for the synthesis of NPs. Among the microorganisms used in biosynthesis, fungi tend to be the most attractive due to their large scaling of biomass. They are not expensive, and they imply easy management for their maintenance and cultivation. The synthesis of metallic NPs from an intracellular or extracellular form has been interesting due to the mechanisms they use. However, in this new fashion of green synthesis, it is necessary to improve aspects such as reproducibility, calculating the rate of nanomaterial obtained, the separation of secondary biomolecules from the fungus and NPs formed, the little stability reported, etc. However, their uses in medicine are attractive due to the biocompatibility they may have concerning other nanomaterials synthesized by other methods that can be toxic to humans.

References

Abdelghany TM (2020) GC/MS analysis of Juniperus procera extract and its activity with silver nanoparticles against Aspergillus flavus growth and aflatoxins production. Biotechnol Rep 27:e00496

Adil SF, Assal ME, Khan M (2015) Biogenic synthesis of metallic nanoparticles and prospects toward green chemistry. Dalton Trans 44:9709–9717

Ahmad A, Mukherjee P, Mandal D, Senapati S, Khan MI, Kumar R, Sastry M (2002) Enzyme mediated extracellular synthesis of CdS nanoparticles by the fungus, *Fusarium oxysporum*. J Am Chem Soc 124:12108–12109

Ahmad A, Senapati S, Khan MI, Kumar R, Ramani R, Srinivas V, Sastry M (2003) Intracellular synthesis of gold nanoparticles by a novel alkalotolerant actinomycete, *Rhodococcus* species. Nanotechnology 14:824–828

Ahmad A, Senapati S, Khan MI et al (2005) Extra-/intracellular biosynthesis of gold nanoparticles by an alkalotolerant fungus Trichothecium sp. J Biomed Nanotechnol 1:47–53

Alani F, Moo-Young M, Anderson W (2012) Biosynthesis of silver nanoparticles by a new strain of Streptomyces sp. compared with Aspergillus fumigatus. World J Microbiol Biotechnol 28:1081–1086

Baker A, Iram S, Syed A, Elgorban AM, Al-Falih MA, Bahkali AH, Khan MS, Kim J (2021) Potentially bioactive fungus mediated silver nanoparticles. Nanomaterials 11:3227

Balaji DS, Basavaraja S, Deshpande R, Mahesh DB, Prabhakar BK, Venkataraman A (2009) Extracellular biosynthesis of functionalized silver nanoparticles by strains of *Cladosporium cladosporioides* fungus. Colloids Surf B Biointerfaces 68:88–92

Bansal V, Rautray D, Ahmad A, Sastry M (2004) Biosynthesis of zirconia nanoparticles using the fungus Fusarium oxysporum. J Mater Chem 14:3303–3305

Bansal V, Rautray D, Bharde A et al (2005) Fungus mediated biosynthesis of silica and titania particles. J Mater Chem 15:2583–2589

Banu NA, Balasubramanian C (2014) Myco-synthesis of silver nanoparticles using Beauveria bassiana against dengue vector, Aedes aegypti (Diptera:Culicidae). Parasitol Res 113:2869–2877

Basavaraja S, Balaji SD, Lagashetty A, Rajasab AH, Venkataraman A (2008) Extracellular biosynthesis of silver nanoparticles using the fungus *Fusarium semitectum*. Mater Res Bull 43:1164–1170

Bhainsa KC, D'Souza SF (2006) Extracellular biosynthesis of silver nanoparticles using the fungus *Aspergillus fumigatus*. Colloids Surf B Biointerfaces 47:160–164

Bharde A, Rautaray D, Bansal V, Ahmad A, Sarkar I, Yusuf SM, Sanyal M, Sastry M (2006) Extracellular biosynthesis of magnetite using fungi. Small 1:135–141

Birla SS, Tiwari VV, Gade AK et al (2009) Fabrication of silver nanoparticles by Phoma glomerata and its combined effect against Escherichia coli Pseudomonas aeruginosa and Staphylococcus aureus. Lett Appl Microbiol 48:173–179

Blackwell M (2011) The fungi: 1, 2, 3 ... 5.1 million species? Am J Bot 98:426–438

Bocate KP, Reis GF, de Souza PC, Junior AGO, Durán N, Nakazato G, Furlaneto MC, de Almeida RS, do Panagio, L.A. (2019) Antifungal activity of silver nanoparticles and simvastatin against toxigenic species of Aspergillus. Int J Food Microbiol 291:79–86

Bowman SM, Free JF (2011) The structure and synthesis of the fungal cell-wall. Bioessays 28:799–808

Castro-Longoria E, Vilchis-Nestor AR, Avalos-Borja M (2011) Biosynthesis of silver, gold and bimetallic nanoparticles using the filamentous fungus *Neurospora crassa*. Colloids Surf B Biointerfaces 83:42–48

Castro-Longoria E, Moreno-Velásquez SD, Vilchis-Nestor AR, Arenas-Berumen E, Avalos-Borja M (2012) Production of platinum nanoparticles and nanoaggregates using *Neurospora crassa*. J Microbiol Biotechnol 22:1000–1004

Chen JC, Lin ZH, Ma XX (2003) Evidence of the production of silver nanoparticles via pretreatment of Phoma sp. 32883 with silver nitrate. Lett Appl Microbiol 37:105–108

Chen G, Yi B, Zeng G et al (2014) Facile green extracellular biosynthesis of CdS quantum dots by white rot fungus Phanerochaete chrysosporium. Colloids Surf B Biointerfaces 117:199–205

Das SK, Das AR, Guha AK (2009) Gold nanoparticles: microbial synthesis and application in water hygiene management. Langmuir 25:8192–8819

Devi LS, Joshi SR (2015) Ultrastructures of silver nanoparticles biosynthesized using endophytic fungi. J Microsc Ultrastruct 3:29–37

Dhanasekar NN, Rahul G, Narayanan KB et al (2015) Green chemistry approach for the synthesis of gold nanoparticles using the fungus Alternaria sp. J Microbiol Biotechnol 25:1129. https://doi.org/10.4014/jmb.1410.10036
Dhillon GS, Brar SK, Kaur S, Verma M (2012) Green approach for nanoparticles biosynthesis by fungi: current trends and applications. Crit Rev Biotechnol 32:49–73
Duran A, Nombela C (2004) Fungal cell-wall biogenesis: building a dynamic interface with the environment. Microbiology 150:3099–3103
Durán N, Marcato PD, Alves OL, de Souza GIH, Esposito E (2005) Mechanistic aspects of biosynthesis of silver nanoparticles by several *Fusarium oxysporum* strains. J Nanobiotechnol 3:8. https://doi.org/10.1186/1477-3155-3-8
Duran N, Marcato PD, Alves OL et al (2005) Mechanistic aspects of biosynthesis of silver nanoparticles by several Fusarium oxysporum strains. J Nanobiotechnol 3:1–8
Duran N, Alves OL, Esposito E et al (2006) Production process of silver nanoparticles stabilized by proteins and antibacterial textile process and of effluent treatment of the effluents produced. Patent PIBr 0605681-4, Brazil
Gericke M, Pinches A (2006) Microbial production of gold nanoparticles. Gold Bull 39:22–28
Gholami-Shabani MH, Akbarzadeh A, Mortazavi M, Emzadeh MK (2013) Evaluation of the antibacterial properties of silver nanoparticles synthesized with Fusarium oxysporum and Escherichia coli. Int J Life Sci Biotechnol Pharma Res 2:333–348
Guilger-Casagrande M, de Lima R (2019) Synthesis of silver nanoparticles mediated by fungi: a review. Front Bioeng Biotechnol 7:287
Hemath NKS, Kumar G, Karthik LB, Rao HKV (2010) Extracellular biosynthesis of silver nanoparticles using the filamentous fungus Penicillium sp. Arch Appl Sci Res 2:161–167
Holan ZR, Volesky B (1995) Accumulation of cadmium, lead and nickel by fungal and wood biosorbents. Appl Biochem Biotechnol 53(2):133–146
Honary S, Barabadi H, Gharaei-Fatahabad E, Naghibi F (2013) Green synthesis of silver nanoparticles induced by the fungus Penicillium citrinum. Trop J Pharma Res 12:7–11
Ingle A, Gade A, Pierrat S et al (2008) Mycosynthesis of silver nanoparticles using the fungus Fusarium acuminatum and its activity against some human pathogenic bacteria. Curr Nanosci 4:141–144
Ingle A, Rai M, Gade A, Bawaskar M (2009) *Fusarium solani*: a novel biological agent for the extracellular synthesis of silver nanoparticles. J Nanopart Res 11:2079–2085
Jain N, Bhargava A, Majumdar S et al (2011) Extracellular biosynthesis and characterization of silver nanoparticles using Aspergillus flavus NJP08: a mechanism perspective. Nanoscale 3:635–641
Khan M, Khan AU, Hasan MA, Yadav KK et al (2021a) Agro-nanotechnology as an emerging field: a novel sustainable approach for improving plant growth by reducing biotic stress. Appl Sci 11:2282
Khan M, Khan AU, Bogdanchikova N, Garibo D (2021b) Antibacterial and antifungal studies of biosynthesized silver nanoparticles against plant parasitic nematode Meloidogyne incognita, plant pathogens *Ralstonia solanacearum* and *Fusarium oxysporum*. Molecules 26:2462
Khan M, Khan AU, Rafatullah M, Alam M, Bogdanchikova N, Garibo D (2022) Search for effective approaches to fight microorganisms causing high losses in agriculture: application of *P. lilacinum* metabolites and mycosynthesised silver nanoparticles. Biomolecules 12:174
Klaus-Joerger T, Joerge R, Olsson E, Granqvist CG (2001) Bacteria as workers in the living factory: metal accumulating bacteria and their potential for material sciences a review. Trends Biotechnol 19:15–20
Kumar SA, Ansary AA, Ahmad A, Khan MI (2007) Extracellular biosynthesis of CdSe quantum dots by the fungus, *Fusarium oxysporum*. J Biomed Nanotechnol 3:190–194
Li G, He D, Qian Y, Guan B, Gao S, Cui Y, Yokoyama K, Wang L (2012) Fungus-mediated green synthesis of silver nanoparticles using Aspergillus terreus. Int J Mol Sci 13:466–476
Madigan MT, Martinko JM (2006) Brock biology of microorganisms, vol 1. Pearson Prentice Hall, Hoboken

Mansoori GA (2005) Molecular-based study of condensed matter in small systems: principles of nanotechnology. World Scientific Publishing, Hackensack
Mehra RK, Winge DR (1991) Metal ion resistance in fungi: molecular mechanisms and their regulated expression. J Cell Biochem 45:30–40
Moghaddam AB, Namvar F, Moniri M, Tahir P, Azizi S, Mohamad R (2015) Nanoparticles biosynthesized by fungi and yeast: a review of their preparation, properties, and medical applications. Molecules 20:16540–16565. https://doi.org/10.3390/molecules200916540
Mukherjee P, Ahmad A, Mandal D, Senapati S, Sainkar SR, Khan MI, Parishcha R, Ajaykumar PV, Alam M, Kumar R et al (2001a) Fungus-mediated synthesis of silver nanoparticles and their immobilization in the mycelial matrix: a novel biological approach to nanoparticle synthesis. Nano Lett 1:515–519
Mukherjee P, Ahmad A, Mandal D et al (2001b) Bioreduction of AuCl4− ions by the fungus Verticillium sp. and surface trapping of the gold nanoparticles formed. Angew Chem Int Ed 40:3585–3588
Mukherjee P, Roy M, Mandal BP et al (2008) Green synthesis of highly stabilized nanocrystalline silver particles by a nonpathogenic and agriculturally important fungus T. asperellum. Nanotechnology 19:103–110
Narayanan KB, Sakthivel N (2010) Biological synthesis of metal nanoparticles by microbes. Adv Colloid Interface Sci 156:1–13
Raheman F, Deshmukh S, Ingle A, Gade A, Rai M (2011) Silver nanoparticles: novel antimicrobial agent synthesized from an endophytic fungus Pestalotia sp. isolated from leaves of Syzygium cumini (L). Nano Biomed Eng 3:174–178
Rai M, Yadav A, Gade A (2008) Current trends in phytosynthesis of metal nanoparticles. Crit Rev Biotechnol 28:277–284
Rai M, Yadav A, Rai Gade A (2009) Silver nanoparticles: as a new generation of antimicrobials. Biotechnol Adv 27:76–83
Rai M, Yadav A, Bridge P, Gade A (2009a) Myconanotechnology: a new and emerging science. In: Rai MK, Bridge PD (eds) Applied mycology. CAB International, New York, pp 258–267
Riddin TL, Gericke M, Whiteley CG (2006) Analysis of the inter- and extracellular formation of platinum nanoparticles by *Fusarium oxysporum* f. sp. *lycopersici* using response surface methodology. Nanotechnology 17:3482–3489
Sanghi R, Verma P (2009) Biomimetic synthesis and characterization of protein capped silver nanoparticles. Bioresour Technol 100:502–504
Sastry M, Ahmad A, Islam Khan M, Kumar R (2003) Biosynthesis of metal nanoparticles using fungi and actinomycete. Curr Sci 85:162–170
Sharma VK, Yngard RA, Lin Y (2009) Silver nanoparticles: green synthesis and their antimicrobial activities. Adv Colloid Interface Sci 145:83–96
Soni N, Prakash S (2012) Synthesis of gold nanoparticles by the fungus Aspergillus niger and its efficacy against mosquito larvae. Rep Parasitol 2:1–7
Sujatha S, Tamilselvi S, Subha K, Panneerselvam A (2013) Studies on biosynthesis of silver nanoparticles using mushroom and its antibacterial activities. Int J Curr Microbiol Appl Sci 2:605–614
Thakkar KN, Mhatre SS, Parikh RY (2010) Biological synthesis of metallic nanoparticles. Nanomedicine 6:257–262
Uddin I, Adyanthaya S, Syed A, Selvaraj K, Ahmad A, Poddar P (2008) Structure and microbial synthesis of sub-10 nm Bi_2O_3 nanocrystals. J Nanosci Nanotechnol 8:3909–3913
Uskokovic V (2008) Nanomaterials and nanotechnologies: approaching the crest of this big wave. Curr Nanosci 4:119–129
Vahabi K, Mansoori GA, Karimi S (2011) Biosynthesis of silver nanoparticles by the fungus Trichoderma reesei. Insciences J 1:65–79
Vala AK, Shah S, Patel R (2014) Biogenesis of silver nanoparticles by marine derived fungus Aspergillus flavus from Bhavnagar coast, gulf of Khambat, India. J Mar Biol Oceanogr 3(1):1–3

Varshney R, Mishra AN, Bhadauria S, Gaur MS (2009) A novel microbial route to synthesize silver nano particles using fungus *Hormoconis resinae*. Dig J Nanomater Biostruct 4:349–355
Vetchinkina EP, Loshchinina EA, Burov AM, Dykman LA, Nikitina VE (2014) Enzymatic formation of gold nanoparticles by submerged culture of the basidiomycete Lentinus edodes. J Biotechnol 182–183:37–45
Yadav A, Kon K, Kratosova G, Duran N, Ingle AP, Rai M (2015) Fungi as an efficient mycosystem for the synthesis of metal nanoparticles: progress and key aspects of research. Biotechnol Lett 37:2099–2120

Chapter 9
What It Takes to Control Plant Pathogenic Fungi Using Chitosan and Chitosan-Based Nanoparticles in the Twenty-First Century

Abdulaziz Bashir Kutawa, Syazwan Afif Mohd Zobir, and Khairulmazmi Ahmad

Abstract The use of nanoparticles is a safe approach that offers a novel control method against different fungal pathogens affecting plants. The negative impacts of chemical fungicides worldwide have compelled the utilization of new strategies to manage plant diseases. The use of nanoparticles in plant protection offers target delivery, strong efficacy, and unique mechanisms of action with low toxic impacts. Chitosan has attracted attention as a result of its desirable qualities and antimicrobial activities. Chitosan nanoparticles alone or when coupled with other compounds (active ingredients) provided a great effect against plant pathogenic fungi. It gives two benefits (disease control and improving the defense mechanisms in the host). Because of the need for natural antimicrobial agents that could minimize the negative effects of chemical pesticides. This work aimed to explore the unique properties of chitosan and the antifungal potentialities of chitosan-based and chitosan nanoparticles alone against several fungal phytopathogens. In addition, the mechanisms of

---

A. B. Kutawa
Department of Plant Protection, Faculty of Agriculture, Universiti Putra Malaysia, Serdang, Malaysia

Department of Plant Science and Biotechnology, Faculty of Life Science, Federal University Dutsin-Ma, Dutsin-Ma, Nigeria

S. A. M. Zobir
Department of Plant Protection, Faculty of Agriculture, Universiti Putra Malaysia, Serdang, Malaysia

K. Ahmad (✉)
Department of Plant Protection, Faculty of Agriculture, Universiti Putra Malaysia, Serdang, Malaysia

Institute Tropical Agriculture and Food Security (ITAFoS), Universiti Putra Malaysia, Serdang, Malaysia
e-mail: khairulmazmi@upm.edu.my

M. Khan, J.-T. Chen (eds.), *Nanoparticles in Plant Biotic Stress Management*,
https://doi.org/10.1007/978-981-97-0851-2_9

action against different fungal pathogens and biosafety are also extensively discussed. The use of chitosan and chitosan-based nanotechnological approaches has been promising for the management of fungal pathogens affecting varieties of plant species; this method is considered safe for the environment and the host plants.

Keywords Biosafety · Disease control · Chitosan · Fungi · Mode of action · Nanotechnology

9.1 Introduction

To lessen reliance on synthetic pesticides, there is a significant interest in creating an alternative plant disease control method. Without a doubt, the most valuable organism for plant growth destruction is the pathogenic fungi of plants. Among the many approaches, inventions supported by nanotechnology have produced desired data mostly by using nanofungicides (Cho et al. 2010; El-Mohamedya et al. 2019). Chitosan is a biopolymer derived from the alkaline deacetylating of chitins (Fig. 9.1). The majority of the protective cuticles of crustaceans including shrimp, crabs, prawns, and lobsters are derivative of chitin, a homopolymer made of β-(1,4) linked N acetylglucosamine. It is extensively used in the domains of agriculture, medicine, pharmaceuticals, and the environment (Abd El-Aziz et al. 2018; Mahmoud et al. 2018). Due to its excellent antifungal effect, broad spectrum in use, and low toxicity for human cells, chitosan offers a number of benefits over other forms of molecules (Youssef et al. 2019).

The need for a safe alternative to manufactured pesticides spurs research into natural antifungal substances like chitosan. The most essential substance in agricultural nanotechnology is chitosan which is non-toxic, antifungal, and biodegradable. According to studies (Meng et al. 2010), chitosan has an antifungal property in nature due to its cationic amino groups which form the biological constituents and its functions. However, its insolubility in water and reduced fungicidal efficacy are the reasons why chitosan molecule has not been mainly used as an antifungal substance (Youssef et al. 2018).

Chitosan's physicochemical properties are being modified to increase its antifungal effectiveness (Meng et al. 2010). O-hydroxy phenylaldehyde thiosemicarbazone chitosan and triethylene diamine dithiocarbamate chitosan are two chemically modified forms of chitosan that have greater antifungal properties than chitosan (El-Mohamedya et al. 2019; Meng et al. 2010). Due to their high permeability to the biological membrane, biodegradability, cost efficiency, wide antifungal potentialities, and non-toxicity to humans, chitosan nanoparticles are preferred for usage in a variety of industries. Due to modified physicochemical qualities like surface area, size, active functional groups, cationic nature, and increased encapsulation potentialities, chitosan nanoparticles have a wide range of biological functions, etc. alone or in combination with other elements (Saharan et al. 2013). Few publications are known on the utilization of chitosan nanoparticles in plant infection control,

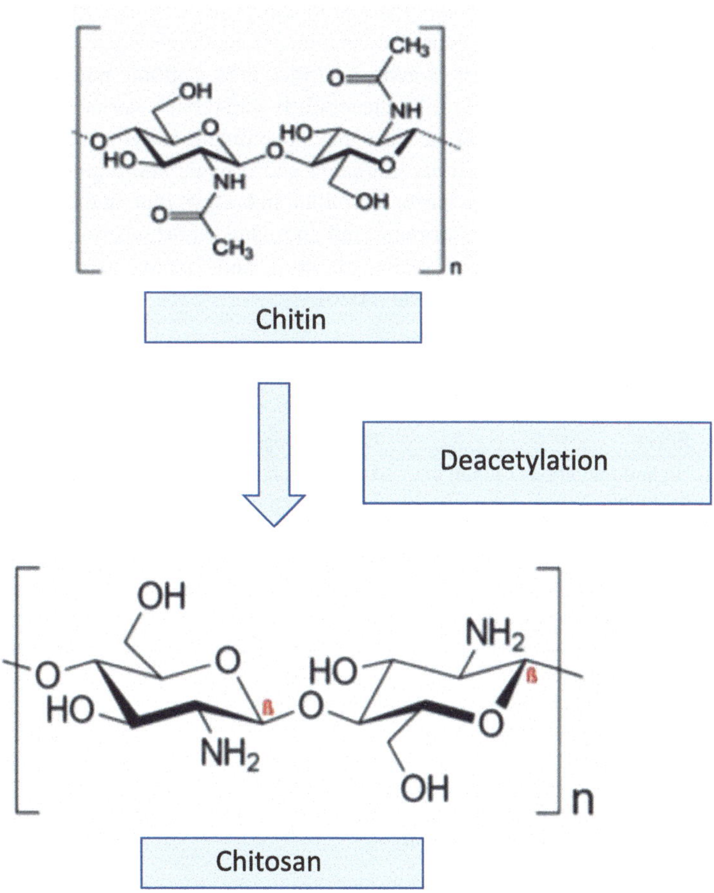

Fig. 9.1 Derivation of chitosan from chitin using the deacetylation method at a temperature of 80 °C

specifically against the fungi despite their potential uses in the agricultural sector. This chapter aimed to focus on the desired properties of chitosan and the antifungal properties of chitosan-based and chitosan nanofungicides against fungal plant pathogens and, more so, to study its overall modes of action for different plant pathogenic fungi and biosafety.

9.2 Definition and Properties of Chitosan

A few researchers have verified that chitosan's capacity to stop pathogen development, which affects spore germination, viability, and sporulation, is what causes it to work against the pathogenic fungi. By the stimulation and inhibition of numerous biochemical processes during the phytopathogenic interactions, chitosan's effect may disrupt pathogen cells or serve as an inducer of defensive responses in the plants (Hassan et al. 2022). The increased surface area, cationic nature, and small size of chitosan are what enable it to successfully carry out a variety of biological activities (Table 9.1). In addition, because they have functional groups in their structures, they easily react with various inorganic and organic compounds (Choudhar et al. 2019). In general, chitosan has been studied in both its pure and functionalized nanomaterial forms with other organic and inorganic substances because it is a highly effective antibacterial, plant elicitor, and plant growth regulator (El-Mohamedya et al. 2019; Devi et al. 2020).

Table 9.1 The fungicidal effects of chitosan alone against several fungal phytopathogens

Chitosan	Effect	Target pathogen	Sources
Chitosan	Highest inhibition (14 mm) of the fungi	*Aspergillus flavus*, *A. niger*, and *A. fumigates*	Debnath et al. (2022)
Chitosan	Complete inhibition at 5.0 g/L (in vitro)	*Alternaria solani*	Ghule et al. (2021)
Chitosan	Inhibition was highest (10.66 mm) *A. flavus*	*Rhizoctonia solani*, *Alternaria alternata*, and *A. flavus*	Shakeel et al. (2018)
Chitosan	Decrease deoxynivalenol and fumonisin formation in irradiated wheat and maize and the growth rate of the pathogens also decrease	*F. verticillioides* and *F. proliferatum*	Zachetti et al. (2019)
Chitosan	Post-emergence seedlings mortality of 7.67% where the seeds were treated with chitosan (1%)	*Colletotrichum capsici*	Akter et al. (2018)
Chitosan	In vitro antifungal activity (50%) inhibition	*Pyricularia oryzae*	Kutawa et al. (2021)
Chitosan	In vitro antifungal activity (50%) inhibition	*Ganoderma boninense*	Maluin et al. (2019a)
Chitosan	Decrease in disease severity of 42.8% and 16.60% after the application of chitosan 1.0 g/L + *T. harzianum* and chitosan (1 g/L)	*Fusarium oxysporum* f. sp. *Radicis lycopersici*	El-Mohamedy et al. (2017)
Chitosan	100% inhibition at a concentration of 5 g/L	*Fusarium oxysporum*, *F. solani*, *Sclerotium rolfsii*, *Rhizoctonia solani*, *Pythium* sp., *Phytophthora* sp., and *Macrophomina phaseolina*	El-Mohamedy et al. (2013)

9.3 Chitosan Nanofungicides for the Control of Fungal Infections in Plant

9.3.1 Antifungal Effects of Chitosan Nanofungicides Against Some Species of Fungi

Chitosan's usefulness in creating chitosan NPs either on its own or after being combined with other organic and inorganic materials has been recognized. To increase their efficacy and prevent environmental toxicity, the synthesized chitosan-based nanocomposites might enable control, targeted, protected, and systemic release of active components (Devi et al. 2020; Chouhan et al. 2017). A smart chitosan-based nano-agri input may be produced by combining chitosan with other various compounds, which might result in precision farming that is both efficient and affordable (Fig. 9.2). This section describes in detail many NMs made of chitosan that may be able to avert the effect of the plant pathogenic fungi (Table 9.2).

Chitosan nanomaterials were investigated due to their biological potentialities. They were tested against several fungi and were found effective in averting the growth of the fungi (Chookhongkha et al. 2012). Chitosan NMs at a concentration (0.6%) had inhibited the growth of mycelia for *Colletotrichum capsici*, *C. gloeosporioides*, *A. niger*, and *Rhizopus* sp. Chitosan nanoparticles have better potentialities than chitosan alone in terms of mycelial growth reduction. Moreover, chitosan combined with chickpea seeds had a good antifungal effect, and this can be based on two facts: (1) Seeds treated with chitosan yield more lignin and phenolic substances. (2) Chitosan can inhibit the growth of the mycelia (Chookhongkha et al. 2012). The chitosan nanoparticles were examined against phytopathogenic fungi (*Macrophomina phaseolina*, *Rhizoctonia solani*, and *Alternaria alternata*) at different concentrations from 0.001 to 0.1% (in vitro). The inhibitory effect of 87.6% was found against *Macrophomina phaseolina* at concentrations of 0.1%. The growth of *R. solani* fungus was inhibited by all the tested concentrations of chitosan nanoparticles (Saharan et al. 2013). In a different research work, a method of preparing chitosan nanoparticles using an anionic protein obtained from *Penicillium oxalicum* was used. These nanoparticles have greatly inhibited the growth and development of *Alternaria solani*, *Pyricularia grisea*, and *F. oxysporum* (Sathiyabama and

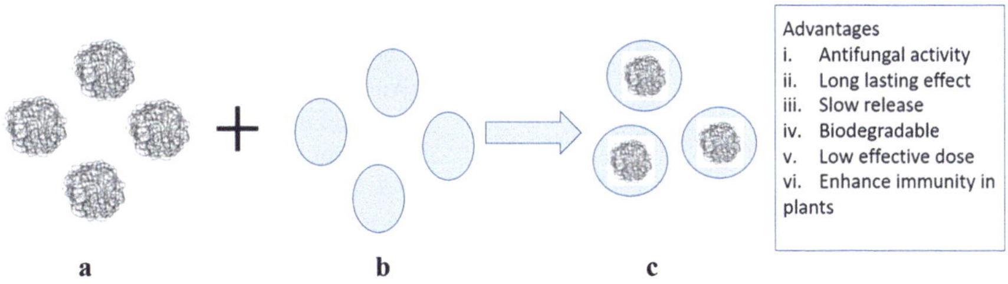

Fig. 9.2 Chitosan reacting with other bioactive compounds to form a nanomaterial with desirable properties: (**a**) chitosan, (**b**) active ingredient, (**c**) formed nanomaterial

Table 9.2 Chitosan coupled with other compounds utilized against several phytopathogenic fungi

Nanomaterials	Particle size (nm)	Polydispersity index (PDI)	Zeta potential (mV)	Effects	Sources
Chitosan dazomet	7	–	–	Antifungal effect against *Pyricularia oryzae*, *Ganoderma boninense* (in vitro)	Maluin et al. (2019a)
Chitosan NPs	89.8	0.22	−37	Antifungal effects against *Alternaria solani*, *Pyricularia grisea*, *Fusarium oxysporum*. It also enhances the germination and growth of chickpea seedling (in vitro)	Sathiyabama and Parthasarathy (2016)
Chitosan NPs	83.3	0.31	−28	In vivo and in vitro fungicidal effects against *Pyricularia oryzae*	Manikandan and Sathiyabama (2016)
Chitosan NPs	180.9	0.31	+45.6	Fungicidal effect against *Fusarium graminearum*	Kheiri et al. (2017)
Cu (II)–chitosan nanogel	220	0.20	+40	Antifungal effect against *Fusarium graminearum* (in vitro)	Brunel et al. (2013)
Cu–chitosan NPs	196.4	0.50	+88	Antifungal effects against *A. alternata*, *Rhizoctonia solani*, and *M. phaseolina* (in vitro)	Saharan et al. (2013)
Cu–chitosan NPs	2.5–25	N/A	N/A	Fungicidal effects of *F. solani*	Devi et al. (2020)
Cu–chitosan NPs	2–3	N/A	N/A	Inhibition of *Sclerotium rolfsii* and *R. solani* (in vitro)	Rubina et al. (2017)
Cu–chitosan NPs	374.3	0.33	+22.6	In vivo and in vitro fungicidal effect on *Curvularia lunata* in corn	Choudhar et al. (2019)
Chitosan–saponin NPs	373.9	1	+31	Antifungal effect of *M. phaseolina*, *A. alternata*, and *R. solani* (in vitro)	Saharan et al. (2013)

Table 9.2 (continued)

Nanomaterials	Particle size (nm)	Polydispersity index (PDI)	Zeta potential (mV)	Effects	Sources
Oleoyl–chitosan NPs	296.9	N/A	N/A	Mycelial and spore germination inhibition of *Verticillium dahliae* (in vitro)	Xing et al. (2017)
Chitosan–Zn NPs	200–300	0.22	+34	In vivo and in vitro antifungal effect of *C. lunata* in corn plants	Choudhar et al. (2019)
Chitosan loaded with salicylic acid NPs	368.7	0.1	+34.1	In vivo and in vitro fungicidal effect of post flowering stalk rot of corn incited by *Fusarium verticillioides*	Kumaraswamy et al. (2019)
Chitosan–silica NPs	110	N/A	N/A	Antifungal effects of *Phomopsis asparagi* (in vitro)	Cao et al. (2016)
Chitosan–*Schinus molle* (CS-EO) essential oil NPs	754	9.1 ± 1.74	N/A	Antifungal effects on the conidia of *Aspergillus parasiticus*	Cao et al. (2016)
Mentha piperita EO in cinnamic acid–chitosan nanogel	N/A	N/A	N/A	Fungicidal effects of *A. flavus* during the storage of tomato fruits	Devi et al. (2020)
Chitosan loaded with *Zataria multiflora* EOs nanoparticles	125–175	N/A	N/A	*Botryticidal* effects of *B. cinerea* in strawberry (in vivo and in vitro)	Mohammadi et al. (2015)
Chitosan boehmite–thyme oil and alumina nanocomposite films	N/A	N/A	N/A	*Monilinia laxa* inhibition during the storage of peach	Cindi et al. (2015)
Chitosan–*Cymbopogon martinii* essential oil	455–480	39.3–37.2	N/A	Antifungal effects of *Fusarium graminearum*, during the storage of corn grains	Kalagatur et al. (2018)
Chitosan–thymol nanoparticles	175 ± 21	37 ± 2.7	0.4 ± 0.1	Antifungal effects on the mycelia of *Botrytis cinerea* in tomato and blueberries during the storage period	Medina et al. (2019)

Parthasarathy 2016). The rate of inhibition for *Fusarium oxysporum ciceri*, *Pyricularia grisea*, and *A. solani* was 87%, 92%, and 72%, respectively (Table 9.2). A seed treatment with chitosan nanoparticles yielded a good morphological effect which included improved vegetative biomass, seed vigor index, and percent germination of chickpea seedlings. The antifungal effect of nanoparticles can be due to their size and highly permeable membrane (Devi et al. 2020; Saharan et al. 2015). Their lower PDI values, higher zeta potential, and small size make the nanoparticles efficient and stable against phytopathogens.

9.4 Chitosan Coupled with Synthetic Fungicidal Active Ingredients for Managing Plant Fungal Diseases

Maluin et al. (2019b) carried out another study in which they created nanofungicides of the three distinct types by entrapping chitosan, dazomet, and hexaconazole: chitosan combined with dazomet, chitosan combined with hexaconazole, and chitosan combined with dazomet and hexaconazole formulations (Maluin et al. 2019b). Furthermore, it was deduced from the results that the antifungal activity will be strong if the particles are tiny (Maluin et al. 2019c). The lowest EC50 for chitosan combined with dazomet and hexaconazole was discovered to be 3.5 ng/mL. The lowest EC50 for chitosan combined with dazomet and chitosan combined with hexaconazole, respectively, was 13.7 ng/mL and 4.6 ng/mL. Additionally, (Maluin et al. 2019a) observed that phenazine nanoformulation has 70.74% suppression of the growth of the fungus *G. boninense* at 1000 μg/mL (Maluin et al. 2019a).

The fungicide was intercalated in zinc–aluminum layered double hydroxide (Al-Zn-LDH) by exchange of ions to create the nanodelivery form of fungicide used by Mustafa et al. (Mustafa and Komatsu 2016). Dazomet and hexaconazole were the fungicides employed, and they were shown to be effective in the control of *G. boninense* (Maluin et al. 2019b). Hexaconazole's controlled release characteristic was achieved in H-Al-Zn-LDH, and the EC50 value was discovered to be 30.0·2.9 ng/mL. H-Al-Zn-LDH and D-Al-Zn-LDH have shown a less phytotoxic impact than traditional fungicides (Maluin et al. 2019c).

9.5 Chitosan Coupled with Natural Products for Managing Plant Fungal Diseases

9.5.1 Fungicidal Effects of Zataria Multiflora Essential Oil in Chitosan

One of the EOs that appears to be a possible natural product for managing fruit post-harvest loss is *Zataria multiflora* essential oils (ZEO). According to Sajed et al. (Sajed et al. 2013), oxygenated monoterpenes make up the majority of the hydro-distilled ZEO's constituents (about 70%), followed by monoterpenes, sesquiterpene, and oxygenated sesquiterpene. In order to preserve fruit quality and reduce fungal rot, EOs utilized volatile chemicals; however, these compounds are rapidly affected by high pressure, temperatures, oxygen, and light. Additionally, they are insoluble in H_2O, necessitating a regulated release for several applications (Devi et al. 2020; Martin et al. 2010). So, to get the most out of employing EOs as antimicrobial compounds, continuous and regulated release is essential.

These compounds' nanoencapsulation techniques may offer a workable and effective solution to some of these issues, such as physical instability. Using chitosan nanoparticles (CSNPs), Mohammadi et al. (Mohammadi et al. 2015) examined the encapsulation of ZEO to increase the stability and antifungal effect against the fungus (*Botrytis cinerea*) responsible for strawberry gray mold infection. ZEO was enclosed in CSNPs using the ionic gelation process, and the transmission electron microscope analysis revealed that the average size was between 125 and 175 nm (Devi et al. 2020). Works on the release of ZEO in vitro showed that this release was sustained and regulated for a period of 40 days. When ZEO was enclosed by CSNPs, it had greater anti-*Botrytis cinerea* activity compared to ZEO (in vitro and in vivo). After 7 days of storage at 4 °C, followed by another period of storage at 20 °C for 2–3 days, strawberries infected with *Botrytis* had dramatically reduced in terms of disease incidence and severity at 1500 ppm. These results demonstrated the use of CSNPs as an EO-controlled release method to increase the antifungal effect (Devi et al. 2020).

9.5.2 Antifungal Activity of Chitosan–Thymol Nanoparticles

The main antibacterial component of the thyme plant is thymol (2 isopropyl-5-methylphenol). Due to its capacity to bind to bacterial proteins and produce cell membrane permeability and disintegration, it has a potent antibacterial effect (Ahmad et al. 2011). Thymol interferes with energy-producing mechanisms, which prevents the cell from recovering (Ahmad et al. 2011). To increase the life span of the food, it may be used as a natural fungicidal material during the packaging of the active (Mirdehghan and Valerob 2017). By incorporating chitosan–thymol nanofungicides made by the ionic exchange process, Medina et al. (Medina et al. 2019) improved the effectiveness of chitosan–quinoa protein films on the enhancement of the life span of tomato and blueberry cherries. They were able to generate NPs with hydrodynamic diameters (175±21 nm) comparable to TEM diameters (153 42 nm). The zeta potential value was 37 2.7 mV, while the PDI value was 0.4 0.1 mV. In several dilutions applied to the PDA bearing the same amounts of active chemicals, the effects of thymol–chitosan nanoparticles (TCNPs), CNPs, and thymol–chitosan (CT) mix on the mycelial growth inhibition were assessed. While CT mix only demonstrated complete inhibition at a higher dose (50% v/v), CTNP formulations showed inhibition of 100% for all the dilutions (10, 25, and 50%) (Medina et al. 2019). At higher concentrations (50% v/v), CNP demonstrated the lowest suppression of mycelial development (74%). The treatment that showed inhibition at the least concentration (10%) was CTNPs.

9.5.3 *Antifungal Activity of Thyme–Tea Tree, Thyme–Pepper Mint, and Chitosan–Thyme–Oregano Essential Oil*

The management of fungal proliferation and contamination in processed food is now being aided by bio-nanocomposite-based packaging that contains EOs obtained from green plants (Hossain et al. 2017). To overcome dose restrictions and enhance the stability of active substances, EOs are more effectively employed in meals when entrapped in suitable delivery methods (Van Long et al. 2016). When EOs are encapsulated at the nanoscale, their bioactivities are improved. They do so via tissue infusion or passive mechanism which allows the EO levels necessary to ensure the antibacterial effect to be reduced (Van Long et al. 2016).

Due to the modest concentrations of the bioactive ingredient used, food flavor, natural scent, and taste remain unchanged (Lu et al. 2016). By encasing EO nanoemulsion, Hossaina et al. (Hossaina et al. 2019) developed cellulose nanocrystals coupled with chitosan (CNCs) antifungal films. For a storage period of 8 weeks at 28 °C, chitosan-based films combined with EO mixtures of oregano, thyme, peppermint, and tea tree reduced the fungal growth (*Aspergillus flavus*, *Aspergillus niger*, *Penicillium chrysogenum*, and *Aspergillus parasiticus*) by 51–77% on the inoculated rice. Thyme–peppermint, thyme–tea tree, and thyme–oregano formulations have z-averages of 69.9 nm, 76.58 nm, and 57.9 nm, respectively, as well as PDI values of 0.25, 0.21, and 0.32 and zeta-potential values of 51, 50, and 53 mV (Hossaina et al. 2019). After 24 h of inoculation, they demonstrated significant inhibition of 83.73 2.55, 75.60 1.27, 87.95, and 6.81% against the fungus *A. niger*. Over the course of 12 weeks of storage, there was a gradual release of volatile chemicals (26%), but the paddy samples wrapped with bioactive films underwent no distinct changes in flavor, odor, and color. The chitosan matrix-incorporated CNCs were crucial in maintaining the release and physicochemical characteristics of the nanocomposite film (Devi et al. 2020; Hossaina et al. 2019).

9.5.4 *Antifungal Effect of Cymbopogon martinii Essential Oils Coupled with Chitosan*

A tropical herb in the Poaceae family, *Cymbopogon martinii*, is often referred to as motia, Indian geranium, or rosha (Kakaraparthi et al. 2015). The chemical components of its EOs include bioactive substances including humulene, geraniol, caryophyllene, linalool, geranyl acetate, limonene, selinenes, etc. (Kalagatur et al. 2018). The antifungal efficacy of *C. martini* essential oils (CMEOs) was examined against the post-harvest *Fusarium graminearum* fungus which is responsible for causing *Fusarium* head blight infection in corn plants (Devi et al. 2020). They discovered that CMEOs had MIC of 421.7±27.14 ppm and minimum fungicidal concentrations of 618.3±79.35 ppm. Craters, vesicles, rough surfaces, and protuberances of macroconidia underwent morphological alteration when exposed to CMEOs as opposed to controls. Fungi died as a result of increased lipid peroxidation and ROS concentration (Kalagatur et al. 2018). The chitosan-entrapped CMEO nanoparticles (CeCMEONPs) had a zeta potential ranging from 37.2 to 39.3 mV and a spherical shape between 455 and 480 nm in size. The bioactive components of CMEOs were effectively conjugated with chitosan and successfully produced CeCMEONPs according to the FTIR analyses (Kalagatur et al. 2018). By gradually releasing the antifungal components of CeCMEONPs, a stable complex structure generated between CMEOs and chitosan improved the lifespan and antifungal properties of CMEOs. In a lab setting, during the course of a 28-day storage period, grains of maize have been utilized as the samples to test the antimycotoxin and antifungal potentialities of CeCMEONPs and CMEOs against *F. graminearum*. The fungal growth was inhibited by CMEOs and CeCMEONPs at 900 and 700 ppm, respectively. CeCMEONPs had superior antimycotoxin and antifungal properties compared to CMEO, which may be explained by the antifungal components of CeCMEONPs' regulated release of antifungal agents over time.

9.5.5 Antifungal Activity of Thyme Oil with Chitosan–Boehmite

Thyme oil (TO) sachets were packaged in PET punnets that were additionally packed with lidding films made of a boehmite–chitosan nanocomposite, according to Cindi et al. (Cindi et al. 2015). They discovered that brown rot severity and incidence were decreased in experimentally infected *Prunus persica* (cv. Kakawa) peach fruits when kept at 25 °C for 5 days. In addition, the incidence of brown rot infection in naturally infected fruits was decreased to 10% when kept at 0.5 °C, 90% RH for a total of 7 days. Linalool (37.6%), caryophyllene (9.47%), and thymol (56.43%) were kept inside the punnet as active substances (Cindi et al. 2015). The taste, appearance, and flavor of natural peach remain the same as such people prefer to buy the fruit packed from commercial punnet which contains thyme oil (sachet) and sealed using boehmite–chitosan film.

9.5.6 Antifungal Activity of Thiadiazole Functionalized Chitosan Derivative

Li et al. (Li et al. 2013) demonstrated that a group of water-soluble derivatives of chitosan, which are 1,3,4-thiadiazole, 2-methyl-1,3,4-thiadiazole, and 2-phenyl-1,3,4-thiadiazole, had strong fungicidal properties against the pathogenic fungi like *Phomopsis asparagi*, *Monilinia fructicola*, and *Colletotrichum lagenarium*. At 1.0 mg/mL, the inhibition was measured to be 31.6% for *C. lagenarium*. When compared to chitosan, the antifungal properties of chitosan derivative showed higher results. With inhibition of 82.5, 65.8, and 75.3% against *M. fructicola*, *P. asparagi*, and *C. lagenarium*, respectively, at 1.00 mg/mL, MTPCTS had the greatest performance among the investigated chitosan derivatives. The hydrophobic moiety and the length of the thiadiazole's alkyl substituent could influence the antifungal effects of the chitosan derivative (Li et al. 2013).

9.5.7 *Chitosan Coupled with Pepper Tree Essential Oils*

The pepper tree which belongs to the family Anacardiaceae also contains EOs with antibacterial qualities (Lopez et al. 2014). The pepper tree contains the chemical components of EOs like FC; - phellandrene, FC; and -pinene, limonene, phellandrene myrcene, and monoterpenes. Its EOs have been shown to be effective against *Fusarium solani* filamentous fungus (Rhouma et al. 2009). The mycelium inhibition of *A. flavus* was observed to be up to 53.5% at 500 ppm of pepper tree EOs. *A. niger*, *A. oryzae*, and *A. japonicus* were all also inhibited significantly by the antifungal agent (Alanis-Garza et al. 2007). Against *A. fumigates*, the EOs had a MIC of >1000 mg/mL (Alanis-Garza et al. 2007).

Chitosan nanomaterials were prepared by Luque-Alcaraz et al. (Luque-Alcaraz et al. 2016) and included pepper tree EOs with a zeta potential of +9.1±1.74 mV and particle size of 754±7.5 nm. They looked at how varied chitosan nanoparticle concentrations in pepper tree EO affected the conidia of *A. parasiticus*. It was discovered that, when compared to controls, all treatments decreased the vitality of the fungal conidia. According to their findings, adding pepper tree EO to chitosan nanocomposites is a substitute that maintains the fungicidal effects of both constituents while reducing the potential for EOs to volatilize and the ensuing loss of their activities (Devi et al. 2020).

9.6 Mechanisms of Action for Chitosan Nanoparticles

As a result, interference that damages the cell wall of the fungus may result in content loss and death. According to their findings, *Pyricularia oryzae* strain Gry hyphae treated with CNP exhibited swollen, cell wall damage, and aberrant structure that resulted in partial loss of contents (Ibrahim et al. 2020a; Pariona et al. 2019; Oussou-Azo et al. 2020). In contrast, *Pyricularia oryzae* strain Gry's cell walls displayed typical structural properties in the untreated control (Fig. 9.3c). Similar to AgNPs, CNPs have been discovered to exhibit antifungal effects on the cell walls of several plant fungal pathogens, including *Fusarium solani*, *F. oxysporum*, *Colletotrichum gloeosporioides*, and *Fusarium graminearum* (Oussou-Azo et al. 2020; Kriti et al. 2020).

It is well known that the conidia of pathogenic fungi play a significant part in infecting and colonizing plants (Khalil et al. 2019; Nicomrat et al. 2017). As a

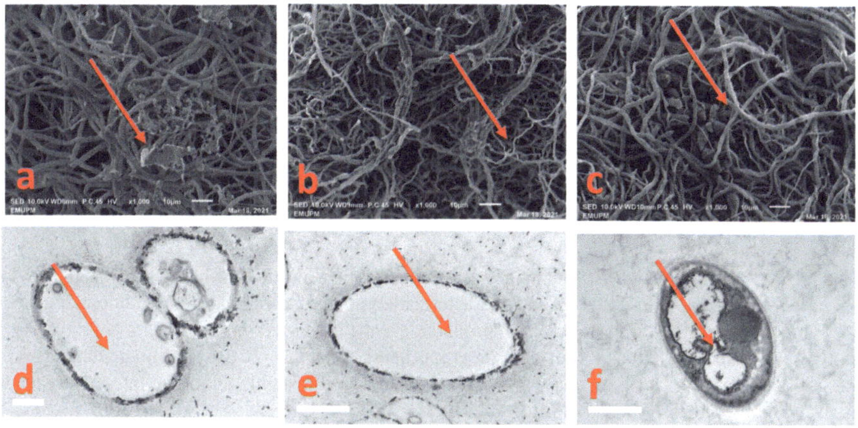

Fig. 9.3 Morphological changes of the treated rice pathogen. (**a**) Complete disruption of *P. oryzae* mycelia after being treated with chitosan hexaconazole agronanofungicides. (**b**) Shrinkage of *P. oryzae* mycelia after being treated with chitosan dazomet agronanofungicides. (**c**) Normal growth of *P. oryzae* mycelia (control). (**d**) Leakage of major cellular organelles after being treated with chitosan hexaconazole agronanofungicides. (**e**) Leakage of major cellular organelles after being treated with chitosan dazomet agronanofungicides. (**f**) The fungal cell was intact without any leakage for the control

result, slowing down the pace of conidial germination will greatly lessen the danger posed by pathogenic fungi in rice. The germ tube development and conidial germination of *P. oryzae* strain Gry could be effectively inhibited by the synthesized CNPs, and the fungicidal effects could increase with the rise in CNP concentrations. In actuality, 83% of spores germinated, whereas the length of the germ tubes in the negative control was 77.63 m (Ibrahim et al. 2020b). Furthermore, our study conducted in 2021 showed that both chitosan hexaconazole and chitosan dazomet agronanofungicides have played a role in disrupting the mycelia and conidia of *P. oryzae* isolated from rice plants (in vitro), while the control was not affected as presented in Fig. 9.3.

9.7 Conclusion and Future Perfectives

Chitosan and chitosan nanoparticles can play a significant role as nanopesticides against different fungal pathogens. Chitosan is a flexible biomaterial with outstanding fungicidal properties. It is easily manipulable using a variety of chemical and physical methods. By combining chitosan's functional groups (−NH2 and −OH), which may be functionalized with both organic and inorganic compounds to expand its range of applications, this biopolymer can serve as a special platform for the production of effective fungicides. Similarly, new fungicides that are promising and eco-friendly with lower doses can be manufactured in this way.

Acknowledgments Our profound gratitude goes to the Malaysian Ministry of Higher Education for the financial support through the Fundamental Research Grant Scheme: FRGS/1/2019/STG03/UPM/02/17. The authors of this work thank the staff of the Department of Plant Protection, Universiti Putra Malaysia, for their immense support.

References

Abd El-Aziz ME, Morsi SM, Salama DM, Abdel-Aziz MS, Abd Elwahed MS, Shaaban EA, Youssef AM (2018) Preparation and characterization of chitosan/polyacrylic acid/copper nanocomposites and their impact on onion production. Int J Biol Macromol 123:856–865

Ahmad A, Khan A, Akhtar F, Yousuf S, Xess I, Khan L, Manzoor N (2011) Fungicidal activity of thymol and carvacrol by disrupting ergosterol biosynthesis and membrane integrity against Candida. Eur J Clin Microbiol Infect Dis 30:41–50

Akter J, Jannat R, Hossain MM, Ahmed JU, Rubayet MT (2018) Chitosan for plant growth promotion and disease suppression against anthracnose in chilli. Int J Environ Agric Biotechnol 3(3):264343

Alanis-Garza BA, Gonzalez-Gonzalez GM, Salazar-Aranda R, de Torres NW, Rivas-Galindo VM (2007) Screening of antifungal activity of plants from the northeast of Mexico. J Ethnopharmacol 114(3):468–471

Brunel F, Gueddari NEE, Moerschbacher BM (2013) Complexation of copper (II) with chitosan nanogels: toward control of microbial growth. Carbohydr Polym 92:1348–1356

Cao L, Zhang H, Cao C, Zhang J, Li F, Huang Q (2016) Quaternized chitosan-capped mesoporous silica nanoparticles as nanocarriers for controlled pesticide release. Nanomaterials (Basel) 6(7):126

Cho Y, Shi R, Borgens RB (2010) Chitosan produces potent neuro protection and physiological recovery following traumatic spinal cord injury. J Exp Biol 213(9):1513–1520

Chookhongkha N, Sopondilok T, & Photchanachai S (2012) Effect of chitosan and chitosan nanoparticles on fungal growth and chilli seed quality. In: International conference on postharvest pest and disease management in exporting horticultural crops-PPDM2012 973. Acta Horticulturae, Bangkok, pp 231–237

Choudhar RC, Kumaraswamy RV, Kumari S, Sharma SS, Pal A, Raliya R, Saharan V (2019) Zinc encapsulated chitosan nanoparticle to promote maize crop yield. Int J Biol Macromol 12:126–135

Chouhan S, Sharma K, Guleria S (2017) Antimicrobial activity of some essential oils—present status and future perspectives. Medicines (Basel) 4(3):58

Cindi MD, Shittu T, Sivakumar D, Bautista-Banos S (2015) Chitosan boehmite-alumina nanocomposite films and thyme oil vapour control brown rot in peaches (*Prunus persica* L.) during postharvest storage. Crop Prot 72:127–131

Debnath D, Samal I, Mohapatra C, Routray S, Kesawat MS, Labanya R (2022) Chitosan: an autocidal molecule of plant pathogenic fungus. Life (Basel) 12(11):1908

Devi KA, Prajapati D, Kumar A, Pal A, Bhagat D, Singh BR, …, Saharan V (2020) Smart nanochitosan for fungal disease control. In: Nanopesticides: from research and development to mechanisms of action and sustainable use in agriculture, pp 23–47

El-Mohamedy RS, Abdel-Kader MM, Abd-El-Kareem F, El-Mougy NS (2013) Inhibitory effect of antagonistic bio-agents and chitosan on the growth of tomato root rot pathogens in vitro. J Agric Technol 9(6):1521–1533

El-Mohamedy RSR, Shafeek MR, Abd El-Samad EEDH, Salama DM, Rizk FA (2017) Field application of plant resistance inducers (PRIs) to control important root rot diseases and improvement growth and yield of green bean (Phaseolus vulgaris L.). Aust J Crop Sci 11(5):496–505

El-Mohamedya RSR, Abd El-Aziz ME, Kamel S (2019) Antifungal activity of chitosan nanoparticles against some plant pathogenic fungi in vitro. Agric Eng Int CIGR J 21:201–209
Ghule MR, Ramteke PK, Ramteke SD, Kodre PS, Langote A, Gaikwad AV et al (2021) Impact of chitosan seed treatment of fenugreek for management of root rot disease caused by Fusarium solani under in vitro and in vivo conditions. 3 Biotech 11:290
Hassan EO, Shoala T, Attia AM, Badr OA, Mahmoud SY, Farrag ES, El-Fiki IA (2022) Chitosan and nano-chitosan for management of Harpophora maydis: approaches for investigating antifungal activity, pathogenicity, maize-resistant lines, and molecular diagnosis of plant infection. J Fungi 8(5):509
Hossain F, Follett P, Salmieri S, Vu KD, Jamshidian M, Lacroix M (2017) Perspectives on essential oil loaded nanodelivery packaging technology for controlling stored cereal and grain pests. In: Nollet LML, Rathore HS (eds) Green pesticides handbook essential oils for pest control. Routledge, London, pp 487–508
Hossaina F, Follettb P, Salmieria S, Vua KD, Fraschinic C, Lacroixa M (2019) Antifungal activities of combined treatments of irradiation and essential oils (EOs) encapsulated chitosan nanocomposite films in *in vitro* and *in situ* conditions. Int J Food Microbiol 295:33–40
Ibrahim E, Zhang M, Zhang Y, Hossain A, Qiu W, Chen Y et al (2020a) Green-synthesization of silver nanoparticles using endophytic bacteria isolated from garlic and its antifungal activity against wheat fusarium head blight pathogen Fusarium graminearum. Nanomaterials 10(2):219
Ibrahim E, Luo J, Ahmed T, Wu W, Yan C, Li B (2020b) Biosynthesis of silver nanoparticles using onion endophytic bacterium and its antifungal activity against rice pathogen Magnaporthe oryzae. J Fungi 6(4):294
Kakaraparthi PS, Srinivas KVNS, Kumar JK, Kumar AN, Rajput DK, Anubala S (2015) Changes in the essential oil content and composition of palmarosa (*Cymbopogon martini*) harvested at different stages and short intervals in two different seasons. Ind Crops Prod 69:348–354
Kalagatur NK, Nirmal Ghosh OS, Sundararaj N, Mudili V (2018) Antifungal activity of chitosan nanoparticles encapsulated with Cymbopogon martinii essential oil on plant pathogenic fungi Fusarium graminearum. Front Pharmacol 9:610
Khalil NM, Abd El-Ghany MN, Rodríguez-Couto S (2019) Antifungal and anti-mycotoxin efficacy of biogenic silver nanoparticles produced by Fusarium chlamydosporum and Penicillium chrysogenum at non-cytotoxic doses. Chemosphere 218:477–486
Kheiri A, Jorf SM, Malihipour A, Saremi H, Nikkhah M (2017) Synthesis and characterization of chitosan nanoparticles and their effect on Fusarium head blight and oxidative activity in wheat. Int J Biol Macromol 102:526–538
Kriti A, Ghatak A, Mandal N (2020) Inhibitory potential assessment of silver nanoparticle on phytopathogenic spores and mycelial growth of Bipolaris sorokiniana and Alternaria brassicicola. Int J Curr Microbiol Appl Sci 9(3):692–699
Kumaraswamy RV, Kumari S, Choudhary RC, Sharma SS, Pal A, Raliya R, Saharan V (2019) Salicylic acid functionalized chitosan nanoparticle: a sustainable biostimulant for plant. Int J Biol Macromol 12:59–69
Kutawa AB, Ahmad K, Ali A, Hussein MZ, Abdul Wahab MA, Adamu A et al (2021) Trends in nanotechnology and its potentialities to control plant pathogenic fungi: a review. Biology 10(9):881
Li Q, Ren J, Dong F, Feng Y, Gu G, Guo Z (2013) Synthesis and antifungal activity of thiadiazole-functionalized chitosan derivatives. Carbohydr Res 373:103–107
Lopez A, Castro S, Andina MJ (2014) Insecticidal activity of microencapsulated *Schinus mole* essential oil. Ind Crop Prod 53:209–216
Lu W, Kelly AL, Miao S (2016) Emulsion-based encapsulation and delivery systems for polyphenols. Trends Food Sci Technol 47:1–9
Luque-Alcaraz AG, Cortez-Rocha MO, Velázquez-Contreras CA, Acosta-Silva AL, Santacruz-Ortega HDC, Burgos-Hernández A, Argüelles-Monal WM, Plascencia-Jatomea M (2016) Enhanced antifungal effect of chitosan/pepper tree (*Schinus molle*) essential oil bionanocomposites on the viability of *Aspergillus parasiticus* spores. J Nanomater 2016:6060137

Mahmoud SH, Salama DM, Abd El-Aziz ME (2018) Effect of chitosan and chitosan nanoparticles on growth, productivity and chemical quality of green snap bean. Biosci Res 15(4):4307–4432

Maluin FN, Hussein MZ, Yusof NA, Fakurazi S, Idris AS, Zainol Hilmi NH, Jeffery Daim LD (2019a) Preparation of chitosan–hexaconazole nanoparticles as fungicide nanodelivery system for combating Ganoderma disease in oil palm. Molecules 24(13):2498

Maluin FN, Hussein MZ, Yusof NA, Fakurazi S, Seman IA, Hilmi NHZ, Daim LDJ (2019b) Enhanced fungicidal efficacy on Ganoderma boninense by simultaneous co-delivery of hexaconazole and dazomet from their chitosan nanoparticles. RSC Adv 9(46):27083–27095

Maluin FN, Hussein MZ, Yusof NA, Fakurazi S, Idris AS, Hilmi NHZ, Jeffery Daim LD (2019c) A potent antifungal agent for basal stem rot disease treatment in oil palms based on chitosan-dazomet nanoparticles. Int J Mol Sci 20(9):2247

Manikandan A, Sathiyabama M (2016) Preparation of chitosan nanoparticles and its effect on detached rice leaves infected with *Pyricularia grisea*. Int J Biol Macromol 84:58–61

Martin A, Varona S, Navarrete A, Cocero MJ (2010) Encapsulation and co-precipitation processes with supercritical fluids: applications with essential oils. Open Chem Eng J 4(1):31

Medina E, Caro N, Abugoch L, Gamboa A, Diaz-Dosque M, Tapia C (2019) Chitosan thymol nanoparticles improve the antimicrobial effect and the water vapour barrier of chitosan-quinoa protein films. J Food Eng 240:191–198

Meng X, Yang L, Kennedy JF, Tian S (2010) Effects of chitosan and oligochitosan on growth of two fungal pathogens and physiological properties in pear fruit. Carbohydr Polym 81(1):70–75

Mirdehghan HS, Valerob D (2017) Bioactive compounds in tomato fruit and its antioxidant activity as affected by incorporation of aloe, eugenol, and thymol in fruit package during storage. Int J Food Prop 20:798–806

Mohammadi A, Hashemi M, Hosseini SM (2015) Nanoencapsulation of *Zataria multiflora* essential oil preparation and characterization with enhanced antifungal activity for controlling *Botrytis cinerea*, the causal agent of gray mould disease. Innov Food Sci Emerg Technol 28:73–80

Mustafa G, Komatsu S (2016) Toxicity of heavy metals and metal-containing nanoparticles on plants. Biochim Biophys Acta 1864(8):932–944

Nicomrat D, Janlapha W, Singkran N (2017) In vitro susceptibility of silver nanoparticles on Thai isolated fungus Pyricularia oryzae from Oryza sativa L. leaves. Appl Mech Mater 866:148–151

Oussou-Azo AF, Nakama T, Nakamura M, Futagami T, Vestergaard MDCM (2020) Antifungal potential of nanostructured crystalline copper and its oxide forms. Nanomaterials 10(5):1003

Pariona N, Mtz-Enriquez AI, Sánchez-Rangel D, Carrión G, Paraguay-Delgado F, Rosas-Saito G (2019) Green-synthesized copper nanoparticles as a potential antifungal against plant pathogens. RSC Adv 9(33):18835–18843

Rhouma A, Daoud HB, Ghanmi S, Salah HB, Romdhane M, Demak M (2009) Antimicrobial activities of leaf extracts of *Pistacia* and *Schinus* species against some plant pathogenic fungi and bacteria. J Plant Pathol 91(2):339–345

Rubina MS, Vasil'kov AY, Naumkin AV, Shtykova EV, Abramchuk SS, Alghuthaymi MA, Abd-Elsalam KA (2017) Synthesis and characterization of chitosan–copper nanocomposites and their fungicidal activity against two sclerotia-forming plant pathogenic fungi. J Nanostructure Chem 7(3):249–258

Saharan V, Mehrotra A, Khatik R, Rawal P, Sharma SS, Pal A (2013) Synthesis of chitosan-based nanoparticles and their in vitro evaluation against phytopathogenic fungi. Int J Biol Macromol 62:677–683

Saharan V, Sharma G, Yadav M, Choudhary MK, Sharma SS, Pal A, Raliya R, Biswas P (2015) Synthesis and in vitro antifungal efficacy of Cu-chitosan nanoparticles against pathogenic fungi of tomato. Int J Biol Macromol 7:346–353

Sajed H, Sahebkar A, Iranshahi M (2013) Zataria multiflora Boiss. (Shirazi thyme)—an ancient condiment with modern pharmaceutical uses. J Ethnopharmacol 145(3):686–698

Sathiyabama M, Parthasarathy R (2016) Biological preparation of chitosan nanoparticles and its in vitro antifungal efficacy against some phytopathogenic fungi. Carbohydr Polym 151:321–325

Shakeel Q, Lyu A, Zhang J, Wu M, Li G, Hsiang T, Yang L (2018) Biocontrol of Aspergillus flavus on peanut kernels using Streptomyces yanglinensis 3-10. Front Microbiol 9:1049
Van Long NN, Joly C, Dantigny P (2016) Active packaging with antifungal activities. Int J Food Microbiol 220:73–90
Xing K, Liu Y, Shen X, Zhu X, Li X, Miao X, Qin S (2017) Effect of O-chitosan nanoparticles on the development and membrane permeability of *Verticillium dahliae*. Carbohydr Polym 165:334–343
Youssef AM, Abdel-Aziz ME, El-Sayed ESA, Abdel-Aziz MS, Abd El-Hakim AA, Kamel S, Turky G (2018) Morphological, electrical & antibacterial properties of trilayered Cs/PAA/PPy bionanocomposites hydrogel based on Fe3O4-NPs. Carbohydr Polym 196:483–493
Youssef AM, Abd El-Aziz ME, Abd El-Sayed ES, Moussa MA, Turky G, Kamel S (2019) Rational design and electrical study of conducting bionanocomposites hydrogel based on chitosan and silver nanoparticles. Int J Biol Macromol 140:886–894
Zachetti VG, Cendoya E, Nichea MJ, Chulze SN, Ramirez ML (2019) Preliminary study on the use of chitosan as an eco-friendly alternative to control Fusarium growth and mycotoxin production on maize and wheat. Pathogens 8(1):29

Chapter 10
The Docking and Physiological Characteristics as Detectors of Nanoparticle's Role in Plant Responses to Biotic Stress

Sozan E. El-Abeid, Mohamed A. M. El-Tabakh, Ahmed Z. I. Shehata, Reham I. AbdelHamid, and Ahmed G. Soliman

Abstract Plant pathogens and other pests that feed on plants are a major cause of crop loss. Historically, many pesticides have been used to control plant pests, even though these chemicals are extremely toxic. However, these chemicals are considered fundamentals of supplying food and reducing crop loss. As a result, plant diseases are one of the biotic stresses that harm plants and productivity in general. Recently, nanomaterial has been utilized as a novel method to reduce the amount of hazardous substances released into the environment. The present chapter aims to shed light on the physiological effects of nanomaterial in terms of hormonal and enzymatic aspects as their role as exterminators for some diseases, in addition to highlighting some innovative new tools used to give a visualization of the nanomaterial-pathogen site of action, interaction, and their stability.

S. E. El-Abeid (✉)
Nanotechnology and Advanced Nano-Materials Laboratory (NANML), Mycology and Disease Survey Research Department, Plant Pathology Research Institute, Agricultural Research Center, Giza, Egypt
e-mail: sozanalabeid@arc.sci.eg

M. A. M. El-Tabakh (✉)
Marine Biology and Ichthyology Branch, Zoology Department, Faculty of Science, Al-Azhar University, Cairo, Egypt
e-mail: dr.m.eltabakh.201@azhar.edu.eg

A. Z. I. Shehata
Zoology Department, Faculty of Science, Al-Azhar University, Cairo, Egypt
e-mail: ahmed.ibrahem84@azhar.edu.eg

R. I. AbdelHamid
Sugar Crops Research Institute, Agricultural Research Center,Giza,Egypt, Giza, Egypt
e-mail: RehamIbrahim@agr.asu.edu.eg

A. G. Soliman
Biotechnology Program, Faculty of Agriculture, Ain Shams University, Cairo, Egypt
e-mail: Ahmedgamal_soliman@agr.asu.edu.eg

M. Khan, J.-T. Chen (eds.), *Nanoparticles in Plant Biotic Stress Management*,
https://doi.org/10.1007/978-981-97-0851-2_10

Keywords Nanomaterials · Physiological effects · Nanocide · Visualization action · Docking

10.1 Introduction

Nanotechnology has recently gained significant attention in the agricultural sector due to its potential to provide novel approaches for controlling and managing plant diseases. Nanoparticles, characterized by their size ranging from 1 to 100 nm, have distinct features that render them well-suited for diverse applications in plant disease control. This topic aims to learn how nanoparticles may be used for the prevention and treatment of plant diseases. Nanoparticles' primary function in preventing plant disease is as delivery vehicles for bioactive compounds. Nanoparticles can encapsulate and safeguard bioactive substances, including fungicides, bactericides, and plant growth regulators, augmenting their stability and effectiveness. Through encapsulation, nanoparticles can provide regulated release mechanisms, facilitating a prolonged and targeted distribution to specific diseased plant tissues (El-Abeid et al. 2020; Mosa et al. 2022).

Furthermore, nanoparticles can potentially enhance the bioavailability and absorption of these bioactive chemicals inside plants. The diminutive dimensions of these particles enable them to readily infiltrate the plant's cuticle or cell walls, enhancing the efficacy of their uptake via the roots or leaves. Improved efficacy against infections and better overall disease control result from increased absorption (El-Abeid et al. 2020). Furthermore, nanoparticles have antibacterial characteristics in addition to their role as transporters for bioactive chemicals. Several studies have shown that certain categories of nanoparticles, including silver nanoparticles, possess various antibacterial properties against diverse diseases. These substances can interfere with the cellular structures of bacteria or fungi, impeding their development and reproductive processes (Choudhary et al. 2022).

The antimicrobial properties shown by nanoparticles provide them a compelling substitute for traditional chemical pesticides, which may potentially harm human well-being and the ecosystem. In addition, nanoparticles can induce the plant's innate defensive responses against pathogens. Plants may be rendered more resistant to disease infection by the induction of systemic acquired resistance (SAR) or the activation of defense-related genes, thus enhancing their ability to combat pathogens (Abdelkhalek et al. 2022). The observed enhancement in the immune system of plants facilitates a more efficient defense against infections and diminishes their vulnerability to subsequent pathogenic assaults. One further functional aspect of nanoparticles in controlling plant diseases is their potential use in diagnostics. Nanoparticles can be functionalized by attaching ligands or antibodies that exhibit specificity toward pathogen-related compounds or antigens. Nanoparticles facilitate the development of expeditious and highly responsive diagnostic instruments, hence facilitating the timely detection and characterization of plant diseases (Benjamin et al. 2023). The early identification of diseases is of utmost importance to promptly adopt control measures and effectively mitigate their spread. In general, nanoparticles possess a diverse array of functional capacities in the regulation and administration of plant diseases. Nanoparticles possess several advantageous characteristics in contemporary agriculture, including their capacity to serve as carriers for bioactive chemicals, boost the bioavailability of these compounds, display antibacterial capabilities, activate plant defense systems, and facilitate diagnostics. Ongoing research and development endeavors in this domain exhibit significant potential for implementing sustainable and efficient disease control approaches in agricultural production.

This chapter aims to provide insights into the physiological impacts of nanomaterial in relation to hormonal and enzymatic factors, particularly their efficacy as agents for combating certain diseases (Fig. 10.1). Additionally, this chapter emphasizes utilizing novel techniques for visualizing the site of action, interaction, and stability of nanomaterials and their relation with pathogens.

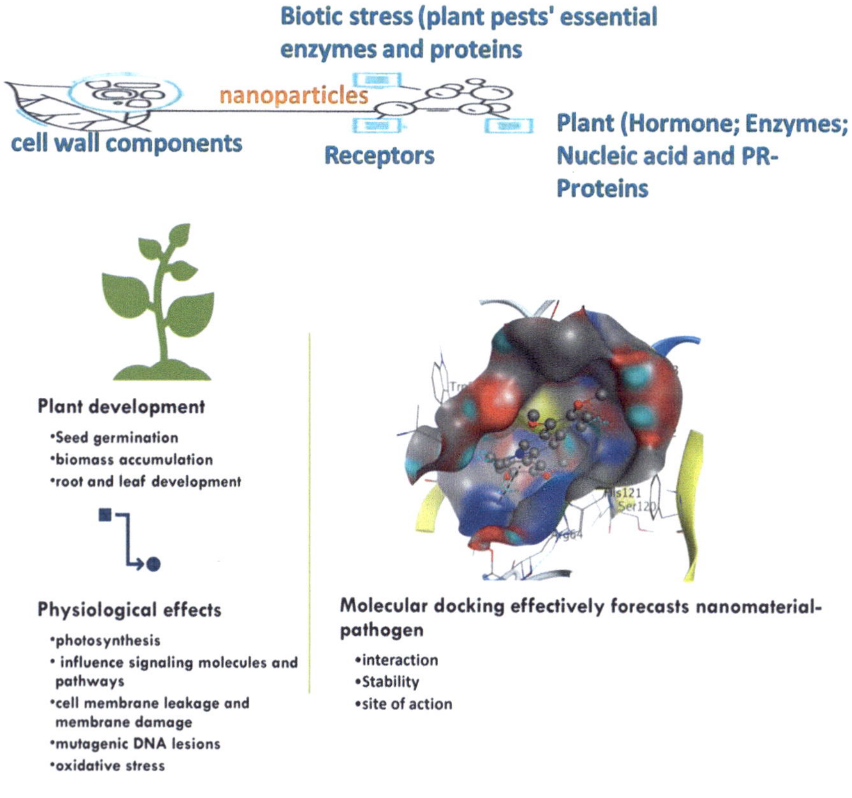

Fig. 10.1 Schematic representation to illustrate the docking and physiological characteristics as detectors of nanoparticle's role in plant responses to biotic stress

10.2 The Functional Roles of Nanoparticles in Controlling and Managing Plant Diseases

Nanoparticles possess the capacity to function as nano-scale insecticides, fertilizers, and nutrition, exhibiting the ability to selectively target specific cellular organelles within both plants and diseases. Nanoparticles have been found to enhance various physiological processes in plants, including metabolism, seed germination, plant development, and nutrient use. The utilization of nanoparticles has the potential to yield two significant advantages: a reduction in the utilization of agrochemicals and a mitigation of the environmental impact associated with current farming practices. One potential mechanism by which nanoparticles can mitigate the pathogenicity and growth of phytopathogens is their interference with many cellular processes, including membrane function, metabolism, respiration, and gene expression. Nanoparticles possess the capacity to enhance a plant's resilience to both biotic and abiotic stresses by modulating its physiological and biochemical responses. The application of nanoparticles in the identification and management of infections can be achieved by the utilization of nano-sensors, nano-carriers, and nano-biocides.

10.3 Their Role in the Regulation of Biotic Stress in Plants

1. Delivery of antimicrobial agents: Nanoparticles may transport drugs to plant infection spots. These nanoparticles can encapsulate or attach antimicrobial compounds such as fungicides, bactericides, or bio-control agents, allowing for targeted and controlled release of these agents to combat plant diseases effectively (Mosa et al. 2022).
2. Enhanced penetration and uptake: Nanoparticles can improve the penetration and uptake of antimicrobial agents into plant tissues. Since they are so small and have such a high surface area to volume ratio, they can readily pierce the plant's cuticle or cell walls, allowing antimicrobial chemicals to be delivered directly to the cells where the infections are hiding out.
3. Controlled release of active compounds: Nanoparticles can be engineered to release antimicrobial compounds in a controlled manner, ensuring a sustained and prolonged effect against plant pathogens. Modifying the surface properties or encapsulating the active compounds within nanoparticles can regulate their release rate, providing long-lasting protection against diseases.
4. Improved stability and bioavailability: Nanoparticles can enhance the stability and bioavailability of antimicrobial agents by protecting them from degradation or inactivation in harsh environmental conditions. They can shield the active compounds from UV radiation, pH variations, enzymatic degradation, or interactions with other molecules in the plant's surroundings.

5. Targeted delivery to specific plant tissues: Nanoparticles can be functionalized with ligands or antibodies that specifically recognize certain plant tissues or pathogen receptors. These allow targeted antimicrobial agent delivery to infected areas while minimizing off-target effects on non-infected tissues.
6. Induction of plant defense mechanisms: Certain nanoparticles have been found to stimulate plants' natural defense mechanisms against pathogens. These nanoparticles can trigger plant immune responses, leading to enhanced production of defense-related molecules such as phytoalexins, pathogenesis-related proteins, or reactive oxygen species that help combat diseases.
7. Enhanced nutrient uptake and growth promotion: Some nanoparticles boost nutrient intake and plant development, making plants more disease-resistant. These nanoparticles can enhance the availability and uptake of essential nutrients, stimulate root development, or modulate plant hormone levels, thereby strengthening plants' overall health and resilience against pathogens (Mosa et al. 2022; El-Abeid et al. 2020).
8. Diagnostic tools for disease detection: Nanoparticles may be used as diagnostic tools for early detection of plant diseases. Functionalized nanoparticles can specifically bind to pathogen-derived molecules or biomarkers in infected plants, enabling rapid and sensitive detection of diseases through techniques such as colorimetric assays or biosensors. Overall, nanoparticles offer promising strategies for controlling and managing plant diseases by providing targeted delivery of antimicrobial agents, enhancing their stability and bioavailability, stimulating plant defense mechanisms, improving nutrient uptake, and enabling early disease detection (Benjamin et al. 2023).

10.4 Nanoparticle Impact on Plant Growth and Development

Nanoparticles have been proven to have dual activity in preventing and treating wilt disease in common bean plants (Tortella et al. 2023; El-Sayed et al. 2023). According to Li et al. (2023a, b), these substances have the potential to function as highly targeted agents, capable of delivering herbicides, nano-pesticide fertilizers, or genes to particular cellular organelles inside plants. According to recent studies by Li et al. (2023a, b) and Wang et al. (2023), using nanoparticles has shown potential to enhance membrane integrity and plant-water connections, mitigating drought stress' effects. Additionally, these nanoparticles have shown the capacity to boost nutrient absorption in plants under drought stress. Additionally, they can provide a novel remedy for the insufficiency of nutrients in crops, stimulate plant growth and advancement, and impede plant disease proliferation (Li et al. 2023a, b; Siddiqui et al. 2015).

Moreover, previous studies have shown that silica nanoparticles (SiNPs) might positively impact the development and growth of plants, particularly the presence of both biotic and abiotic stressors. Silicon nanoparticles (SiNPs) can partially improve leaf and stem development (Mosa et al. 2022) by regulating plant hormone metabolism and soluble sugar levels. According to Li et al. (2023a, b), SiNPs can potentially enhance wheat development via direct or indirect mechanisms, such as the elevation of auxin (IAA) and fructose levels.

10.5 What Is the Mechanism Behind Nanoparticles' Effect on Plant Growth?

The unique physicochemical features of nanomaterials provide significant prospects in the agricultural domain. The physiological, morphological, and genotoxic alterations that occur when nanoparticles interact with plants are important in using nanotechnology for agriculture. The effects of nanoparticles on plant growth and development may differ depending on the characteristics of the nanomaterial, the application technique, and the type of plant used, as shown by the studies. The effects of nanoparticles on plants' physiology, biochemistry, and molecular processes have recently been studied. Particular attention has been paid to the risks and benefits of utilizing nanomaterials on economically important crops and their absorption, translocation, and biotransformation (Nair 2016). Nanoparticles have been identified as beneficial agents for enhancing plant growth, facilitating developmental processes, and providing protective effects. They confer specificity in the administration of pesticides, improve nutrition availability, manage pathogenicity, boost photosynthetic capacity, and increase the germination rate. However, there have been recorded cases of nanoparticle toxicity and bioaccumulation, leading to a number of drawbacks. Therefore, it is essential to have a clear awareness of the benefits and drawbacks of nanoparticles and to do extensive research into their varied properties.

The effects of nanoscale ZnO on peanut (*Arachis hypogaea*) plants were studied, and it was shown that 1000 ppm ZnO had a positive effect on germination, seedling vigor, plant growth, blooming, chlorophyll content, pod yield, and root development (Pawar and Laware 2018). Inhibitory effects were seen, however, at doses beyond 2000 ppm (Pawar and Laware 2018). Biocompatibility studies have shown that zinc oxide nanoparticles (ZnO-NPs) pose no harm to living organisms. ZnO-NPs have been found in previous studies to improve seed germination and boost plant growth. The antibacterial characteristics discovered in these nanoparticles may also aid in disease prevention and plant defense.

Throughout their evolution, ZnO nanoparticles (NPs) have been shown to have positive and negative effects on plant growth and metabolism, according to several studies. Specific features of the nanoparticles themselves, as well as the physiological qualities of the host plant, affect the absorption, translocation, and storage of zinc oxide nanoparticles (ZnO-NPs) by plants. It is theorized that ZnO exerts its effects by causing an increase in membrane lipid peroxidation via the production of reactive oxygen species. This, in turn, reduces cell viability by allowing reducing sugars, DNA, and proteins to flow out of the membrane. Nanoparticles have been found to have deleterious effects on vegetation. The results of this study demonstrate that nonpolar molecules (NPs) may be absorbed, transported, and accumulated inside terrestrial plant species. Some types of edible plants have been found to contain nanoparticles (NPs), which may reduce agricultural yields and pose health hazards to humans. Understanding how nanoparticles (NPs) move through plants and how they exert their toxicity is crucial for any risk assessment to be credible. Several stages of plant development, including expansion, germination, biomass accumulation, and root and leaf development, may be negatively impacted by an abundance of nanoparticles.

However, depending on the plant species, the kind of nanoparticles utilized, and their dosage, the impacts of NPs might be either helpful or destructive. The physiological effects of nanoparticles on plants led to increased seed germination rates and total biomass or grain yield. The NPs also improved the plant's resistance to biotic and abiotic stresses. The molecular-level processes are crucial to the plant's biological functions. In Ali et al. 2021, Ali and coworkers did research.

10.6 Nanoparticle-Microbe Interactions Inside Plant System

Various factors, including nanoparticle characteristics, application method, and plant species, are thought to influence how nanoparticles influence plant growth and development, as Agrahari and Dubey (2020) reported. In almost every way, nanoparticles have helped plants flourish and thrive while keeping them safe. Nanoparticles boost photosynthetic capacity and germination rate, decrease pathogen prevalence, and improve the specificity with which pesticides are delivered (2022). Rasheed et al. found that the effects of various nanoparticles on plant development were very variable (Rasheed et al. 2022).

The green pigment chlorophyll captures the light energy necessary for photosynthesis. It is found in the thylakoid membrane of chloroplasts and is the main pigment used in photosynthesis. Light is captured by chlorophyll and converted into chemical energy, which is then used to synthesize glucose from carbon dioxide and water. The process of photosynthesis may be affected positively or negatively by nanoparticles. Nanoparticles may alter the ability of plants to take in solar energy. Up-conversion nanoparticles, for instance, may transform infrared light into visible light, which solar cells can use to generate more electricity. Up-conversion nanoparticles may be used to increase the amount of energy harvested from solar panels. These nanoparticles take in infrared radiation and output visible light. Not all nanoparticles have the same impact on carotenoid levels, and the photosynthetic mechanism reacts differently when nanoparticles are present.

This compound stimulates the reaction in the plant's light-harvesting complex, and it also inhibits the electron transport system and changes the activity of RuBisCo, carbonic anhydrase, or phosphoenolpyruvate carboxylase (PEP) enzymes, all of which work together to halt metabolism (Poddar et al. 2020). Nanoparticles may have both negative (decreased chlorophyll content and electron transport rate) and positive (increased chlorophyll content, RuBisCo enzyme activity, PSII performance, CO_2 harvesting, and a broader chloroplast photoabsorption spectrum) effects on photosynthesis (Ghorbanpour et al. 2021). Researchers found that when SWCNTs were exposed to photosynthetic-pigment-protein assemblies in an aqueous environment, the steady-state chlorophyll fluorescence in the thylakoid membrane-biohybrid systems was quenched, and the average fluorescence lifetimes were shortened. The impact was unrelated to alterations in the structure and function of photosynthetic membranes at the macro level.

In photosynthetic organisms, nanoparticles may influence signaling molecules and pathways. For instance, single-walled carbon nanotubes (SWCNT) boosted the activity of plant signaling chemicals like nitric acid and facilitated the quick transit of electrons, leading to a threefold increase in photosynthetic production in the chloroplasts (Ghorbanpour et al. (2021)). Muhammad et al. (2021) thoroughly explain how the presence of various nanoparticles affects photosynthesis and how the photosynthetic process reacts to those nanoparticles.

Leakage of photosynthetic electrons toward the nanotubes, most likely at the level of the PSII acceptor site, is suggested by the reduction of the fluorescence signal in thylakoid membrane-biohybrid systems (Lambreva et al. 2023).

Titanium dioxide (TiO_2) nanoparticles increase the chlorophyll content of plants like tomatoes and oilseeds, among other beneficial benefits. Arabidopsis, spinach, and tomato all had their chlorophyll content raised, their RuBisCo (Ribulose-1,5-bisphosphate carboxylase/oxygenase) activity boosted, and their net photosynthesis stimulated by a foliar spray of TiO_2 NPs (Chen et al. 2017). TiO_2 NPs were shown to improve photosynthesis, growth rate, and solar energy absorption in spinach plants by increasing leaf chlorophyll content and influencing the thylakoid membrane (Karvar et al. 2022). Adil et al. found that chlorophyll a and b and total content concentrations rose in response to increasing concentrations of ZnO NPs, from 0.06 to 0.12 g (Adil et al. 2022). In contrast to the positive effects of ZnO NPs on seedling growth and chlorophyll and carotenoid content, Pb severely slowed their development (Hussain et al. 2021). By increasing plant growth and enhancing chlorophyll and carotenoid levels, ZnO NPs reduce Pb-induced stress (Hussain et al. 2021). Using differential modulation of photosynthesis and anti-oxidative defense mechanisms, ZnO-NPs may protect plant seedlings from Cd and Pb stress by reducing HMs-induced phytotoxicity and boosting physiochemical activity Alhammad et al. (2023).

It is unclear how nanoparticles compromise the structure and function of photosynthetic membranes. Various investigations have proven that polyvinyl chloride (PVC) nanoparticles cause cell membrane leakage and membrane damage (Li et al. 2022). An in-depth understanding of nanoparticles' interactions with chloroplast and its structural components (e.g., thylakoid membranes), signaling molecules, and pathways is necessary because increased photosynthetic efficiency is a potential impact of NPs on photosynthetic organisms of major economic and ecological significance (e.g., crops and algae).

Polyvinyl chloride (PVC) nanoparticles can cause mutagenic DNA lesions, generate reactive oxygen species, destruct cellular membranes, enhance membrane lipid peroxidation, and thus inhibit the metabolism and growth of plants (Ahmed et al. 2021). The negative effect of NPs on plant growth is mainly caused by excessive NPs (Gao et al. 2023).

DNA damage in plants caused by nanoparticles may be mutagenic. Examples include the first proof that manufactured nanoparticles may harm DNA in plant cells, published by a team from NIST and UMass Amherst (Atha et al. 2012). DNA damage from oxidative stress is the primary source of genotoxicity from NPs. Plants may experience oxidative stress due to nanoparticle production of reactive oxygen species (ROS). In addition to causing damage to DNA and RNA, ROS may also destroy proteins and enzymes inside a cell. TiO_2 nanoparticles stimulate the RuBisCo enzyme. To create reactive oxygen species (ROS), nanoparticles release metal ions and interact with water molecules (Yu et al. 2020; Abdal Dayem et al. 2017). Carbon nanomaterial, copper oxide nanoparticles, and metal-based nanoparticles, including gold, iron, cerium, copper, and titanium, may all produce reactive oxygen species (ROS) (Wen et al. 2020; Zhang et al. 2021; Sicwetsha et al. 2021).

10.7 Reducing Pathogen Infection by Affecting Their Receptors

Germline-encoded pathogen recognition receptors (PRRs) identify pathogen-associated molecular patterns (PAMPs). Macrophages and dendritic cells have these receptors, two types of immune cells. Until more specialized adaptive immunity is created, innate immunity serves a critical function as the body's initial line of defense. Therefore, reducing pathogen infection by affecting their receptors is a promising approach to combat infectious diseases. However, it is important to note that this is an active area of research, and there are no approved therapies yet that target PRRs to reduce pathogen infection (Jang et al. 2015).

Nanoparticles have been proposed as a promising approach to reduce pathogen infection by affecting their receptors. In particular, bio-receptor-functionalized nanoparticles effectively detect pathogenic bacteria and microbial biomolecules. Antibodies, oligonucleotide aptamers, proteins, enzymes, nano-zymes (chromogens), cells, antigens, microbes, bio-mimic substances, and other tiny molecules are all examples of bio-receptors that may be attached to these nanoparticles. Pathogen-associated molecular patterns (PAMPs) are recognized by bio-receptors, initiating a reaction that ultimately eliminates the pathogen (Lin et al. 2021). These nanoparticles are very useful for detecting harmful bacteria and other microbial macromolecules. Binding to target pathogens causes a change in the nanoparticles' refractive index, which in turn causes a shift in the surface plasmon resonance peak. Some research has shown that harmful bacteria and biomolecules may be detected visually, rapidly, and with a high intensity of colorimetry at a very low analytic concentration. The selectivity and specificity of a bio-receptor-functionalized nanoparticle, as well as the binding relationship displayed by the nanoparticle, bio-receptor, and analytes to create a bio-sensing complex, determine how well it functions as a detection component (Daramola et al. 2022).

It's worth noting that this is a dynamic field of study and that, yet, no medicines have been authorized that specifically target PRRs in an effort to lessen pathogen infection. There are two types of chemicals that pattern recognition receptors (PRRs) may identify: PAMPs, which are molecules associated with microbial pathogens, and DAMPs, which are molecules linked with host cell components that are produced after cell injury or death. The microbe-specific molecules that a given PRR recognizes are called pathogen-associated molecular patterns (PAMPs) and include bacterial carbohydrates (such as lipopolysaccharide or LPS, mannose), nucleic acids (such as bacterial or viral DNA or RNA), bacterial peptides (flagellin, microtubule elongation factors), peptidoglycans and lipoteichoic acids (from Gram-positive bacteria), N-formylmethionine, lipoproteins, and fungal glucans and chitin. Uric acid and extracellular ATP are damage-associated molecular patterns (DAMPs) that serve as endogenous stress signals.

PRRs are classified as localized in the membrane or the cytoplasm based on their location in the cell. The Toll-like receptors (TLRs) and the C-type lectin receptors are two examples of membrane-bound PRRs (CLRs). Many PRRs, such as NOD-like (NLRs) and RIG-I-like receptors, are in the cytoplasm (RLRs). Toll-like receptors mediate cytokine generation and release and activate other host defense processes in response to pathogen-associated molecular patterns detected either extracellularly or endosomal (Amarante-Mendes et al. 2018).

10.8 The Role of Nanoparticles in Regulating Gene Expression for Biotic Stress in Plants

It has been discovered that nanoparticles may regulate biotic stress in plants by modulating gene expression. Gene expression refers to producing a protein or other gene product from genetic instructions. In biotic stress response and defense, nanoparticles may be used to control gene expression (Kandhol et al. 2022).

The pathogenesis-related (PR) gene is an important player in plant immunity. Proteins having antimicrobial characteristics are encoded by a collection of genes called PR genes, which are activated in response to pathogen assault. These proteins are essential for preventing the proliferation of harmful microorganisms (Zribi et al. 2021; Jain and Khurana 2018).

Increased defensive responses against pathogens in plants have been attributed to nanoparticles' ability to stimulate the expression of PR genes. Silver and titanium nanoparticles, for example, have been shown in studies to stimulate the overexpression of PR genes in plants. Due to this upregulation, the enhanced synthesis of PR proteins is one factor in the plant's resistance to infections (Pérez-Labrada et al. 2020).

The precise mechanisms by which nanoparticles regulate gene expression in plants are still being investigated. However, it is believed that nanoparticles can interact with plant cell receptors and signaling molecules, triggering a cascade of molecular events that ultimately result in gene expression changes. Reactive oxygen species (ROS) produced by nanoparticles inside plant cells may serve as signaling molecules to turn on genes involved in defense (Thwala et al. 2013; Vannini et al. 2013; Jiang et al. 2017; Tripathi et al. 2017).

It's worth noting that the nanoparticle type, concentration, and length of exposure may all affect how they affect gene expression. Nanoparticle-mediated gene expression modification and its impact on plant defense against pathogens need more study to completely grasp the underlying processes. Nanoparticles may affect plant gene expression, boosting the plant's defensive response against pathogens by, among other things, upregulating genes involved in pathogenesis. One of the functional functions of nanoparticles in the management and prevention of plant diseases is the regulation of gene expression.

10.8.1 Multiple Crop Types Were Studied to See How Nanoparticles Affected Gene Expression and Plant Disease Resistance

Tomato (*Solanum lycopersicum*)

Silver nanoparticles (AgNPs) were studied for their effect on gene expression and disease resistance in tomato plants in a study by Rai et al. (2009). They analyzed the expression of important defense-related genes after treating tomato plants with AgNPs. In this study, we found that AgNP treatment induced an increase in the expression of many genes involved in plant defense against pathogens, known as pathogenesis-related (PR) genes. Increased defensive responses in tomato plants may be attributed to this upregulation. In addition, the treated tomato plants showed enhanced resistance to the early blight-causing fungus *Alternaria solani*. Disease severity was reduced, and pathogen development was stymied in plants treated with AgNP.

Sugar Beet (*Beta vulgaris*)

Khodakovskaya et al. (2012) investigated the effects of carbon nanotubes (CNTs) on sugar beet plants and their defense response against the fungal pathogen *Cercospora beticola*, which causes *Cercospora* leaf spot disease. This study revealed that applying CNTs to sugar beet plants resulted in the upregulation of defense-related genes associated with pathogen recognition and defense signaling pathways. These genes included those involved in synthesizing antimicrobial compounds and reinforcing cell walls. Moreover, CNT-treated sugar beet plants exhibited increased resistance against *Cercospora* leaf spot disease. The CNT treatment reduced disease severity and inhibited the growth and colonization of pathogens on plant leaves.

Arabidopsis Plant Studies

Vivancos et al. (2015) investigated the role of silicon nanoparticles (SiNPs) in enhancing resistance against the fungal pathogen *Fusarium oxysporum* in *Arabidopsis* plants. The researchers observed that SiNP treatment increased the expression of defense-related genes associated with the SA signaling pathway. This upregulation of defense gene expression resulted in improved resistance against fungal pathogens. This study demonstrated the potential of SiNPs to modulate gene expression and enhance plant defense mechanisms in *Arabidopsis*.

Sugarcane (*Saccharum officinalis*)

The effects of copper nanoparticles (CuNPs) on sugarcane development and disease control were studied by Elmer and White (2016). They tested the effects of CuNPs on sugarcane plants by measuring the expression of defense-related genes. Genes involved in pathogen detection and defense signaling were shown to be elevated after treatment with CuNP. The activation of plant defense mechanisms against pathogens is reflected in this increase. In addition, sugarcane plants treated with CuNP showed increased resistance to the bacterial pathogen *Xanthomonas albilineans*, the causative agent of sugarcane leaf scald disease. The severity of the sickness was lessened, and the spread of the infection was stifled after treatment with CuNP.

In summary, increased expression of defense genes involved in fending off diseases, known as pathogenesis-related (PR) genes, has been linked to plant exposure to nanoparticles. Silver nanoparticles (AgNPs), carbon nanotubes (CNTs), and copper nanoparticles (CuNPs) have all been proven in studies on tomato, sugar beet, *Arabidopsis*, and sugarcane to upregulate defense-related genes. In tomato plants, AgNP treatment led to the upregulation of PR genes and improved resistance against the fungal pathogen *Alternaria solani*. Similarly, in sugar beet plants, the CNT treatment upregulated defense-related genes associated with pathogen recognition and defense signaling pathways, resulting in enhanced resistance against the fungal pathogen *Cercospora beticola*. Silicon nanoparticles (SiNPs) applied to *Arabidopsis* plants boost the plant's resistance to the fungal disease *Fusarium oxysporum* by increasing the expression of defense-related genes linked with the salicylic acid (SA) signaling pathway, while sugarcane plants are more resistant to the bacterial pathogen *Xanthomonas albilineans*, which causes leaf scald disease, after receiving a treatment with CuNP.

The precise mechanisms by which nanoparticles regulate gene expression in plants are still being investigated. However, it is believed that nanoparticles interact with plant cell receptors and signaling molecules and generate reactive oxygen species (ROS), triggering molecular cascades that result in changes in gene expression. Overall, these studies demonstrate the potential of nanoparticles to enhance plant defense against pathogens by modulating gene expression. Further research is needed to fully understand the mechanisms underlying nanoparticle-mediated gene expression modulation and its impact on plant defense mechanisms.

10.8.2 Plant Hormone Inducers

Nanoparticles have become promising tools in plant science because of their ability to modulate gene expression and influence plant defense responses. Their interaction with plant cells at the molecular level, potentially involving specific genes and receptors, is crucial for harnessing their full potential in controlling and managing plant diseases (Khalid et al. 2022). While the exact details of nanoparticle-cell interactions in plants are still unknown, research has provided insights into the general processes involved. Nanoparticles interact with receptors on the surface of plant cells, triggering signaling cascades that can lead to changes in gene expression. These interactions can activate or mediate various signaling pathways within plant cells, such as hormone signaling or reactive oxygen species signaling (Khan et al. 2019).

Nanoparticles such as silver nanoparticles (AgNPs), titanium dioxide nanoparticles (TiO_2 NPs), and silicon nanoparticles (SiNPs) have been found to interact with plant hormone signaling pathways, potentially influencing plant growth and defense responses. The following are some general observations.

Silver Nanoparticles (AgNPs)

Auxins, gibberellins, and ethylene are only some plant hormones whose production and signaling are affected by AgNPs. These nanoparticles may affect hormone synthesis in plants, which can either increase or decrease hormone production (Tripathi et al. 2022).

Titanium Dioxide Nanoparticles (TiO_2 NPs)

Auxins and abscisic acid are two plant hormones that TiO_2 NPs might affect during biosynthesis and transduction (ABA). Increased synthesis of auxins, hormones essential for plant growth and development, has been linked to their presence (Tripathi et al. 2022).

Silicon Nanoparticles (SiNPs)

SiNPs have been reported to modulate hormone signaling pathways, including those associated with abscisic acid (ABA) and salicylic acid (SA). SiNP treatment enhances ABA signaling, which is involved in plant stress responses and drought tolerance. While these observations suggest that nanoparticles can potentially influence plant hormone signaling, it is important to note that the specific mechanisms and effects may vary depending on nanoparticle type, concentration, and plant species.

10.9 Receptor-Mediated Interactions

Nanoparticles may alter gene expression in plants by interacting with particular receptors on the surfaces of plant cells. Plant defensive responses are mediated by various membrane-bound receptors, some of which are involved in pattern recognition (Swartzwelter et al. 2021).

10.9.1 Signaling Pathways and Gene Expression

Nanoparticles can potentially activate or modulate various signaling pathways within plant cells, leading to changes in gene expression. These pathways may involve hormone signaling, reactive oxygen species (ROS) signaling, or other defense-related signaling cascades (Hasanuzzaman et al. 2021).

10.9.2 Defense-Related Genes

Exposure to nanoparticles may affect the expression of genes involved in plant defense. These genes can include pathogenesis-related (PR) genes, which encode antimicrobial proteins, genes involved in synthesizing secondary metabolites, reinforcement of cell walls, and other defense mechanisms (Khoshru et al. 2023).

10.9.3 Some Examples of Bio-receptor-Functionalized Nanoparticles

Antibody-functionalized gold nanoparticles: These nanoparticles have been shown to effectively detect and kill bacteria such as *Escherichia coli* and *Staphylococcus aureus*.

Aptamer-functionalized magnetic nanoparticles: These nanoparticles have been shown to effectively detect and kill bacteria such as *Salmonella typhimurium*.

Protein-functionalized silver nanoparticles: These nanoparticles have been shown to effectively detect and kill bacteria such as *Pseudomonas aeruginosa* (Mocan et al. 2017).

Enzyme-functionalized silica nanoparticles: These nanoparticles have been shown to effectively detect and kill bacteria such as *Staphylococcus aureus*.

Nanoparticle-based detection and killing of bacteria rely on bio-receptors functionalized on the surface of nanoparticles detecting pathogen-associated molecular patterns (PAMPs) (Mocan et al. 2017). Binding to target pathogens causes a change in the nanoparticles' refractive index, which in turn causes a shift in the surface plasmon resonance peak. This change is detected by optical sensors that can rapidly detect pathogens (Selvarajan et al. 2020).

In addition to detection, bio-receptor-functionalized nanoparticles can also kill bacteria. The bio-receptors recognize PAMPs on the pathogen's surface and trigger a response that leads to the destruction of the pathogen. For example, antibody-functionalized gold nanoparticles can bind to bacterial cells through specific interactions between antibodies and bacterial cell surface proteins or carbohydrates (Tian et al. 2021). Once bound, these nanoparticles can induce oxidative stress or generate heat, leading to bacterial cell death (Tian et al. 2021). Similarly, aptamer-functionalized magnetic nanoparticles can bind to bacterial cells through specific interactions between aptamers and bacterial cell surface proteins or carbohydrates (Tian et al. 2021). Once bound, these nanoparticles can induce mechanical stress or generate heat, leading to bacterial cell death. This is a developing field of study, and there are currently no FDA-approved medicines specifically targeting PRRs to lessen pathogen infection. Further study is required to evaluate the safety and effectiveness of these nanoparticles for clinical usage.

10.10 Effective Nanocides in Literature and the Market

Recent articles show that metal nanoparticles can be used as nanocides to control plant fungal diseases and create environmentally friendly fungicides that selectively activate their pathogen-suppressing properties only in the chosen fungal pathogen (Table 10.1). Nanocides effectively control plant fungal infection, thus providing a targeted approach to plant disease management that minimizes the impact on the surrounding environment and non-target organisms. This innovative use of nanotechnology in agriculture could revolutionize the way we protect crops from fungal pathogens, ensuring food security while maintaining ecological balance.

Nanoguard is a commercially available product that employs the usage of silver nanoparticles to effectively eradicate odor-causing microorganisms that may potentially interact with various types of fabric materials. The silver particles exhibit precision design, possess non-ionic properties, and are characterized by their non-toxic nature. The efficacy of nanoguard-treated fabrics in eradicating various bacterial strains has been substantiated by independent testing, demonstrating a 99.99% elimination rate upon contact.

Nanozymes refer to nanomaterial-based artificial enzymes that imitate the catalytic regions found in real enzymes or possess multivalent components to facilitate various reactions (Gomaa 2022). The subsequent instances illustrate the application of nanomaterials in nanozymes: Metal oxides are compounds that consist of a metal element bonded with oxygen. These chemicals are ubiquitous in the natural world and have many potential uses. Nanomaterials made from metal oxides like cerium oxide and iron oxide have been the subject of much study to mimic the activity of several naturally occurring enzymes (Wang et al. 2016). Due to their intrinsic catalytic properties, carbon-based nanomaterials such as fullerene and its derivatives,

Table 10.1 Effective nanocides against plant pathogens in literature and the market

Compound	The target	Type of nanoparticles	References
Nanoguard	Bacteria, viruses, fungi	Product-based–nano-silver	Tran et al. (2022)
Nanozyme	Enhance the effectiveness of enzyme	– Carbon quantum dots, carbon nanotubes, and graphene oxide – Au, Ag, Pt, PD, Rh, Ru, and Ir	Jeelani et al. (2020) Wong et al. (2021) Wang et al. (2023)
Nanoagri	Pesticides and fertilizers	Such as carbon, silica dioxide, iron oxide, titanium dioxide	Nazir et al. (2020), https://aradbranding.com/en/nano-agri-products-agriculture-upsc/
Chitosan-copper nanoparticles	Damping off fungi	Chitosan copper	Vanti et al. (2020)
Silver nanoparticles as a fungicide	Against *Sclerotium rolfsii* on wheat plants	Silver	Desai et al. (2022)
Copper-graphene nanocomposite	Fusarium wilt disease and Root rot	Graphene copper	El-Abeid et al. (2020)
A delivery fungicide in mesoporous silica nanoparticles	Fusarium crown and root rot in tomato	Mesoporous silica Chitosan and fungicide	Mosa et al. (2022)
ZnO nanoparticles as a nanofungicide	*Magnaporthe grisea*	ZnO	Ghamari et al. (2022)

carbon quantum dots, carbon nanotubes, and graphene oxide show great promise as non-metallic nanozymes (Wong et al. 2021). Also, gold (Au), silver (Ag), platinum (Pt), palladium (PD), rhodium (Rh), ruthenium (Ru), and iridium (Ir) are only a few of the metals that make up metallic nanozyme (Wang et al. 2023).

10.11 Molecular Docking Effectively Forecasts the Nanomaterial-Pathogen Site of Action, Interaction, and Stability

The binding interactions between ligands (nanomaterials) and targets may be predicted using a computer method called molecular docking (pathogen biomolecules). Molecular docking finds the optimal binding mode by simulating interactions between the ligand and the target in various conformations and orientations (Fig. 10.1). It employs scoring methods to assess the stability of ligand-target complexes and estimate binding affinities (Lin et al. 2020). To rationally develop and find new insecticides, understanding their biomolecular interactions by molecular docking is an interesting scaffold. Insecticide complexes' binding energy, free energy, and stability may be estimated using data gained via docking. Because it is performed before the experimental section of any inquiry, molecular docking can prove the viability of any insecticidal action. Enzymes may be used as a predictor of insecticide binding capabilities to nucleic acid, and molecular docking has transformed the study of interactions between tiny chemicals (ligands) and protein targets. This is why entomologists are often looking into the method of interaction between nucleic acid and insecticides in an attempt to demonstrate the harmful impact mechanism of insecticides at the molecular level (Mehrotra et al. 2013; Agarwal et al. 2014). The experimental methods are made accessible to determine the true binding mode of the complex if the docking algorithm predicts the indicated interaction (Dar and Mir 2017). As a result, researchers can create improved pesticides. This information may aid in identifying structural modifications to pesticides that will result in sequence-/structure-specific binding (Holt et al. 2008).

Due to their novel characteristics and possible therapeutic uses, nanomaterials have emerged as intriguing options for controlling plant pests. Understanding the molecular connections between nanomaterials and disease is crucial for developing individualized and effective treatment methods. To understand the underlying workings, interactions, and stability of nanomaterials at the nanoscale, the computational method of molecular docking is essential (Li et al. 2023a, b).

Molecular docking allows for the atomic-level study of how nanomaterials interact with plant pests' essential enzymes and proteins. It reveals unique binding locations inside timeless essential processes, which sheds light on the therapeutic potential of nanomaterials. Nanomaterials with increased affinity and specificity may be designed by docking them onto pathogen biomolecular targets, which helps discover critical residues involved in binding interactions (Krishnani et al. 2022).

The method of action of nanomaterials may be deduced from the binding poses and interaction patterns inside pathogen biomolecular targets, which are revealed by molecular docking. It sheds light on the mechanisms through which nanomaterials interfere with pathogen activity, enzymes, and cellular functions. To stop viruses from entering cells or replicating, docking studies may show how nanoparticles interact with surface proteins. Docking also reveals the binding mechanisms of nanomaterials to cell wall components or key enzymes, which may kill bacteria or prevent them from releasing virulence factors (Lokhande et al. 2023).

To create successful treatments, evaluating the stability and dynamics of nanomaterial-pathogen complexes is essential. By revealing the stability and strength of connections, molecular docking helps improve nanomaterial development. Surface changes, functionalization, and other methods may improve stability, toxicity, and targeting. The size, shape, surface charge, and functional groups of nanomaterials are all factors in their therapeutic effectiveness, and molecular docking studies help choose the best ones (Majumder et al. 2023).

Experimental validation of molecular docking results is common practice to verify predicted interactions and assess therapeutic potential. Binding affinities, kinetics, and thermodynamics of nanomaterial-pathogen interactions may be verified using surface plasmon resonance, isothermal titration calorimetry, or fluorescence spectroscopy. Nanomaterial-based therapies' effectiveness, biodistribution, and safety characteristics may be gleaned from cellular and animal research (Malla et al. 2023).

Molecular docking is useful for learning about the mode of action of insecticides such as synthetic chemical compounds, plant extracts, microbial insecticides, nanoparticles, etc., which have often been used against plant pests over many years. The order Coleoptera has been the focus of less research on the effects of pesticides on insects than any other order of insects. From this vantage point, the following examples are adequate. In Domínguez-Arrizabalaga et al. 2020, Domínguez-Arrizabalaga et al. investigated the efficacy of *Bacillus thuringiensis* as a microbial pesticide agent against many agro-forestry pest species, including Coleoptera and Lepidoptera. Insects of the Coleoptera, Lepidoptera, Hemiptera, and Hymenoptera orders, as well as mites, are killed by *Bacillus thuringiensis* because of the inclusions, called parasporal crystalline inclusions, that are formed of proteins (Cry and Cyt) (Schnepf et al. 1998). In addition, while *B. thuringiensis* is in its vegetative development phase, it secretes toxins into the medium. The vegetative insecticidal protein (Vip) and the secretory insecticidal protein (Sip) are two examples of these poisons (Sip) (Palma et al. 2014).

Since the active toxin contacts a specific receptor in the midgut epithelial cells, where an oligomeric pre-pore structure is generated, and since changes in midgut receptors are the first step in the development of insect resistance, molecular docking may shed light on the mode of action of a wide range of toxins (Jurat-Fuentes and Crickmore 2017). Most important coleopteran families, such as Chrysomelidae, Curculionidae, Scarabaeidae, Tenebrionidae, etc., have been demonstrated to be susceptible to the Cry3Aa, Cry3Ba, Cry3Bb, and Cry3Ca proteins (Domínguez-Arrizabalaga et al. 2020). Although heterologous competition assays show that each protein may bind at a different site, it has been shown that Cry3Ba, Cry3Aa, and Cry3Ca all share the same receptor (Rausell et al. 2004). Also, Cry3Bb, Cry3Ca, and Cry7Aa proteins competed in the same binding sites in sweet potato weevil, *Cylas puncticollis*; thus, the three proteins may become ineffective because of mutation in the midgut receptors (Hernández-Martínez et al. 2014). Ochoa-Campuzano et al. (2007) reported that Cry3Aa in *Leptinotarsa decemlineata* is considered a receptor of ADAM metalloprotease, and this binding interaction improves Cry3Aa pore formation.

To connect pyridylpyrazole to functional groups like alkanes, benzene rings, and heterocycles, a carboxylate group is used as a "linker." Yang et al. (2022) created new RyRs insecticides comprising an N-pyridylpyrazole moiety. The majority of the pieces came from the structures of common agricultural chemicals. Insecticidal properties against *Plutella xylostella* and *Spodoptera frugiperda* were tested after the design and synthesis of 44 new pyridylpyrazole carboxylate derivatives. The potential target of these compounds was also discussed, with compounds G34 and G35 being related to ryanodine receptors (RyRs). Crystal structures of the N-terminal domain of *P. xylostella* RyR were used to study how chemicals G34 and G35 dock with RyR. Consistency between the molecules G34 and G35 and the RyR N-terminal domain gap is possible. Both the aniline portion of G34 and the acetanilide portion of G35, which are surrounded by the residues A/Val-73, A/Arg-74, A/Asn-102, B/

Glu-155, B/Tyr-175, and A/Val-73, A/Gly-101, B/Pro-148, B/Ile-166, and B/Tyr-175, respectively, reach deep into the hydrophobic pocket. Also, residues A/Val-73, A/Gly-101, A/Asn-102, and B/Tyr-175 encircled chlorantraniliprole. One hydrogen bond was established between compound G34 and the B/Tyr-175 residue at a distance of 2.2, while three hydrogen bonds were formed between compound G35 and the A/Val-73, A/Gly-101, and B/Pro-148 residues at distances of 1.9, 3.5, and 3.0. Evidence for a possible insecticidal mechanism against *P. xylostella* is provided by the ability of compounds G34 and G35 to fix in the N-terminal domain of RyR. One possible explanation for G34's diminished insecticidal activity against *P. xylostella* is that it has less hydrogen bond interactions in its docking complex (binding energy = −5.65 kcal.mol^{-1}) than G35 (binding energy = −5.94 kcal.mol^{-1}). Chlorantraniliprole bound to the protein receptor with a lower binding energy (−7.03 kcal.mol^{-1}) than the chemical G35 (binding energy = −5.94 kcal.mol^{-1}), presumably because it established four hydrogen bonds with the receptor's A/Val-73, A/Gly-101, A/Asn-102, and B/Tyr-175 residues. This may explain why compound G35 has less insecticidal activity against *P. xylostella* than chlorantraniliprole. Due to the lack of crystal structures for cowpea aphids, *Aphis craccivora*, a docking investigation was performed using X-ray crystal structures of acetylcholine (protein AChBP) from *Lymnaea stagnalis* (PDB 2ZJU) (Alanazi et al. 2022). With an RMSD of 1.3201 and a binding energy of −8.17 kcal/mol, the cocrystallized ligand imidacloprid was redocked inside the AChBP binding area. H-bond lengths and binding energy were evaluated when the target compounds were docked within the AChBP and H-bonding amino acid residues. The binding score of Compound 9 to the AChBP site was −8.51 kcal/mol, which is quite close to that of imidacloprid. In addition, the quinoxaline moiety connected with Tyr385 through arene-H interactions, while the azomethine N created H-bonds with Trp143. In contrast, the thiazolidine quinoxaline derivative 15 formed three hydrogen bonds: (1) Trp143 with CN, (2) Thr144 with quinoxaline NH, and (3) Cys187 with thiazolidino S. This compound also exhibited arene-H interactions with the thiazolidino moiety. The thiazolidinone C=O established a hydrogen bond with Trp143, making compound 16 the most active chemical targeting cowpea aphid nymphs (with a binding score of −10.54 kcal/mol).

Another compound that affects insect acetylcholinesterase is 2, 3-dimethylmaleic anhydride. CASTp study of the projected model of AChE revealed three significant pockets, designated CP1 (CASTp Pocket 1), CP2 (CASTp Pocket 2), and CP3 (CASTp Pocket 3), against the rice weevil, Sitophilus oryzae. Most of the residues that make up the catalytic triad (SER111, GLU237, and HIS351) and the substrate binding site (GLY29, GLY30, GLY31, GLU110, SER111, ALA112, VAL115, SER266, LEU270, ASN271, VAL306, GLY352, and ILE355) of the modeled AChE can be found in CP1. Using the AutoDock 4.2 software, we performed docking analysis in two different ways: first, with the whole protein and, then, with only the region of interest. The initial docking simulations helped us identify and verify putative ligand binding locations. Cluster analysis was used to determine the resulting numerous conformations of AChE and 2,3-dimethylmaleic anhydride complexes. Four of the most populous clusters were selected for further study, denoted below as AC1, AC2, AC3, and AC4. Three of the four clusters, AC1, AC3, and AC4, were found to validate the three main CASTp pockets, CP1, CP2, and CP3, as possible ligand binding sites. The last cluster, AC2, didn't share any features with the others; therefore, it was singled out as a separate and distinct candidate for a binding site. Within CP1, a cluster of AC1 forms close to the region allocated for substrate binding. Consequently, the binding site for future study was validated jointly by considering the AC1 and CP1 sites. In addition, fresh targeted docking simulations were run on the validated binding location. The second round of docking studies yielded a docking conformation with a binding energy of 5.25 kcal/mol, and this structure was chosen as the complex's typical structure. Protein and ligand interact with one another on the level of 153 individual atomic atoms. In addition, 2,3-dimethyl maleic anhydride formed hydrogen bonds with CYS198 and GLU199 at 1.81 and 2.17, respectively, from these amino acids. Individual amino acids and the protein's collective interaction energies (IE) with the ligand were also studied. Both electrostatic (IE = −18.50 kcal/mol) and van der Waals (IE = −19.16 kcal/mol) interactions contributed equally to the IE between the protein and ligand. It was determined that the interaction energy of the established residues is less than −2.0 kcal/mol. The amino acids GLU199, GLU191, and PHE202 were identified as van der Waal interaction contributors, whereas the amino acids PHE241, PHE200, and TYR244 were identified as polar or electrostatic interaction contributors (Singh et al. 2017). Furthermore, GPI-anchored alkaline phosphatases (ALP) are also important for the Cry3Aa binding to Tenebrio molitor brush border membrane vesicles (BBMV) and are highly expressed when larvae are exposed to Cry3Aa (Zúñiga-Navarrete et al. 2013).

10.12 Advantages and Disadvantages of Docking

Molecular docking is a computational technique used in drug discovery and various other scientific fields to predict the interactions between molecules, such as small molecules (ligands) and biological macromolecules (receptors), like proteins or nucleic acids. When applying molecular docking to study interactions between nanomaterials and plant receptors, there are both advantages and disadvantages to consider:

10.12.1 Advantages

Cost-effective: Molecular docking is a cost-effective and time-saving method compared to experimental techniques. It can significantly reduce the time and resources required for the initial screening of potential interactions.

High Throughput: It allows for the screening of a large number of nanomaterials and plant receptors in a relatively short amount of time, making it suitable for virtual high-throughput screening.

Hypothesis Generation: Molecular docking can provide insights into potential binding sites, binding modes, and affinity rankings for various nanomaterials, aiding in the generation of hypotheses for further experimental validation.

Customization: The method is highly customizable, allowing researchers to adjust parameters and incorporate additional information like energy calculations, solvation effects, and flexibility of molecules to improve accuracy.

Reduction of Lab Work: By narrowing down the list of potential candidates, molecular docking can reduce the amount of experimental work required, saving time and resources (Van Dijk et al. 2005; Orihuela and Ungerfeld 2019).

10.12.2 Disadvantages

Simplification: Molecular docking relies on simplifications and assumptions, such as rigid receptor structures and static conformations, which may not accurately represent the dynamic nature of biological systems. In addition from previous work, molecular docking technique gives a low binding affinity which does not suggest the real interaction may be due to the presence of unfavorable interactions and only a few significant interactions were detected between nanomaterial and amino acid pocket (Rarey et al. 1996).

Accuracy and False Positives: Docking results can produce false positives and false negatives due to limitations in force field accuracy, scoring functions, and the neglect of important factors like water-mediated interactions and induced-fit effects (Lill 2011).

Parameter Sensitivity: The accuracy of molecular docking is highly dependent on the choice of parameters, scoring functions, and force fields, which can introduce variability in results. This limitation increases due to there is a limited number of tools that can be used to preform docking for nanomaterial in addition to the previous knowledge needed to add atom parameters to the successful docking process (Jain 2007; Befort et al. 2021).

Data Availability: The accuracy of molecular docking heavily relies on the availability of high-quality structural data for both nanomaterials and plant receptors. A lack of accurate structural information can lead to unreliable predictions.

Limited Information: Docking can only provide information about binding affinity and binding modes but cannot predict the biological activity or toxicity of nanomaterials accurately (Lin et al. 2020).

Another challenge is due to the complexity of biological systems. In the case of plant receptors, the complexity of plant signaling pathways, post-translational modifications, and allosteric effects may not be adequately captured by molecular docking alone (Sheik Amamuddy et al. 2020).

Validation: Docking results should always be validated through experimental methods, such as X-ray crystallography or binding assays, which can be time-consuming and costly (Ross et al. 2012).

In summary, molecular docking can be a valuable tool for predicting interactions between nanomaterials and plant receptors, but it should be used cautiously, considering its limitations and the need for experimental validation. It can help guide research efforts by providing a starting point for further investigation but should not be solely relied upon for drawing definitive conclusions about the biological activity or behavior of nanomaterials in complex systems.

10.13 Conclusions

Nanoparticles play a significant role in the agricultural domain due to their unique physicochemical features. They interact with plants, causing physiological, morphological, and genotoxic changes. The effects of nanoparticles on plant growth and development vary depending on the nanomaterial's characteristics, application technique, and plant type. They have been identified as beneficial agents for enhancing plant growth, facilitating developmental processes, and providing protective effects. However, there have been recorded cases of nanoparticle toxicity and bioaccumulation, leading to drawbacks. Nanoparticles can also positively impact several stages of plant development, including expansion, germination, biomass accumulation, and root and leaf development. The physiological effects of nanoparticles on plants led to increased seed germination rates and total biomass or grain yield. Within plant systems, nanoparticles can affect photosynthesis. Some nanoparticles may alter the ability of plants to take in solar energy, while others can have both negative and positive effects on photosynthesis. Nanoparticles have the potential to influence signaling molecules and pathways in photosynthetic organisms. However, they can also cause cell membrane leakage and membrane damage, leading to mutagenic DNA lesions and oxidative stress.

Pattern recognition receptors (PRRs) can recognize microbial infections and host cell-linked DAMPs. Membrane-bound or cytoplasm-bound PRRs generate cytokines and defend the host. By altering germline-encoded pathogen recognition receptors, nanoparticles may inhibit pathogen infection. These receptors detect pathogen-associated molecular patterns (PAMPs) and are the body's first defense against infections. PRR-targeted pathogen infection treatments are not authorized. The connections between nanoparticles, chloroplasts, signaling molecules, and pathways and the possible deleterious consequences of NPs on plant development and metabolism need more study.

Nanoparticles have been found to regulate gene expression for biotic stress in plants by modulating gene expression. PR genes, which encode antimicrobial proteins, are essential for preventing the proliferation of harmful microorganisms. Nanoparticles can stimulate the expression of PR genes, leading to increased defensive responses against pathogens in plants. Silver and titanium nanoparticles have been shown to stimulate the overexpression of PR genes in plants, enhancing the synthesis of PR proteins and reducing disease severity. The precise mechanisms by which nanoparticles regulate gene expression in plants are still being investigated, but it is believed that they interact with plant cell receptors and signaling molecules, triggering a cascade of molecular events that ultimately result in gene expression changes. Reactive oxygen species (ROS) produced by nanoparticles inside plant cells may serve as signaling molecules to turn on genes involved in defense.

Nanoparticles have become promising tools in plant science due to their ability to modulate gene expression and influence plant defense responses. Their interaction with plant cells at the molecular level, potentially involving specific genes and receptors, is crucial for harnessing their full potential in controlling and managing plant diseases. Receptor-mediated interactions between nanoparticles and plant hormones have been observed, with silver nanoparticles (AgNPs), titanium dioxide nanoparticles (TiO_2 NPs), and silicon nanoparticles (SiNPs) potentially influencing plant growth and defense responses. Bio-receptor-functionalized nanoparticles, such as antibody-functionalized gold nanoparticles, aptamer-functionalized magnetic nanoparticles, protein-functionalized silver nanoparticles, and enzyme-functionalized silica nanoparticles, have been shown to effectively detect and kill bacteria. However, further research is needed to fully understand the mechanisms underlying nanoparticle-mediated gene expression modulation and its impact on plant defense mechanisms. Nanocides are effective in controlling plant-fungal infections by selectively activating their pathogen-suppressing properties. These nanozymes, made from metal oxides, carbon-based nanomaterials, and metals like gold, silver, platinum, palladium, rhodium, ruthenium, and iridium, have shown a 99.99% elimination rate upon contact.

Molecular docking is a computer method that predicts the binding interactions between nanomaterials and pathogens, allowing researchers to develop and find new insecticides. This method helps estimate the binding energy, free energy, and stability of insecticide complexes, proving the viability of any insecticidal action. Molecular docking has transformed the study of interactions between tiny chemicals and protein targets, allowing entomologists to demonstrate the harmful impact mechanism of insecticides at the molecular level. Nanomaterials have emerged as a promising option for controlling plant pests due to their novel characteristics and potential therapeutic uses. Understanding the molecular connections between nanomaterials and diseases is crucial for developing effective treatment methods. Molecular docking is essential for studying nanomaterials' interactions with plant pests' essential enzymes and proteins, revealing unique binding locations, and revealing the therapeutic potential of nanomaterials. Docking studies help design nanomaterials with increased affinity and specificity by docking them onto pathogen biomolecular targets. The mode of action of nanomaterials can be deduced from their binding poses and interaction patterns inside pathogen biomolecular targets. This helps in determining the mechanisms through which nanomaterials interfere with pathogen activity, enzymes, and cellular functions. Experimental validation of molecular docking results is common practice to verify predicted interactions and assess therapeutic potential. Molecular docking is useful for learning about the mode of action of insecticides, such as synthetic chemical compounds, plant extracts, microbial insecticides, and nanoparticles.

References

Abdal Dayem A, Hossain MK, Lee SB, Kim K, Saha SK, Yang GM et al (2017) The role of reactive oxygen species (ROS) in the biological activities of metallic nanoparticles. Int J Mol Sci 18(1):120

Abdelkhalek A, Király L, Al-Mansori ANA, Younes HA, Zeid A, Elsharkawy MM, Behiry SI (2022) Defense responses and metabolic changes involving phenylpropanoid pathway and PR genes in squash (Cucurbita pepo L.) following Cucumber mosaic virus infection. Plants, 11(15):1908.

Adil M, Bashir S, Bashir S, Aslam Z, Ahmad N, Younas T et al (2022) Zinc oxide nanoparticles improved chlorophyll contents, physical parameters, and wheat yield under salt stress. Front Plant Sci 13:932861

Agarwal S, Jangir DK, Mehrotra R, Lohani N, Rajeswari M (2014) A structural insight into major groove directed binding of nitrosourea derivative nimustine with DNA: a spectroscopic study. PLoS One 9:e104115. https://doi.org/10.1371/journal.pone.0104115

Agrahari S, Dubey A (2020) Nanoparticles in plant growth and development. In: Biogenic nanoparticles and their use in agro-ecosystems, pp 9–37

Ahmed B, Rizvi A, Ali K, Lee J, Zaidi A, Khan MS, Musarrat J (2021) Nanoparticles in the soil–plant system: a review. Environ Chem Lett 19:1545–1609

Alanazi MA, Arafa WAA, Althobaiti IO, Altaleb HA, Bakr RB, Elkanzi NAA (2022) Green design, synthesis, and molecular docking study of novel quinoxaline derivatives with insecticidal potential against *Aphis craccivora*. ACS Omega 7(31):27674–27689. https://doi.org/10.1021/acsomega.2c03332

Alhammad BA, Ahmad A, Seleiman MF (2023) Nano-hydroxyapatite and ZnO-NPs mitigate Pb stress in maize. Agronomy 13(4):1174

Ali S, Mehmood A, Khan N (2021) Uptake, translocation, and consequences of nanomaterials on plant growth and stress adaptation. J Nanomater 2021:1–17

Amarante-Mendes GP, Adjemian S, Branco LM, Zanetti LC, Weinlich R, Bortoluci KR (2018) Pattern recognition receptors and the host cell death molecular machinery. Frontiers in immunology, 9:417707.

Atha DH, Wang H, Petersen EJ, Cleveland D, Holbrook RD, Jaruga P et al (2012) Copper oxide nanoparticle mediated DNA damage in terrestrial plant models. Environ Sci Technol 46(3):1819–1827

Befort BJ, DeFever RS, Tow GM, Dowling AW, Maginn EJ (2021) Machine learning directed optimization of classical molecular modeling force fields. J Chem Inf Model 61(9):4400–4414

Benjamin SR, de Lima F, Nascimento VAD, de Andrade GM, Oriá RB (2023) Advancement in paper-based electrochemical biosensing and emerging diagnostic methods. Biosensors 13(7):689

Chen J, Wang X, Pan D, Lyu S, Wei X, Wang C (2017) Titanium as a beneficial element for crop production

Choudhary A, Singh S, Ravichandiran V (2022) Toxicity, preparation methods and applications of silver nanoparticles: an update. Toxicol Mech Methods 32(9):650–661

Dar AM, Mir S (2017) Molecular docking: approaches, types, applications and basic challenges. J Anal Bioanal Tech 8:356. https://doi.org/10.4172/2155-9872.1000356

Daramola OB, Omole RK, Akinwale IV, Otuyelu FO, Akinsanola BA, Fadare TO et al (2022) Bioreceptors functionalized nanoparticles: a resourceful sensing and colorimetric detection tool for pathogenic bacteria and microbial biomolecules. Front Nanotechnol 4:885803

Desai AS, Singh A, Edis Z, Haj Bloukh S, Shah P, Pandey B, Bhagat N (2022) An in vitro and in vivo study of the efficacy and toxicity of plant-extract-derived silver nanoparticles. Journal of Functional Biomaterials, 13(2):54.

Domínguez-Arrizabalaga M, Villanueva M, Escriche B, Ancín-Azpilicueta C, Caballero P (2020) Insecticidal activity of *Bacillus thuringiensis* proteins against coleopteran pests. Toxins 12:430. https://doi.org/10.3390/toxins12070430

El-Abeid SE, Ahmed Y, Daròs JA, Mohamed MA (2020) Reduced graphene oxide nanosheet-decorated copper oxide nanoparticles: a potent antifungal nanocomposite against fusarium root rot and wilt diseases of tomato and pepper plants. Nanomaterials 10(5):1001

Elmer W, White JC (2016) The use of metallic oxide nanoparticles to enhance growth of tomatoes and eggplants in disease-infested soil or soilless medium. Environ Sci Nano 3(5):1072. https://doi.org/10.1039/C6EN00146G

El-Sayed ESR, Mohamed SS, Mousa SA, El-Seoud MAA, Elmehlawy AA, Abdou DA (2023) Bifunctional role of some biogenic nanoparticles in controlling wilt disease and promoting growth of common bean. AMB Express 13(1):41

Gao M, Chang J, Wang Z, Zhang H, Wang T (2023) Advances in transport and toxicity of nanoparticles in plants. J Nanobiotechnol 21(1):75

Ghamari R, Ahmadikhah A, Tohidfar M, Bakhtiarizadeh MR (2022) RNA-Seq analysis of Magnaporthe grisea transcriptome reveals the high potential of ZnO nanoparticles as a nanofungicide. Front Plant Sci 13:896283

Ghorbanpour M, Movahedi A, Hatami M, Kariman K, Bovand F, Shahid MA (2021) Insights into nanoparticle-induced changes in plant photosynthesis. Photosynthetica 59(4):570–586

Gomaa EZ (2022) Nanozymes: a promising horizon for medical and environmental applications. J Clust Sci 33(4):1275–1297

Hasanuzzaman M, Raihan MRH, Masud AAC, Rahman K, Nowroz F, Rahman M, Nahar K, Fujita M (2021) Regulation of reactive oxygen species and antioxidant defense in plants under salinity. Int J Mol Sci 22:9326. https://doi.org/10.3390/ijms22179326

Hernández-Martínez P, Vera-Velasco NM, Martínez-Solís M, Ghislain M, Ferré J, Escriche B (2014) Shared binding sites for the *Bacillus thuringiensis* proteins Cry3Bb, Cry3Ca, and Cry7Aa in the African sweet potato pest *Cylas puncticollis* (Brentidae). Appl Environ Microbiol 80:7545–7550. https://doi.org/10.1128/AEM.02514-14

Holt PA, Chaires JB, Trent JO (2008) Molecular docking of intercalators and groove-binders to nucleic acids using Autodock and Surflex. J Chem Inf Model 48:1602–1615. https://doi.org/10.1021/ci800063v

Hussain F, Hadi F, Rongliang Q (2021) Effects of zinc oxide nanoparticles on antioxidants, chlorophyll contents, and proline in Persicaria hydropiper L. and its potential for Pb phytoremediation. Environ Sci Pollut Res 28:34697–34713

Jain AN (2007) Surflex-Dock 2.1: robust performance from ligand energetic modeling, ring flexibility, and knowledge-based search. J Comput Aided Mol Des 21:281–306

Jain D, Khurana JP (2018) Role of pathogenesis-related (PR) proteins in plant defense mechanism. In: Singh A, Singh IK (eds) Molecular aspects of plant-pathogen interaction. Springer, Singapore, pp 265–281

Jang JH, Shin HW, Lee JM, Lee HW, Kim EC, Park SH (2015) An overview of pathogen recognition receptors for innate immunity in dental pulp. Mediat Inflamm 2015:794143

Jeelani PG, Mulay P, Venkat R, Ramalingam C (2020) Multifaceted application of silica nanoparticles. A review. Silicon 12:1337–1354

Jiang HS, Yin LY, Ren NN, Zhao ST, Li Z, Zhi Y et al (2017) Silver nanoparticles induced reactive oxygen species via photosynthetic energy transport imbalance in an aquatic plant. Nanotoxicology 11:157–167. https://doi.org/10.1080/17435390.2017.1278802. [PubMed] [CrossRef] [Google Scholar]

Jurat-Fuentes JL, Crickmore N (2017) Specificity determinants for Cry insecticidal proteins: insights from their mode of action. J Invertebr Pathol 142:5–10. https://doi.org/10.1016/j.jip.2016.07.018

Kandhol N, Jain M, Tripathi DK (2022) Nanoparticles as potential hallmarks of drought stress tolerance in plants. Physiol Plant 174(2):e13665. https://doi.org/10.1111/ppl.13665

Karvar M, Azari A, Rahimi A, Maddah-Hosseini S, Ahmadi-Lahijani MJ (2022) Titanium dioxide nanoparticles (TiO2-NPs) enhance drought tolerance and grain yield of sweet corn (Zea mays L.) under deficit irrigation regimes. Acta Physiol Plant 44(2):14

Khalid MF, Iqbal Khan R, Jawaid MZ, Shafqat W, Hussain S, Ahmed T, Rizwan M, Ercisli S, Pop OL, Alina Marc R (2022) Nanoparticles: the plant saviour under abiotic stresses. Nanomaterials 12:3915. https://doi.org/10.3390/nano12213915

Khan MR, Adam V, Rizvi TF, Zhang B, Ahamad F, Jośko I, Zhu Y, Yang M, Mao C (2019) Nanoparticle-plant interactions: two-way traffic. Small 15:e1901794. https://doi.org/10.1002/smll.201901794

Khodakovskaya M et al (2012) Carbon nanotubes induce growth enhancement of tobacco cells. ACS Nano 6(3):2128–2135. https://doi.org/10.1021/nn204643g

Khoshru B, Mitra D, Joshi K, Adhikari P, Rion MSI, Fadiji AE et al (2023) Decrypting the multifunctional biological activators and inducers of defense responses against biotic stresses in plants. Heliyon 9:e13825. https://doi.org/10.1016/j.heliyon.2023.e13825

Krishnani KK, Boddu VM, Chadha NK, Chakraborty P, Kumar J, Krishna G, Pathak H (2022) Metallic and non-metallic nanoparticles from plant, animal, and fisheries wastes potential and valorization for application in agriculture. Environ Sci Pollut Res 29(54):81130–81165

Lambreva MD, Akhtar P, Sipka G, Margonelli A, Lambrev PH (2023) Fluorescence quenching in thylakoid membranes induced by single-walled carbon nanotubes. Photochem Photobiol Sci 22:1625–1635

Li M, Zhang Y, Li C, Lin J, Li X (2022) Polyvinyl chloride nanoparticles affect cell membrane integrity by disturbing the properties of the multicomponent lipid bilayer in Arabidopsis thaliana. Molecules 27(18):5906

Li DF, Tang Q, Yang MF, Xu HM, Zhu MZ, Zhang Y et al (2023a) Plant-derived exosomal nanoparticles: potential therapeutic for inflammatory bowel disease. Nanoscale Adv 5(14):3575–3588

Li Y, Xi K, Liu X, Han S, Han X, Li G et al (2023b) Silica nanoparticles promote wheat growth by mediating hormones and sugar metabolism. J Nanobiotechnol 21(1):2

Lill MA (2011) Efficient incorporation of protein flexibility and dynamics into molecular docking simulations. Biochemistry 50(28):6157–6169

Lin X, Li X, Lin X (2020) A review on applications of computational methods in drug screening and design. Molecules 25(6):1375

Lin W, Zhang J, Xu JF, Pi J (2021) The advancing of selenium nanoparticles against infectious diseases. Front Pharmacol 12:682284

Lokhande KB, Pawar SV, Madkaiker S, Nawani N, Venkateswara SK, Ghosh P (2023) High throughput virtual screening and molecular dynamics simulation analysis of phytomolecules against BfmR of Acinetobacter baumannii: anti-virulent drug development campaign. J Biomol Struct Dyn 41(7):2698–2712

Majumder S, Eckersall PD, George S (2023) Bovine mastitis: examining factors contributing to treatment failure and prospects of nano-enabled antibacterial combination therapy. ACS Agric Sci Technol 3(7):562–582

Malla BA, Ali A, Maqbool I, Dar NA, Ahmad SB, Alsaffar RM, Rehman MU (2023) Insights into molecular docking and dynamics to reveal therapeutic potential of natural compounds against P53 protein. J Biomol Struct Dyn 41:8762–8781

Mehrotra R, Jangir DK, Agarwal S, Ray B, Singh P et al (2013) Interaction studies of anticancer drug lomustine with calf thymus DNA using surface enhanced Raman spectroscopy. MAPAN 28:273–277. https://doi.org/10.1007/s12647-013-0086-5

Mocan T, Matea CT, Pop T, Mosteanu O, Buzoianu AD, Puia C et al (2017) Development of nanoparticle-based optical sensors for pathogenic bacterial detection. J Nanobiotechnol 15:1–14

Mosa MA, El-Abeid SE, Khalifa MMA, Elsharouny TH, El-Baz SM, Ahmed AY (2022) Smart pH responsive system based on hybrid mesoporous silica nanoparticles for delivery of fungicide to control fusarium crown and root rot in tomato. J Plant Pathol 104(3):979–992

Muhammad I, Shalmani A, Ali M, Yang QH, Ahmad H, Li FB (2021) Mechanisms regulating the dynamics of photosynthesis under abiotic stresses. Front Plant Sci 11:615942

Nair R (2016) Effects of nanoparticles on plant growth and development. In: Plant nanotechnology: principles and practices, pp 95–118

Nazir R, Ayub Y, Tahir L. (2020) Green-nanotechnology for precision and sustainable agriculture. In: Biogenic nano-particles and their use in agro-ecosystems, pp 317–357

Ochoa-Campuzano C, Real MD, Martínez-Ramírez AC, Bravo A, Rausell C (2007) An ADAM metalloprotease is a Cry3Aa Bacillus thuringiensis toxin receptor. Biochem Biophys Res Commun 362:437–442. https://doi.org/10.1016/j.bbrc.2007.07.197

Orihuela A, Ungerfeld R (2019) Tail docking in sheep (Ovis aries): a review on the arguments for and against the procedure, advantages/disadvantages, methods, and new evidence to revisit the topic. Livestock Sci 230:103837

Palma L, Muñoz D, Berry C, Murillo J, Caballero P (2014) *Bacillus thuringiensis* toxins: an overview of their biocidal activity. Toxins 6:3296–3325. https://doi.org/10.3390/toxins6123296

Pawar VA, Laware SL (2018) Seed priming a critical review. Int J Sci Res Biol Sci 5(5):94–101

Pérez-Labrada F, Hernández-Hernández H, López-Pérez MC, González-Morales S, BenavidesMendoza A, Juárez-Maldonado A (2020) Chapter 13—nanoparticles in plants: morphophysiological, biochemical, and molecular responses. In: Tripathi DK, Singh VP, Chauhan DK, Sharma S, Prasad SM, Dubey NK, Ramawat N (eds) Plant life under changing environment. Academic, Cambridge, pp 289–322

Poddar K, Sarkar D, Sarkar A (2020) Nanoparticles on photosynthesis of plants: effects and role. In: Green nanoparticles: synthesis and biomedical applications, pp 273–287

Rai M et al (2009) Silver nanoparticles as a new generation of antimicrobials. Biotechnol Adv 27:76–83. https://doi.org/10.1016/j.biotechadv.2008.09.002

Rarey M, Wefing S, Lengauer T (1996) Placement of medium-sized molecular fragments into active sites of proteins. J Comput Aided Mol Des 10:41–54

Rasheed A, Li H, Tahir MM, Mahmood A, Nawaz M, Shah AN et al (2022) The role of nanoparticles in plant biochemical, physiological, and molecular responses under drought stress: a review. Front Plant Sci 13:976179

Rausell C, García-Robles I, Sánchez J, Muñoz-Garay C, Martínez-Ramírez AC, Real MD, Bravo A (2004) Role of toxin activation on binding and pore formation activity of the *Bacillus thuringiensis* Cry3 toxins in membranes of *Leptinotarsa decemlineata* (Say). Biochim Biophys Acta 1660:99–105. https://doi.org/10.1016/j.bbamem.2003.11.004

Ross GA, Morris GM, & Biggin PC (2012) Rapid and accurate prediction and scoring of water molecules in protein binding sites. PloS one, 7(3):e32036.

Schnepf E, Crickmore N, Van Rie J, Lereclus D, Baum J, Feitelson J, Zeigler DR, Dean DH (1998) *Bacillus thuringiensis* and its pesticidal crystal proteins. Microbiol Mol Biol Rev 62:775–806. https://doi.org/10.1128/mmbr.62.3.775-806.1998

Selvarajan V, Obuobi S, Ee PLR (2020) Silica nanoparticles—a versatile tool for the treatment of bacterial infections. Front Chem 8:602

Sheik Amamuddy O, Veldman W, Manyumwa C, Khairallah A, Agajanian S, Oluyemi O et al (2020) Integrated computational approaches and tools for allosteric drug discovery. Int J Mol Sci 21(3):847

Sicwetsha S, Mvango S, Nyokong T, Mashazi P (2021) Effective ROS generation and morphological effect of copper oxide nanoparticles as catalysts. J Nanopart Res 23:1–18

Siddiqui MH, Al-Whaibi MH, Firoz M, Al-Khaishany MY (2015) Role of nanoparticles in plants. In: Nanotechnology and plant sciences: nanoparticles and their impact on plants, pp 19–35

Singh KD, Labala RK, Devi TB, Singh NI, Chanu HD, Sougrakpam S, Nameirakpam BS, Sahoo D, Rajashekar Y (2017) Biochemical efficacy, molecular docking and inhibitory effect of 2, 3-dimethylmaleic anhydride on insect acetylcholinesterase. Sci Rep 7(1):12483. https://doi.org/10.1038/s41598-017-12932-0

Swartzwelter BJ, Mayall C, Alijagic A, Barbero F, Ferrari E, Hernadi S, Michelini S, Navarro Pacheco NI, Prinelli A, Swart E, Auguste M (2021) Cross-species comparisons of nanoparticle interactions with innate immune systems: a methodological review. Nanomaterials (Basel) 11(6):1528. https://doi.org/10.3390/nano11061528

Thwala M, Musee N, Sikhwivhilu L, Wepener V (2013) The oxidative toxicity of Ag and ZnO nanoparticles towards the aquatic plant Spirodela punctata and the role of testing media parameters. Environ Sci Process Impacts 15:1830–1843. https://doi.org/10.1039/c3em00235g

Tian EK, Wang Y, Ren R, Zheng W, Liao W (2021) Gold nanoparticle: recent progress on its antibacterial applications and mechanisms. J Nanomater 2021:2501345

Tortella G, Rubilar O, Pieretti JC, Fincheira P, de Melo Santana B, Fernández-Baldo MA et al (2023) Nanoparticles as a promising strategy to mitigate biotic stress in agriculture. Antibiotics (Basel) 12(2):338

Tran NN, Le TNQ, Pho HQ, Tran TT, Hessel V (2022) Nanofertilizers and nanopesticides for crop growth. In: Plant and nanoparticles. Singapore, Springer Nature Singapore, pp 367–394

Tripathi DK, Singh S, Singh S, Srivastava PK, Singh VP, Singh S et al (2017) Nitric oxide alleviates silver nanoparticles (AgNps)-induced phytotoxicity in Pisum sativum seedlings. Plant Physiol Biochem 110:167–177. https://doi.org/10.1016/j.plaphy.2016.06.015. [PubMed] [CrossRef] [Google Scholar]

Tripathi D, Singh M, Pandey-Rai S (2022) Crosstalk of nanoparticles and phytohormones regulate plant growth and metabolism under abiotic and biotic stress. Plant Stress, 100107

Van Dijk AD, Boelens R, Bonvin AM (2005) Data-driven docking for the study of biomolecular complexes. FEBS J 272(2):293–312

Vannini C, Domingo G, Onelli E, Prinsi B, Marsoni M, Espen L et al (2013) Morphological and proteomic responses of Eruca sativa exposed to silver nanoparticles or silver nitrate. PLoS One 8:e68752. https://doi.org/10.1371/journal.pone.0068752

Vanti GL, Masaphy S, Kurjogi M, Chakrasali S, Nargund VB (2020) Synthesis and application of chitosan-copper nanoparticles on damping off causing plant pathogenic fungi. Int J Biol Macromol 156:1387–1395

Vivancos J, Labbé C, Menzies JG, Bélanger RR (2015) Silicon-mediated resistance of Arabidopsis against powdery mildew involves mechanisms other than the salicylic acid(SA)-dependent defence pathway. Mol Plant Pathol 16(6):572–582. https://doi.org/10.1111/mpp.12213

Wang X, Guo W, Hu Y, Wu J, Wei H, Wang X, …, Wei H (2016) Metal oxide-based nanomaterials for nanozymes. In: Nanozymes: next wave of artificial enzymes, pp 57–91

Wang X, Xie H, Wang P, Yin H (2023) Nanoparticles in plants: uptake, transport and physiological activity in leaf and root. Materials 16(8):3097

Wen T, Liu J, He W, Yang A (2020) Nanomaterials and reactive oxygen species (ROS). In: Nanotechnology in regenerative medicine and drug delivery therapy, pp 361–387

Wong EL, Vuong KQ, Chow E (2021) Nanozymes for environmental pollutant monitoring and remediation. Sensors 21(2):408

Yang S, Peng H, Zhu J, Zhao C, Xu H (2022) Design, synthesis, insecticidal activities, and molecular docking of novel pyridylpyrazolo carboxylate derivatives. J Heterocyclic Chem 59:1366–1375. https://doi.org/10.1002/jhet.4476

Yu Z, Li Q, Wang J, Yu Y, Wang Y, Zhou Q, Li P (2020) Reactive oxygen species-related nanoparticle toxicity in the biomedical field. Nanoscale Res Lett 15(1):115

Zhang L, Zhu C, Huang R, Ding Y, Ruan C, Shen XC (2021) Mechanisms of reactive oxygen species generated by inorganic nanomaterials for cancer therapeutics. Front Chem 9:630969

Zribi I, Ghorbel M, Brini F (2021) Pathogenesis related proteins (PRs): from cellular mechanisms to plant defense. Curr Protein Pept Sci 22:396–412. https://doi.org/10.2174/1389203721999201231212736

Zúñiga-Navarrete F, Gómez I, Peña G, Bravo A, Soberón M (2013) A Tenebrio molitor GPI-anchored alkaline phosphatase is involved in binding of *Bacillus thuringiensis* Cry3Aa to brush border membrane vesicles. Peptides 41:81–86. https://doi.org/10.1016/j.peptides.2012.05.019

Chapter 11
The Physiological Impacts of Nanoparticle Size, Morphology, and Concentration on the Phytopathogens Causing Plant Biotic Stress

Khaled Sayed-Ahmed and Yasser M. Shabana

Abstract This chapter focuses on the mechanisms of nanoparticles (NPs) against various phytopathogens (PHPs) responsible for plant biotic stress. Several environmental factors negatively affect plants, leading to a significant decline in crop productivity. These factors include biotic and abiotic stresses. Plant biotic stress is caused by diverse PHPs, including fungi, bacteria, viruses, nematodes, oomycetes, and parasitic plants. NPs have the potential to reduce the infection of PHPs by using lower doses and exhibiting higher efficiency than traditional pesticides. Additionally, NPs, as ultra-small particles, can target viral particles, which may provide an effective approach to controlling plant viruses. Therefore, the use of nanotechnology as an advanced approach to combat the growth of PHPs has attracted researchers' interest. However, a series of studies is required to evaluate the physiological impacts of NPs on PHPs and plants and the potential toxicity of these particles to human health in advance. PHPs attack plants through various mechanisms and cause diseases in leaves, roots, stems, fruits, and vascular systems. In this regard, the effects of nanoparticle size, morphology, and concentration on the growth of PHPs are discussed in this chapter to review the best potential applications of nanotechnology as a sustainable means of managing plant biotic stress.

K. Sayed-Ahmed
Department of Agricultural Biotechnology, Faculty of Agriculture, Damietta University, New Damietta, Egypt

The Center for Excellence in Research of Advanced Agricultural Sciences, Faculty of Agriculture, Damietta University, New Damietta, Egypt
e-mail: drkhaled_1@du.edu.eg

Y. M. Shabana (✉)
Plant Pathology Department, Faculty of Agriculture, Mansoura University, El-Mansoura, Egypt

M. Khan, J.-T. Chen (eds.), *Nanoparticles in Plant Biotic Stress Management*,
https://doi.org/10.1007/978-981-97-0851-2_11

Keywords Nanoparticles · Phytopathogens · Plant biotic stress · Size · Morphology · Concentration

11.1 Introduction

In recent years, the field of nanotechnology has emerged as a promising avenue for addressing plant biotic stress caused by phytopathogens (PHPs). The study of the physiological impacts of nanotechnology on PHPs requires a multidisciplinary knowledge encompassing chemistry, physics, biology, materials science, and other fields to explore the impact of nanoparticles (NPs) based on their size, concentration, and shape, on the growth of harmful PHPs (Rastogi et al. 2018; Tiwari et al. 2018). NPs have garnered significant interest in agriculture due to their promising chemical, physical, and biological properties as sustainable alternatives to traditional and harmful pesticides in combatting PHPs (Bhagyaraj and Oluwafemi 2018; Liu et al. 2023). The utilization of nanotechnology is crucial in addressing the challenge of PHPs developing resistance to conventional agricultural pesticides, which often leads to increased chemical doses. Consequently, NPs, either individually or in combination, can be employed through diverse agricultural practices to combat PHPs (Consolo et al. 2020; Akpinar et al. 2021).

The efficacy of NPs is governed by their size, shape, concentration, surface chemistry, and structure. The shape and size of NPs play a critical role in successful attachment and penetration into microbe cells (Schwab et al. 2016). Regarding structure, metallic structures, in particular, can contribute to the release of ions, which serve as the primary mechanism of toxicity against PHPs. Surface reactivity and coating provide NPs with the appropriate charge and interaction for targeting PHPs. However, the activity of NPs against PHPs typically depends on the sensitivity of the particular PHP species to the specific type of NPs used. Additionally, the dosage of NPs employed in PHP control plays a significant role in their effectiveness (Kanakari and Dendrinou-Samara 2023).

Understanding the physiological impacts of nanoparticle size, morphology, and concentration on PHPs is crucial for developing effective strategies for managing plant biotic stress. By harnessing the unique properties of NPs, such as their size-dependent behavior and shape-specific interactions, researchers can design targeted approaches to combat phytopathogenic infections. This knowledge opens up new possibilities for the development of innovative and sustainable nanotechnology-based solutions to enhance crop protection and improve agricultural productivity in the face of biotic stress challenges.

11.2 Nanotechnology

11.2.1 Fundamentals of Nanotechnology

Nanotechnology is a multidisciplinary field that involves the manipulation and control of materials at the nanoscale (typically 1–100 nm) to create new properties and functionalities. Nanomaterials (NMs) can be classified into different categories based on their structures, such as one-, two-, and three-dimensional NMs (Wan et al. 2016; Tiwari et al. 2018). Common types of NMs include composite-based, carbon-based, inorganic-based, and organic-based NMs. The unique physical and chemical characteristics of NMs are influenced by factors like size, morphology, size distribution, agglomeration state, aspect ratio, surface area, and crystallinity. NMs exhibit properties such as unique optical features resulting from the interaction between their outer electrons and light, magnetic characteristics based on electron charges and magnetic spins, and mechanical properties like hardness and elasticity. Nanotechnology has found applications in various fields, including industry, agriculture, and medicine, by enhancing catalytic activity, selectivity, and reactivity in chemical reactions. Many NPs show higher selectivity and reactivity than their bulk counterparts, maximizing the product and minimizing the required time for the chemical reaction due to the small NPs size and high specific area (Mori and Shitara 1994; Häkkinen et al. 2003).

NPs can be prepared using several techniques, which can be summarized into two main approaches: bottom-up and top-down (Birnbaum and Pique 2011). In the top-down approach, the bulk is broken down into tiny particles with one or all of their dimensions at the nanoscale (Balasooriya et al. 2017). In this approach, different mechanisms, such as attrition, electrochemical explosion, laser ablation, milling, nanolithography, and chemical etching, are used to prepare NPs. In the bottom-up approach, NPs are synthesized as smaller particles, such as atoms and/or molecules, using several techniques, including spinning, biological methods, plasma, sol–gel methods, pyrolysis, atomic condensation, and vapor deposition. A bottom-up approach in preparing NPs is preferred in many cases. NPs synthesized using the bottom-up approach generally exhibit greater surface energy, smaller volume, larger surface area, lower imperfections, more monodispersity, and spatial confinement than those prepared using the top-down approach.

11.2.2 Characterization Techniques

Characterizing NMs requires the use of various techniques to analyze their properties. Some common characterization techniques include the following:

- *Scanning Electron Microscopy (SEM):* SEM is used to measure the size, size distribution, and morphology of NMs. It does not require the transmission of an electron beam and can be used for bulk samples (Ma 2016).
- *Transmission Electron Microscopy (TEM):* TEM involves focusing an electron beam on a thin sample to produce high-resolution micrographs that reveal the size, lattice type, and shape of NMs (Carter 2009). It is also useful for characterizing the crystalline structure of specific sections of materials (Anderson et al. 2013).
- *X-ray Diffraction (XRD):* XRD is used to determine the crystallinity nature of NMs and can also provide information about their size (Kumar et al. 2022).
- *Raman Spectroscopy:* Raman spectroscopy is used to examine the composition of materials at the nanoscale, both qualitatively and quantitatively. It can detect small variations in the composition of NMs (Shipp et al. 2017; Smith and Dent 2019; Worsfold et al. 2019).
- *UV/Vis Spectroscopy:* UV/vis spectroscopy is employed to characterize NMs by detecting their surface plasmon resonance (SPR), which provides information about their size and stability (Torres-Rivero et al. 2021; Abbas 2019).
- *Energy-Dispersive X-ray Spectroscopy (EDX):* EDX is an analytical technique used to determine the chemical composition of NM surfaces based on the unique electronic structure of each element (Srivastava 2012; Kalaivani 2021).
- *Atomic Force Microscopy (AFM):* AFM is used to identify the structural information of NMs, providing 3D visualization capabilities and data on size, surface texture, morphology, and roughness. It can be used in controlled conditions or even liquid dispersions (Yang et al. 2007; Chen and Xu 2020).
- *Confocal Laser Microscopy (CLSM):* CLSM produces clear images of specimens, even those that may appear blurred under a traditional microscope. It provides three-dimensional images related to the distribution of NMs at a micrometer resolution.
- *Selected Area Diffraction Pattern (SADP):* SADP is an instrument used in TEM to collect crystallographic data from small regions, typically around 1 nm, to characterize the architecture of NMs (Wang 2011; Zou et al. 2017).

These characterization techniques help researchers understand the size, morphology, composition, and other properties of NMs, enabling the development and optimization of nanotechnology applications.

11.3 Phytopathogens (PHPs)

Phytopathogens, or PHPs, are organisms that cause diseases in plants. They have a significant impact on agricultural productivity and can result in substantial economic losses. The widespread occurrence of PHPs poses a global challenge in the field of agriculture.

PHPs are particularly concerning because they affect crops that are of high economic importance, such as staple food crops. The damage caused by these pathogens often leads to decreased yields and reduced quality of agricultural produce. To manage these diseases, farmers often resort to the extensive and continuous use of pesticides (Al-Ani and Furtado 2020), which can have environmental and health implications. In addition to the direct impact on crop production, specific PHPs are capable of producing mycotoxins. Mycotoxins are toxic compounds produced by certain species or subgenera of fungi. These mycotoxins can contaminate crops and pose a threat to livestock, humans, poultry, and fish when consumed. They can cause various health issues and have detrimental effects on animal and human health. Efforts to manage and control PHPs are crucial to ensuring food security, protecting the environment, and safeguarding public health (Schasteen 2023).

11.3.1 Types of PHPs

PHPs can be classified into various categories based on the type of organism causing the plant disease (Horst 2013). The main types of PHPs include the following:

- *Fungi:* Fungal PHPs are one of the most common types and cause significant damage to crops. They can infect various plant parts and produce toxins that affect plant development and cause disease symptoms (Peng et al. 2021).
- *Viruses:* Plant viruses are small infectious agents that can cause diseases in plants. They typically invade plant cells and disrupt normal cellular functions, leading to symptoms such as stunting, discoloration, and deformities.
- *Nematodes:* Nematodes are microscopic roundworms that can be parasitic to plants. They feed on plant tissues and can cause extensive damage to roots, leading to reduced plant growth and yield (Jones et al. 2013).
- *Bacteria:* Phytopathogenic bacteria infect plants and cause diseases such as bacterial blight, bacterial wilt, and soft rot. They can enter plant tissues through wounds or natural openings and disrupt normal plant functions (Aguilar-Marcelino et al. 2020).
- *Phytoplasmas:* Phytoplasmas are specialized bacteria that are transmitted by insect vectors. They infect the phloem tissues of plants and can cause a wide range of symptoms, including stunting, abnormal growth, and yellowing of leaves (Namba 2019).
- *Parasitic Higher Plants:* Some higher plants are parasitic and can infect other plants, leading to negative effects on the host plants' growth and development (Hegenauer et al. 2017).
- *Mollicutes:* Mollicutes are bacteria that lack a cell wall and can parasitize plants. They are among the smallest self-replicating organisms and can cause diseases in various crops (Kahane et al. 1993).

- *Parasitic Green Algae:* Certain species of green algae can infect plants and cause foliar diseases, particularly in greenhouse environments or warm humid climates (Pombert et al. 2014; Joubert and Rijkenberg 1971).
- *Protozoa:* Protozoa are single-celled eukaryotic organisms that can feed on organic matter, including plant debris or other microorganisms. Some protozoa can cause diseases in plants (Bertrand et al. 2015).
- *Viroids:* Viroids are infectious pathogens consisting of small, circular RNA molecules. They can infect plants and interfere with normal plant growth and development (Kovalskaya and Hammond 2014).

11.3.2 Mechanisms of Phytopathogen Attack

PHPs employ various mechanisms to attack and infect plants. These mechanisms include the following:

1. *Toxin Production:* Many PHPs, such as fungi, produce toxins that can directly damage plant cells or interfere with plant metabolism. These toxins can inhibit nucleic acid and protein synthesis, disrupt cellular processes, and induce cell death (Shang et al. 2016).
2. *Infection and Colonization:* PHPs, including bacteria and fungi, can enter plant tissues through wounds or natural openings. They then colonize and multiply within the plant, leading to the destruction of plant cells and tissues (Tampakaki et al. 2009).
3. *Disruption of Plant Physiology:* PHPs can interfere with plant physiology by draining nutrients from the host plant, disrupting nutrient transport systems, and altering plant hormone levels. These disruptions can have detrimental effects on plant growth and development.
4. *Molecular Interactions:* Some PHPs, particularly bacteria and fungi, secrete proteins called effector proteins that interact with host plant cells. These effectors manipulate host plant defense responses, allowing the pathogen to evade detection and establish an infection.
5. *Induction of Reactive Oxygen Species (ROS):* Certain pathogens, such as fungi, can induce the production of reactive oxygen species (ROS) in plant cells. ROS are highly reactive molecules that can cause oxidative damage to plant cells and contribute to disease development.
6. *Alteration of Signaling Pathways:* PHPs can interfere with plant signaling pathways, including defense signaling pathways, to suppress plant immune responses. This allows the pathogen to evade recognition and establish an infection (Dyakov and Zinovyeva 2007; Zeilinger et al. 2016).

Understanding the mechanisms of PHP attacks is crucial for developing effective strategies to manage plant diseases and reduce crop losses.

11.4 Plant Biotic Stress Induced by PHPs

Biotic stress in plants can be induced by various types of PHPs, including fungi, bacteria, viruses, insects, nematodes, and other pathogens. These pathogens can cause significant damage to plants, leading to changes in plant physiology and various negative effects on plant growth and development.

Physiological Changes: PHP infection can result in a range of physiological changes in host plants. These changes may include alterations in plant growth patterns, such as early flowering or reduced biomass accumulation. PHPs can also impact reproductive processes, leading to minimized seed sets. Additionally, plants may accumulate specific metabolites as a defense response to combat the detrimental effects of pathogen attack (Kovalchuk 2021).

Cellular and Molecular Responses: When plants experience biotic stress, a suite of cellular and molecular processes is triggered. Plants activate various defense mechanisms to combat the harmful impacts of pathogens and maintain their survival. These responses involve complex signaling pathways and gene expression changes to enhance plant resistance (Iqbal et al. 2021).

Limitations on Plant Growth: Biotic stress, including that induced by fungal and bacterial PHPs, can significantly impact crop productivity. Environmental factors and the presence of pathogens can restrict the complete development of host plants throughout their growth period. These limitations can lead to reduced crop yields and economic losses.

Plant Adaptation Strategies: Plants have evolved unique biological, molecular, and physiological strategies to cope with adverse conditions and promote their growth. However, continuous stress can overwhelm their defense systems and limit their ability to cope with environmental challenges (Chitnis et al. 2020).

Endophytes as a Potential Solution: Endophytes, which are fungi and bacteria that reside within the tissues of plants, have gained attention as potential alternatives for disease management and enhancing plant resistance to PHPs and environmental stressors. Endophytes can stimulate plant defense responses, influence the accumulation of secondary metabolites, activate antioxidant enzymes, and positively regulate gene expression related to plant protection.

Challenges in Stress Tolerance: Traditional methods, including genetic engineering, physical interventions, and chemical treatments, have limitations in providing long-term stress tolerance in plants. Many of these methods are not environmentally friendly. Therefore, the use of endophytes as a means to enhance plant resistance and resilience to PHPs and environmental stressors is being explored as a more sustainable and eco-friendly approach (Zheng et al. 2021; Biswas et al. 2016).

By understanding the mechanisms of biotic stress and exploring innovative strategies like utilizing endophytes, researchers aim to develop effective and sustainable solutions to mitigate the negative impacts of PHPs on plant health and agricultural productivity.

11.5 Abiotic Stress

Abiotic stress refers to the negative impact of non-living factors or environmental conditions on living organisms, including plants. These stressors are typically physical or chemical in nature and include factors such as salinity, drought, extreme temperatures (cold or heat stress), flooding, nutrient deficiencies or toxicities, high light intensity, and heavy metal pollution (Ben-Ari and Lavi 2012).

Impact on Plant Health: Abiotic stress can have significant adverse effects on plant health and productivity. It disrupts normal physiological processes and can lead to growth inhibition, reduced photosynthesis, wilting, leaf discoloration, reduced nutrient uptake, impaired reproductive development, and even plant death. Abiotic stress can also make plants more susceptible to diseases and pests.

Common Abiotic Stresses: Drought and salinity are among the most prevalent and economically significant abiotic stresses affecting crop productivity worldwide. Drought stress occurs when plants experience inadequate water availability, while salinity stress refers to the presence of high salt concentrations in the soil, which can hinder water uptake by plants.

Genetic Responses: The genetic responses to abiotic stress are complex and involve multiple genes and molecular pathways. Unlike biotic stress, which often involves specific interactions with pathogens, abiotic stress affects plants through various mechanisms. Identifying and controlling the genetic factors involved in abiotic stress responses can be challenging due to their multigenic nature.

Plant Defense Mechanisms: Plants have developed various defense mechanisms to cope with abiotic stress. These include structural adaptations like a waxy cuticle to reduce water loss, the accumulation of compatible solutes (osmolytes) to maintain cellular osmotic balance, the production of stress-related proteins (molecular chaperones) to protect against protein denaturation, and the activation of antioxidant systems to mitigate the damage caused by reactive oxygen species.

Signaling Molecules: Abiotic stress triggers the production and signaling of various molecules within plants. These include reactive oxygen species (ROS), stress hormones (e.g., abscisic acid), polyamines, gasotransmitters (e.g., nitric oxide), calcium ions, and phytochromes (light receptors). These molecules play crucial roles in signal transduction and the activation of stress-responsive genes.

Cellular Damage: Abiotic stress can cause cellular damage in plants. Reactive oxygen species generated under stress conditions can lead to oxidative stress, causing damage to proteins, lipids, and DNA. Osmotic stress, particularly dehydration, can disrupt cellular water balance, leading to membrane injury and impaired cellular functions (He et al. 2018).

Efforts are focused on understanding the mechanisms of abiotic stress tolerance in plants and developing strategies to enhance plant resilience. These include breeding for stress-tolerant crop varieties, improving water and nutrient management practices, implementing precision agriculture techniques, and utilizing biotechnological approaches to enhance stress tolerance in crops. By mitigating the impact of abiotic stress on plants, agricultural productivity can be improved, and food security can be ensured in challenging environmental conditions.

11.6 Physiological and Physical Effects of NPs on the PHPs

11.6.1 Antimicrobial Activities of NPs Against the PHPs

Antifungal Activities

The use of NPs to control phytopathogenic fungi holds great promise in agricultural fields. Until now, several metal NPs, such as Ag, Se, Cu, Mg, Fe, and Ni NPs, have been developed and investigated as potential antifungal agents. Among these, Ag-NPs have been extensively studied for their antifungal properties against several pathogenic fungi in agriculture, followed by Cu-NPs (Cruz-Luna et al. 2021). However, the high cytotoxicity associated with synthesized Ag-NPs limits their application in agriculture, particularly on edible plant parts for controlling fungal PHPs. Therefore, there is a need for other effective and low-cytotoxicity MNPs as antifungal agents, leading to the exploration of specific MNPs such as Se-NPs, which exhibit significant potential in combating phytopathogenic fungi (Garza-García et al. 2021). In this regard, Lazcano-Ramírez et al. (2023) evaluated the antifungal activities of Se-NPs synthesized through plant-mediated methods against *Fusarium oxysporum* and *Colletotrichum gloeosporioides*. They found that Se-NPs at different concentrations (0–1.7 mg/mL) exhibited antifungal activity against both PHPs at concentrations ranging from 0.25 to 1.7 mg/mL.

Selenium, a semi-metallic element belonging to the chalcogen family, was previously considered toxic but has since been recognized for its vital roles in various organisms. It exists naturally in different oxidation states, both inorganic and organic forms. Se-NPs synthesized using mycosynthesis have been investigated as potential antifungal agents, effectively combating pearl millet downy mildew disease in greenhouse conditions. They inhibit the sporulation of pathogenic fungi on the host plant leaves and enhance localized and systemic resistance against a wide range of biotic stresses (Joshi et al. 2019). The antifungal activity of Se-NPs may be attributed to the presence of ABC transporters, which are involved in secreting cell wall-degrading enzymes and antibiotics. Furthermore, Daniel et al. (2023) reported that ZnO-NPs exhibit antifungal activity against pathogenic fungi through various mechanisms. The most significant processes include the generation of reactive oxygen species (ROS), DNA damage, and lysosomal instability. ZnO-NPs achieve their antifungal activity by generating reactive oxygen species, which disrupt various metabolic pathways and interfere with fungal growth.

In summary, NPs, particularly Se-NPs and ZnO-NPs, demonstrate promising antifungal activities against phytopathogenic fungi associated with plant biotic stress. Exploring these NPs as potential alternatives to conventional fungicides offers the opportunity to control fungal PHPs in agriculture effectively. However, further research is necessary to optimize the formulation, concentration, and application methods of these NPs to ensure their efficacy while minimizing any potential negative effects on the environment and non-target organisms.

Antibacterial Activities

Substances used as antibacterial agents may be natural, semi-synthetic, or synthetic. While these substances have played a crucial role in managing infections caused by bacterial PHPs, their extensive use has led to the development of bacterial resistance, posing a significant challenge to their continued effectiveness and sustainability. To address this issue, nanotechnology offers a promising alternative by replacing traditional antibacterial agents and potentially mitigating PHP resistance (Rudramurthy et al. 2016).

The antibacterial activity of several NPs has been evaluated against numerous plant pathogenic bacteria using different methods (i.e., disc diffusion assay). Among the plant pathogenic bacteria, rod-shaped bacilli are the most common, infecting different parts of the host plant and causing functional inhibition. Some plant pathogenic bacteria, such as *Erwinia* and *Agrobacterium* species, degrade the cell wall or inhibit specific hormones in the host plant (Gautam and Kumar 2020). The most NPs type used in these studies was Ag-NPs, followed by Cu-NPs. In this respect, Doolotkeldieva et al. (2022) investigated the effects of the prepared Ag-NPs, Cu-NPs, and their mixture at different concentrations on the bacterial growth of *Erwinia amylovora*, *Pectobacterium carotovorum*, and *Pseudomonas syringae*. They found that *P. syringae* was the most sensitive to the tested NPs. The antibacterial activities of the fabricated MnO-NPs were also evaluated against *E. amylovora*, *Erwinia persicina*, and *Acetobacter pasteurianus*. It was found that *E. persicina* displayed greater resistance to the MnO-NPs, while *E. amylovora* showed higher sensitivity compared to the other tested bacteria (El-Moslamy et al. 2023).

In summary, the antibacterial activities of NPs against plant pathogenic bacteria can be attributed to several key processes. These include disrupting the cell membrane of the bacteria, inducing the formation of reactive oxygen species (ROS), facilitating the entry of NPs into bacterial cells through their membranes, and altering genetic characteristics through interactions between NPs, proteins, and DNA (Wang et al. 2017), as shown in Fig. 11.1. These mechanisms collectively contribute to the antimicrobial effects of NPs, offering potential strategies for controlling bacterial PHPs and combating resistance. However, further research is needed to optimize NP formulations, concentrations, and delivery methods to ensure their efficacy while minimizing any potential adverse effects on the environment and non-target organisms.

Antiviral Activities

Plant pathogenic viruses are responsible for massive losses in crop productivity, accounting for a substantial portion of the damage caused by PHPs. Plant viral diseases are challenging to control due to the diversity, mutations, recombination, and rapid evolution of viruses, which limit management options. Nanotechnology has emerged as a novel approach to address these challenges in the management and diagnosis of plant pathogenic viruses (Dutta et al. 2022). The field of

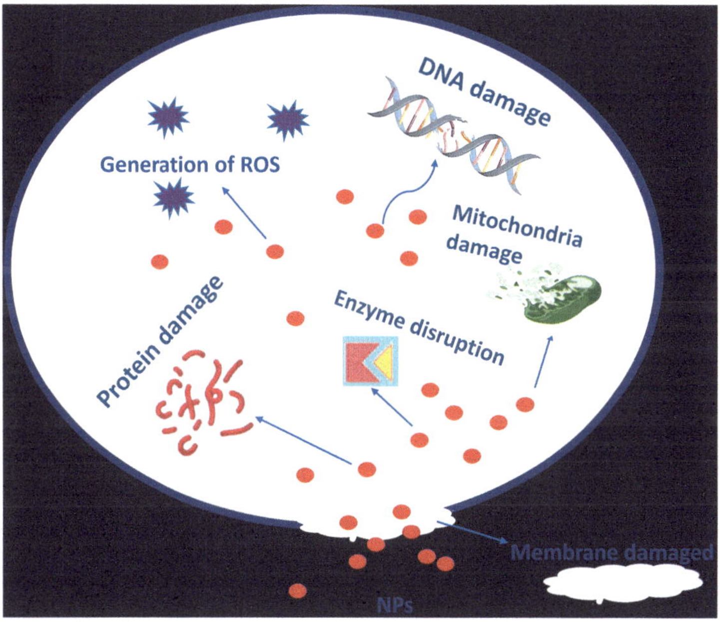

Fig. 11.1 Mechanisms of action of NPs against phytopathogens (PHPs). They include mitochondrial damage, generation of reactive oxygen species (ROS), and remarkable changes in genetic materials causing damage to DNA and thus disruption of protein synthesis as a result of the intracellular release of nanoparticles (NPs) in PHPs cells

"nanophytovirology" focuses on utilizing nanotechnology techniques to manage viral diseases in plants. Various NPs possess unique physicochemical characteristics that allow them to interact with plant pathogenic viruses, their vectors, and plants in beneficial ways. However, the investigation of the antiviral activity of NPs against plant pathogenic viruses is still in its early stages, requiring in-depth studies on factors such as application timing, dosage, and mode of action. While the mode of action of MNPs against plant pathogenic viruses is not yet fully understood, their potential as antiviral agents is being explored (Singh et al. 2022).

Several types of NPs have shown promising effects in combating plant pathogenic viruses. Ag-NPs are commonly used as antiviral agents due to their ability to reduce the infectivity and local lesion formation caused by viruses like the *Tomato mosaic virus* (Noha et al. 2018). Au-NPs have been shown to disrupt coat proteins, leading to the elimination of *Barley yellow dwarf virus* infections and dissociation of *Barley yellow mosaic virus* particles (Singh et al. 2022). ZnO-NPs have been found to decrease the invasion speed of *Tobacco mosaic virus*, resulting in a significant reduction in the incidence and severity of viral diseases (Hao et al. 2017). TiO_2-NPs and Fe_2O_3-NPs have demonstrated the ability to reduce viral replication and infection by the *Turnip mosaic virus* (Hao et al. 2018). Additionally, SiO_2-NPs limit viral infection through several mechanisms, such as a reduction in the speed of virus invasion and suppression of virus accumulation (Singh et al. 2022).

These examples highlight the potential of NPs as antiviral agents against plant pathogenic viruses. However, further research is needed to understand their precise mode of action, optimize their application protocols, and evaluate their efficacy and safety in different crops and environmental conditions. The development of NPs for effective viral disease management holds great promise for minimizing losses caused by plant pathogenic viruses in economically important crops.

Physiological Effects of NPs Against Other PHPs

In addition to the potential of NPs against plant pathogenic fungi, bacteria, and viruses, they have also demonstrated promising effects against other PHPs such as plant-parasitic nematodes and oomycetes.

Plant-parasitic nematodes cause a drastic decrease in crop quality and yield. SiO_2-NPs have been utilized to combat various plant-parasitic nematodes. Application of SiO_2-NPs at a concentration of 0.1 g/L resulted in a notable reduction in nematode infection in carrots by 40% and a remarkable decrease in the number of galls by 30% compared to the untreated control. Foliar spray of SiO_2-NPs at a concentration of 0.2 g/L reduced the infection of *Meloidogyne incognita* in eggplant, leading to a 37.92% improvement in dry shoot weight and a 70.42% increase in chlorophyll content (Khan et al. 2022). Cu-doped-ZnO-NPs and Cu-NPs were tested against *Meloidogyne javanica*. In this respect, Cu-doped-ZnO-NPs exhibited higher efficacy against *M. javanica* with an EC_{50} value of 2.60 μg/mL compared to Cu-NPs (EC_{50} = 25 μg/mL). The tested NPs inhibited the growth of *M. javanica* without notable phytotoxicity (Tryfon et al. 2022). Bio-Ag-NPs at a concentration of 100 μg/mL have been shown to decrease nematode mortality and activity, as well as inhibit egg hatching and larval movement (Li et al. 2023).

Downy mildew, caused by oomycetes, can result in significant crop yield losses. SiO_2-NPs have been investigated for their effects on downy mildew-infected Thompson seedless grapevines. Spraying SiO_2-NPs at a concentration of 150 ppm resulted in the overexpression of defense-related genes and transcription factors associated with the jasmonate and ethylene-responsive pathways. Disease severity was reduced by up to 81.5% with this treatment (Li et al. 2023). Cu-NPs have also shown potential as antimicrobial agents against the harmful effects of *Phytophthora sojae*. Cu-NPs induce oxidative stress, leading to disruptions in mycelial cells, carbon metabolism, intracellular structure, amino acid metabolism, and organic acid metabolic processes (Li et al. 2023).

These examples demonstrate the potential of NPs in controlling other PHPs, including plant-parasitic nematodes and oomycetes. However, further research is needed to optimize NP formulations, concentrations, and application methods for different PHPs and crop systems while ensuring their safety and minimizing any potential adverse effects.

11.6.2 The Effect of NPs Size

The size of NPs is a crucial factor that influences their antimicrobial activities. Different NP sizes can exhibit varying degrees of antimicrobial efficacy. Several studies have investigated the effect of NP size on their antimicrobial properties against bacteria and fungi.

It has been observed that the antibacterial activity of ZnO increases as the size of the particles decreases. This property has sparked interest among researchers due to the unique properties of ZnO-NPs at the nanoscale (Şahin et al. 2021). The variation in NP size can be attributed to changes in the ionic radius and electronegativity of the host metal, such as zinc, which may lead to the formation of nanorods (Munawar et al. 2021). Similarly, the antifungal activity of Ag-NPs against the phytopathogenic fungus *Fusarium oxysporum* f. sp. *radicis-lycopersici* (FORL) was evaluated at different nanosizes (3, 5, 8, and 10 nm). The smallest Ag-NPs showed the highest antifungal activity, indicating a size-dependent effect (Akpinar et al. 2021). Nano-chitosan particles also exhibit size-dependent effects, where smaller particles achieve higher dispersibility and efficacy than large particles (Xing et al. 2015).

In general, NPS exhibit various characteristics compared to their counterparts owing to the fact that the decrease in NPs size gradually maximizes the surface/volume ratio. The physical and chemical properties of the materials and metals differ depending on their size. The antimicrobial activities of different sizes of Ag-NPs were evaluated against *Bipolaris sorokiniana* and *Magnaporthe grisea*. Their efficacy against various types of PHPs, such as soil-borne fungi, was evaluated. Ag-NPs ranging from 20 to 30 nm showed better penetration within treated plant tissue than large particles (Jo et al. 2009). In addition to chemical stability and dispersity, the Ag-NP size is another crucial factor affecting their antimicrobial activities. Ag-NPs exhibited a rapid aggregation, which led to a significant increase in their size, causing highly variable antimicrobial activities against the same pathogen based on the size of the tested NPs.

In this respect, Baker et al. (2005) studied the effect of the size of Ag-NPS on their antimicrobial activities. Ag-NPs with two varied sizes ranged from 5 to 70 nm and from 50 to greater than 100 nm with average values of 15 and 75 nm, respectively. The results obtained from this study revealed that smaller Ag-NPs (approximately 15 nm) showed higher antibacterial activity than those around 75 nm. Moreover, it was found that Ag-NPs with sizes ranging from 1 to 10 nm can penetrate the bacterial cell membrane and have highly reactive {111} facets and electronic effects as well as a large surface area. Therefore, Ag-NPs with smaller sizes are more effective against bacterial strains, ultimately causing cell death at the end (Tang and Zheng 2018). Regarding CuO-NPs, their antibacterial activities against both gram-positive and gram-negative bacteria were found to be variable and size-dependent. The smallest CuO-NPs around 20 nm showed a higher inhibitory effect than those with a larger size (Azam et al. 2012). On the other hand, the dependence of the minimum inhibitory concentration of the prepared ZnO-NPs for various bacterial strains on their sizes was studied. It was observed that the use of ZnO-NPs with sizes of up to 10 nm was not effective. At these average sizes of ZnO-NPs, NPs distribution by size is complex and not usually narrow. ZnO-NPs in small size are prone to aggregation (Kumar et al. 2013).

ZnO-NPs at varied concentrations in the range of 0.125–1.00 mg/mL showed outstanding inhibition against *P. syringae* growth. In contrast, bulk ZnO exhibited lower inhibition at the same concentration. ZnO-NPs at a concentration of 0.5 mg/mL completely inhibited *P. syringae* growth, whereas approximately 30% of bacterial cells were viable after treatment with bulk ZnO at the same concentration, indicating a significant effect of the size of the antimicrobial agents against PHPs (Fan et al. 2023).

11.6.3 The Effect of NPs Shape

The shape of NPs is another important factor that influences their antimicrobial activities. Different NP shapes can exhibit varying degrees of antimicrobial efficacy. Several studies have investigated the effect of NP shape on their antimicrobial properties against bacteria and fungi.

Triangular magnetic NPs were found to be more antibacterial than spherical or rod-shaped MNPs. The sharp ends present in triangular MNPs enable them to penetrate the cell membrane more easily, leading to enhanced antibacterial activity (Siddiquah et al. 2018). Similarly, the antibacterial activities of Ag-NPs with different shapes (spherical, rod-shaped, and triangular) were evaluated, and truncated triangular Ag-NPs exhibited the highest activity, followed by spherical and rod-shaped Ag-NPs (Tang and Zheng 2018).

Flower-shaped ZnO-NPs, which may reach large sizes of up to 3 μm, showed antimicrobial activity against both gram-positive and gram-negative bacteria. Spherical ZnO-NPs exhibited antimicrobial activity regardless of the type of inhibited organism. Hexagonal ZnO-NPs displayed higher bactericidal activity against specific bacteria compared to triangular ZnO-NPs. Thorn-like ZnO-NPs demonstrated significant inhibition of bacteria and fungi, providing both antifungal and antibacterial activities (Siddiquah et al. 2018). Polygonal Cu-NPs with an average size of 345 nm exhibited higher antifungal activity (up to 46%) against *Fusarium oxysporum* compared to spherical Cu-NPs with a smaller size of around 278 nm at the same concentration. The polygonal shape of Cu-NPs, with a greater surface-area-to-volume ratio, enhanced their antifungal activity (Mohamed 2018). The shape-dependent antifungal activity of ZnO-NPs in hexagonal rods and lamellar platelets was investigated against *Colletotrichum gloeosporioides* and *Fusarium* spp. The platelet-shaped ZnO-NPs at approximately 246 nm with an average thickness of 48 nm exhibited higher antifungal activity and minimized the growth of *Fusarium solani* by up to 65%. This suggests that the selectivity of the interaction between platelet-shaped ZnO-NPs and pathogenic fungi contributes to their enhanced antifungal activity (Mosquera-Sánchez et al. 2020).

Overall, the shape of NPs plays a significant role in determining their antimicrobial activities. Different shapes can affect the penetration ability, surface area, and interaction with microorganisms, leading to variations in antimicrobial efficacy. However, it is important to note that the optimal shape may vary depending on the specific NPs, target microorganisms, and environmental conditions. Further research is needed to fully understand the shape-dependent effects of NPs and to optimize their applications in antimicrobial treatments.

11.6.4 The Impact of NPs Dose

Doolotkeldieva et al. (2022) found that the antibacterial activities of the tested Ag-NPs, Cu-NPs, and a mixture of them at different concentrations on the bacterial growth of *E. amylovora*, *P. carotovorum*, and *P. syringae* increased with increasing NPs concentration. The antifungal activity of Ag-NPs against *Fusarium oxysporum* was evaluated at different concentrations (12.5–100 ppm) of the NPs to study the effect of NPs concentration on their antifungal activity. Mycelium growth abilities were minimized by approximately 50, 75, and 90% as a result of Ag-NPs (3 nm) treatment at 25, 37.5, and 50 ppm, respectively (Akpinar et al. 2021). In the case of ZnO-NPs in the range of 20–70 nm, by increasing the concentration of ZnO-NPs up to 15 mM for 6 days, a great inhibition of *Colletotrichum* sp. up to 96% was observed as a result of the formation of groups of hyphal structures and loss in the continuity of some hyphae (Mosquera-Sánchez et al. 2020). Bafghi et al. (2022) investigated the effect of the Se-NPs concentration on their antifungal activities against *Aspergillus flavus*. They found that *A. flavus* was resistant to the effects of Se-NPs prepared at low concentrations of 0.065 and 0.125 μg/mL. On the other hand, at high concentrations ranging from 0.5 to 64 μg/mL, *A. flavus* was sensitive to the antifungal effect of the Se-NPs.

Ag-NPs were tested against *Alternaria alternata* at different concentrations to study the effect of NPs concentration on fungal growth. In this respect, the minimum inhibitory concentration (MIC) and minimum fungicidal concentration (MFC) were determined. MFC is the lowest concentration causing 99.9% death of the inoculum, while MIC is the lowest concentration that results in a reduction in PHP growth compared to that of the control. The MIC and MFC of the tested Ag-NPs were found at 125 ppm and 500 ppm, respectively. Determination of the MIC is necessary to determine the appropriate dose of NPs. High concentrations of NPs may lead to undesirable harm to the surrounding environment (Bhatti et al. 2014; Dashora et al. 2022). The antibacterial activity of ZnO-NPs at concentrations ranging from 0.125 to 1 mg/mL was evaluated against *P. syringae* growth. It was found that their antibacterial activity increased with increasing NPs concentration (Fan et al. 2023).

In summary, the morphology (size and shape), type, and surface chemistry of the prepared NPs are known to be the factors that most influence the effectiveness of NPs against PHPs. The morphology of NPs contributes to their ability to bind and

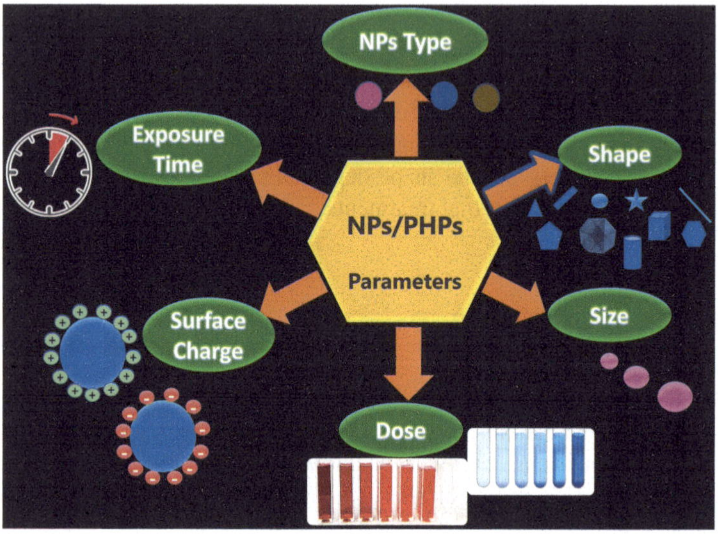

Fig. 11.2 Physiological and physical parameters of NPs against PHPs. The shape of the prepared NPs significantly affects their antimicrobial activity. NPs can be formed in several shapes, including spherical, rod, triangular, cubic, cylindrical, hexagonal, polygonal, octahedral, needle, star, and pentagonal. The dose of NPs used in the treatment and the exposure time of PHPs to these NPs play a crucial role in the control of PHPs. The efficacy of NPs differs from one type to another and depends on the size and surface charge (positive or negative) of the prepared NPs

penetrate PHP cells. Various types of NPs with different structures play a crucial role in the release of ions responsible for the toxicity mechanism in PHPs. The charge of NPs depends on the surface reactivity and surface coating. The surface charge influences the behavior of NPs during the reaction and their binding to their target. In addition, the effectiveness of NPs against PHPs relies on additional parameters such as the effective dose and duration of PHPs exposure to NPs, as shown in Fig. 11.2. However, these parameters vary from one PHP to another, depending on their sensitivity to the NPs used in their control (Kanakari and Dendrinou-Samara 2023).

11.7 Advantages of NPs Compared to Conventional Pesticides

Liu et al. (2023) evaluated the environmental implications of CuO-NPs used as pesticides and compared them to conventional pesticides for tomato disease control. They revealed that CuO-NPs showed more than tenfold antimicrobial performance compared to conventional $CuSO_4$. CuO-NPs exhibited remarkable advantages in terms of overall environmental performance (reduced 67.8%), human health (reduced 69.6%), ecosystem quality (reduced 53.2%), and renewable energy (reduced 64.7%) per unit efficacy in comparison with conventional $CuSO_4$ pesticides.

Many conventional agrochemicals threaten the environment and pose serious risks to human health. The use of nanotechnology in agriculture to improve agricultural practices has been increasing in recent years. Some promising properties of nanoformulations may minimize the cytotoxicity of NPs on aquatic life compared with their conventional counterparts. The overall toxicity of nanoformulations to non-target aquatic species is usually less than that of their conventional counterparts. However, some nanoformulations may show higher toxicity than bulk materials and lower toxicity than ionic formulations. Moreover, organic nanopesticides show no general trend in their overall toxicity (Zhang and Goss 2022).

The antifungal activities of S-NPs and green-synthesized Cu-NPs at different concentrations (5–100 μg/mL) were evaluated against *Sclerotinia sclerotiorum* and *Botrytis cinerea* and compared to $CuSO_4$ (4000 μg/mL), micro sulfur (MS) at a concentration of 1000 μg/mL, and the conventional fungicide Topsin-M 70 WP (1000 μg/mL). Their antifungal activities were evaluated in vivo and in vitro at varied temperature levels of 10 and 20 °C on cucumber fruits. The results obtained from this investigation showed that S-NPs and Cu-NPs showed the highest antifungal activities among all treatments. The cytotoxicity levels of S-NPs and Cu-NPs were evaluated in Vero and WI-38 cells. It was found that S-NPs resulted in less cytotoxicity than Cu-NPs (Sadek et al. 2022).

11.8 Conclusions

Nanotechnology has been introduced as an alternative and sustainable way to fight PHPs. The use of NPs to combat the harmful impact of PHPs on crop productivity and quality has been investigated and has shown promising results. Biotic stress can be induced by numerous PHPs, including fungi, bacteria, viruses, insects, nematodes, and other PHPs, which cause massive damage to the host plant. On the other hand, abiotic stress is induced by non-living stressors, such as inappropriate temperature and other ecological extremes, in addition to salinity and drought, which are the main causes of crop productivity loss worldwide. The anti-phytopathogenic mechanisms of NPs against PHPs can be summarized as follows: generation of ROS, DNA damage, lysosomal instability, transfer of NPs into the bacterial cells through their membranes, and changing genetic characteristics due to the interactions of the NPs with proteins and DNA. Ag-NPs are the most commonly used antimicrobial agents due to their ability to release ROS, which significantly reduces the infectivity and severity of PHPs. Additionally, plant pathogenic viruses account for 47% of the total losses induced by PHPs, and more than 50% of plant diseases are caused by viruses because they are usually unavoidable or uncontrollable. The effects of the size, shape, and dose of NPs on the infectivity of PHPs were investigated. It was found that the NPs concentration was the most effective factor affecting the efficacy of NPs, followed by their size and shape.

Several studies have compared NPs as potential anti-phytopathogenic and conventional pesticides. In general, NPs showed a higher efficacy against the tested PHPs and lower cytotoxicity compared to conventional agrochemicals. In addition, PHPs did not show resistance to the tested NPs, as in the case of conventional chemical compounds in agricultural fields. Therefore, the use of NPs as an alternative way to fight PHPs minimizes the required amounts of agrochemicals. Further studies on the effect of the size, shape, and dose of NPs on their antimicrobial activities are important to maximize the applicability of NPs as high-potential anti-phytopathogenic agents.

References

Abbas Q (2019) Understanding the UV–Vis spectroscopy for nanoparticles. J Nanomater Mol Nanotechnol 8(3):1–3

Aguilar-Marcelino L, Mendoza-de-Gives P, Al-Ani LKT, López-Arellano ME, Gómez-Rodríguez O, Villar-Luna E, Reyes-Guerrero DE (2020) Using molecular techniques applied to beneficial microorganisms as biotechnological tools for controlling agricultural plant pathogens and pest. In: Molecular aspects of plant beneficial microbes in agriculture. Academic Press, pp 333–349

Akpinar I, Unal M, Sar T (2021) Potential antifungal effects of silver nanoparticles (AgNPs) of different sizes against phytopathogenic *Fusarium oxysporum* f. sp. *radicis-lycopersici* (FORL) strains. SN Appl Sci 3(4):506

Al-Ani LKT, Furtado EL (2020) The effect of incompatible plant pathogens on the host plant. In: Molecular aspects of plant beneficial microbes in agriculture. Academic Press, pp 47–57

Anderson W, Kozak D, Coleman VA, Jämting ÅK, Trau M (2013) A comparative study of submicron particle sizing platforms: accuracy, precision and resolution analysis of polydisperse particle size distributions. J Colloid Interface Sci 405:322–330

Azam A, Ahmed AS, Oves M, Khan MS, Memic A (2012) Size-dependent antimicrobial properties of CuO nanoparticles against Gram-positive and-negative bacterial strains. Int J Nanomedicine:3527–3535

Bafghi MH, Nazari R, Darroudi M, Zargar M, Zarrinfar H (2022) The effect of biosynthesized selenium nanoparticles on the expression of CYP51A and HSP90 antifungal resistance genes in *Aspergillus fumigatus* and *Aspergillus flavus*. Biotechnol Prog 38(1):e3206

Baker C, Pradhan A, Pakstis L, Pochan DJ, Shah SI (2005) Synthesis and antibacterial properties of silver nanoparticles. J Nanosci Nanotechnol 5(2):244–249

Balasooriya ER, Jayasinghe CD, Jayawardena UA, Ruwanthika RWD, Mendis de Silva R, Udagama PV (2017) Honey mediated green synthesis of nanoparticles: new era of safe nanotechnology. J Nanomater 2017

Ben-Ari G, Lavi U (2012) Marker-assisted selection in plant breeding. In: Plant biotechnology and agriculture. Academic Press, pp 163–184

Bertrand JC, Caumette P, Lebaron P, Matheron R, Normand P, Ngando TS (eds) (2015) Environmental microbiology: fundamentals and applications. Springer, Dordrecht, The Netherlands, pp 659–753

Bhagyaraj SM, Oluwafemi OS (2018) Nanotechnology: the science of the invisible. In: Synthesis of inorganic nanomaterials. Woodhead Publishing, pp 1–18

Bhatti HN, Khan SS, Khan A, Rani M, Ahmad VU, Choudhary MI (2014) Biotransformation of monoterpenoids and their antimicrobial activities. Phytomedicine 21(12):1597–1626

Birnbaum AJ, Pique A (2011) Laser induced extraplanar propulsion for three-dimensional microfabrication. Appl Phys Lett 98(13):134101

Biswas T, Kalra A, Mathur AK, Lal RK, Singh M, Mathur A (2016) Elicitors' influenced differential ginsenoside production and exudation into medium with concurrent Rg3/Rh2 panaxadiol induction in *Panax quinquefolius* cell suspensions. Appl Microbiol Biotechnol 100:4909–4922
Carter DBWCB (2009) Transmission electron microscopy a textbook for materials science. Springer Publication
Chen J, Xu K (2020) Applications of atomic force microscopy in materials, semiconductors, polymers, and medicine: a minireview. Instrum Sci Technol 48(6):667–681
Chitnis VR, Suryanarayanan TS, Nataraja KN, Prasad SR, Oelmüller R, Shaanker RU (2020) Fungal endophyte-mediated crop improvement: the way ahead. Front Plant Sci 11:561007
Consolo VF, Torres-Nicolini A, Alvarez VA (2020) Mycosinthetized Ag, CuO and ZnO nanoparticles from a promising *Trichoderma harzianum* strain and their antifungal potential against important phytopathogens. Sci Rep 10(1):20499
Cruz-Luna AR, Cruz-Martínez H, Vásquez-López A, Medina DI (2021) Metal nanoparticles as novel antifungal agents for sustainable agriculture: current advances and future directions. J Fungi 7(12):1033
Daniel AI, Keyster M, Klein A (2023) Biogenic zinc oxide nanoparticles: a viable agricultural tool to control plant pathogenic fungi and its potential effects on soil and plants. Sci Total Environ 165483
Dashora A, Rathore K, Raj S, Sharma K (2022) Synthesis of silver nanoparticles employing *Polyalthia longifolia* leaf extract and their in-vitro antifungal activity against phytopathogen. Biochem Biophys Rep 31:101320
Doolotkeldieva T, Bobusheva S, Zhasnakunov Z, Satybaldiev A (2022) Biological activity of Ag and Cu monometallic nanoparticles and Ag-Cu bimetallic nanocomposites against plant pathogens and seeds. J Nanomater 2022
Dutta P, Kumari A, Mahanta M, Biswas KK, Dudkiewicz A, Thakuria D et al (2022) Advances in nanotechnology as a potential alternative for plant viral disease management. Front Microbiol 13:935193
Dyakov YT, Zinovyeva SV (2007) Horizontal pathosystem: parasite attack factors. In: Comprehensive and molecular phytopathology. Elsevier, pp 139–159
El-Moslamy SH, Yahia IS, Zahran HY, Kamoun EA (2023) Novel biosynthesis of MnO NPs using Mycoendophyte: industrial bioprocessing strategies and scaling-up production with its evaluation as anti-phytopathogenic agents. Sci Rep 13(1):2052
Fan G, Xiao Q, Li Q, Xia Y, Feng H, Ma X et al (2023) Antimicrobial mechanisms of ZnO nanoparticles to phytopathogen *Pseudomonas syringae*: damage of cell envelope, suppression of metabolism, biofilm and motility, and stimulation of stomatal immunity on host plant. Pesticide Biochem Physiol 194:105455
Garza-García JJ, Hernández-Díaz JA, Zamudio-Ojeda A, León-Morales JM, Guerrero-Guzmán A, Sánchez-Chiprés DR et al (2021) The role of selenium nanoparticles in agriculture and food technology. Biol Trace Elem Res:1–21
Gautam AK, Kumar S (2020) Techniques for the detection, identification, and diagnosis of agricultural pathogens and diseases. In: Natural remedies for pest, disease and weed control. Academic Press, pp 135–142
Häkkinen H, Abbet S, Sanchez A, Heiz U, Landman U (2003) Structural, electronic, and impurity-doping effects in nanoscale chemistry: supported gold nanoclusters. Angew Chem Int Ed 42(11):1297–1300
Hao Y, Cao X, Ma C, Zhang Z, Zhao N, Ali A et al (2017) Potential applications and antifungal activities of engineered nanomaterials against gray mold disease agent *Botrytis cinerea* on rose petals. Front Plant Sci 8:1332
Hao Y, Yuan W, Ma C, White JC, Zhang Z, Adeel M et al (2018) Engineered nanomaterials suppress Turnip mosaic virus infection in tobacco (*Nicotiana benthamiana*). Environ Sci Nano 5(7):1685–1693
He M, He CQ, Ding NZ (2018) Abiotic stresses: general defenses of land plants and chances for engineering multistress tolerance. Front Plant Sci 9:1771

Hegenauer V, Körner M, Albert M (2017) Plants under stress by parasitic plants. Curr Opin Plant Biol 38:34–41

Horst RK (ed) (2013) Westcott's plant disease handbook. Springer Science & Business Media

Iqbal Z, Iqbal MS, Hashem A, Abd-Allah EF, Ansari MI (2021) Plant defense responses to biotic stress and its interplay with fluctuating dark/light conditions. Front Plant Sci 12:631810

Jo YK, Kim BH, Jung G (2009) Antifungal activity of silver ions and nanoparticles on phytopathogenic fungi. Plant Dis 93(10):1037–1043

Jones JT, Haegeman A, Danchin EG, Gaur HS, Helder J, Jones MG et al (2013) Top 10 plant-parasitic nematodes in molecular plant pathology. Mol Plant Pathol 14(9):946–961

Joshi SM, De Britto S, Jogaiah S, Ito SI (2019) Mycogenic selenium nanoparticles as potential new generation broad spectrum antifungal molecules. Biomol Ther 9(9):419

Joubert JJ, Rijkenberg FHJ (1971) Parasitic green algae. Annu Rev Phytopathol 9(1):45–64

Kahane I, Adoni A, Tully JG (1993) Biology of mollicutes. In: Rapid diagnosis of mycoplasmas. Springer Science & Business Media, pp 3–14

Kalaivani K (2021) A study on the synthesis and characterization methods of Nanomaterials. Mater Today: Proc 45:3551–3555

Kanakari E, Dendrinou-Samara C (2023) Fighting phytopathogens with engineered inorganic-based nanoparticles. Materials 16(6):2388

Khan M, Siddiqui ZA, Parveen A, Khan AA, Moon IS, Alam M (2022) Elucidating the role of silicon dioxide and titanium dioxide nanoparticles in mitigating the disease of the eggplant caused by *Phomopsis vexans*, *Ralstonia solanacearum*, and root-knot nematode *Meloidogyne incognita*. Nanotechnol Rev 11(1):1606–1619

Kovalchuk I (2021) Transgenerational genome instability in plants. In: Genome stability. Academic Press, pp 659–678

Kovalskaya N, Hammond RW (2014) Molecular biology of viroid–host interactions and disease control strategies. Plant Sci 228:48–60

Kumar KM, Mandal BK, Naidu EA, Sinha M, Kumar KS, Reddy PS (2013) Synthesis and characterisation of flower shaped zinc oxide nanostructures and its antimicrobial activity. Spectrochim Acta A Mol Biomol Spectrosc 104:171–174

Kumar M, Ranjan R, Dandapat S, Srivastava R, Sinha MP (2022) XRD analysis for characterization of green nanoparticles: a mini review. Global J Pharmacy Pharmaceut Sci 10(1):555779

Lazcano-Ramírez HG, Garza-García JJ, Hernández-Díaz JA, León-Morales JM, Macías-Sandoval AS, García-Morales S (2023) Antifungal activity of selenium nanoparticles obtained by plant-mediated synthesis. Antibiotics 12(1):115

Li T, Li Y, Qin P, Wei G, Chen C (2023) Comparing the inhibitory effects of CuO-rGO, CuO NPs, and $CuCl_2$ on the oomycete *Phytophthora sojae*: insights from phenotypic and transcriptomic analyses. Environ Sci Nano 10(9):2299–2311

Liu P, Ren Z, Ding W, Kong D, Hermanowicz SW, Huang Y (2023) Comparative environmental impact assessment of copper-based nanopesticides and conventional pesticides. ACS Agric Sci Technol 3(7):593–600

Ma Y (2016) Characterization of nanomaterials in nanotoxicological analyses. Toxicol Nanomater:1–21

Mohamed EA (2018) Non-dependency of in-vitro fungicidal efficiency of copper nanoparticles against *Fusarium oxysporum* upon particle size. J Plant Pathol Microbiol 9(2)

Mori S, Shitara Y (1994) Tribochemical activation of gold surface by scratching. Appl Surf Sci 78(3):269–273

Mosquera-Sánchez LP, Arciniegas-Grijalba PA, Patiño-Portela MC, Guerra-Sierra BE, Muñoz-Florez JE, Rodríguez-Páez JE (2020) Antifungal effect of zinc oxide nanoparticles (ZnO-NPs) on *Colletotrichum* sp., causal agent of anthracnose in coffee crops. Biocatal Agric Biotechnol 25:101579

Munawar T, ur Rehman MN, Nadeem MS, Mukhtar F, Manzoor S, Ashiq MN, Iqbal F (2021) Facile synthesis of Cr-Co co-doped CdO nanowires for photocatalytic, antimicrobial, and supercapacitor applications. J Alloys Compd 885:160885

Namba S (2019) Molecular and biological properties of phytoplasmas. Proc Japan Acad Ser B 95(7):401–418
Noha K, Bondok AM, El-Dougdoug KA (2018) Evaluation of silver nanoparticles as antiviral agent against ToMV and PVY in tomato plants. Sciences 8(01):100–111
Peng Y, Li SJ, Yan J, Tang Y, Cheng JP, Gao AJ et al (2021) Research progress on phytopathogenic fungi and their role as biocontrol agents. Front Microbiol 12:670135
Pombert JF, Blouin NA, Lane C, Boucias D, Keeling PJ (2014) A lack of parasitic reduction in the obligate parasitic green alga *Helicosporidium*. PLoS Genet 10(5):e1004355
Rastogi A, Singh P, Haraz FA, Barhoum A (2018) Biological synthesis of nanoparticles: an environmentally benign approach. In: Fundamentals of nanoparticles. Elsevier, pp 571–604
Rudramurthy GR, Swamy MK, Sinniah UR, Ghasemzadeh A (2016) Nanoparticles: alternatives against drug-resistant pathogenic microbes. Molecules 21(7):836
Sadek ME, Shabana YM, Sayed-Ahmed K, Abou Tabl AH (2022) Antifungal activities of sulfur and copper nanoparticles against cucumber postharvest diseases caused by *Botrytis cinerea* and *Sclerotinia sclerotiorum*. J Fungi 8(4):412
Şahin B, Soylu S, Kara M, Türkmen M, Aydin R, Çetin H (2021) Superior antibacterial activity against seed-borne plant bacterial disease agents and enhanced physical properties of novel green synthesized nanostructured ZnO using *Thymbra spicata* plant extract. Ceram Int 47(1):341–350
Schasteen CS (2023) Safety of food and beverages: oilseeds, legumes and derived products. In: Reference module in food science. Elsevier. https://doi.org/10.1016/B978-0-12-822521-9.00159-3
Schwab F, Zhai G, Kern M, Turner A, Schnoor JL, Wiesner MR (2016) Barriers, pathways and processes for uptake, translocation and accumulation of nanomaterials in plants–critical review. Nanotoxicology 10(3):257–278
Shang Y, Xiao G, Zheng P, Cen K, Zhan S, Wang C (2016) Divergent and convergent evolution of fungal pathogenicity. Genome Biol Evol 8(5):1374–1387
Shipp DW, Sinjab F, Notingher I (2017) Raman spectroscopy: techniques and applications in the life sciences. Adv Opt Photon 9(2):315–428
Siddiquah A, Hashmi SS, Mushtaq S, Renouard S, Blondeau JP, Abbasi R et al (2018) Exploiting in vitro potential and characterization of surface modified zinc oxide nanoparticles of *Isodon rugosus* extract: their clinical potential towards HepG2 cell line and human pathogenic bacteria. EXCLI J 17:671
Singh R, Kuddus M, Singh PK, Choden D (2022) Nanotechnology for nanophytopathogens: from detection to the management of plant viruses. Biomed Res Int 2022
Smith E, Dent G (2019) Modern Raman spectroscopy: a practical approach. John Wiley & Sons
Srivastava R (2012) Synthesis and characterization techniques of nanomaterials. Int J Green Nanotechnol 4(1):17–27
Tampakaki AP, Hatziloukas E, Panopoulos NJ (2009) Plant pathogens, bacteria. In: Encyclopedia of microbiology, 3rd edn. Elsevier, pp 655–677
Tang S, Zheng J (2018) Antibacterial activity of silver nanoparticles: structural effects. Adv Healthc Mater 7(13):1701503
Tiwari P, Srivastava M, Mishra R, Ji G, Prakash R (2018) Economic use of waste *Musa paradisiaca* peels for effective control of mild steel loss in aggressive acid solutions. J Environ Chem Eng 6(4):4773–4783
Torres-Rivero K, Bastos-Arrieta J, Fiol N, Florido A (2021) Metal and metal oxide nanoparticles: an integrated perspective of the green synthesis methods by natural products and waste valorization: applications and challenges. Comp Anal Chem 94:433–469
Tryfon P, Kamou NN, Ntalli N, Mourdikoudis S, Karamanoli K, Karfaridis D et al (2022) Coated Cu-doped ZnO and Cu nanoparticles as control agents against plant pathogenic fungi and nematodes. NanoImpact 28:100430
Wan J, Lacey SD, Dai J, Bao W, Fuhrer MS, Hu L (2016) Tuning two-dimensional nanomaterials by intercalation: materials, properties and applications. Chem Soc Rev 45(24):6742–6765
Wang CC (ed) (2011) Laser scanning: theory and applications. BoD–Books on Demand

Wang L, Hu C, Shao L (2017) The antimicrobial activity of nanoparticles: present situation and prospects for the future. Int J Nanomedicine:1227–1249

Worsfold P, Townshend A, Poole CF, Miró M (2019) Encyclopedia of analytical science. Elsevier

Xing K, Zhu X, Peng X, Qin S (2015) Chitosan antimicrobial and eliciting properties for pest control in agriculture: a review. Agron Sustain Dev 35:569–588

Yang H, Wang Y, Lai S, An H, Li Y, Chen F (2007) Application of atomic force microscopy as a nanotechnology tool in food science. J Food Sci 72(4):R65–R75

Zeilinger S, Gupta VK, Dahms TE, Silva RN, Singh HB, Upadhyay RS et al (2016) Friends or foes? Emerging insights from fungal interactions with plants. FEMS Microbiol Rev 40(2):182–207

Zhang Y, Goss GG (2022) Nanotechnology in agriculture: Comparison of the toxicity between conventional and nano-based agrochemicals on non-target aquatic species. J Hazard Mater 439:129559

Zheng R, Li S, Zhang X, Zhao C (2021) Biological activities of some new secondary metabolites isolated from endophytic fungi: a review study. Int J Mol Sci 22(2):959

Zou Y, Celli A, Zhu H, Elmahdy A, Cao Y, Hui X, Maibach H (2017) Confocal laser scanning microscopy to estimate nanoparticles' human skin penetration in vitro. Int J Nanomed:8035–8041

Chapter 12
Nanoparticles in Clonal Micropropagation and Plant Stress Control

Tatiana A. Grodetskaya, Olga A. Fedorova, Konstantin V. Zhuzhukin, Peter M. Evlakov, and Alexander T. Eprintsev

Abstract The use of nanoparticles (NPs) helps to solve a large number of problems in various fields. Biotechnology is one of the developing industries that has the opportunity to improve the methodological base and resulting products thanks to the use of NPs. Despite a large number of studies and significant advances in this area, the question of the safety and feasibility of using NPs in the biotechnology of various plants, and in particular forest crops, remains open. This review is aimed at systematizing the available research results on the use of NPs to combat unfavorable environmental conditions in agriculture and forestry, the use of plant in vitro technology to increase the percentage of sterile explants, improve survival rate and morphogenesis, protect against stress, as well as stimulation of secondary metabolites synthesis. The concentrations of NPs, their toxic effects, and prospects are also discussed.

Keywords Nanoparticles · Plant stress · Biotechnology · In vitro · Agriculture · Forestry · Toxic effects

---

T. A. Grodetskaya · O. A. Fedorova · P. M. Evlakov (✉)
Research Institute of Innovative Technologies of the Forestry Complex, Laboratory of PCR Analysis, Voronezh State University of Forestry and Technologies Named After G. F. Morozov, Voronezh, Russia
e-mail: tatyana.pokusina@yandex.ru; fed-olga78@mail.ru; peter.evlakov@yandex.ru

K. V. Zhuzhukin
Department of Chemistry, Voronezh State University of Forestry and Technologies Named After G.F. Morozov, Voronezh, Russia
e-mail: kinkon18@yandex.ru

A. T. Eprintsev
Department of Biochemistry and Cell Physiology, Voronezh State University, Voronezh, Russia
e-mail: bc366@bio.vsu.ru

M. Khan, J.-T. Chen (eds.), *Nanoparticles in Plant Biotic Stress Management*,
https://doi.org/10.1007/978-981-97-0851-2_12

12.1 Introduction

Nanotechnology is currently an actively developing area of research, and interest in it is growing and spreading in the branches of many sciences, including medicine, chemical, food and pharmaceutical industries, agriculture, forestry, and biotechnology. According to some estimates, the consumption of nanoparticles by various industries is up to 11.5 million tons per year (Thounaojam et al. 2021). Interest in the use of nanomaterials in biotechnology is due to their unique physical and chemical properties, namely, nanosize, high specific surface, melting point, wettability, and thermal and electrical conductivity, which can be controlled in the interests of a researcher, which gives them an undoubted advantage over their bulk counterparts (Mahajan et al. 2022). The relative cheapness of nanoparticle-based preparations is determined by their high activity and, at the same time, low consumption. Nanosizes allow particles to freely pass through the cell membrane, which allows them to affect living organisms at extremely low doses. Penetrating the cell, nanoparticles regulate the processes of cell division, respiration and metabolism, the functioning of enzyme systems, and the activity of genes. A detailed study of nanoparticles has opened up the prospect of their use in medicine and biotechnology as antiseptic agents and for targeted delivery of medicinal components and in agriculture and forestry to reduce the effects of adverse environmental factors on plants. At the same time, the high biological activity of nanoparticles makes it possible to achieve the most positive effects at the lowest possible concentrations, while an insignificant increase in the concentration of nanoparticles, by the standards of full-sized molecules, can cause toxic effects in living organisms and significantly disrupt the balance in the environment. It is necessary to prevent the uncontrolled use of nanomaterials to minimize negative effects, which requires a careful approach to their study, including planning experiments and interpreting the data obtained. This review is aimed at summarizing and analyzing knowledge in the field of using nanoparticles in plant biotechnology to improve growth characteristics and combat adverse environmental factors, including phytopathogens.

12.2 Obtaining Nanoparticles

There are several ways to obtain nanoparticles. Among them are physical vapor deposition, in which molecules are deposited on the surface of a substrate at low temperatures, and hydrothermal synthesis, which involves the production of nanoparticles by heating a solution to high temperatures and pressures used to obtain metal and nanoparticles and their oxides. Chemical vapor deposition is a method implying gaseous reactants to be deposited on the surface of a substrate under the action of an electric field. This method makes it possible to obtain nanoparticles of various materials, such as metals, oxides, and semiconductors. The sol-gel method is a process for obtaining nanoparticles by gelling a sol. This method is often used to synthesize ceramic nanoparticles and other materials. Laser ablation involves the production of nanoparticles using laser radiation. This method is used for the synthesis of metal nanoparticles in gaseous media. Aerodynamic sputtering will allow nanoparticles to be synthesized from solutions or melts by sputtering onto a substrate. This method can be used to obtain nanoparticles of various materials. One of the methods for obtaining nanoparticles is "biosynthesis" or "green synthesis," which involves the use of microorganisms and plant extracts (Mohanta et al. 2020) for the synthesis of nanosized particles from metal ions. This technology is positioned as environmentally friendly in comparison with physical and chemical methods of synthesis, where toxic compounds are released as by-products.

12.3 Plant Biotechnology

Tissue culture is one of the most actively developing areas of plant biotechnology. It has found its application in the mass production of cultivated agricultural and woody plants for the creation of plantations for various purposes and the cultivation of medicinal plants for use in the pharmaceutical industry, as well as for germplasm conservation and cryopreservation. Clonal micropropagation is a method of mass asexual reproduction of plants in tissue and cell culture, which allows mass replication of plants in laboratory conditions in a relatively short period of time and can be performed continuously regardless of climate and season. The advantage of the technology is the production of a genetically homogeneous planting material, completely identical to the original specimen, which makes it possible to preserve and increase rare and valuable plant genotypes. Tissue culture opens up prospects for genetic manipulation to create improved varieties, as well as the production of biologically active compounds (Singh and Ahamad 2022). Plants are a source of carbohydrates, proteins, trace elements, secondary metabolites, glycosides, phenolic compounds, and antioxidants valuable for the food and drug industry, the content of which can be regulated to a certain extent by changing the composition of the nutrient medium. Clonal micropropagation is widely used to obtain agricultural, forestry, medicinal, and ornamental plants, including the preservation and reproduction of plants that are difficult to propagate by conventional methods. Creation and enhancement of a collection of germplasm of rare and endangered plant species is a particularly urgent task in connection with the growing threats of climate change and natural disasters in the form of droughts and fires.

In vitro cloning involves the use of an artificial nutrient medium containing all the necessary vitamins and microelements necessary for successful tissue differentiation and regeneration and the development of a full-fledged plant. The regeneration of a whole plant from its part is a complex and multi-stage process that requires the plant to have an appropriate regenerative potential and create a favorable microclimate. The selection of lighting, temperature, maintaining 100% humidity, and the presence of absolute sterility are indispensable conditions for a successful result.

12.3.1 Nanoparticles in In vitro Plant Technology

The biotechnological use of nanoparticles implies a wide range of applications, such as fertilizers, growth promoters, smart pesticides, remediators, and biosensors for determining pollution and degradation of contaminants (Majeed et al. 2020). One of the most promising areas is the use of nanoparticles in in vitro cloning technology. Among the numerous aspects of the use of nanoparticles in tissue culture are nutrition or sterilization, as well as stimulation of the development of stress resistance, while one does not exclude the other. The combined action of different types of nanoparticles, the use of nanocomposites, or the simultaneous use of hormones and growth stimulants can enhance the positive effect of nanoparticles on cultivated plants. Nanoparticles can reduce the negative impact of abiotic and biotic stress in plants in vitro, stimulating their regeneration potential; triggering the synthesis of secondary metabolites, protective proteins, and antioxidant enzymes; and reducing active radicals and ROS (Fig. 12.1).

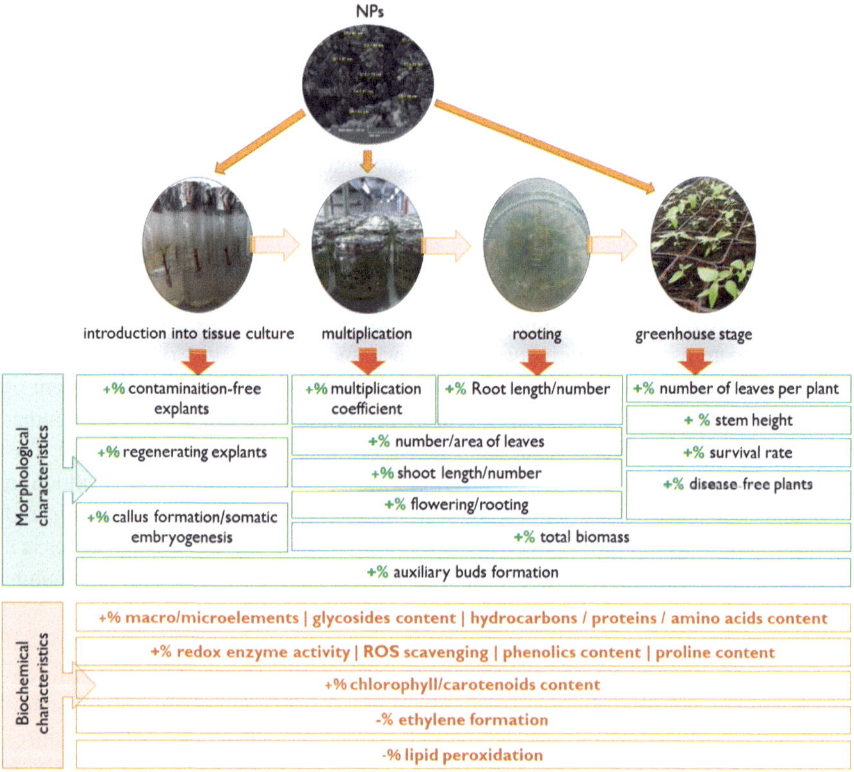

Fig. 12.1 The impacts on morphological and biochemical characteristics of in vitro plants on different stages of micropropagation

Combining nanoparticles with various types of sanitizing agents will help develop an efficient protocol to fight infection in tissue culture. The advantage of nanoparticles is their prolonged action and the possibility of use in microdoses at all stages of in vitro cultivation, which enhances the immune response in plants, on the one hand, and suppresses infection, on the other hand. They are widely used at different concentrations and at different stages of in vitro cultivation, depending on the type of plants used. The choice of the chemical composition of nanoparticles, concentration, and method of treatment (surface treatment of seeds (Prażak et al. 2020), plant organs (Cai et al. 2020; Hussain et al. 2019; Pérez-Labrada et al. 2019; Satti et al. 2021), introduction into the cultivation medium, treatment of the substrate (Hernández-Hernández et al. 2019)) will depend on goals settled by the researcher. At the same time, the shape, size, reactivity, type of coating, and level of aggregation of nanoparticles will significantly affect the growth and development of plants. The mechanism and form of the effect of nanoparticles also depend on the method of their application. The surface treatment promotes the absorption of nanoparticles through stomata and their further adsorption in mesophyll cells (El-Shetehy et al. 2021), increasing the activity of antioxidant enzymes (Pérez-Labrada et al. 2019).

Sterilizing Effect of Nanoparticles in Culture In vitro

The introduction of plants into culture in vitro requires compliance with a number of basic requirements, one of which is the maintenance of asepsis. For introduction into culture, fragments of a shoot with an apical or lateral meristem are used which are the infected material. The nutrient medium contains all the micro- and macroelements and sugars necessary for the development of the plant, as a result of which it is also favorable for the growth of microorganisms that develop faster than the explants themselves (Mahendran et al. 2019). The use of washing solutions helps to get rid of external infection but often does not help to suppress the development of the internal microflora of plants. Introduction into culture in vitro is accompanied by infection of explants with pathogens of bacterial and fungal origin. Fungal phytopathogens are less demanding on the nutrient medium and cause more damage when cultivated in vitro. Infection of the nutrient medium can cause necrosis and lead to the complete death of explants (Fig. 12.2).

Fig. 12.2 Infection of poplar explants with pathogens of fungal origin

Primary sterilization includes washing the explants in running water, followed by washing in a solution containing surfactants. Next, explants are introduced onto nutrient media that have undergone autoclaving. Despite the observance of all protection measures, the percentage of infected plants can remain high. The fight against infection includes a number of methods, including the use of surfactants, chlorine-containing compounds, and antibiotics. Despite the significant effectiveness of the use of the above methods, they have a number of limitations, of which toxicity to humans and plants is one of the most significant. Antibiotics can largely solve the problem of in vitro contamination of plants, but they cause resistance in microorganisms and negatively affect plant immunity, inhibiting their growth and development (Reed et al. 1997).

Sterilization of the initial plant material for introduction into in vitro culture is the first and main step that determines the results of in vitro cloning to the greatest extent. The choice of disinfectant depends on the size, age, type, and condition of the parent plant, as well as cultivation conditions. The level of contamination of the introduced plants directly depends on the origin of the plant material. Explants taken from open environments are most susceptible to both external and internal infection. At the same time, under controlled greenhouse conditions, the percentage of infection can be significantly reduced. These features can significantly affect the degree of infection of explants in culture in vitro.

When obtaining an aseptic tissue culture, an important role in sterilization belongs to the selection of disinfecting agents, the effectiveness of concentrations, and the duration of treatment to remove the infection and obtain regenerating viable explants. To sterilize the material, various compounds are used with different concentrations and exposure times, while the degree of their disinfecting effect also varies. The main sterilizing agents can be chemical agents such as hydrogen peroxide (H_2O_2), mercury (II) chloride ($HgCl_2$), calcium hypochlorite [$Ca(ClO)_2$], silver nitrate ($AgNO_3$), sodium hypochlorite, and antibiotics. The longer in time and higher in concentration of the influence of the disinfectant, the greater the effect can be achieved during sterilization. At the same time, high concentrations of chemical agents cause growth inhibition, reduced enzyme activity, mutations, and necrotic changes in explants. A decrease in plant immunity under the action of chemical agents leads to an inability to fight external and internal infection, which leads to an increase in the percentage of contamination and death of explants.

Nanoparticles can be a promising alternative to the above agents for combating contamination. The use of nanoparticles in culture in vitro helps to effectively eliminate the contamination and maintain viability and even has a stimulating effect on the growth and development of explants under optimal conditions. Ag, ZnO, CuO, TiO_2, Fe_2O_3, Co, MgO, Al_2O_3, Ni, and SiO_2 nanoparticles are most often used for explant disinfection. Ag nanoparticles can be used both for surface sterilization of explants and for their introduction into the culture medium (Tung et al. 2021a). It was also shown the possibility of replacing $HgCl_2$ и $Ca(ClO)_2$ with CuNPs in surface disinfection of petioles, inflorescence, and stem nodes explants of ex vitro tuberous begonias for introduction into culture (Bao et al. 2022).

Explant multiplication is the critical step in propagation in vitro. The risk factor at this stage may be the lack of immunity in explants and the risk of their infection when transferred to a new nutrient medium. The presence of nanoparticles in the cultivation medium can help stimulate plant immunity to protect them from the effects of pathogenic microorganisms.

Rooting of explants is an important process for the formation of a whole plant in tissue culture. This stage is determined largely by the presence of internal and added hormonal factors. At the same time, callus formation is undesirable, which can be provoked by a high content of mineral salts in the nutrient medium. The use of plant nutrients in a nanoform can contribute to solving this problem due to the need to use significantly lower concentrations of nanoparticles. The addition of Ag (Hegazi et al. 2021; Tung et al. 2022; Zia et al. 2020), Si-based (Khafri et al. 2023), and ZnO NPs (Al-Mayahi 2021; Awad 2020) to the cultivation medium contributed to the acceleration of root regeneration in various plants.

Transfer to greenhouse conditions can be accompanied by a loss of up to 50% of in vitro plants due to a sharp change in cultivation conditions, moisture regime, temperature, and lighting. At the same time, the use of nanopreparations can simplify the transition of regenerants to independent mineral nutrition and photosynthesis in greenhouse conditions and accelerate their growth after ex vitro transplantation. Strawberry microclones treated with silver nanoparticles under in vitro conditions had a higher survival rate when adapted under greenhouse conditions compared to control plants (Tung et al. 2021b).

Influence of Nanoparticles on Plant Survival and Morphogenesis In vitro

Nanoparticles in culture in vitro have a complex effect on the plant and microflora. The fight against infection can be accompanied by the presence of the effect of stimulating the growth processes of plants in tissue culture conditions. The undoubted advantage of using nanoparticles in tissue culture has been shown in a large number of works, where a positive effect of nanoparticles on the processes of survival, callogenesis, organogenesis, an increase in the multiplication factor, stimulation of shoot growth, root growth, and leaf formation in various plants was revealed. To stimulate the processes of morphogenesis and increase the biomass and multiplication factor of plants, both metallic and non-metallic nanoparticles are used (Table 12.1).

NPs help to increase the survival rate, improve the health of plants, and reduce the effect of vitrification (Tung et al. 2022). The use of NPs contributed to an increase in the multiplication rate, acceleration of shoot development (El-Mahdy and Elazab 2020; Sorcia-Morales et al. 2021), an increase in the length of the stem and roots and the number of roots per plant (El-Mahdy and Elazab 2020), and an increase in the content of chlorophyll (El-Mahdy and Elazab 2020; Elsayh et al. 2022).

Silver is a promising agent for reducing contamination, stimulating growth and development processes, and stimulating the development of explant stress

Table 12.1 Improvement of morphological and biochemical traits of plants with the application of nanoparticles in vitro

Nanoparticles	Plant	Effect	Reference
Ag	*Psidium friedrichsthalianum* (O. Berg) Nied.	Multiplication factor, number/area of leaves	Andújar et al. (2020)
	Olea europaea L.	Bud formation, root length/number of roots, number/area of leaves	Hegazi et al. (2021)
	Dianthus caryophyllus L.	Root length/number of roots, activity of antioxidant enzymes	Zia et al. (2020)
	Fragaria × ananassa Duch.	Biomass, shoot development, root length/number of roots	Tung et al. (2021b)
	Passiflora edulis Sims f. edulis	Shoot development, flowering/fruiting, chlorophyll/carotenoid content	Phong et al. (2022)
	Vanilla planifolia Jacks. ex Andrews	Biomass, shoot development, macro-/microelement composition, activity of antioxidant enzymes, phenolics content, lipid oxidation	Spinoso-Castillo et al. (2017)
	Gaillardia pulchella Foug cv. 'Torch Yellow'	Biomass, shoot development, hydrocarbons/proteins/amino acids, chlorophyll/carotenoid content	Manokari et al. (2023e)
	Rosa hybrida L. 'Baby Love'	Biomass, shoot development, multiplication factor, chlorophyll/carotenoid content	Ha et al. (2020)
	Tecomella undulata (Roxb.) Seem.	Shoot development	Sarmast et al. (2015)
	Gerbera jamesonii L.	Shoot development, multiplication factor	Tung et al. (2022)
	Cucumis metuliferus E. Meyer ex. Naudin	Root length/number of roots	Lai et al. (2022)
	Phoenix dactylifera L.	Shoot development, root length/number of roots, number/area of leaves, proline content, hydrocarbons/proteins/amino acids, chlorophyll/carotenoids content	Elsayh et al. (2022)
	Populus × canescens (Ait) Smith.	Shoot development, root length/number of roots	Vasyukova et al. (2021)

ZnO	*Phoenix dactylifera* L.	Root length/number of roots, macro/microelement composition, antioxidant enzyme activity, hydrocarbons/proteins/amino acids	Al-Mayahi (2021); Awad (2020)
	Stevia rebaudiana Bertoni	Biomass, root length/number of roots, antioxidant enzyme activity, glycoside content	Ahmad et al. (2020a)
	Vigna radiata L.	Phenol content, glycoside content	Iqbal et al. (2022)
	Punica granatum L.	Biomass, shoot development, chlorophyll/carotenoid content	El-Mahdy and Elazab (2020)
	Fragaria x ananassa Duch.	Antioxidant enzyme activity, proline content	Abu Zeid et al. (2023)
	Thymus spp.	Phenolic content	Mosavat et al. (2019)
	Zataria multiflora Boiss	Phenolic content	
	Triticum aestivum L.	Callus formation	Nalci et al. (2019)
	Olea europaea L.	Number of shoots, green fresh and total dry weigh, chlorophyll a and b content, soluble protein, carotenoids, anthocyanins, total phenol content, the radical scavenging activity	Regni et al. (2022)
	Salvia officinalis L.	Biomass, shoot length, number of shoots, total chlorophyll, proline content, antioxidant enzyme activity	Alenezi et al. (2022)
	Cydonia oblonga Mill.	Shoot length and the number of leaves	Farhadi et al. (2023)
Ti	*Sequoia sempervirens* L.	Shoot number and length, root percentage and root number, total chlorophyll, total phenolic and flavonoid contents	El-Sayed et al. (2021)
TiO_2	*Arundinaria pygmaea* L.	Decrease in ROS concentration, degree of peroxidation and soluble protein, increase in antioxidant activity, chlorophyll content, and total biomass	Emamverdian et al. (2021)
TiS3 nanoribbons	*B. pubescens* Ehrh.	Germination initiation, rhizogenesis activation	Zakharova et al. (2021)
	Populus alba × *Populus tremula*	Germination initiation, rhizogenesis activation	
CuO	*Vigna radiata* L.	Phenol content, glycoside content	Iqbal et al. (2022)
	Stevia rebaudiana Bertoni	Biomass, root length/number of roots, antioxidant enzyme activity, glycoside content	Ahmad et al. (2020a)
	Triticum aestivum L.	Callus formation	Nalci et al. (2019)
	Rosa hybrida L. 'Baby Love'	Biomass, root length/number of roots, chlorophyll/ carotenoid content, reduced ethylene	Ha et al. (2020)
	Gerbera jamesonii L.	Root length/number of roots	Tung et al. (2022)

(continued)

Table 12.1 (continued)

Nanoparticles	Plant	Effect	Reference
Fe-based	*Triticum aestivum* L.	Callus formation	Nalci et al. (2019)
	Sequoia sempervirens L.	Antioxidant enzyme activity	El-Sayed et al. (2021)
MgO	*Vigna unguiculata* L.	Callus formation, shoot multiplication	Koçak et al. (2023)
Sibased	*Prunus cerasifera* L.	Root length/number of roots	Khafri et al. (2023)
	Cydonia oblonga Mill.	Number of axillary shoots	Farhadi et al. (2023)
	Lavandula officinalis L.	Multiplication factor	Khattab et al. (2022)
	Vitex trifolia L.	Shoot proliferation, rooting efficiency, biomass production, foliar micro-structural traits	Cokul Raj et al. (2023b)
	Thunbergia erecta (Benth.) T. Anderson	Rooting efficiency, shoot amplification, foliar micro-structural traits	Manokari et al. (2023c)
	Cymbopogon citratus (DC.) Stapf	Shoot proliferation, shoot length, robust leaf, and biomass	Manokari et al. (2023b)
	Vitex negundo L.	Flower bud induction	Cokul Raj et al. (2023a)
	Dioscorea pentaphylla L.	Production of mini-tubers	Manokari et al. (2023a)
	Santalum album L.	Micro-morphological structures of the shoots	Manokari et al. (2023d)

resistance in tissue culture. It was shown that silver nanoparticles had microbicidal activity and simultaneously stimulated the development of shoots, leaves, and roots under in vitro conditions (Andújar et al. 2020). The use of silver nanoparticles contributed to the increase in biomass and shoot length in gaillardia (Manokari et al. 2023e), vanilla (Spinoso-Castillo et al. 2017), rose (Ha et al. 2020), and garden strawberry (Tung et al. 2021b), the latter also showing an increase in root development, which has also been shown in olive (Hegazi et al. 2021), carnation (Zia et al. 2020), kiwano (Lai et al. 2022), and gray poplar (Vasyukova et al. 2021).

The use of ZnO NPs contributed to an increase in biomass in stevia (Ahmad et al. 2020a) and pomegranate (El-Mahdy and Elazab 2020), roots in stevia (Ahmad et al. 2020a), and dates (Al-Mayahi 2021; Awad 2020).

CuO and Co nanoparticles contributed to an increase in biomass and root growth in stevia (Ahmad et al. 2020a) and rose (Ha et al. 2020), respectively. Exposure to MgO nanoparticles improved callus formation, and Mg NPs increased shoot multiplication in cowpea plants in vitro (Koçak et al. 2023).

The regeneration potential of explants is determined by their multiplication factor. An increase in this parameter was noted when using silver nanoparticles in guava (Andújar et al. 2020), rose, and gerbera (Tung et al. 2022; Tung et al. 2021b), when using MWCNTs in sugar cane (Sorcia-Morales et al. 2021) and ZnO NPs in pomegranate (El-Mahdy and Elazab 2020).

Hormones are used to stimulate the processes of survival, growth, and development of microshoots in the cultivation medium. Cytokinins such as meta-topolin are important in vitro growth promoters (Lai et al. 2022). The simultaneous use of nanoparticles can enhance the positive effect of the use of media based on hormonal preparations. The use of Si nanoparticles in a medium containing meta-topolin and indole-3-acetic (IAA) acid contributed to the formation of the microstructure of tubers in *Dioscorea pentaphylla* L. (Manokari et al. 2023a). The addition of nanosilver to the culture medium containing cytokinins had a positive effect on the formation of buds, leaves (Hegazi et al. 2021), and root regeneration (Hegazi et al. 2021; Lai et al. 2022) in plants in vitro. Adding silver nanoparticles to a medium containing benzylamino purine (BA), gibberellic acid (GA_3), and 2,3-naphaleneacetic acid (NAA), had a positive effect on the development of shoots and roots and contributed to an increase in the content of chlorophylls, carotenoids, and proline (Elsayh et al. 2022). The use of MS medium containing 2,4-dichlorophenoxyacetic acid (2,4-D) and kinetin (Kin) and supplemented with ZnO nanoparticles contributed to an increase in the concentration of secondary metabolites in thyme callus (Mosavat et al. 2019). MS medium containing benzylaminopurine, IAA and enriched with Si nanoparticles had a positive effect on shoot proliferation, rooting efficiency, formation of leaf microstructure, and flower bud induction in various plants (Cokul Raj et al. 2023b; Manokari et al. 2023a, b; c, d, e).

Stimulation of the Formation of Secondary Metabolites When Using Nanoparticles In vitro

Plants are a source of synthesis of secondary metabolites necessary for plants themselves to protect against adverse effects, as well as for human use in commercial activities due to their high biological activity and invaluable medicinal properties. However, the production of secondary metabolites using in vivo and in vitro cultures remains low and does not meet commercial demand (Inam et al. 2023). Plants in culture in vitro are deliberately subjected to various influences to stimulate the synthesis of beneficial compounds. Nanoparticles have a good application as elicitors in the technology of obtaining tannins, sugars, proteins, enzymes, glycosides, and phenolates, since they have a stimulating effect and exclude toxic effects at low concentrations. The impact of nanoparticles causes the formation of signaling molecules and ROS that promote transcriptional control and regulation of secondary metabolism (Inam et al. 2023). The use of ZnO nanoparticles contributed to an increase in the content of macroelements N, P, K, and S in date palm (Al-Mayahi 2021; Awad 2020) and an increase in the production of phenols and glycosides in mung bean (Iqbal et al. 2022); the content of glycosides and steviates also increased in stevia (Ahmad et al. 2020a). The use of ZnO nanoparticles stimulated the formation of thymol and carvacrol from callus cultures of thyme and zataria (Mosavat et al. 2019). Silver nanoparticles stimulated the accumulation of hydrocarbons and proline (Elsayh et al. 2022) in the date palm. The use of silver and titanium oxide nanoparticles helped to stimulate the formation of aloin in *Aloe vera*, the important secondary metabolite used in medicine and cosmetology (Álvarez et al. 2019).

Development of Stress Resistance When Using Nanoparticles in Culture In Vitro

The use of nanoparticles has also been noted to reduce stress in various plants under in vitro conditions. The influence of salt stress can significantly reduce the adaptive capacity of plants, while the use of nanoparticles under in vitro conditions significantly reduces the negative effects caused by salinity. Thus, the use of ZnO NPs under salt stress conditions contributed to a decrease in the accumulation of toxic Na^+ and Cl^- ions and Na^+/K^+ ratios, an increase in the absorption of K^+ ions and proline content, and an increase in activity of antioxidant enzymes in strawberries (Abu Zeid et al. 2023). The impact of ZnO nanoparticles under salt stress contributed to an increase in the number of roots, biomass, content of chlorophyll, soluble protein and carotenoids, anthocyanins, phenols, and removal of active radicals from the olive (Regni et al. 2022). Biomass, shoot length, number of shoots, total chlorophyll, proline content, and activity of antioxidant enzymes also increased in sage (Alenezi et al. 2022).

12.3.2 Silver Nanoparticles in Plant Biotechnology

For thousands of years, silver has been known for its antimicrobial properties and has been used to store food, water, and wine (Xu et al. 2020). Even before the discovery of antibiotics, silver was used in wound dressings, creams, and as an antibacterial coating on medical devices. Currently, ionic and colloidal forms of silver are used as additives in medical immunostimulating drugs, to modify the surface of implants for sterilization (Yin et al. 2020). Silver nanofluids have been applied in solar collectors, cooling and air circulation systems, and energy management in architectural design to provide natural convection and cavities (Yavari et al. 2022), improve the performance of photovoltaic systems (Abdelrazik et al. 2019), and for disinfection and desalination of water (Parsa et al. 2020).

Silver nanoparticles are widely used in medicine, food and industrial production, and biotechnology due to their unique physicochemical and biological properties (Bruna et al. 2021; Xu et al. 2020). They have higher capacitance and higher surface area than bulk silver (Yaqoob et al. 2020). The advantage of using silver nanoparticles is their antimicrobial properties, which make it possible to fight even multiresistant bacteria (Bruna et al. 2021).

The use of silver nanoparticles in tissue culture is the most common due to their antimicrobial, antifungal, and antiviral properties. Nanosizes of 1–100 nm allow them to easily penetrate the cell membrane of pathogens, disintegrating it and destroying intracellular ATP, disrupting cell division processes, and provoking oxidative stress, protein disfunction, and DNA damage, which leads to damage to the pathogen cell (Anees Ahmad et al. 2020). At the same time, the high efficiency of using silver nanoparticles is observed at low concentrations of 0.1–100 mg/L, depending on the type of plant. Ag NPs were used for surface sterilization of basil at a concentration of 10, 50, and 100 mg/L for 5, 10, 20, 30, and 60 min (Adebomojo and AbdulRahaman 2020). Treatment of 250 ppm Ag NPs of 4-week-old ex vitro leaves showed optimal sterilization results for chrysanthemums. Treatment of unsterilized medium with 4 ppm nanoparticles showed 100% sterility (Tung et al. 2021a). Silver NPs obtained by the method have shown their effectiveness against gram-positive and gram-negative bacteria (Aritonang et al. 2019) including antibiotic-resistant strains (Bruna et al. 2021) as well as against viruses (Sharma et al. 2019).

An important feature of silver nanoparticles in plant agriculture and biotechnology is the ability to stimulate growth processes, accumulate biomass and secondary metabolites, and activate defense pathways in plants to adapt to abiotic and biotic stress. Silver NPs have a positive effect on germination and stimulate the development and activity of the antioxidant system of cultivated plants under in vitro conditions. The use of Ag nanoparticles stimulated the growth and multiplication of in vitro clones (Khattab et al. 2022), bud formation (Hegazi et al. 2021), regeneration of shoots and roots (Elsayh et al. 2022), increased leaf area (Andújar et al. 2020), accumulation of chlorophyll a and b (Ha et al. 2020;

Cokul Raj et al. 2023b), organization of the stomatal complex (Manokari et al. 2023a), and the formation of flowers and fruits (Phong et al. 2022). Treatment with silver nanoparticles increased the activity of ascorbate peroxidase and superoxide dismutase (Tung et al. 2022), lipid oxidation (Spinoso-Castillo et al. 2017), and the content of phytohormones, abscisic acid, gibberellin A_3, IAA (Sadak 2019), zeatin and ethylene (Cuong et al. 2023), phenolic substances (Bello-Bello et al. 2017), chlorophylls (Castro-González et al. 2019; Sadak 2019), and carotenoids and carbohydrates (Elsayh et al. 2022). Silver nanoparticles increased the chemical and biological activity of plant extracts (Khattab et al. 2022).

12.3.3 TiO_2 NPs

Titanium is a versatile metal that finds numerous applications in both technology and biotech fields due to its unique optical, thermal, electrical, and magnetic properties. It has found its application in photocatalysis, the functioning of electrochromic and electronic devices, as well as sensory devices (Irshad et al. 2021). Titanium's exceptional strength-to-weight ratio makes it ideal for aerospace components. Titanium is biocompatible, resistant to corrosion, strong, and durable, making it ideal for medical implants, bioreactors, electronic devices, and sports equipment.

Titanium dioxide (TiO_2) nanoparticles are utilized in diagnostic assays, biosensors, and drug delivery systems in biotechnology for their unique chemical and optical properties. They are promising materials for use in plant biotechnology having unique properties such as photocatalytic activity, antibacterial activity, and the ability to improve plant photosynthesis. The photocatalytic activity of TiO_2 allows it to remove organic substances such as pesticides, insecticides, and other pollutants that can negatively affect plant growth and development. In addition, TiO_2 can be used to combat bacterial infections that cause plant diseases. Improving plant photosynthesis with TiO_2 is also a promising direction in biotechnology. TiO_2 NPs can be used to increase the efficiency of photosynthesis, which can lead to higher yields and improved product quality. TiO_2 NPs had a positive effect on bamboo plants exposed to heavy metals by reducing the concentration of ROS, the degree of peroxidation, and soluble protein and increasing antioxidant activity, chlorophyll content, and total biomass (Emamverdian et al. 2021). Titanium oxide nanoparticles contributed to an increase in the number and length of shoots, the number of roots, and the total chlorophyll, phenol, and flavonoid content in sequoia (El-Sayed et al. 2021).

Despite the many positive properties, the use of TiO_2 NPs in plant biotechnology should be carried out with caution to avoid negative consequences for the environment and human health. More research and testing on the safety and efficacy of TiO_2 NPs in biotechnological processes are needed.

12.3.4 Si Nanoparticles

Silicon is the second most abundant element on the planet. Si helps reduce transpiration losses in plants and increases resistance to various diseases and abiotic stresses (Feizi 2023). Silicon nanoparticles penetrate plant coats and are transported in plants through aquaporins and cause a cascade of morphophysiological changes, regulating the work of photosynthetic genes. Si nanoparticles have demonstrated a protective role under stress conditions by affecting the water status in plants, the metabolism of reactive oxygen species, the enzymatic profile, and the regulation of transcription of stress resistance genes in plants (Mukarram et al. 2022; Saha et al. 2021) and a positive effect on the rhizosphere microbiome (Rajput et al. 2021). The use of silicon nanoparticles contributed to an increase in the mass of roots under the influence of drought in barley (Ghorbanpour et al. 2020); an increase in the content of proline, soluble sugars, chlorophylls, and carotenoids and the activity of antioxidant enzymes; and a decrease in the concentration of hydrogen peroxide in strawberries (Zahedi et al. 2023). The use of silicon nanoparticles can also be effective in protection against salinity (González-García et al. 2021; Junedi et al. 2023) and exposure to heavy metals, including Pb, Cd, and As (Bakhtiari et al. 2023; Fatemi et al. 2021; González-Moscoso et al. 2022), and insects (Bapat et al. 2020). Si nanoparticles may be more effective in reducing Cd uptake than conventional silicon (Rizwan et al. 2021). Both Si and Si-NP can increase the concentration of mineral nutrients, protein, and some amino acids in plants (Riaz et al. 2022). Amorphous silicon nanoparticles are of interest for use in tissue culture because they are cheap and biodegradable (Cokul Raj et al. 2023a). Under in vitro conditions, it plays an important role by stimulating the synthesis of secondary metabolites and stabilizing the structure of plant tissues (Manokari et al. 2023c). The use of Si-based nanoparticles in tissue culture contributed to an increase in multiplication in *Lavandula officinalis* L. (Khattab et al. 2022), regeneration of lateral roots in *Prunus cerasifera* L. (Khafri et al. 2023), and lateral shoots in *Cydonia oblonga* Mill (Farhadi et al. 2023). Manokari et al. showed that the use of silicon nanoparticles improved shoot proliferation, rooting efficiency, biomass production, and leaf microstructural organization in *Vitex trifolia* L., *Thunbergia erecta* (Benth.) T. Anderson, and *Cymbopogon citratus* (DC.) Stapf (Cokul Raj et al. 2023b; Manokari et al. 2023b; Manokari et al. 2023c). The use of Si nanoparticles in culture in vitro contributed to the formation of flower buds in *Vitex negundo* L. (Cokul Raj et al. 2023a).

12.3.5 ZnO Nanoparticles

ZnO NPs have found their application in various industries, including agriculture, forestry, photocatalysis, medicine, food and cosmetics, and other activities related to their antibacterial properties (Zeghoud et al. 2022). ZnO nanoparticles can act as an alternative to traditional antimicrobial and antifungal drugs, showing activity against gram-positive and gram-negative bacteria and unicellular and multicellular fungi (Lashin et al. 2023). ZnO NPs have pesticidal activity and increase the yield of plants, so they have found their application in agriculture.

Zinc nanoparticles provoke the formation of ROS, which contributes to the destruction of microbial cells. The antimicrobial properties of zinc nanoparticles depend on their size, with higher activity observed in smaller particles (Kaushik et al. 2019). Zn is a natural antioxidant, protecting cell membranes, mitochondria, and DNA from ROS damage (Zhuzhukin et al. 2023). Zn is one of the most important trace elements necessary for plant growth and development and provides the most important processes of photosynthesis and phytochrome synthesis, being a structural and cofactor element of various proteins and enzymes and maintaining membrane integrity by stimulating the antioxidant system (Zhuzhukin et al. 2023). Despite the sufficient content in the soil, the bioavailability of this trace element for plants remains low, which leads to deficiency and affects vegetative growth and plant yield. The use of zinc in nanoform helps to solve the problem of its bioavailability in soil due to its small size, higher specific surface area, and higher solubility (Zhuzhukin et al. 2023).

The use of zinc oxide nanoparticles in tissue culture contributed to the stimulation of rooting, an increase in biomass, the content of glycosides in stevia (Ahmad et al. 2020a), and the content of secondary metabolites in callus cultures in *Thymus* species and *Zataria multiflora* Boiss (Mosavat et al. 2019). ZnO nanoparticles increased the content of macro- and microelements and stimulated callus and root regeneration in *Phoenix dactylifera* L. (Al-Mayahi 2021). An increase in the activity of antioxidant enzymes under the influence of zinc oxide nanoparticles was also shown in various plants (Abu Zeid et al. 2023; Ahmad et al. 2020a; Al-Mayahi 2021). The use of ZnO nanoparticles in tissue culture contributed to the reduction of salt stress by increasing the activity of antioxidant enzymes, reducing free radicals, and accumulating secondary metabolites, while stimulation of growth parameters was observed in various plants (Abu Zeid et al. 2023; Alenezi et al. 2022; Regni et al. 2022).

12.3.6 Carbon Nanotubes

Carbon nanotubes (CNTs) are an allotropic form of sp^2-hybridized carbon that has a cylindrical structure with a diameter ranging from a few nanometers to tens of nanometers. Carbon-coated nanoparticles can enter plant cells and move to remote positions through the vascular system (Corredor et al. 2009). The organization of cylinders helps to divide CNTs into two types, single-walled CNTs (SWCNTs) and multi-walled CNTs (MWCNTs). They have unique properties such as high strength, thermal properties, electrical conductivity, and chemical stability, which make them promising for use in various fields, including agriculture and plant biotechnology.

The use of SWCNTs accelerated the growth of maize seminal roots by upregulating gene expression (Yan et al. 2013). A low concentration (20 mg/L) of SWCNTs and MWCNTs was found to increase leaf growth and development, chlorophyll and gibberellin content, and chloroplast development-associated and cell size-associated gene expression in rice seedlings (Zhang et al. 2017). The use of functionalized SWCNTs helped to promote the growth and cellular metabolism of the in vitro plants of *Rubus adenotrichos* Schltdl (Flores et al. 2014). There are some studies on the effect of SWCNTs on salt stress and drought in plants, which reliably show that the use of nanomaterials contributes to the protection of plants from these negative consequences (Hatami et al. 2017).

Over the past decade, significant strides have been made in the exploration of the impact of MWCNTs on plants. MWCNTs, characterized by their relatively easier and more cost-effective production when compared to SWCNTs, have emerged as key players in this research domain. Remarkable findings have emerged, showcasing enhanced growth in a variety of plant species. Broccoli (Martínez-Ballesta et al. 2016), chickpea (Martinez-Ballesta et al. 2020), barley, soybean, and corn (Lahiani et al. 2013) have exhibited augmented growth patterns when exposed to MWCNTs. This phenomenon can be attributed to the MWCNTs' ability to penetrate cell walls, engendering the creation of novel pores. These openings facilitate increased water absorption, thereby contributing to heightened germination rates (Yousefi et al. 2017). Furthermore, it has been observed that MWCNTs possess the capacity to upregulate the expression of genes associated with critical processes like cell division, cell wall formation, and water transport. This genetic modulation results in the remarkable growth enhancement of tobacco cells (Khodakovskaya et al. 2012). In a parallel revelation, MWCNTs were found to infiltrate seed coats, bolstering water uptake mechanisms in tomato plants (Khodakovskaya et al. 2009).

The intricate interplay between CNTs and plants results in a multifaceted array of physiological and morphological transformations. These alterations are contingent upon several factors, including the CNT type, size, concentration, the specific plant species involved, and the developmental stage of the plants (Khodakovskaya et al. 2013). This dynamic interaction orchestrates modifications in the expression of diverse genes within the plant's genome. For instance, in a study denoted as (Yan et al. 2013), researchers documented a notable upregulation of genes associated with seminal root development in maize when subjected to SWCNTs. Similarly, the exposure of tobacco cells and tomato tissues to MWCNTs resulted in a conspicuous upregulation of gene expression (Khodakovskaya et al. 2012, 2013). Furthermore, the application of fullerenes induced upregulation in genes associated with oxidative stress, salt tolerance, water deprivation response, wound healing, and defense against pathogens in *Arabidopsis* plants (Landa et al. 2012).

12.4 The Use of Nanoparticles in Agriculture

Agriculture is the main source of food for most countries in the world. Fertilization is necessary to obtain high yields, but their uptake by plants remains extremely low. In this regard, nanofertilizers are widely used at present. It has been shown that hydroxyapatite nanoparticles can be used as nanofertilizers in tomato cultivation and do not exhibit a phytotoxic effect (Marchiol et al. 2019).

The main problem of obtaining a high yield is the presence of unfavorable conditions, such as drought, salinization, the presence of heavy metals in the soil, and the presence of phytopathogens. Nanoparticles have shown their effectiveness in combating both abiotic and biotic stress factors; therefore, special attention is paid to their study in relation to crops. The effectiveness of the use of both metal-containing and non-metal nanoparticles on soils contaminated with heavy metals (Cd, Pb, and As) has been shown for various agricultural plants. Thus, nanoparticles reduced the negative impact of heavy metals on wheat plants (Hussain et al. 2019; Manzoor et al. 2021), soybean (Ahmad et al. 2020b), and coriander (Fatemi et al. 2021) by reprogramming the morphological and physiological status and reducing the absorption of heavy metal by plants, while biomass, the content of photosynthetic pigments, and the activity of antioxidant enzymes increased (Hussain et al. 2019).

Salinity and drought significantly limit the growth of plants in the field. Here nanoparticles proved to be a universal tool for reducing the consequences of abiotic stress, while the yield of treated plants increased significantly. Nanoparticles regulated the content of photosynthetic pigments (Manzoor et al. 2021) and contributed to maintaining the Na^+/K^+ balance of cells (Liu et al. 2021) when exposed to salinity. Zn, B, Si, and zeolite NPs increased growth and productivity, nutrient concentrations, proline, gibberellic acid, protein, carbohydrates, and antioxidant enzymes in potatoes (Mahmoud et al. 2019). Metal-containing nanoparticles reduced the negative effects of drought by stimulating the expression of drought-associated genes (Linh et al. 2020). Under greenhouse conditions, treatment with Zn-containing nanoparticles contributed to an increase in the biomass and rate of photosynthesis in coffee plants (Rossi et al. 2019).

12.5 The Mechanism of Uptake and Relocation of Nanoparticles in Plants

High interest in the use of nanoparticles, their large-scale industrial production, and their use in various sectors of the economy inevitably lead to their mass release into the environment, soils, and water resources. The possibility of penetration of nanoparticles into plants depends on the size and charge of nanoparticles, as well as on the type of plant and the stage of its growth. Nanoparticles penetrate plants through the leaf cuticle and stomata, the root system, and sometimes wounds and roots origin when applied to the soil and foliar spraying with fertilizers or pesticides, as well as when using an artificial nutrient medium (Inam et al. 2023; Mishra et al. 2022). For terrestrial plant parts, the cuticle is the primary barrier that screens out nanoparticles larger than 5 nm; nanoparticles up to 20 nm in size can penetrate through stomata, be transported along the phloem, and accumulate in the leaves, fruits, and seeds. Larger nanoparticles have been reported to penetrate natural barriers by expanding existing pores or forming larger ones (Inam et al. 2023). Penetrating through the root and overcoming the barrier in the form of the endoderm Caspar strip, the nanoparticles reach the vascular cylinder and are further distributed through the xylem to all organs and tissues (Thounaojam et al. 2021). The cell wall is an important barrier to the passage of nanoparticles. Overcoming the cell wall, nanoparticles reach the cell membrane, the internalization of which occurs through the induction of new pores with the help of ion carrier substances or with the help of membrane-bound carrier proteins (Inam et al. 2023). When entering a cell, nanoparticles interact with various organelles and can significantly change cell metabolism, provoke ROS production, and regulate gene transcription. One of the possible mechanisms of influence on the plant organism is the stimulation of the expression of regulatory genes. It has been shown that Ag NPs can influence the expression of ethylene production genes. A decrease in the expression of TuACS, which contains light-dependent, hormone-dependent, and stress-dependent cis-regulatory elements in the promoter region, under the influence of nanosilver, contributed to delayed aging and increased survival in *Tecomella undulata* under in vitro conditions (Sarmast et al. 2015).

Nanoparticles can be formed in the soil through natural processes (Huang et al. 2022). Forest fires provoke the formation of a certain amount of CNTs, which are products of combustion, which fall into the soil and are absorbed by plants. MWCNTs can act as signaling molecules in plants, initiating the synthesis of secondary metabolites (Samadi et al. 2020). The presence of an increased content of carbon nanotubes in soil can be perceived by woody plants as a signal to occupy a free area (Fig. 12.3).

12.6 The Use of Nanoparticles in Forestry

A significant amount of research on in vitro cultivation is focused on agricultural, ornamental, and medicinal, usually herbaceous or shrubby plants. At the same time, there are not many reports on forest tree species.

The importance of forests in commercial and ecological terms cannot be overestimated. The forest is a source of wood and lumber, a raw material for the production of paint and varnish products and fuel. Forest plants, mushrooms, and berries serve as a source of food for humans and animals. The forest is the main source of oxygen on the planet and a pool for the decarbonization of the environment. At the same time, the problem of forest loss is the most acute, including due to global climate change toward droughts and the spread of forest fires. Forests annually disappear over large areas as a result of economic activities (logging, construction of

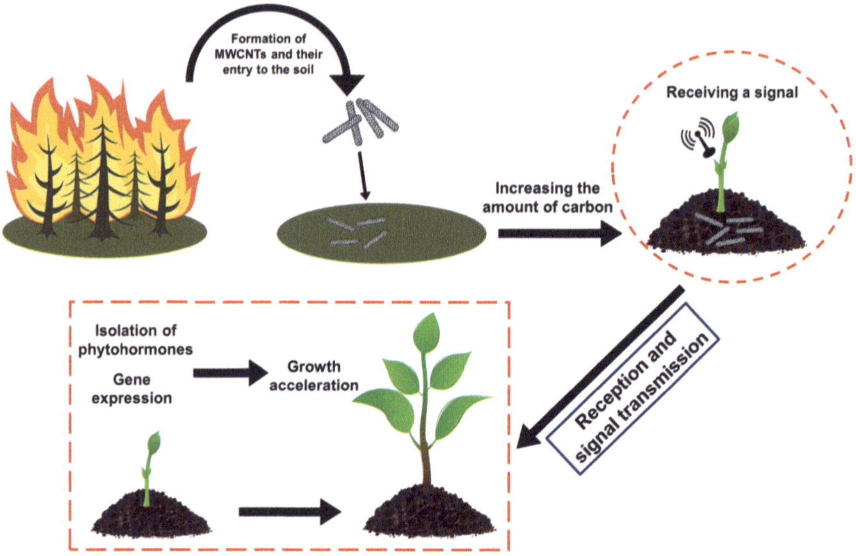

Fig. 12.3 Possible influence of MWCNTs on plant growth acceleration (Zhuzhukin et al. 2023)

roads and other infrastructure, expansion of settlements, reduction in the resistance of secondary forests to insects, diseases, etc.). At the same time, due to the long period of rotation of forest tracts, there is no possibility of their rapid restoration.

An urgent task at the world level is the transition to a low-carbon economy, which implies the widespread use of such decarbonization tools as forest climate projects. Carbon (climate) neutrality is a major challenge faced by most of the countries that contribute the most to anthropogenic greenhouse gas emissions. At the same time, achieving real carbon neutrality in most cases is impossible without increasing the absorption capacity of natural ecosystems, which can be achieved, among other things, through projects to reduce greenhouse gas emissions in the field of forestry and increase the absorption capacity of forests (forest climate projects). At the same time, the creation of new nanobiotechnological approaches to obtain high-quality planting material for fast-growing forest crops can be an effective tool. Woody plants (Burlacu et al. 2019) such as eucalyptus (Sawalha et al. 2021) can be used to produce metal nanoparticles. Nanoparticles, in turn, accelerate growth, promote the accumulation of chlorophyll, and stimulate photosynthesis (Dong et al. 2019; Siddiqui et al. 2020; Wang et al. 2022), which opens up the potential for intensifying CO_2 uptake and its further deposition by plants.

The main candidate species for such projects will be highly productive and resistant forms, well adapted to the conditions of a particular region. Such species in the forest-steppe zone can be birch, poplar, and oak. Birch can be the main crop due to its high biological stability and productivity in the specific forest conditions of the steppe and forest-steppe, where in many cases birch (*Betula* sp.) should be preferred over conifers. Birch is widely distributed in these areas under natural conditions. An important role will be played by selection-valuable varieties and hybrids, bred and adapted to different growing conditions. One of the priorities in use remains with representatives of the poplar genus (*Populus* L.), which are the fastest-growing species of the temperate zone. The use of nanoparticles can improve the growth parameters of woody plants. In our work, we showed that MWCNTs caused an increase in the length and diameter of the shoot in *Betula pubescens* Ehrh. and *B. pendula* Roth. propagated in vitro and transferred to a greenhouse (Zhuzhukin et al. 2023).

As part of the implementation of climate projects, the "cost of error" when using low-quality planting material is as high as possible, since the period for reaching the estimated productivity is 20–30 years or more, and low tree growth rates can ultimately significantly reduce the economic feasibility of the project. One of the problems of forest climate projects is the presence of limiting factors in the form of the impact of adverse environmental conditions. In connection with global climate change, there is an expansion of the habitat of phytopathogenic microorganisms in the world to the northern and eastern territories. This leads to a deterioration in the quality of planting material of cultivated plants, including woody ones. For seedlings and young plants, diseases that cause rotting of the roots and the death of needles and shoot stems are especially dangerous. Fungi from the genus *Fusarium* and *Alternaria* cause rot of seedlings of coniferous and deciduous species (lodging of seedlings). Plants that are in the juvenile phase of development are mostly affected, which is especially important when microclones are transferred to the substrate. Diagnosis of fusariosis and alternariosis is difficult because these diseases often occur in a latent form. A variety of methods can be used to control fungal diseases of woody plants, including chemicals, biological fungicides, and cultural practices. At the same time, it has been shown that nanoparticles of metals and their oxides are active against fungal phytopathogens that infect woody plants. We showed that the use of CuO nanoparticles contributed to a decrease in the development of colonies of phytopathogenic fungi *Alternaria alternata* and *Fusarium avenaceum*, while *Fusarium oxysporum* was not significantly affected (Grodetskaya et al. 2022).

The use of copper nanoparticles contributed to a decrease in the reproduction of the complex of the beetle *Euwallacea fornicatus* and a symbiotic fungus, which can serve as a new strategy for the treatment of pathologies caused by pests of woody plants (Cruz et al. 2021).

Bacterial plant diseases also cause great lesions that can be general, causing the death of the entire plant or its parts. It can appear on the roots (root rot) and in the vascular system (vascular diseases); to be local, is limited to the disease of individual parts or organs of the plant; and also manifests itself on parenchymal tissues (parenchymal diseases such as rot, spotting, burns) or may be mixed. A special place is occupied by bacterioses associated with the appearance of neoplasms (tumors). The causative agents of bacterioses are mainly non-spore-bearing bacteria from *Pseudomonadaceae* family and others. Infection of trees occurs through various damages to the bark and cambium, including damage by insects, which further leads to the rapid development of infection and death of plants. In this regard, timely diagnosis and development of methods for adequate control in the early stages of disease development are of particular importance. The main problem is the difficulty of early detection of diseases and the use of pesticides when their effect is already reduced due to the neglect of the disease. In addition, the choice of pesticide depends on the type of pathogen and the severity of the disease and may induce resistance after prolonged use. Currently, the problem of phytopathogenic infections is being solved by increasing the use of pesticides, fungicides, and bactericides that have a negative impact on the environment and human health. The identified problems indicate the need to search for effective and environmentally friendly substances to regulate the number of phytopathogenic bacteria and fungi that infect woody plants. Copper oxide nanoparticles loaded onto the surfaces of graphene oxide sheets (GO-Cu NPs) (Li et al. 2017b), silver-chitosan nanocomposite (Shahryari et al. 2020), poly(DL-lactide-co-glycolide acid) (PLGA) copolymer-based biopolymeric NPs (Fortunati et al. 2016), Ag (Jiang et al. 2022), and ZnO (Fan et al. 2023) nanoparticles were found to be effective against different strains of *P. syringae*.

A promising task is the development of smart pesticides based on nanomaterials and their complexes, which solve several major problems at once. These are the possibility of targeted delivery to the source of infection, the use of deliberately small doses in comparison with conventional drugs, the impact on a wide range of pathogens, and the impact on a pathogenic organism and on the treated plant. Moreover, if fungicides and antibiotics often have a negative effect on plants, nanoparticles, on the contrary, stimulate the immune response of plants, thereby facilitating the fight against phytopathogens.

12.6.1 Use of NPs in In Vitro Culture of Forest Plants

Preservation and enhancement of the protective and resource potential of forests and prevention of deforestation and degradation are the main task of sustainable forest management. One of the possible ways to combat deforestation, as well as to intensify the production of wood as a raw material for the timber processing industry, is plantation cultivation using selectively improved fast-growing tree species (Gureev et al. 2021). Biotechnological approaches based on the cultivation of organs and tissues of perennial plants in vitro, on artificial nutrient media, under controlled aseptic conditions, will allow full use of breeding achievements and intensive cultivation of high-quality planting material of valuable genotypes of tree crops for plantation forestry (Lebedev et al. 2020). Despite numerous difficulties, to date, there are more than 200 species of woody plants from 40 families that have been propagated in vitro. Techniques and methods have been developed for clonal micropropagation of birch, poplar, aspen, and adult trees of Karelian birch (*Betula pendula* Roth var. carelisa Merkl.); juvenile material of English oak and Scots pine and poplar haploids were obtained from anther callus; the first regenerated plants of these species were obtained. Systems of reproduction and preservation in vitro of ornamental, subtropical, stone fruit, fruit, and essential oil crops have been created and improved. Despite the significant results achieved in tissue culture of woody plants, there is ample room for improvement in these methods in terms of industrial production using nanoparticles. Nanomaterials are a promising tool for improving the quality of sterilization, increasing stress resistance, and increasing productivity in the culture of forest plants. The most common types of nanoparticles used in this field are metal (Au, Ag, Cu, Fe) and metal oxide NPs such as titanium dioxide (TiO_2), zinc oxide (ZnO), copper oxide (CuO), iron oxide (Fe_2O_3), and magnesium oxide (MgO). TiS_3 nanoribbons had a sterilizing and stimulating effect at the initiation stage and acted as an activator of rhizogenesis at the rooting stage in downy birch and white poplar × aspen hybrid (Zakharova et al. 2021). Titanium and iron oxide nanoparticles had a positive effect on morphometric parameters, the activity of antioxidant enzymes, and the synthesis of secondary metabolites during in vitro cultivation of sequoia (El-Sayed et al. 2021). Nanosilver is one of the promising agents for combating the contamination of woody plants during in vitro cultivation. The use of silver NPs together with NaOCl contributed to the complete elimination of contamination and stimulation of the development of shoots and roots in graying poplar (Vasyukova et al. 2021). The use of silver nanoparticles at a high concentration (50 ppm) contributed to the inhibition of the growth of white rot fungus colonies in beech (Piętka et al. 2022). The use of 150 ppm silver nanoparticles contributed to the inhibition of phytopathogenic fungi *Rhizoctonia solani* and *Fusarium oxysporum* f. sp. lycopersici in vitro (Purohit et al. 2022).

12.7 The Use of Nanoparticles in the Fight Against Plant Pathogens

Agricultural production is associated with a constant threat of damage by phytopathogens, yield losses from which in some years amount to 30–50%. Excessive and not always justified use of fungicides harms plants and the environment. The introduction of new technologies based on the use of nanoparticles in the agricultural industry can reduce the negative consequences of the use of synthetic fungicides, as well as reduce the damage to cultivated plants by phytopathogens.

Boron, manganese, molybdenum, copper, zinc, cobalt, and iodine are of great importance in plant nutrition, protection against diseases and pests, as well as the formation of the crop and its quality. These elements are called trace elements since the content of most of them ranges from thousandths to hundred-thousandths of a percent. Trace elements are involved in many physiological and biochemical processes in plants. Trace elements Cu, Mn, and Zn are of decisive importance for the activation of enzymes and the synthesis of biomolecules involved in plant protection. Currently, in the field of agricultural biotechnology, there has been a tendency to use nanoparticles of metals and their oxides, which exhibit increased biological activity including antifungal activity, due to unique physicochemical characteristics such as size, surface area, and Ilyin's charge. Scientists from the All-Russian

Research Institute of Potato Farming have developed a preparation based on nanoparticles of the vital elements of iron, zinc, copper, and molybdenum to protect potatoes from diseases common in the Moscow region of Russia, late blight, alternariosis, and rhizoctoniosis, and evaluated its effectiveness. A decrease in the prevalence and development of diseases during the growing season with the use of pre-sowing treatment with metal nanoparticles in the composition of the film from 2.2 to 4.2 times was established. Such a decrease in potato diseases during pre-sowing treatment of tubers with a nanopreparation leads to an increase in potato yield by 25.7% and marketable tubers by 11.7% compared with the control.

The largest amount of scientific research is devoted to the development of ways to use silver nanoparticles in agricultural practice (Jo et al. 2009; Li et al. 2017; Mansoor et al. 2021; Singh and Ahamad 2022). As we know, silver is one of the most powerful natural antiseptics. For the first time, a pesticide containing silver was registered in the USA in 1954, and at present, there are about 110 preparations based on silver. Zerebra Agro, Serebromedin, Zeromix Alpha, and others have been registered and successfully tested in Russia as growth stimulant seeds of crops with fungicidal properties.

One of the theories explaining the mechanism of action of silver on microorganisms is the adsorption theory. According to this theory, the microbial cell loses its viability as a result of the interaction of electrostatic forces between the cells of microorganisms that carry a negative charge and positively charged silver ions. Silver reacts with the cell membrane of microorganisms and blocks its ability to transfer oxygen into the cell, which leads to "suffocation" of the microorganism and its death. Another theory of the toxicity of silver nanoparticles is partly explained by the production of reactive oxygen species (ROS), which leads to apoptosis of cells of the phytopathogenic organism (Kim and Ryu 2013).

In an in vitro study on cultures of phytopathogenic fungi *Rhizoctonia solani*, *Macrophomina phaseolina*, *Sclerotinia sclerotiorum*, *Trichoderma harzianum*, and *Pythium aphanidermatum*, the effect of silver nanoparticles at concentrations of 6, 8, 10, 12, 14, and 16 mg/L (Mahdizadeh et al. 2015) was tested. The most sensitive concentration of 6 mg/L was established and further investigated in a greenhouse. The sensitivity of fungi to AgNPs decreased in the series *P. aphanidermatum* > *S. sclerotiorum* > *M. phaseolina* > *R. solani* > *T. harzianum*. The minimal negative effect of *Trichoderma harzianum* fungus nanoparticles indicates the fact of partial or complete absence of inhibition of a representative of beneficial microflora. Based on the research, it was concluded that AgNPs could potentially be used to control plant pathogens in the field.

Other scientists have shown a significant antifungal effect of silver nanoparticles on plant pathogens belonging to the genera *Alternaria*, *Fusarium*, *Botrytis*, *Cladosporium*, *Pythium*, *Didymella*, etc. at a higher concentration of 100 mg/L (Kim et al. 2012). The reason why the work turned out to be so overestimated compared to the rest of the toxic concentrations is not clear. Perhaps this is due to the different shapes of the studied nanoparticles, as well as differences in their sizes. It has been established that the smaller the size of the nanoparticles, the greater the antifungal activity they possess. So, in the works of (Soltani Nejad et al. 2017;

Madbouly et al. (2017); Win et al. 2020), when studying silver nanoparticles 10–30 nm in size, higher inhibitory effects were obtained against a number of phytopathogenic fungi than when studying larger 40–70 nm nanoparticles (Kasprowicz et al. 2010; Krishnaraj et al. 2012; Valsalam et al. 2019). This is because smaller nanoparticles penetrate or destroy the pathogen cell membrane faster and thus combine fungal hyphae and mycelium and inactivate these pathogens.

In a study (Kim et al. 2012), the antifungal activity of Ag NPs against the ragweed fungus *Raffaelea* sp., which is responsible for the mortality of a large number of oak trees in Korea, depended on their form: CV, AT, and PR. Fungal growth in the presence of silver nanoparticles was significantly inhibited in a dose-dependent manner. Microscopic observations have shown that silver nanoparticles have a detrimental effect not only on fungal hyphae but also on the germination of conidia.

The effect of silver nanocomposites at a concentration of 0.000625% (6.25 mg/L) on the causative agent of bacterial potato disease *Clavibacter michiganensis* subsp. sepedonicus causes ring rot (Perfileva et al. 2021). Polysaccharides, arabinogalactan, starch, and humic substances were used as stabilizing matrices. The design of nanoparticles in a matrix is advantageous since in matrices (nanocomposites) toxic nanoparticles are densely packed into a plant-safe matrix of natural origin and are able to be released under the action of bacterial exoenzymes, killing them. The experiments performed have shown that during the incubation of bacteria with silver nanocomposites in various matrices, the bacteria *Clavibacter michiganensis* subsp. sepedonicus experience stress, which subsequently leads to their death, which is confirmed by data obtained using vital dyes. The observed effect of reducing the growth of bacteria by 40% and changing their morphology after incubation with silver nanocomposites is probably associated with local destructive changes in the bacterial cell wall.

No less popular at present is copper, which in the nanodispersed state has important biological properties, in particular, antimicrobial, antiviral, antibacterial, anticancer, and fungicidal, but at the same time is characterized by relatively low toxicity (Borkow and Gabbay 2009; Bramhanwade et al. 2016). The antifungal effect of Cu-chitosan nanocomposites at concentrations of 30, 60, and 100 μg/mL against two anastomotic groups of *Rhizoctonia solani*, the causative agent of cotton root rot, was studied (Abd-Elsalam et al. 2017). The results showed that Cu-chitosan nanocomposite in combination with *Trichoderma* can suppress *R. solani* cotton seedling disease in vivo. Evaluation of *R. solani* in a greenhouse with a *Trichoderma* strain showed a synergistic inhibitory effect. Light micrographs of mycelium, Cu-chitosan nanocomposites, showed the destruction of hyphae structures.

Using biogenic synthesis, Jazan University scientists obtained copper nanoparticles, the size of which CuNP varied from 6 to 26 nm (Abboud 2021). The antifungal activity of CuNP was evaluated by testing against three phytopathogenic fungi including *Fusarium oxysporum*, *Fusarium solani*, and *Curvularia curvulatat* with growth inhibition of 86.25%, 32.92%, and 68.42%, respectively, at 200 mg/L. *F. oxysporum* was more affected by CuNP, followed by *C. curvulatat* and *F. solani*.

It was noted that the highest antifungal activity against *Alternaria alternata* and *Botrytis cinerea* was shown by spherical copper nanoparticles (d ~ 38 nm) (Ouda

2014). Perhaps this is because spherical nanoparticles have a high penetrating ability into phytopathogenic organisms. Other forms have also been found, such as faceted ones with sizes in the 200–500 nm length range, which have shown high efficiency against *F. solani*, *Neofusicoccum* sp., and *F. oxysporum* (Pariona et al. 2019).

Another equally important factor inhibiting the growth of phytopathogenic fungi is the concentration of Cu nanoparticles. Using nanoparticles at concentrations of 0.1, 0.25, and 0.5 mg/L against wilt pathogen *F. oxysporum*, it was shown that a low concentration (0.1 mg/L) contributed to strong oxidative stress in the mycelium of the fungus and an average (0.25 mg/L) higher concentration (0.5 mg/L) showed antifungal activity (Hermida-Montero et al. 2019). Studies (Malandrakis et al. 2019) also established a dose-dependent effect of copper nanoparticles on *B. cinerea*, *A. alternata*, *M. fructicola*, *C. gloeosporioides*, *F. solani*, *F. oxysporum*, and *V. dahliae*. Tested high concentrations of 50, 100, 500, and 1000 mg/L showed a toxic effect on phytopathogens to some extent, and the highest concentration (1000 mg/L) inhibited all types of studied fungi. In addition, the fungicidal activity of the ionic form of copper ($CuSO_4$) was investigated in the studies, which, as studies have shown, turned out to be less toxic against phytopathogens than copper nanoparticles. One of the indisputable facts that serve to protect the use of nanoparticles in agriculture is their non-toxicity, in contrast to the ionic form of metals, in relation to representatives of beneficial soil microorganisms: *Rhizobium* spp., *Trichoderma harzianum*, etc.

Of the other nanoparticles, there are works on the use of Zn NPs in medical practice (Hernández-Sierra et al. 2008), but data on their relation to plant phytopathogens are rare. The mechanisms of ZnO nanoparticle's action on the phytopathogen *Pseudomonas syringae* pv. Tabaci. have been investigated. It was established that under the influence of nanoparticles (0.5 mg/mL), there are global changes in the expression of genes of *Pseudomonas syringae* pv. Tabaci encoding processes such as protein secretion; signal transducer activity; chemotaxis; taxis; degradation of valine, leucine, and isoleucine; biofilm formation; etc. At the same time, ZnO nanoparticles (0.5 mg/mL) potentially improved plant growth and immunity due to enzymatic activity and global analysis of the molecular response. In addition, changes in gene expression in the ABA signaling pathway, ABA concentration, and stomatal aperture all suggest that ZnO NPs can specifically stimulate stomatal immunity, which is important for protection against bacterial infection (Fan et al. 2023).

The work (Zabrieski et al. 2015) showed the effect of zinc nanoparticles at a concentration of 50, 100, 250, and 500 mg/L on representatives of the genus *Pythium* in experiments in vitro. A significant antifungal effect was noted for all tested concentrations. However, in the field and greenhouse experiments, soil pH, as well as the presence of chelators and materials for absorbing Zn, may change that influences the effectiveness of even high concentrations of nanoparticles, which must be taken into account when recommending nanoparticles as a means of controlling and treating various diseases.

Of the non-metals, it should be noted the activity of selenium nanoparticles encapsulated in a starch matrix against the bacterium that causes potato ring rot

Clavibacter sepedonicus in potato plants in vitro (Perfileva et al. 2021). Se nanocomposites exhibit bactericidal and antibiofilm activity against the bacterium under study. The presence of the ability to destroy bacterial biofilms is very important since the ability to form films makes bacteria invulnerable to the action of antibacterial components. Due to their extremely small size and high reactivity, Se nanoparticles can quickly penetrate biofilm materials and cell membranes. In addition, stimulation of plant growth and development was noted. In experiments on plants in vitro, the nanocomposite reduced the negative effect of potato infection with *Clavibacter sepedonicus* due to the enhancement of the protective functions of plants. The absence of selenium accumulation in potato tissues after its treatment with nanoparticles was established. The results obtained allow us to consider Se/Cr nanocomposites as an effective and environmentally friendly agent for stimulating the development of agricultural plants due to targeted low-dose delivery of antimicrobial nanoselenium biocomposites to bacterial phytopathogens.

Thus, a review of multiple sources confirms that metal and non-metal nanoparticles are an excellent option for the control and treatment of various diseases of agronomic importance. At the same time, it is necessary to investigate their possible health risks due to their uncontrolled use, release into the natural environment, and dangerous consequences.

The presence of phytopathogens is the main threat to plant development. Currently, an alternative to pesticides is the use of nanoparticles. Metal NPs are excellent antimicrobial agents and are of particular importance in agricultural applications (Castillo-Henríquez et al. 2020). Copper oxide nanoparticles have been shown to have antibacterial activity by destroying bacterial cells (Asghar and Asghar 2020) and also reducing the growth of fungal colonies (Grodetskaya et al. 2022). In plants, nanoparticles can be an effective tool in the fight against infection of various types and localizations. The main defense mechanism can be systemic acquired resistance. Si-based NPs activated systemic acquired resistance in *Arabidopsis* through exposure to salicylic acid (El-Shetehy et al. 2021). Treatment with metal nanoparticles contributed to the normalization of water content, membrane index, protein and phenols (Satti et al. 2021), increased growth, enzymatic and non-enzymatic activity (Sofy et al. 2021), increased chlorophyll content (Hernández-Hernández et al. 2019; Li et al. 2021), vitamin C, glutathione, flavonoids (Hernández-Hernández et al. 2019), activation of antioxidant defense (Ghorbanpour et al. 2020), reduction of oxidative stress (Li et al. 2017; Sofy et al. 2021), an increase in the content of salicylic acid, and the activity of PR genes (Cai et al. 2020) in various plants.

12.8 Concentration and Toxic Effects of Nanoparticles

The ever-increasing need for the use of nanoparticles may provoke their uncontrolled use. Metal nanoparticles are usually obtained using toxic chemical-reducing agents that pose a threat to the environment (Mustapha et al. 2022). Getting into the soil from wastewater containing waste from food, cosmetics, paint and varnish, and other industries, nanoparticles interact with soil microflora and are absorbed and transformed by plants, thus being included in the trophic chain. The problem lies in the lack of knowledge of the mechanism of accumulation and removal of nanomaterials from biological objects and their impact on the quality of life and human health. The toxic effect can also be enhanced due to the increased activity and reactivity of nanoparticles compared to bulk materials.

A concentration-dependent effect from the use of nanoparticles in tissue culture has been established. The unique properties of nanoparticles, which are useful at low concentrations, can certainly play a negative role for living organisms when certain values are reached. The selection of the concentration of nanoparticles for plant treatment and modification of media in culture in vitro can vary within certain limits, but a significant number of works indicate that they do not exceed several hundred mg/L of the substrate. It has been shown that different concentrations of nanoparticles have different effects on morphometric and physiological parameters of plants. It was shown that the addition of 2.5 mg/L ZnO nanoparticles to the WPM medium contributed to the best proliferation rate, while an increase in nanoparticles to 5 mg/L promoted an increase in the growth rate of shoots and biomass in pomegranate. At the same time, increasing the concentration to 10 mg/L did not have a positive effect on the growth of pomegranate (El-Mahdy and Elazab 2020). The use of 5 mg/L titanium oxide nanoparticles caused an increase in chlorophylls, phenols, and flavonoid content in sequoia, while an increase to 10 mg/L promoted shoot and root growth (El-Sayed et al. 2021). The use of 1 mg/L SiO_2 NPs and 2.5 mg/L ZnO NPs in the cultivation medium improved the proliferation and growth rate of quince rootstock explants (Farhadi et al. 2023). The use of 6 and 18 mg/L ZnO nanoparticles caused an increase in biomass, root development, synthesis of carotenoids, chlorophylls, anthocyanins and phenols, protein accumulation, and reduction of free radicals in olive (Regni et al. 2022), and 100 and 150 mg/L nanoparticles stimulated the formation of secondary metabolites in thyme and *Zataria multiflora* (Mosavat et al. 2019). Treatment at 10 and 15 mg/L reduced the negative effects of salt stress in sage (Alenezi et al. 2022) and strawberry (Abu Zeid et al. 2023), respectively.

If low concentrations have a positive effect on the morphometric and biochemical characteristics of plants and stimulate the activity of antioxidant enzymes, then an increase in concentration can lead to toxic effects. An increase in the concentration of Ag nanoparticles above 10 mg/L in the cultivation medium can provoke a toxic effect, which is expressed in a decrease in the biomass (Zia et al. 2020) and the number of roots (de Timoteo et al. 2019). Zinc oxide nanoparticles are among the most toxic and can induce oxidative stress. High concentrations of ZnO NPs provoke the formation of ROS, which leads to damage to chromosomes and mitochondria (Zeghoud et al. 2022). While some studies showed low uptake of CNTs by plants, the absence of oxidative stress, and morphological and biochemical changes (Larue et al. 2012), a number of works report on their toxicity for biological objects.

Wang et al. (2014) noted that carboxylated MWCNTs caused oxidative stress, imbalance in nutrient content, and damage in *Vicia faba* L. leaves. The use of carbon nanoparticles resulted in flowering delay and yield decrease in rice (Lin et al. 2009). Nanotubes inhibited root elongation in tomato and lettuce (Cañas et al. 2008). Researchers have reported reduced biomass in squash (Stampoulis et al. 2009), reduced root length in lettuce plants (Lin and Xing 2007), and DNA damage in *Allium cepa* plants (Ghosh et al. 2011). Decreased growth, uptake, and translocation have been observed in soybean plants when treated with MWCNTs at a concentration of 10–50 mg/L (Zhai et al. 2015). The treatment with MWCNTs, fullerene, and reduced graphene oxide resulted in cell deformation, an increase in antioxidant enzyme activity in rice, and a dramatic change in the bacterial community, which showed the need to control the release of nanomaterials into the environment (Hao et al. 2018).

12.9 Conclusion

The production of nanoparticles is growing every year with the expansion of their areas of application. Nanomaterials are actively used in biotechnology to increase sterility, improve plant growth and development, increase productivity, and develop resistance to heavy metals, drought and salinity. In in vitro culture, including woody plants, they have found their use as sterilizing agents, growth and immunity stimulants, and elicitors for the synthesis of useful compounds and micro-/macroelements, supplementing or replacing the use of hormones in the cultivation medium. The antiviral, antibacterial, and antifungal properties of nanoparticles are extremely useful in controlling plant pathogens and developing resistance to diseases and pests. The use of nanoparticles will help to significantly reduce the loss of crops and the death of forest plant species due to pests. Despite natural barriers, nanoparticles penetrate plant cells with relative ease and provoke a change in metabolism, acting as second messengers and regulating the expression of key genes. Nanoparticles hold great promise for use in plant biotechnology and natural ecosystems. Some of their amounts are formed in natural conditions during the oxidation of metals, the electrolysis of solutions, and when various substances, such as water or salt, are exposed to ultraviolet radiation, as a result of fires in burners.

At the same time, the use of nanoparticles in plant biotechnology requires caution and compliance with certain safety rules, since they can have a negative impact on plant and human health. The synthesis of nanoparticles is accompanied by the production of toxic compounds that are hazardous to the environment. In addition, even insignificant concentrations of nanoparticles on the scale of full-sized molecules can have a toxic effect on flora and fauna, polluting the soil and wastewater. Therefore, before using nanoparticles, careful research and evaluation of their impact on plants and the environment are necessary.

References

Abboud MAA (2021) A novel biological approach to copper nanoparticles synthesis: characterization and its application against phytopathogenic fungi, 29 November 2021, Preprint (Version 2). Research Article. Available at Research Square. https://doi.org/10.21203/rs.3.rs-125001/v2

Abdelrazik AS, Al-Sulaiman FA, Saidur R (2019) Optical behavior of a water/silver nanofluid and their influence on the performance of a photovoltaic-thermal collector. Sol Energy Mater Sol Cells 201:110054. https://doi.org/10.1016/j.solmat.2019.110054

Abd-Elsalam KA, Vasil'kov AY, Said-Galiev EE et al (2017) Bimetallic blends and chitosan nanocomposites: novel antifungal agents against cotton seedling damping-off. Eur J Plant Pathol. https://doi.org/10.1007/s10658-017-1349-8

Abu Zeid IM, Mohamed FH, Metwali EMR (2023) Responses of two strawberry cultivars to NaCl-induced salt stress under the influence of ZnO nanoparticles. Saudi J Biol Sci 30:103623. https://doi.org/10.1016/j.sjbs.2023.103623

Adebomojo A, AbdulRahaman A (2020) Surface sterilization of Ocimum seeds and tissues with biosynthesized nanosilver and its effects on callus induction. IOP Conf Ser Mater Sci Eng 805:012024. https://doi.org/10.1088/1757-899X/805/1/012024

Ahmad MA, Javed R, Adeel M et al (2020a) Engineered ZnO and CuO nanoparticles Ameliorate morphological and biochemical response in tissue culture regenerants of Candyleaf (*Stevia rebaudiana*). Molecules 25:1356. https://doi.org/10.3390/molecules25061356

Ahmad P, Alyemeni MN, Al-Huqail AA et al (2020b) Zinc oxide nanoparticles application alleviates arsenic (As) toxicity in soybean plants by restricting the uptake of as and modulating key biochemical attributes, antioxidant enzymes, ascorbate-glutathione cycle and glyoxalase system. Plan Theory 9:825. https://doi.org/10.3390/plants9070825

Alenezi NA, Al-Qurainy F, Tarroum M et al (2022) Zinc oxide nanoparticles (ZnO NPs), biosynthesis, characterization and evaluation of their impact to improve shoot growth and to reduce salt toxicity on salvia officinalis in vitro cultivated. Processes 10:1273. https://doi.org/10.3390/pr10071273

Al-Mayahi AMW (2021) The effect of humic acid (HA) and zinc oxide nanoparticles (ZnO-NPS) on in vitro regeneration of date palm (Phoenix dactylifera L.) cv. Quntar. Plant Cell Tissue Organ Culture (PCTOC) 145:445–456. https://doi.org/10.1007/s11240-021-02020-7

Álvarez SP, Tapia MAM, Vega MEG et al (2019) Nanotechnology and plant tissue. Culture:333–370

Andújar I, González N, García-Ramos JC et al (2020) Argovit™ silver nanoparticles reduce contamination levels and improve morphological growth in the in vitro culture of Psidium friedrichsthalianum (O. Berg). Nied SN Appl Sci 2:2110. https://doi.org/10.1007/s42452-020-03948-9

Anees Ahmad S, Sachi Das S, Khatoon A et al (2020) Bactericidal activity of silver nanoparticles: a mechanistic review. Mater Sci Energy Technol 3:756–769. https://doi.org/10.1016/j.mset.2020.09.002

Aritonang HF, Koleangan H, Wuntu AD (2019) Synthesis of silver nanoparticles using aqueous extract of medicinal plants' (*Impatiens balsamina* and *Lantana camara*) fresh leaves and analysis of antimicrobial activity. Int J Microbiol 2019:1–8. https://doi.org/10.1155/2019/8642303

Asghar MA, Asghar MA (2020) Green synthesized and characterized copper nanoparticles using various new plants extracts aggravate microbial cell membrane damage after interaction with lipopolysaccharide. Int J Biol Macromol 160:1168–1176. https://doi.org/10.1016/j.ijbiomac.2020.05.198

Awad K (2020) In vitro assessment of ZnO nanoparticles on Phoenix dactylifera L. micropropagation. Sci J King Faisal Univ. https://doi.org/10.37575/b/agr/2000

Bakhtiari M, Raeisi Sadati F, Raeisi Sadati SY (2023) Foliar application of silicon, selenium, and zinc nanoparticles can modulate lead and cadmium toxicity in sage (Salvia officinalis L.) plants by optimizing growth and biochemical status. Environ Sci Pollut Res 30:54223–54233. https://doi.org/10.1007/s11356-023-25959-w

Bao HG, Tung HT, Van HT et al (2022) Copper nanoparticles enhanced surface disinfection, induction and maturation of somatic embryos in tuberous begonias (Begonia × tuberhybrida

Voss) cultured in vitro. Plant Cell Tissue Organ Culture (PCTOC) 151:385–399. https://doi.org/10.1007/s11240-022-02360-y

Bapat G, Zinjarde S, Tamhane V (2020) Evaluation of silica nanoparticle mediated delivery of protease inhibitor in tomato plants and its effect on insect pest Helicoverpa armigera. Colloids Surf B Biointerfaces 193:111079. https://doi.org/10.1016/j.colsurfb.2020.111079

Bello-Bello JJ, Chavez-Santoscoy RA, Lecona-Guzmán CA et al (2017) Hormetic response by silver nanoparticles on in vitro multiplication of sugarcane (*Saccharum* spp. Cv. Mex 69-290) using a temporary immersion system. Dose-Response 15:155932581774494. https://doi.org/10.1177/1559325817744945

Borkow G, Gabbay J (2009) Copper, an ancient remedy returning to fight microbial, fungal and viral infections. Curr Chem Biol 3:272–278. https://doi.org/10.2174/2212796810903030272

Bramhanwade K, Shende S, Bonde S et al (2016) Fungicidal activity of Cu nanoparticles against Fusarium causing crop diseases. Environ Chem Lett 14:229–235. https://doi.org/10.1007/s10311-015-0543-1

Bruna T, Maldonado-Bravo F, Jara P, Caro N (2021) Silver nanoparticles and their antibacterial applications. Int J Mol Sci 22:7202. https://doi.org/10.3390/ijms22137202

Burlacu E, Tanase C, Coman N-A, Berta L (2019) A review of bark-extract-mediated green synthesis of metallic nanoparticles and their applications. Molecules 24:4354. https://doi.org/10.3390/molecules24234354

Cai L, Cai L, Jia H et al (2020) Foliar exposure of Fe3O4 nanoparticles on Nicotiana benthamiana: evidence for nanoparticles uptake, plant growth promoter and defense response elicitor against plant virus. J Hazard Mater 393:122415. https://doi.org/10.1016/j.jhazmat.2020.122415

Cañas JE, Long M, Nations S et al (2008) Effects of functionalized and nonfunctionalized single-walled carbon nanotubes on root elongation of select crop species. Environ Toxicol Chem 27:1922. https://doi.org/10.1897/08-117.1

Castillo-Henríquez L, Alfaro-Aguilar K, Ugalde-Álvarez J et al (2020) Green synthesis of gold and silver nanoparticles from plant extracts and their possible applications as antimicrobial agents in the agricultural area. Nano 10:1763. https://doi.org/10.3390/nano10091763

Castro-González CG, Sánchez-Segura L, Gómez-Merino FC, Bello-Bello JJ (2019) Exposure of stevia (Stevia rebaudiana B.) to silver nanoparticles in vitro: transport and accumulation. Sci Rep 9:10372. https://doi.org/10.1038/s41598-019-46828-y

Cokul Raj M, Manokari M, Arumugam N et al (2023a) Silicon nanoparticles mediated in vitro flowering and study of Pollen viability in Vitex negundo L. SILICON 15:4861–4871. https://doi.org/10.1007/s12633-023-02397-4

Cokul Raj M, Manokari M, Dey A et al (2023b) Improvements in morphometric and structural traits of Vitex trifolia L. to exogenous application of nano-silicon in vitro. S Afr J Bot 160:613–621. https://doi.org/10.1016/j.sajb.2023.07.049

Corredor E, Testillano PS, Coronado M-J et al (2009) Nanoparticle penetration and transport in living pumpkin plants: in situ subcellular identification. BMC Plant Biol 9:45. https://doi.org/10.1186/1471-2229-9-45

Cruz LF, Cruz JC, Carrillo D et al (2021) In-vitro evaluation of copper nanoparticles as a potential control agent against the fungal symbionts of the invasive ambrosia beetle Euwallacea fornicatus. Crop Prot 143:105564. https://doi.org/10.1016/j.cropro.2021.105564

Cuong DM, Mai NTN, Tung HT et al (2023) Positive effect of silver nanoparticles in micropropagation of Limonium sinuatum (L.) Mill. 'White'. Plant Cell Tissue Organ Culture (PCTOC). https://doi.org/10.1007/s11240-023-02488-5

de Timoteo CO, Paiva R, dos Reis MV et al (2019) Silver nanoparticles in the micropropagation of Campomanesia rufa (O. Berg) Nied. Plant Cell Tissue Organ Culture (PCTOC) 137:359–368. https://doi.org/10.1007/s11240-019-01576-9

Dong R, Li Y, Li W et al (2019) Recent developments in luminescent nanoparticles for plant imaging and photosynthesis. J Rare Earths 37:903–915. https://doi.org/10.1016/j.jre.2019.04.001

El-Mahdy MT, Elazab DS (2020) Impact of zinc oxide nanoparticles on pomegranate growth under in vitro conditions. Russ J Plant Physiol 67:162–167. https://doi.org/10.1134/S1021443720010045

El-Sayed IM, Salama WH, Salim RG et al (2021) Relevance of nanoparticles on micropropagation, antioxidant activity and molecular characterization of Sequoia sempervirens L. Plant. Jordan J Biol Sci 14:373–382. https://doi.org/10.54319/jjbs/140225

Elsayh SAA, Arafa RN, Ali GA et al (2022) Impact of silver nanoparticles on multiplication, rooting of shoots and biochemical analyses of date palm Hayani cv. by in vitro. Biocatal Agric Biotechnol 43:102400. https://doi.org/10.1016/j.bcab.2022.102400

El-Shetehy M, Moradi A, Maceroni M et al (2021) Silica nanoparticles enhance disease resistance in Arabidopsis plants. Nat Nanotechnol 16:344–353. https://doi.org/10.1038/s41565-020-00812-0

Emamverdian A, Ding Y, Mokhberdoran F et al (2021) Different physiological and biochemical responses of bamboo to the addition of TiO2 NPs under heavy metal toxicity. Forests 12:759. https://doi.org/10.3390/f12060759

Fan G, Xiao Q, Li Q et al (2023) Antimicrobial mechanisms of ZnO nanoparticles to phytopathogen Pseudomonas syringae: damage of cell envelope, suppression of metabolism, biofilm and motility, and stimulation of stomatal immunity on host plant. Pestic Biochem Physiol 194:105455. https://doi.org/10.1016/j.pestbp.2023.105455

Farhadi S, Ma'mani L, Kermani MJ et al (2023) Rice husk-derived biogenic silica nanoparticles and zinc oxide nanoparticles as nano-additives for improving in vitro quince rootstock propagation. Plant Cell Tissue Organ Culture (PCTOC). https://doi.org/10.1007/s11240-023-02556-w

Fatemi H, Esmaiel Pour B, Rizwan M (2021) Foliar application of silicon nanoparticles affected the growth, vitamin C, flavonoid, and antioxidant enzyme activities of coriander (Coriandrum sativum L.) plants grown in lead (Pb)-spiked soil. Environ Sci Pollut Res 28:1417–1425. https://doi.org/10.1007/s11356-020-10549-x

Feizi S (2023) Role of nanomaterials in plant cell and tissue culture. In: Nanomaterial interactions with plant cellular mechanisms and macromolecules and agricultural implications. Springer International Publishing, Cham, pp 359–397

Flores D, Chacón R, Alvarado L et al (2014) Effect of using two different types of carbon nanotubes for Blackberry (Rubus adenotrichos) in vitro plant rooting, growth and histology. Am J Plant Sci 05:3510–3518. https://doi.org/10.4236/ajps.2014.524367

Fortunati E, Rescignano N, Botticella E et al (2016) Effect of poly(dl-lactide-co-glycolide) nanoparticles or cellulose nanocrystals-based formulations on Pseudomonas syringae pv. tomato (Pst) and tomato plant development. J Plant Dis Protect 123:301–310. https://doi.org/10.1007/s41348-016-0036-x

Ghorbanpour M, Mohammadi H, Kariman K (2020) Nanosilicon-based recovery of barley (*Hordeum vulgare*) plants subjected to drought stress. Environ Sci Nano 7:443–461. https://doi.org/10.1039/C9EN00973F

Ghosh M, Chakraborty A, Bandyopadhyay M, Mukherjee A (2011) Multi-walled carbon nanotubes (MWCNT): induction of DNA damage in plant and mammalian cells. J Hazard Mater 197:327–336. https://doi.org/10.1016/j.jhazmat.2011.09.090

González-García Y, Cárdenas-Álvarez C, Cadenas-Pliego G et al (2021) Effect of three nanoparticles (Se, Si and Cu) on the bioactive compounds of bell pepper fruits under saline stress. Plan Theory 10:217. https://doi.org/10.3390/plants10020217

González-Moscoso M, Martínez-Villegas N, Cadenas-Pliego G, Juárez-Maldonado A (2022) Effect of silicon nanoparticles on tomato plants exposed to two forms of inorganic arsenic. Agronomy 12:2366. https://doi.org/10.3390/agronomy12102366

Grodetskaya TA, Evlakov PM, Fedorova OA et al (2022) Influence of copper oxide nanoparticles on gene expression of birch clones in vitro under stress caused by phytopathogens. Nanomaterials 12:864. https://doi.org/10.3390/nano12050864

Gureev AP, Mashkina OS, Shabanova EA et al (2021) Study of the amount of oxidative damage to mitochondrial and chloroplast DNA in clones of white poplar (Populus alba L.) during long-term in vitro cultivation for 26 years. Plant Mol Biol 106:479–489. https://doi.org/10.1007/s11103-021-01157-5

Ha NTM, Manh Do C, Hoang TT et al (2020) The effect of cobalt and silver nanoparticles on overcoming leaf abscission and enhanced growth of rose (Rosa hybrida L. 'Baby Love') plantlets cultured in vitro. Plant Cell Tissue Organ Culture (PCTOC) 141:393–405. https://doi.org/10.1007/s11240-020-01796-4

Hao Y, Ma C, Zhang Z et al (2018) Carbon nanomaterials alter plant physiology and soil bacterial community composition in a rice-soil-bacterial ecosystem. Environ Pollut 232:123–136. https://doi.org/10.1016/j.envpol.2017.09.024

Hatami M, Hadian J, Ghorbanpour M (2017) Mechanisms underlying toxicity and stimulatory role of single-walled carbon nanotubes in Hyoscyamus niger during drought stress simulated by polyethylene glycol. J Hazard Mater 324:306–320. https://doi.org/10.1016/j.jhazmat.2016.10.064

Hegazi ES, Yousef A, Abd Allatif A et al (2021) Effect of silver nanoparticles, medium composition and growth regulators on in vitro propagation of picual olive cultivar. Egypt J Chem 64(12):6961–6969. https://doi.org/10.21608/ejchem.2021.78774.3853

Hermida-Montero LA, Pariona N, Mtz-Enriquez AI et al (2019) Aqueous-phase synthesis of nanoparticles of copper/copper oxides and their antifungal effect against Fusarium oxysporum. J Hazard Mater 380:120850. https://doi.org/10.1016/j.jhazmat.2019.120850

Hernández-Hernández H, Quiterio-Gutiérrez T, Cadenas-Pliego G et al (2019) Impact of selenium and copper nanoparticles on yield, antioxidant system, and fruit quality of tomato plants. Plan Theory 8:355. https://doi.org/10.3390/plants8100355

Hernández-Sierra JF, Ruiz F, Cruz Pena DC et al (2008) The antimicrobial sensitivity of Streptococcus mutans to nanoparticles of silver, zinc oxide, and gold. Nanomedicine 4:237–240. https://doi.org/10.1016/j.nano.2008.04.005

Huang D, Dang F, Huang Y et al (2022) Uptake, translocation, and transformation of silver nanoparticles in plants. Environ Sci Nano 9:12–39. https://doi.org/10.1039/D1EN00870F

Hussain A, Ali S, Rizwan M et al (2019) Responses of wheat (Triticum aestivum) plants grown in a Cd contaminated soil to the application of iron oxide nanoparticles. Ecotoxicol Environ Saf 173:156–164. https://doi.org/10.1016/j.ecoenv.2019.01.118

Inam M, Attique I, Zahra M et al (2023) Metal oxide nanoparticles and plant secondary metabolism: unraveling the game-changer nano-elicitors. Plant Cell Tissue Organ Culture (PCTOC). https://doi.org/10.1007/s11240-023-02587-3

Iqbal Z, Javad S, Naz S et al (2022) Elicitation of the in vitro cultures of selected varieties of Vigna radiata L with zinc oxide and copper oxide nanoparticles for enhanced phytochemicals production. Front Plant Sci 13:908532. https://doi.org/10.3389/fpls.2022.908532

Irshad MA, Nawaz R, Ur Rehman MZ et al (2021) Synthesis, characterization and advanced sustainable applications of titanium dioxide nanoparticles: a review. Ecotoxicol Environ Saf 212:111978. https://doi.org/10.1016/j.ecoenv.2021.111978

Jiang L, Xiang S, Lv X et al (2022) Biosynthesized silver nanoparticles inhibit Pseudomonas syringae pv. tabaci by directly destroying bacteria and inducing plant resistance in Nicotiana benthamiana. Phytopathol Res 4:43. https://doi.org/10.1186/s42483-022-00148-8

Jo Y-K, Kim BH, Jung G (2009) Antifungal activity of silver ions and nanoparticles on phytopathogenic fungi. Plant Dis 93:1037–1043. https://doi.org/10.1094/PDIS-93-10-1037

Junedi MA, Mukhopadhyay R, Manjari KS (2023) Alleviating salinity stress in crop plants using new engineered nanoparticles (ENPs). Plant Stress 9:100184. https://doi.org/10.1016/j.stress.2023.100184

Kasprowicz MJ, Kozioł M, Gorczyca A (2010) The effect of silver nanoparticles on phytopathogenic spores of *Fusarium culmorum*. Can J Microbiol 56:247–253. https://doi.org/10.1139/W10-012

Kaushik M, Niranjan R, Thangam R et al (2019) Investigations on the antimicrobial activity and wound healing potential of ZnO nanoparticles. Appl Surf Sci 479:1169–1177. https://doi.org/10.1016/j.apsusc.2019.02.189

Khafri AZ, Zarghami R, Ma'mani L, Ahmadi B (2023) Enhanced efficiency of in vitro rootstock micro-propagation using silica-based nanoparticles and plant growth regulators in myrobalan 29C (Prunus cerasifera L.). J Plant Growth Regul 42:1457–1471. https://doi.org/10.1007/s00344-022-10631-3

Khattab S, El Sherif F, AlDayel M et al (2022) Silicon dioxide and silver nanoparticles elicit antimicrobial secondary metabolites while enhancing growth and multiplication of Lavandula officinalis in-vitro plantlets. Plant Cell Tissue Organ Culture (PCTOC) 149:411–421. https://doi.org/10.1007/s11240-021-02224-x

Khodakovskaya M, Dervishi E, Mahmood M et al (2009) Carbon nanotubes are able to penetrate plant seed coat and dramatically affect seed germination and plant growth. ACS Nano 3:3221–3227. https://doi.org/10.1021/nn900887m

Khodakovskaya MV, de Silva K, Biris AS et al (2012) Carbon nanotubes induce growth enhancement of tobacco cells. ACS Nano 6:2128–2135. https://doi.org/10.1021/nn204643g

Khodakovskaya MV, Kim B-S, Kim JN et al (2013) Carbon nanotubes as plant growth regulators: effects on tomato growth, reproductive system, and soil microbial community. Small 9:115–123. https://doi.org/10.1002/smll.201201225

Kim S, Ryu D-Y (2013) Silver nanoparticle-induced oxidative stress, genotoxicity and apoptosis in cultured cells and animal tissues. J Appl Toxicol 33:78–89. https://doi.org/10.1002/jat.2792

Kim SW, Jung JH, Lamsal K et al (2012) Antifungal effects of silver nanoparticles (AgNPs) against various plant pathogenic fungi. Mycobiology 40:53–58. https://doi.org/10.5941/MYCO.2012.40.1.053

Koçak R, Okcu M, Haliloğlu K et al (2023) Magnesium oxide nanoparticles: an influential element in Cowpea (Vigna unguiculata L Walp). Tissue Culture Agron 13:1646. https://doi.org/10.3390/agronomy13061646

Krishnaraj C, Ramachandran R, Mohan K, Kalaichelvan PT (2012) Optimization for rapid synthesis of silver nanoparticles and its effect on phytopathogenic fungi. Spectrochim Acta A Mol Biomol Spectrosc 93:95–99. https://doi.org/10.1016/j.saa.2012.03.002

Lahiani MH, Dervishi E, Chen J et al (2013) Impact of carbon nanotube exposure to seeds of valuable crops. ACS Appl Mater Interfaces 5:7965–7973. https://doi.org/10.1021/am402052x

Lai CS, Kho YH, Chew BL et al (2022) Organogenesis of Cucumis metuliferus plantlets under the effects of LEDs and silver nanoparticles. S Afr J Bot 148:78–87. https://doi.org/10.1016/j.sajb.2022.04.020

Landa P, Vankova R, Andrlova J et al (2012) Nanoparticle-specific changes in Arabidopsis thaliana gene expression after exposure to ZnO, TiO_2, and fullerene soot. J Hazard Mater 241–242:55–62. https://doi.org/10.1016/j.jhazmat.2012.08.059

Larue C, Pinault M, Czarny B et al (2012) Quantitative evaluation of multi-walled carbon nanotube uptake in wheat and rapeseed. J Hazard Mater 227–228:155–163. https://doi.org/10.1016/j.jhazmat.2012.05.033

Lashin I, Hasanin M, Hassan SAM, Hashem AH (2023) Green biosynthesis of zinc and selenium oxide nanoparticles using callus extract of Ziziphus spina-christi: characterization, antimicrobial, and antioxidant activity. Biomass Convers Biorefinery 13:10133–10146. https://doi.org/10.1007/s13399-021-01873-4

Lebedev VG, Lebedeva TN, Chernodubov AI, Shestibratov KA (2020) Genomic selection for forest tree improvement: methods, achievements and perspectives. Forests 11:1190. https://doi.org/10.3390/f11111190

Li J, Sang H, Guo H et al (2017) Antifungal mechanisms of ZnO and Ag nanoparticles to *Sclerotinia homoeocarpa*. Nanotechnology 28:155101. https://doi.org/10.1088/1361-6528/aa61f3

Li Y, Yang D, Cui J (2017b) Graphene oxide loaded with copper oxide nanoparticles as an antibacterial agent against Pseudomonas syringae pv. tomato. RSC Adv 7:38853–38860. https://doi.org/10.1039/C7RA05520J

Li M, Zhang P, Adeel M et al (2021) Physiological impacts of zero valent iron, Fe3O4 and Fe2O3 nanoparticles in rice plants and their potential as Fe fertilizers. Environ Pollut 269:116134. https://doi.org/10.1016/j.envpol.2020.116134

Lin D, Xing B (2007) Phytotoxicity of nanoparticles: inhibition of seed germination and root growth. Environ Pollut 150:243–250. https://doi.org/10.1016/j.envpol.2007.01.016
Lin S, Reppert J, Hu Q et al (2009) Uptake, translocation, and transmission of carbon nanomaterials in rice plants. Small 5(10):1128–1132. https://doi.org/10.1002/smll.200801556
Linh TM, Mai NC, Hoe PT et al (2020) Metal-based nanoparticles enhance drought tolerance in soybean. J Nanomater 2020:1–13. https://doi.org/10.1155/2020/4056563
Liu J, Li G, Chen L et al (2021) Cerium oxide nanoparticles improve cotton salt tolerance by enabling better ability to maintain cytosolic K+/Na+ ratio. J Nanobiotechnol 19:153. https://doi.org/10.1186/s12951-021-00892-7
Madbouly AK, Abdel-Aziz MS, Abdel-Wahhab MA (2017) Biosynthesis of nanosilver using *Chaetomium globosum* and its application to control *Fusarium* wilt of tomato in the greenhouse. IET Nanobiotechnol 11:702–708. https://doi.org/10.1049/iet-nbt.2016.0213
Mahajan S, Kadam J, Dhawal P et al (2022) Application of silver nanoparticles in in-vitro plant growth and metabolite production: revisiting its scope and feasibility. Plant Cell Tissue Organ Culture (PCTOC) 150:15–39. https://doi.org/10.1007/s11240-022-02249-w
Mahdizadeh V, Safaie N, Khelghatibana F (2015) Evaluation of antifungal activity of silver nanoparticles against some phytopathogenic fungi and Trichoderma harzianum. J Crop Protect 4(3):291–300. Retrieved from URL: http://jcp.modares.ac.ir/article-3-12407-en.html
Mahendran D, Geetha N, Venkatachalam P (2019) Role of silver nitrate and silver nanoparticles on tissue culture medium and enhanced the plant growth and development. In: In vitro plant breeding towards novel agronomic traits. Springer Singapore, Singapore, pp 59–74
Mahmoud AWM, Abdeldaym EA, Abdelaziz SM et al (2019) Synergetic effects of zinc, boron, silicon, and zeolite nanoparticles on confer tolerance in potato plants subjected to salinity. Agronomy 10:19. https://doi.org/10.3390/agronomy10010019
Majeed N, Panigrahi KCS, Sukla LB et al (2020) Application of carbon nanomaterials in plant biotechnology. Mater Today Proc 30:340–345. https://doi.org/10.1016/j.matpr.2020.01.618
Malandrakis AA, Kavroulakis N, Chrysikopoulos CV (2019) Use of copper, silver and zinc nanoparticles against foliar and soil-borne plant pathogens. Sci Total Environ 670:292–299. https://doi.org/10.1016/j.scitotenv.2019.03.210
Manokari M, Dey A, Faisal M et al (2023a) In vitro tuberization using silicon nanoparticles and short-term cold storage of mini-tubers of Dioscorea pentaphylla L. BioNanoScience. https://doi.org/10.1007/s12668-023-01185-z
Manokari M, Cokul Raj M, Dey A et al (2023b) Structural alterations of Cymbopogon citratus (DC.) Stapf leaves and roots caused by silicon nanoparticles during in vitro propagation. Ind Crop Prod 197:116648. https://doi.org/10.1016/j.indcrop.2023.116648
Manokari M, Cokul Raj M, Dey A et al (2023c) Silicon nanoparticles moderated morphometric deficiencies by improving micro-morpho-structural traits in Thunbergia erecta (Benth.) T. Anderson. SILICON 15:5415–5427. https://doi.org/10.1007/s12633-023-02451-1
Manokari M, Dey A, Faisal M et al (2023d) Silicon nanoparticles (SiNPs) positively affect morpho-structural differentiation in micropropagated plantlets of Santalum album L. SILICON. https://doi.org/10.1007/s12633-023-02558-5
Manokari M, Raj MC, Dey A et al (2023e) Silver nanoparticles improved morphogenesis, biochemical profile and micro-morphology of Gaillardia pulchella Foug cv. 'Torch Yellow'. Plant Cell Tissue Organ Culture (PCTOC). https://doi.org/10.1007/s11240-023-02502-w
Mansoor S, Zahoor I, Baba TR et al (2021) Fabrication of silver nanoparticles against fungal pathogens. Front Nanotechnol 3:679358. https://doi.org/10.3389/fnano.2021.679358
Manzoor N, Ahmed T, Noman M et al (2021) Iron oxide nanoparticles ameliorated the cadmium and salinity stresses in wheat plants, facilitating photosynthetic pigments and restricting cadmium uptake. Sci Total Environ 769:145221. https://doi.org/10.1016/j.scitotenv.2021.145221
Marchiol L, Filippi A, Adamiano A et al (2019) Influence of hydroxyapatite nanoparticles on germination and plant metabolism of tomato (Solanum lycopersicum L.): preliminary evidence. Agronomy 9:161. https://doi.org/10.3390/agronomy9040161

Martínez-Ballesta MC, Zapata L, Chalbi N, Carvajal M (2016) Multiwalled carbon nanotubes enter broccoli cells enhancing growth and water uptake of plants exposed to salinity. J Nanobiotechnol 14:42. https://doi.org/10.1186/s12951-016-0199-4

Martinez-Ballesta MC, Chelbi N, Lopez-Zaplana A, Carvajal M (2020) Discerning the mechanism of the multiwalled carbon nanotubes effect on root cell water and nutrient transport. Plant Physiol Biochem 146:23–30. https://doi.org/10.1016/j.plaphy.2019.11.008

Mishra D, Tripathi S, Pant S, Chaturvedi P (2022) Nanomaterials and plant tissue culture: developmental path and contradictory facts. In: Commercial scale tissue culture for horticulture and plantation crops. Springer Nature Singapore, Singapore, pp 153–172

Mohanta YK, Biswas K, Jena SK et al (2020) Anti-biofilm and antibacterial activities of silver nanoparticles synthesized by the reducing activity of phytoconstituents present in the Indian medicinal plants. Front Microbiol 11:1143. https://doi.org/10.3389/fmicb.2020.01143

Mosavat N, Golkar P, Yousefifard M, Javed R (2019) Modulation of callus growth and secondary metabolites in different *Thymus* species and *Zataria multiflora* micropropagated under ZnO nanoparticles stress. Biotechnol Appl Biochem 66:316–322. https://doi.org/10.1002/bab.1727

Mukarram M, Petrik P, Mushtaq Z et al (2022) Silicon nanoparticles in higher plants: uptake, action, stress tolerance, and crosstalk with phytohormones, antioxidants, and other signalling molecules. Environ Pollut 310:119855. https://doi.org/10.1016/j.envpol.2022.119855

Mustapha T, Misni N, Ithnin NR et al (2022) A review on plants and microorganisms mediated synthesis of silver nanoparticles, role of plants metabolites and applications. Int J Environ Res Public Health 19:674. https://doi.org/10.3390/ijerph19020674

Nalci OB, Nadaroglu H, Pour AH et al (2019) Effects of ZnO, CuO and γ-Fe3O4 nanoparticles on mature embryo culture of wheat (Triticum aestivum L.). Plant Cell Tissue Organ Culture (PCTOC) 136:269–277. https://doi.org/10.1007/s11240-018-1512-8

Ouda SM (2014) Antifungal activity of silver and copper nanoparticles on two plant pathogens, Alternaria alternata and Botrytis cinerea. Res J Microbiol 9:34–42. https://doi.org/10.3923/jm.2014.34.42

Pariona N, Mtz-Enriquez AI, Sánchez-Rangel D et al (2019) Green-synthesized copper nanoparticles as a potential antifungal against plant pathogens. RSC Adv 9:18835–18843. https://doi.org/10.1039/C9RA03110C

Parsa SM, Rahbar A, Koleini MH et al (2020) A renewable energy-driven thermoelectric-utilized solar still with external condenser loaded by silver/nanofluid for simultaneously water disinfection and desalination. Desalination 480:114354. https://doi.org/10.1016/j.desal.2020.114354

Pérez-Labrada F, López-Vargas ER, Ortega-Ortiz H et al (2019) Responses of tomato plants under saline stress to Foliar application of copper nanoparticles. Plan Theory 8:151. https://doi.org/10.3390/plants8060151

Perfileva AI, Tsivileva OM, Nozhkina OA et al (2021) Effect of natural polysaccharide matrix-based selenium nanocomposites on Phytophthora cactorum and rhizospheric microorganisms. Nanomaterials (Basel) 11(9):2274. https://doi.org/10.3390/nano11092274

Phong TH, Hieu T, Tung HT et al (2022) Silver nanoparticles: a positive factor for in vitro flowering and fruiting of purple passion fruit (Passiflora edulis Sim f. edulis). Plant Cell Tissue Organ Culture (PCTOC) 151:401–412. https://doi.org/10.1007/s11240-022-02361-x

Piętka J, Adamczuk A, Zarzycka E et al (2022) The application of copper and silver nanoparticles in the protection of fagus sylvatica wood against decomposition by Fomes fomentarius. Forests 13:1724. https://doi.org/10.3390/f13101724

Prażak R, Święciło A, Krzepiłko A et al (2020) Impact of Ag nanoparticles on seed germination and seedling growth of green beans in normal and chill temperatures. Agriculture 10:312. https://doi.org/10.3390/agriculture10080312

Purohit A, Sharma R, Shiv Ramakrishnan R et al (2022) Biogenic synthesis of silver nanoparticles (AgNPs) using aqueous leaf extract of Buchanania lanzan spreng and evaluation of their antifungal activity against phytopathogenic fungi. Bioinorg Chem Appl 2022:1–9. https://doi.org/10.1155/2022/6825150

Rajput VD, Minkina T, Feizi M et al (2021) Effects of silicon and silicon-based nanoparticles on rhizosphere microbiome. Plant Stress Growth Biol (Basel) 10:791. https://doi.org/10.3390/biology10080791

Reed BM, Mentzer J, Tanprasert P, Yu X (1997) Internal bacterial contamination of micropropagated hazelnut: identification and antibiotic treatment. In: Pathogen and microbial contamination management in micropropagation, pp 169–174

Regni L, Del Buono D, Micheli M et al (2022) Effects of biogenic ZnO nanoparticles on growth, physiological, biochemical traits and antioxidants on olive tree in vitro. Horticulturae 8:161. https://doi.org/10.3390/horticulturae8020161

Riaz S, Hussain I, Parveen A et al (2022) Silicon and nano-silicon in plant nutrition and crop quality. In: Silicon and nano-silicon in environmental stress management and crop quality improvement. Elsevier, pp 277–295

Rizwan M, Ali S, Ur Rehman MZ et al (2021) Effects of nanoparticles on trace element uptake and toxicity in plants: a review. Ecotoxicol Environ Saf 221:112437. https://doi.org/10.1016/j.ecoenv.2021.112437

Rossi L, Fedenia LN, Sharifan H et al (2019) Effects of foliar application of zinc sulfate and zinc nanoparticles in coffee (Coffea arabica L.) plants. Plant Physiol Biochem 135:160–166. https://doi.org/10.1016/j.plaphy.2018.12.005

Sadak MS (2019) Impact of silver nanoparticles on plant growth, some biochemical aspects, and yield of fenugreek plant (Trigonella foenum-graecum). Bull Natl Res Cent 43:38. https://doi.org/10.1186/s42269-019-0077-y

Saha G, Mostofa MG, Rahman MM, Tran L-SP (2021) Silicon-mediated heat tolerance in higher plants: a mechanistic outlook. Plant Physiol Biochem 166:341–347. https://doi.org/10.1016/j.plaphy.2021.05.051

Samadi S, Saharkhiz MJ, Azizi M et al (2020) Multi-walled carbon nanotubes stimulate growth, redox reactions and biosynthesis of antioxidant metabolites in Thymus daenensis celak. in vitro. Chemosphere 249:126069. https://doi.org/10.1016/j.chemosphere.2020.126069

Sarmast MK, Niazi A, Salehi H, Abolimoghadam A (2015) Silver nanoparticles affect ACS expression in Tecomella undulata in vitro culture. Plant Cell Tissue Organ Culture (PCTOC) 121:227–236. https://doi.org/10.1007/s11240-014-0697-8

Satti SH, Raja NI, Javed B et al (2021) Titanium dioxide nanoparticles elicited agro-morphological and physicochemical modifications in wheat plants to control Bipolaris sorokiniana. PLoS One 16:e0246880. https://doi.org/10.1371/journal.pone.0246880

Sawalha H, Abiri R, Sanusi R et al (2021) Toward a better understanding of metal nanoparticles, a novel strategy from eucalyptus plants. Plan Theory 10:929. https://doi.org/10.3390/plants10050929

Shahryari F, Rabiei Z, Sadighian S (2020) Antibacterial activity of synthesized silver nanoparticles by sumac aqueous extract and silver-chitosan nanocomposite against Pseudomonas syringae pv. syringae. J Plant Pathol 102:469–475. https://doi.org/10.1007/s42161-019-00478-1

Sharma V, Kaushik S, Pandit P et al (2019) Green synthesis of silver nanoparticles from medicinal plants and evaluation of their antiviral potential against chikungunya virus. Appl Microbiol Biotechnol 103:881–891. https://doi.org/10.1007/s00253-018-9488-1

Siddiqui H, Ahmed KBM, Sami F, Hayat S (2020) Silicon nanoparticles and plants: current knowledge and future perspectives. Sustain Agric Rev 41:129–142

Singh R, Ahamad S (2022) Integration of nanotechnology in plant tissue culture. Curr Nanosci 18:604–610. https://doi.org/10.2174/1573413717666211015115351

Sofy AR, Sofy MR, Hmed AA et al (2021) Ameliorating the adverse effects of tomato mosaic tobamovirus infecting tomato plants in Egypt by boosting immunity in tomato plants using zinc oxide nanoparticles. Molecules 26:1337. https://doi.org/10.3390/molecules26051337

Soltani Nejad M, Bonjar GHS, Khatami M et al (2017) *In vitro* and *in vivo* antifungal properties of silver nanoparticles against *Rhizoctonia solani* , a common agent of rice sheath blight disease. IET Nanobiotechnol 11:236–240. https://doi.org/10.1049/iet-nbt.2015.0121

Sorcia-Morales M, Gómez-Merino FC, Sánchez-Segura L et al (2021) Multi-walled carbon nanotubes improved development during in vitro multiplication of sugarcane (Saccharum spp.) in a semi-automated bioreactor. Plan Theory 10:2015. https://doi.org/10.3390/plants10102015

Spinoso-Castillo JL, Chavez-Santoscoy RA, Bogdanchikova N et al (2017) Antimicrobial and hormetic effects of silver nanoparticles on in vitro regeneration of vanilla (Vanilla planifolia Jacks. ex Andrews) using a temporary immersion system. Plant Cell Tissue Organ Culture (PCTOC) 129:195–207. https://doi.org/10.1007/s11240-017-1169-8

Stampoulis D, Sinha SK, White JC (2009) Assay-dependent phytotoxicity of nanoparticles to plants. Environ Sci Technol 43:9473–9479. https://doi.org/10.1021/es901695c

Thounaojam TC, Meetei TT, Devi YB et al (2021) Zinc oxide nanoparticles (ZnO-NPs): a promising nanoparticle in renovating plant science. Acta Physiol Plant 43:136. https://doi.org/10.1007/s11738-021-03307-0

Tung HT, Bao HG, Cuong DM et al (2021a) Silver nanoparticles as the sterilant in large-scale micropropagation of chrysanthemum. In Vitro Cell Dev Biol Plant. https://doi.org/10.1007/s11627-021-10163-7

Tung HT, Thuong TT, Cuong DM et al (2021b) Silver nanoparticles improved explant disinfection, in vitro growth, runner formation and limited ethylene accumulation during micropropagation of strawberry (Fragaria × ananassa). Plant Cell Tissue Organ Culture (PCTOC) 145:393–403. https://doi.org/10.1007/s11240-021-02015-4

Tung HT, Nguyen PLH, Van Lich T et al (2022) Enhanced shoot and plantlet quality of Gerbera (Gerbera jamesonii Revolution Yellow) cultivar on medium containing silver and cobalt nanoparticles. Sci Hortic 306:111445. https://doi.org/10.1016/j.scienta.2022.111445

Valsalam S, Agastian P, Arasu MV et al (2019) Rapid biosynthesis and characterization of silver nanoparticles from the leaf extract of Tropaeolum majus L. and its enhanced in-vitro antibacterial, antifungal, antioxidant and anticancer properties. J Photochem Photobiol B 191:65–74. https://doi.org/10.1016/j.jphotobiol.2018.12.010

Vasyukova I, Gusev A, Zakharova O et al (2021) Silver nanoparticles for enhancing the efficiency of micropropagation of gray poplar (Populus × canescens Aiton. Sm.). IOP Conf Ser Earth Environ Sci 875:012053. https://doi.org/10.1088/1755-1315/875/1/012053

Wang C, Liu H, Chen J et al (2014) Carboxylated multi-walled carbon nanotubes aggravated biochemical and subcellular damages in leaves of broad bean (Vicia faba L.) seedlings under combined stress of lead and cadmium. J Hazard Mater 274:404–412. https://doi.org/10.1016/j.jhazmat.2014.04.036

Wang L, Ning C, Pan T, Cai K (2022) Role of silica nanoparticles in abiotic and biotic stress tolerance in plants: a review. Int J Mol Sci 23:1947. https://doi.org/10.3390/ijms23041947

Win TT, Khan S, Fu P (2020) Fungus- (Alternaria sp.) mediated silver nanoparticles synthesis, characterization, and screening of antifungal activity against some phytopathogens. J Nanotechnol 2020:1–9. https://doi.org/10.1155/2020/8828878

Xu L, Wang Y-Y, Huang J et al (2020) Silver nanoparticles: synthesis, medical applications and biosafety. Theranostics 10:8996–9031. https://doi.org/10.7150/thno.45413

Yan S, Zhao L, Li H et al (2013) Single-walled carbon nanotubes selectively influence maize root tissue development accompanied by the change in the related gene expression. J Hazard Mater 246–247:110–118. https://doi.org/10.1016/j.jhazmat.2012.12.013

Yaqoob AA, Umar K, Ibrahim MNM (2020) Silver nanoparticles: various methods of synthesis, size affecting factors and their potential applications—a review. Appl Nanosci 10:1369–1378. https://doi.org/10.1007/s13204-020-01318-w

Yavari R, Abed AM, Akbari OA et al (2022) Numerical study of flow and free convection of water/silver nanofluid in a circular cavity influenced by hot fluid flow. Prog Nucl Energy 153:104378. https://doi.org/10.1016/j.pnucene.2022.104378

Yin IX, Zhang J, Zhao IS et al (2020) The antibacterial mechanism of silver nanoparticles and its application in dentistry. Int J Nanomedicine 15:2555–2562. https://doi.org/10.2147/IJN.S246764

Yousefi S, Kartoolinejad D, Naghdi R (2017) Effects of priming with multi-walled carbon nanotubes on seed physiological characteristics of Hopbush (*Dodonaea viscosa* L.) under drought stress. Int J Environ Stud 74:528–539. https://doi.org/10.1080/00207233.2017.1325627

Zabrieski Z, Morrell E, Hortin J et al (2015) Pesticidal activity of metal oxide nanoparticles on plant pathogenic isolates of Pythium. Ecotoxicology 24:1305–1314. https://doi.org/10.1007/s10646-015-1505-x

Zahedi SM, Hosseini MS, Fahadi Hoveizeh N et al (2023) Comparative morphological, physiological and molecular analyses of drought-stressed strawberry plants affected by SiO2 and SiO2-NPs foliar spray. Sci Hortic 309:111686. https://doi.org/10.1016/j.scienta.2022.111686

Zakharova OV, Gusev AA, Muratov DS et al (2021) Titanium trisulfide nanoribbons affect the Downy Birch and Poplar × Aspen hybrid in plant tissue culture via the emission of hydrogen sulfide. Forests 12:713. https://doi.org/10.3390/f12060713

Zeghoud S, Hemmami H, Ben Seghir B et al (2022) A review on biogenic green synthesis of ZnO nanoparticles by plant biomass and their applications. Mater Today Commun 33:104747. https://doi.org/10.1016/j.mtcomm.2022.104747

Zhai G, Gutowski SM, Walters KS et al (2015) Charge, size, and cellular selectivity for multiwall carbon nanotubes by maize and soybean. Environ Sci Technol 49:7380–7390. https://doi.org/10.1021/acs.est.5b01145

Zhang H, Yue M, Zheng X et al (2017) Physiological effects of single- and multi-walled carbon nanotubes on rice seedlings. IEEE Trans Nanobioscience 16:563–570. https://doi.org/10.1109/TNB.2017.2715359

Zhuzhukin KV, Evlakov PM, Grodetskaya TA et al (2023) Effect of multi-walled carbon nanotubes on the growth and expression of stress resistance genes in birch. Forests 14:163. https://doi.org/10.3390/f14010163

Zia M, Yaqoob K, Mannan A et al (2020) Regeneration response of carnation cultivars in response of silver nanoparticles under in vitro conditions. Vegetos 33:11–20. https://doi.org/10.1007/s42535-019-00074-9

Chapter 13
Seed Priming and Spraying with Zinc Oxide and Copper Oxide Nanoparticles Mitigate Crop Diseases and Induce Resistance Against Pathogens

Deeksha Kashyap, Aiman Parveen, and Lukman Ahamad

Abstract Nanotechnology promotes resistance against plant pathogens and provides environmentally acceptable solutions for managing plant diseases. The treatment of plant diseases in diverse crops has drawn researchers' attention to significant advancements in the synthesis of nanoparticles and modified nanoparticles. For the treatment of plant diseases by inducing resistance, foliar spray and seed priming with zinc oxide (ZnO-NPs) and copper oxide (CuO-NPs) nanoparticles were used. On a variety of crops, nanoparticles were also utilized to promote plant development and offer protection against plant diseases, albeit their efficacy varied depending on the plant and the dose used. Therefore, the use of these nanoparticles can be expected to serve as a long-term effective measure for controlling pathogenic fungi, bacteria, and plant parasitic nematodes.

Keywords Nanoparticles · Seed priming · Foliar application · Plant pathogens · Disease management

13.1 Introduction

The agricultural sector is facing a number of difficulties, including continued climate change, declining soil fertility, a lack of micro- and macronutrients, and excessive chemical fertilizer and pesticide use that is damaging the soil ecology. However, since then, the world's population has been growing, driving up food demand at the same time. Therefore, by increasing agricultural yield, repairing and improving soil quality, and safeguarding crops from biotic and abiotic threats, nanotechnology has significantly aided in the development of sustainable agriculture (Khan et al. 2021).

D. Kashyap · A. Parveen · L. Ahamad (✉)
Department of Botany, Aligarh Muslim University, Aligarh, Uttar Pradesh, India

M. Khan, J.-T. Chen (eds.), *Nanoparticles in Plant Biotic Stress Management*,
https://doi.org/10.1007/978-981-97-0851-2_13

Biotic stresses induced by insect pests, plant diseases, and plant-parasitic nematodes are major constraints in enhancing food and nutritional security. Around 40% of food crops are lost due to plant diseases and pests annually (Khan et al. 2021; Khan and Khan 2021). Therefore, their management is of paramount importance. Traditional pest and disease management strategies have mainly focused on utilizing traditional practices and bio-control agents and largely on deploying chemicals. Henceforth, the quest for a safe, eco-friendly, and effective novel management approach is on. The development of nanotechnology has offered hope for pest and disease reduction in several ways. A wide range of nano-formulations based on silver nanoparticles (Ag-NPs), copper oxide nanoparticles (CuO-NPs), zinc oxide nanoparticles (ZnO-NPs), chitosan nanoparticles, and titanium dioxide (TiO_2-NPs) have been synthesized and formulated for use in crop protection (Singh et al. 2023). These nanoparticles are synthesized through various methods (physical, chemical, and biological) and induce plant resistance against pathogens and pests by strengthening the physical barriers, improving plant photosynthesis, and activating plant defense mechanisms. In, addition, NPs can also upregulate the expression of defense-related genes by increasing defense compounds and activating the expression of defense-related genes (Nawaz et al. 2023). Siddiqui et al. (2019) studied that spraying of graphene oxide and zinc oxide nanoparticles showed antimicrobial activity toward fungal, bacterial, and nematode pathogens. In addition, these NPs also played a significant role in controlling the fungal, bacterial, and nematode diseases in carrot plants under greenhouse conditions.

In this chapter, we have discussed the role of zinc oxide and copper oxide nanoparticles in the management of fungal, bacterial, and nematode diseases/pathogens on different crop plants.

13.2 Role of Zinc Oxide and Copper Oxide Nanoparticles against Fungal Pathogens/Diseases in Crop Plants

13.2.1 Role of ZnO Nanoparticles in Plant Disease Management

According to Sabir et al. (2014), zinc oxide nanoparticles (ZnO-NPs) exhibit exceptional optical, physical, and antibacterial qualities. When it comes to their role against fungal pathogens or fungal diseases, ZnO NPs have shown promising antifungal properties. The ZnO-NPs have a potential role in managing fungal diseases/pathogens in various crops. They can inhibit the growth of fungi responsible for diseases such as rusts, smuts, blights, and others, which can lead to significant yield losses. ZnO-NPs typically disrupt cellular processes, distort fungal hyphae, and prevent the growth of fungus. In addition, ZnO-NPs induce oxidative stress by generating reactive oxygen species (ROS) within fungal cells. The oxidative stress can lead to cell damage and death (He et al. 2011; Khan and Siddiqui 2018).

According to Alghuthaymi et al. (2015), nanoparticles (NPs) may offer an option for managing plant diseases since they may shield plants from pathogens. According to

Sabir et al. (2014), zinc oxide nanoparticles (ZnO NPs) can improve agricultural productivity and control plant diseases. According to Prasad et al. (2012), zinc oxide nanoparticles with a mean particle size of 25 nm stimulate plant development, seed germination, and seedling vigor and effectively increase stem and root growth. Numerous research proposed using myco-synthesized ZnO-NPs as a substitute for synthetic fungicides in order to boost growth parameters and metabolic aspects in addition to curing the *Fusarium* wilt disease in grown eggplant (Abdelaziz et al. 2022). Additionally, ZnO-NPs have an impact on proline content, growth, photosynthetic pigments, and fungal infections brought on by *Alternaria* and *F. oxysporum* f. sp. *lycopersici* on tomato plants (Parveen and Siddiqui 2021). ZnO-NPs' capacity to prevent direct fungal development by altering the mycelia in growth and by eliminating mycotoxins such as fusaric acid allows them to control fungal plant diseases brought on by *F. oxysporum* (Yehia and Ahmed 2013). According to Ramachandrappa et al. (2019), increased levels of lipid peroxidation, ROS, and fluctuating ergosterol concentration may be to blame for ZnO-NPs' effectiveness as an antifungal and anti-mycotoxin against *Fusarium* sp.

13.2.2 Role of Copper Oxide Nanoparticles (CuO-NPs) in Plant Disease Management

When used in different ways as inhibitors with effective antimicrobial effects, copper oxide nanoparticles (CuO-NPs), which are simple, affordable, stable, and readily accessible copper compounds with a high surface area and crystal morphology, are suggested to suppress fungal diseases (Devipriya and Roopan 2017; Borgatta et al. 2018; Elmer et al. 2018).

The CuO-NPs have shown potential in various applications, including their role in combating fungal diseases. They can be utilized to control fungal infections in many crops. They can be applied as fungicides to protect plants from diseases caused by fungi, thus reducing the need for traditional chemical fungicides. CuO-NPs can inhibit fungal growth by disrupting cell membranes, interfering with metabolic processes, and generating reactive oxygen species (ROS) that damage fungal cells. ROS can damage the cell membranes, proteins, and genetic material of microorganisms, including fungi, leading to their inhibition or death.

To effectively control *F. oxysporum* and *Verticillium dahliae* infections on tomato and eggplant plants grown in disease-infested soil, Elmer et al. (2018) used CuO and other metal oxide nanoparticles (AlO, FeO, MnO, and ZnO) through foliar spray. They were less fungitoxic than the corresponding reagents applied in bulk (Elmer and White 2016; Malandrakis et al. 2019). Another study found that, when applied topically, Cu NPs and CuO-NPs exhibited potent fungi toxic activity against a number of foliar and soil-borne plant diseases, including *Podosphaera pannosa* and *Botrytis cinerea* (Hao et al. 2019). Giannousi et al. (2013) discovered that compared to Cu or Cu_2O alone, Cu/Cu_2O composite NPs showed higher sustained fungistatic activity on tomato late blight in vivo. We concluded that CuO-NPs interact with meristematic cells, activating metabolic pathways that promote the growth characteristics (Tables 13.1 and 13.2).

Table 13.1 Effects of ZnO and CuO-NPs on fungal pathogens or diseases in different crops

Nanoparticles	Pathogen/disease	Effect	Reference
ZnO	*Alternaria dauci* *Rhizoctonia solani*	Foliar spray of NPs improved the plant growth attributes and chlorophyll and carotenoid contents and reduced the disease indices (leaf blight and crown rot) in carrot plants	Ahamad and Siddiqui (2021)
ZnO	*Fusarium solani* *A. dauci*	NPs application was effective in increasing growth and chlorophyll, carotenoid and proline contents in carrots. In addition, leaf blight and root rot indices were reduced by *A. dauci* and *F. solani*, respectively	Siddiqui et al. (2019)
ZnO	*F. oxysporum*	In vitro treatments of ZnO NPs inhibited the fungus mycelial growth and the sporulation. However, ZnO-NPs increased the plant height with a decrease in disease severity and incidence in tomato plants under greenhouse conditions	González et al. (2021)
ZnO	*F. oxysporum*	ZnO-NPs exhibited in vitro antifungal activity against *F. Oxysporum*. However, in vivo, ZnO NPs suppressed *fusarium* wilt disease in cultivated *Solanum melongena* L. by decreasing the disease severity	Abdelaziz et al. (2022)
ZnO	*A. brassicae*	Cytomorphological studies by TEM and SEM have shown the complete disintegration of cell wall, cell membrane, and cytoplasmic content at 200 ppm concentration of Tb-ZnO NPs	Dhiman et al. (2022)
ZnO	*Colletotrichum* sp.	ZnO-NPs significantly inhibited % fungal growth by causing a loss in the continuity of hyphae and the formation of groups of hyphal structures	Mosquera-Sánchez et al. (2020)
CuO-NPs	*F. oxysporum* f. sp. *niveum*	CuO-NPs are significantly effective against root fungal disease in watermelon	Borgatta et al. (2018)
M-CuO-NPs	*A. brassicae*	NPs significantly inhibited the growth of *A. Brassicae* with NPs at 200 ppm dose. SEM studies also confirmed the morphological changes in hyphae and affected the sporulation pattern	Gaba et al. (2022)
Copper oxide nanoparticles (CNP)	*A. tenuissima*, *A. alternata*, *F. equiseti*, *F. oxysporum*, *F. solani*	NPs inhibited the growth of plant pathogenic fungi in different media at different concentrations	Kasana et al. (2020)
Cu_2ONPs	*F. solani*	NPs had a greater inhibitory effect on *F. solani* growth over control. In addition, NPs considerably reduced the disease incidence of the root rot pathogen in cucumber plants	Kamel et al. (2022)

Table 13.2 Effects of ZnO and CuO-NPs on nematodes and nematode diseases in different crops

Nanoparticles	Nematode	Effect on disease management	Reference
ZnO	*M. incognita*	Plants sprayed with 200 ppm NPs resulted in a higher increase in plant growth and a higher reduction in galling and nematode multiplication on eggplant	Khan and Siddiqui (2018)
ZnO	*M. incognita*	Foliar spray of ZnO NPs at 200 mg L^{-1} caused the maximum reduction in nematode multiplication and root galling of beetroot	Khan and Siddiqui (2021)
CuO	*M. incognita* *M. javanica*	In a controlled pot experiment, it was observed that the formation of galls on melon plants was effectively suppressed, and there was a notable reduction in the population density of nematodes in the soil	Ch et al. (2019)
ZnO and GO	*M. javanica*	ZnO NPs at a concentration of 0.10 mg ml^{-1} caused the highest reduction in galling and nematode multiplication and GO NPs at 0.05 mg ml^{-1} the least on carrot	Siddiqui et al. (2019)
ZnO	*M. incognita*	Spray of ZnO NPs on lentil plants reduced galling and nematode multiplication	Siddiqui et al. (2018)
ZnO	*C. elegans*	The nematode population was significantly reduced by the use of ZnO	Gupta et al. (2015)
CuO	*M. incognita*	CuO-NPs caused significant increases in the growth parameters of cowpeas and caused a reduction in galling, size of galls, egg mass, and root-knot indices	Tauseef et al. (2021)
CuO	*M. incognita*	The pot experiment demonstrated a significant reduction in root infection caused by *M. Incognita* through the application of CuO-NPs. Furthermore, the growth of chickpea plants was notably enhanced, and their physiological attributes, including chlorophyll and carotenoid content, were positively influenced by the presence of CuO-NPs	Khan et al. (2022)
CuNPs ZnO	*M. javanica*	Both NPs cause a reduction in galling and population of *M. javanica* in lettuce	González et al. (2021)

The CuO and ZnO-NPs show promise in combating fungal diseases on various crops; their practical application requires further research, including field trials under different environmental conditions. Local factors, crop types, and specific fungal pathogens will influence the effectiveness of ZnO and CuO-NPs in plant disease management.

13.3 Role of Zinc Oxide and Copper Oxide Nanoparticles against Nematode Pathogens/Diseases in Crop Plants

Nanoparticles are currently defined as materials having one or more dimensions in the range of 1–100 nm (Sekhon 2014). The utilization of nanotechnology in various aspects of plant disease management, diagnosis, and genetic transformations is still in its early stages. Naturally occurring nanoparticles exhibit irregular shapes and sizes, whereas engineered nanoparticles tend to be more uniform and can take on distinct forms, including spherical, sheet-like, rod-like, or even complex structures such as multi-walled carbon nanotubes and dendrimers resembling branching tree patterns. Nanoparticles (NPs) offer valuable contributions to disease management due to their compact size, expansive surface area, and heightened reactivity. They find applications as bactericides, fungicides, nematicides, and nanofertilizers (Elmer and White 2018). Embracing NPs is recognized as an environmentally friendly and secure alternative for plant disease management (Mishra et al. 2017). Zinc oxide nanoparticles (ZnO-NPs) have gained attention for their potential role in combating nematode pathogens and diseases in crop plants. The unique properties of ZnO-NPs, such as their small size and high reactivity, make them a promising candidate for managing nematode-related issues in agricultural settings. Here are some more details about their role:

13.3.1 Mechanisms of Action

ZnO-NPs have exhibited multiple mechanisms of action against nematodes. They can disrupt the nematode cuticle and affect its integrity, causing dehydration and ultimately leading to nematode mortality (Khan and Siddiqui 2021). ZnO-NPs can also interfere with the nematode's sensory and mobility functions, reducing its ability to locate host plants and move within the soil (Dimkpa et al. 2013). These mechanisms collectively contribute to the reduction of nematode infestations.

13.3.2 Inhibition of Egg Hatching and Larval Development

ZnO-NPs have been reported to hinder nematode egg hatching and inhibit the development of larvae. This can lead to reduced nematode populations in the soil, thereby decreasing the potential damage to crop plants (Nair et al. 2010). ZnO-NPs reduce the infection of *Meloidogyne incognita* (root-knot nematode) in beetroot (Khan and Siddiqui 2021).

13.3.3 Plant Defense Enhancement

ZnO-NPs have the potential to enhance plant defense mechanisms against nematode attacks. They can trigger the production of reactive oxygen species (ROS) and activate defense-related genes, leading to an increased resistance of plants to nematode infections (Khan and Siddiqui 2021).

13.3.4 Direct and Indirect Effects

ZnO-NPs have both direct effects on nematodes and indirect effects on plants. Their antimicrobial properties directly impact nematode survival and reproduction. Indirectly, they enhance plant growth and vigor, making the plants more resilient to nematode infections (Sabir et al. 2014).

13.3.5 Environmental Considerations

While the potential of ZnO-NPs in nematode management is promising, their environmental impact needs careful consideration. Research is ongoing to ensure that the use of nanoparticles does not cause unintended consequences in the ecosystem, including soil health and non-target organisms.

Copper oxide nanoparticles (CuO-NPs) stand out among various metal oxide nanoparticles due to their versatile applications in electronics, air and liquid filtration, ceramics, wood preservation, textiles, bioactive coatings, skincare products, films, lubricant oils, and inks (White et al. 2006). Notably, research indicates that exposure to CuO-NPs leads to their absorption, accumulation, and movement within mung bean and wheat plants (Lee et al. 2008). Copper oxide nanoparticles (CuO-NPs) have emerged as a potential tool for combating nematode pathogens and diseases in crop plants due to their antimicrobial properties and versatile applications. Here's more information about their role:

13.3.6 Antimicrobial Action

CuO-NPs possess inherent antimicrobial properties that can be effective against a wide range of microorganisms, including nematodes. The *Spirodela polyrhiza* (greater duckweed) plant was used to create copper oxide nanoparticles. A 400 nm or so ultraviolet spectroscopy absorbance was employed to characterize the nanoparticles. The EDX method was used to determine if copper and oxygen were present in the biogenic nanoparticles. According to data from scanning and transmission electron microscopy (SEM and TEM), copper oxide nanoparticles (CuO-NPs) are spherically shaped, uniformly distributed, and have diameters in the 100 nm range. The antibacterial, antifungal, and nematicidal efficacy of the produced copper oxide nanoparticles against plant pests and pathogens was evaluated. Mycelium growth of the phytopathogenic fungus *Phomopsis vexans* was reduced by 45–50% (Khan et al. 2023).

CuO particles killed *Staphylococcus aureus* and *Bacillus subtilis* more effectively than *Pseudomonas aeruginosa* and *E. coli*, according to Thiruvengadam et al. (2019) who made this observation. Extremely effective antibacterial activity was shown by CuO-NPs made from gum karaya against *S. aureus* and *E. coli* (Khani et al. 2018; Azam et al. 2012). The size of the NPs is thought to affect efficiency (Nezhad et al. 2012). CuO NPs isolated from the aquatic noxious plant *Eichhornia crassipes* were discovered to suppress the growth of *Aspergillus niger* and *Fusarium culmorum* (Vanathi et al. 2016). Studies have demonstrated that CuO-NPs can inhibit the growth, reproduction, and activity of nematodes through mechanisms such as disrupting the nematode cuticle, affecting their mobility, and interfering with feeding behavior (Ch et al. 2019).

13.3.7 Nematode Mortality

CuO-NPs have been found to induce nematode mortality by damaging the nematode cuticle, leading to dehydration and disruption of physiological processes. The nanoparticles can also alter the nematode's sensory functions and hinder its ability to locate host plants, further reducing nematode infestations (Khan et al. 2023).

13.3.8 Inhibition of Egg Hatching and Larval Development

CuO-NPs have been shown to impede nematode egg hatching and larval development. This disrupts the nematode life cycle and can lead to a decrease in nematode populations, subsequently minimizing the damage caused to crop plants (Khan et al. 2023).

13.3.9 Enhancement of Plant Defense

Copper holds a crucial position as a micronutrient for plants, serving dual roles in promoting growth and bolstering defense mechanisms against plant diseases. In response to pathogen attacks, the plant synthesizes three key proteins vital for host defense: plastocyanins, peroxidases, and multi-Cu oxidases. These proteins are essential for mounting effective defense responses and rely on the availability of copper to function optimally (Evans et al. 2007). Similar to other nanoparticles, CuO-NPs may trigger the plant's defense mechanisms against nematode attacks. The nanoparticles can induce the production of reactive oxygen species (ROS) and activate genes involved in plant defense responses, contributing to increased resistance against infections (Badawy et al. 2021).

13.3.10 Environmental Implications

While CuO-NPs show promise for nematode management, their use should be carefully evaluated to ensure minimal environmental impact. Studies are ongoing to determine the potential effects of nanoparticles on soil ecosystems, non-target organisms, and soil health (Khan et al. 2022).

13.4 Conclusion

Nanotechnology can support agriculture and alleviate environmental challenges by producing nanopesticides and managing plant diseases. So, the foliar spray and seed priming of zinc oxide and copper oxide nanoparticles in managing fungal, bacterial, and nematode pathogens and diseases in various crop plants hold significant potential. Their multifaceted mechanisms of action; ability to inhibit fungi, bacteria, and nematode development; and capacity to enhance plant defenses make them an exciting area of research for sustainable and environmentally friendly disease management strategies (Fig. 13.1).

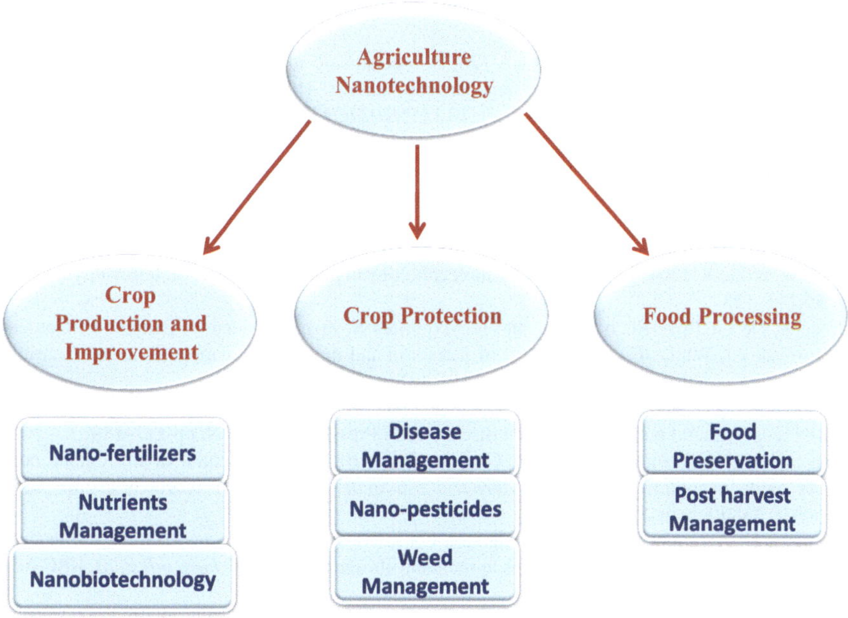

Fig. 13.1 Nanotechnology in agriculture

References

Abdelaziz AM, Salem SS, Khalil AM, El-Wakil DA, Fouda HM, Hashem AH (2022) Potential of biosynthesized zinc oxide nanoparticles to control fusarium wilt disease in eggplant (*Solanum melongena*) and promote plant growth. Biometals 35(3):601–616

Ahamad L, Siddiqui ZA (2021) Effects of silicon dioxide, zinc oxide and titanium dioxide nanoparticles on *Meloidogyne incognita*, *Alternaria dauci* and *Rhizoctonia solani* disease complex of carrot. Exp Parasitol 230:108176. https://doi.org/10.1016/j.exppara.2021.108176

Alghuthaymi MA, Almoammar H, Rai M, Said-Galiev E, Abd-Elsalam KA (2015) Myconanoparticles: synthesis and their role in phytopathogens management. Biotechnol Biotechnol Equip 29(2):221–236

Azam A, Ahmed AS, Oves M, Khan MS, Memic A (2012) Size-dependent antimicrobial properties of CuO nanoparticles against gram-positive and -negative bacterial strains. Int J Nanomed 7:3527–3535

Badawy AA, Abdelfattah NAH, Salem SS, Awad MF, Fouda A (2021) Efficacy assessment of biosynthesized copper oxide nanoparticles (CuO-NPs) on stored grain insects and their impacts on morphological and physiological traits of wheat (*Triticum aestivum* L.). Plant Biol 10:233

Borgatta J, Ma C, Hudson-Smith N, Elmer W, Plaza Perez CD, De La Torre-Roche R, Zuverza-Mena N, Haynes CL, White JC, Hamers RJ (2018) Copper based nanomaterials suppress root fungal disease in watermelon (*Citrullus lanatus*): role of particle morphology, composition and dissolution behavior. ACS Sustain Chem Eng 6(11):14847–14856

Ch G, Ntalli N, Menkissoglu-Spiroudi U, Dendrinou-Samara C (2019) Essential metal-based nanoparticles (copper/iron NPs) as potent Nematicidal agents against *Meloidogyne* spp. J Nanotechnol Res 1(2):44–58

Devipriya D, Roopan SM (2017) *Cissus quadrangularis* mediated ecofriendly synthesis of copper oxide nanoparticles and its antifungal studies against *aspergillus Niger, aspergillus flavus*. Mater Sci Eng C 80:38–44

Dhiman S, Varma A, Prasad R, Goel A (2022) Mechanistic insight of the antifungal potential of green synthesized zinc oxide nanoparticles against *Alternaria brassicae*. J Nanomater 2022:7138843. 13 pages. https://doi.org/10.1155/2022/7138843

Dimkpa CO, McLean JE, Britt DW, Anderson AJ (2013) Antifungal activity of ZnO nanoparticles and their interactive effect with a biocontrol bacterium on growth antagonism of the plant pathogen *fusarium graminearum*. Biometals 26(6):913–924

Elmer WH, White JC (2016) The use of metallic oxide nanoparticles to enhance growth of tomatoes and eggplants in disease infested soil or soilless medium. Environ Sci Nano 3(5):1072–1079

Elmer W, White JC (2018) The future of nanotechnology in plant pathology. Annu Rev Phytopathol 56:111–133

Elmer W, De La Torre-Roche R, Pagano L, Majumdar S, Zuverza-Mena N, Dimkpa C, Gardea-Torresdey J, White JC (2018) Effect of metalloid and metal oxide nanoparticles on fusarium wilt of watermelon. Plant Dis 102(7):1394–1401

Evans I, Solberg E, Huber DM (2007) Copper and plant disease. In: Datnoff LE, Elmer WH, Huber DN (eds) Mineral nutrition and plant disease. APS Press, St. Paul, MN, pp 177–188

Gaba S, Rai AK, Varma A, Prasad R, Goel A (2022) Biocontrol potential of mycogenic copper oxide nanoparticles against *Alternaria brassicae*. Front Chem 10:966396. https://doi.org/10.3389/fchem.2022.966396

Giannousi K, Avramidis I, Dendrinou-Samara C (2013) Synthesis, characterization and evaluation of copper based nanoparticles as agrochemicals against *Phytophthora infestans*. RSC Adv 3(44):21743–21752

González MAM, HernándezJuárez A, BetancourtGalindo R, OchoaFuentes YM, Valdez ALA, LimónCorona ML (2021) Antifungal activity of zinc oxide nanoparticles in *fusarium oxysporum Solanum lycopersicum* pathosystem under controlled conditions. J Phytopathol 169(9):533–544

Gupta S, Kushwah T, Vishwakarma A, Yadav S (2015) Optimization of ZnO-NPs to investigate their safe application by assessing their effect on soil nematode *Caenorhabditis elegans*. Nanoscale Res Lett 10:303. https://doi.org/10.1186/s11671-015-1010-4

Hao Y, Fang PH, Ma CX, White JC, Xiang ZQ, Wang HT, Zhang Z, Rui YK, Xing BS (2019) Engineered nanomaterials inhibit *Podosphaera pannosa* infection on rose leaves by regulating phytohormones. Environ Res 170:1–6

He L, Liu Y, Mustapha A, Lin M (2011) Antifungal activity of zincoxide nanoparticles against *Botrytis cinerea* and *Penicillium expansum*. Microbiol Res 166:207–215

Kamel SM, Elgobashy SF, Omara RI, Derbalah AS, Abdelfatah M, El-Shaer A, Al-Askar AA, Abdelkhalek A, Abd-Elsalam KA, Essa T, Kamran M (2022) Antifungal activity of copper oxide nanoparticles against root rot disease in cucumber. J Fungi 8:911. https://doi.org/10.3390/jof8090911

Kasana RC, Singh SK, Kumar P (2020) Antifungal activity of copper oxide nanoparticles against the fungal pathogens isolated from arid environment of Jodhpur. Int J Curr Microbiol App Sci 9(11):901–909

Khan M, Khan AU (2021) Plant parasitic nematodes effectors and their crosstalk with defense response of host plants: a battle underground. Rhizosphere 17:100288

Khan M, Siddiqui ZA (2018) Zinc oxide nanoparticles for the management of *Ralstonia solanacearum, Phomopsis vexans* and *Meloidogyne incognita* incited disease complex of eggplant. Indian Phytopathol 71(3):355–364

Khan MR, Siddiqui ZA (2021) Role of zinc oxide nanoparticles in the management of disease complex of beetroot (*Beta vulgaris* L.) caused by *Pectobacterium betavasculorum, Meloidogyne incognita* and *Rhizoctonia solani*. Hortic Environ Biotechnol 62:225–241. https://doi.org/10.1007/s13580-020-00312-z

Khan M, Khan AU, Hasan MA, Yadav KK (2021) Agro-nanotechnology as an emerging field: a novel sustainable approach for improving plant growth by reducing biotic stress. Appl Sci 11:2282

Khan A, BaniMfarrej MF, Danish M, Shariq M, Khan MF, Ansari MS, Hashem M, Alamri S, Ahmad F (2022) Synthesized copper oxide nanoparticles via the green route act as antagonists to pathogenic root-knot nematode, *Meloidogyne incognita*. Green Chem Lett Rev 15(3):491–507

Khan M, Khan AA, Parveen A, Min K, Yadav VK, Khan AU, Alam M (2023) Mitigating the growth of plant pathogenic bacterium, fungi, and nematode by using plant-mediated synthesis of copper oxide nanoparticles (CuO-NPs). Green Chem Lett Rev 16(1):2177520

Khani R, Roostaei B, Bagherzade G, Moudi M (2018) Green synthesis of copper nanoparticles by fruit extract of Ziziphus spina-christi (L.) Willd: application for adsorption of triphenylmethane dye and antibacterial assay. J Mol Liq 255:541–549

Lee WM, An YJ, Yoon H, Kweon HS (2008) Toxicity and bioavailability of copper nanoparticles to the terrestrial plants mung bean (*Phaseolus radiatus*) and wheat (*Triticum aestivum*): plant agar test for water-insoluble nanoparticles. Environ Toxicol Chem 27:1915–1921

Malandrakis AA, Kavroulakis N, Chrysikopoulos CV (2019) Use of copper, silver and zinc nanoparticles against foliar and soil-borne plant pathogens. Sci Total Environ 670:292–299

Mishra S, Keswani C, Abhilash PC, Fraceto LF, Singh HB (2017) Integrated approach of agri-nanotechnology: challenges and future trends. Front Plant Sci 8:471. https://doi.org/10.3389/fpls.2017.00471

Mosquera-Sánchez LP, Arciniegas-Grijalba PA, Patiño-Portela MC, Guerra-Sierra BE, Muñoz-Florez JE, Rodríguez-Páez JE (2020) Antifungal effect of zinc oxide nanoparticles (ZnO-NPs) on *Colletotrichum* sp., causal agent of anthracnose in coffee crops. Biocatal Agric Biotechnol 25:101579. https://doi.org/10.1016/j.bcab.2020.101579

Nair R, Varghese SH, Nair BG, Maekawa T, Yoshida Y, Kumar DS (2010) Nanoparticulate material delivery to plants. Plant Sci 179(3):154–163

Nawaz A, Rehman HU, Usman M, Wakeel A, Shahid MS, Alam S, Sanaullah M, Atiq M, Farooq M (2023) Nanobiotechnology in crop stress management: an overview of novel applications. Discover Nano 18:74. https://doi.org/10.1186/s11671-023-03845-1

Nezhad MH, Alamshahi L, Panjehkeh N (2012) Biocontrol efficiency of medicinal plants against Pectobacterium carotovorum, Ralstonia solanacearum and Escherichia coli. Open Conf Proc J 3:46–51

Parveen A, Siddiqui ZA (2021) Zinc oxide nanoparticles affect growth, photosynthetic pigments, proline content and bacterial and fungal diseases of tomato. Arch Phytopathol Plant Protect 54(17–18):1519–1538

Prasad T, Sudhakar P, Sreenivasulu Y, Latha P, Munaswamy V, Reddy RK, Sreeprasad TS, Sajanlal PR, Pradeep T (2012) Effect of nanoscale zinc oxide particles on the germination, growth and yield of Peanut. J Plant Nutr 35(6):905–927

Ramachandrappa LT, Kalagatur NK, Mohan CD et al (2019) Biofabrication of zinc oxide nanoparticles with Syzygium aromaticum flower buds extract and finding its novel application in controlling the growth and mycotoxins of fusarium graminearum. Front Microbiol 10:1244

Sabir S, Arshad M, Chaudhari SK (2014) Zinc oxide nanoparticles for revolutionizing agriculture: synthesis and applications. Sci World J 2014:925494. https://doi.org/10.1155/2014/925494

Sekhon BS (2014) Nanotechnology in Agri-food production: an overview. Nanotechnol Sci Appl 7:31–53

Siddiqui ZA, Khan A, Khan MR, Abd-Allah EF (2018) Effects of zinc oxide nanoparticles (ZnO NPs) and some plant pathogens on the growth and nodulation of lentil (*Lens culinaris Medik.*). Acta Phytopathol Entomol Hungarica 53(2):195–211

Siddiqui ZA, Parveen A, Ahmad L, Hashem A (2019) Effects of graphene oxide and zinc oxide nanoparticles on growth, chlorophyll, carotenoids, proline contents and diseases of carrot. Sci Hortic 249:374–382

Singh AK, Mishra KK, Jeevan B, Paschapur AU, Verma G, Kant L (2023) Nanotechnology in biotic stress management future challenges and opportunities. Biotic Stress Manage Crop Plants Using Nanomater 9:9-6

Tauseef A, Gupta J, Rehman A, Uddin I (2021) Differential response of cowpea towards the CuO nanoparticles under Meloidogyne incognita stress. S Afr J Bot 139:175–182

Thiruvengadam M, Chung IM, Gomathi T, Ansari MA, Khanna VG, Babu V, Rajakumar G (2019) Synthesis, characterization and pharmacological potential of green synthesized copper nanoparticles. Bioprocess Biosyst Eng 42:1769–1777

Vanathi P, Rajiv P, Sivaraj R (2016) Synthesis and characterization of Eichhornia mediated copper oxide nanoparticles and assessing their antifungal activity against plant pathogens. Bull Mater Sci 39:1165–1170

White B, Yin M, Hall A, Le D, Stolbov S, Rahman T, Turro N, O'Brien S (2006) Complete CO oxidation over Cu_2O nanoparticles supported on silica gel. Nano Lett 6(9):2095–2098

Yehia RS, Ahmed OF (2013) In vitro study of the antifungal efficacy of zinc oxide nanoparticles against *fusarium oxysporum* and *Penicillium expansum*. Afr J Microbiol Res 7(19):1917–1923

Chapter 14
Carbon-Based Nanoparticles: Graphene Oxide (GO)-Biotic Stress Minimizers in Plants

Zill-e-Huma Aftab, Tehmina Anjum, Waheed Akram, Muhammad Faisal Shafiq, and Hamza Rafiq

Abstract Agriculture has been a major backbone of the world's economy since ancient civilizations. In recent years agricultural productions have gained a remarkable boost due to excessive usage of chemical pesticides and fertilizers, hence killing the pathogens (*Verticillium*, *Alternaria*, *Fusarium*, *Rhizoctonia*, *Sclerotium*, *Macrophomina*, *Xanthomonas*, *Erwinia*, *Agrobacterium*, aphids, whiteflies, psyllids, etc.) at a significant rate, but these chemicals are dramatically hazardous to the environment at the same time, leading to the appearance of microbial pathogens with increased virulence which attacks the crop plant and deteriorates yield many folds. The encounter of all these problems lies in the development of some sustainable techniques or tools, warranting the significant check on fungal pathogenesis and post-harvest management. In the past two decades, nanotechnology has emerged as a revolutionary solution to manage disease effectively. Nanotechnology can effectively deliver agrochemicals like nutrients, fungicides, and growth regulators with increased efficiency and reduced toxic effects. Nowadays carbon-based nanomaterials specifically graphene oxide (GO) have become a hotspot for researchers/scientists due to their excellent physicochemical properties, as the best remedy for agricultural and environmental problems. Reduced GO is a better carrier for CuO, ZnO, and other nanoparticles as proved by research. Both alone GO and coupled with other nanoparticles have shown splendid potential toward control of *Pseudomonas syringae*, *Ralstonia solanacearum*, *Fusarium graminearum*, and *Fusarium oxysporum* in different crops. Accompanied by antimicrobial potential, GO has been reported to have magnificent effects in the promotion of plant growth, altering metabolic activities (physiology, biochemistry, cytological structures, and gene expressions), hence improving plant health and enabling it to restrict pathogenic activities of plant pathogens.

Zill-e-Huma Aftab · T. Anjum (✉) · W. Akram · M. F. Shafiq · H. Rafiq
Department of Plant Pathology, Faculty of Agricultural Sciences, University of the Punjab, Lahore, Pakistan

M. Khan, J.-T. Chen (eds.), *Nanoparticles in Plant Biotic Stress Management*,
https://doi.org/10.1007/978-981-97-0851-2_14

Keywords Graphene oxide · Nanoparticles · Plants · Disease management

14.1 Introduction

Nanomaterials are molecules ranging from 1 to 100 nm in size, and graphene is an atomically dense two-dimensional crystal made up of single-layered carbon atoms. Graphene oxide (GO) has a monolayer graphene with a basic thickness of 2–3 nm, slightly thicker than simple graphene, as it comprises oxygen-containing functional groups, which change the surface of the GO sheet and make it thicker. Hence GO-based nanomaterials have unique chemical and physical properties, capturing attention for application in diverse fields. The structural formation of graphene and grapheme-based materials (Fig. 14.1) is made up of a single layer of independent

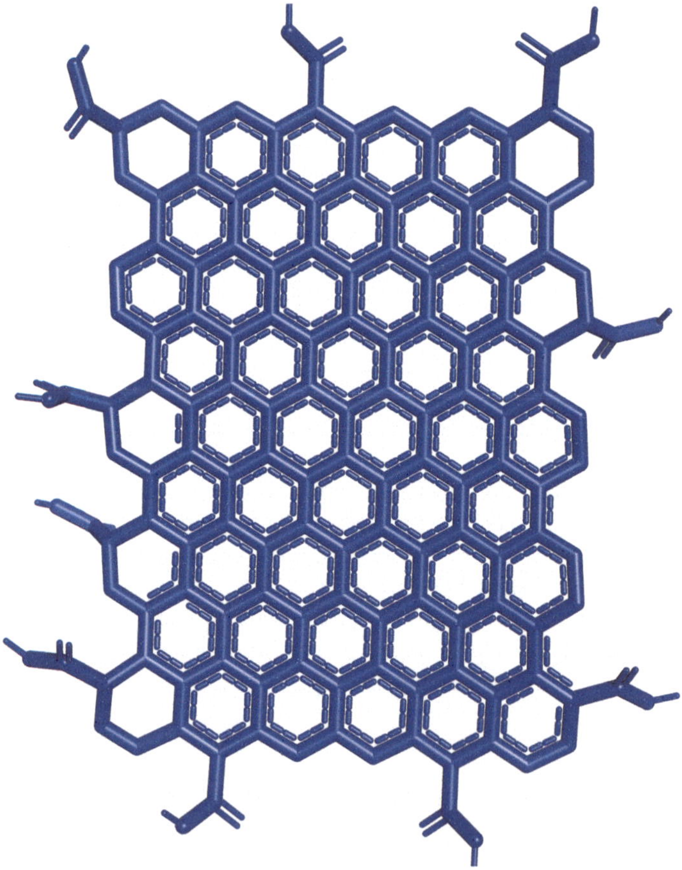

Fig. 14.1 The graphene oxide molecular structure

sp2 hybrid carbon atoms. GO is a two-dimensional carbonaceous material with a hexagonal honeycomb crystal lattice.

Graphene oxide has emerged as a fine hard nanomaterial, having a sheet thickness of about 0.43 nm. Every carbon atom is tightly bonded with the three adjoining carbon atoms in graphene due to the σ bond. The other p electrons get engaged preferably in a π bond with the adjacent carbon atoms with a right angle to the graphene plane. The graphene structure is very durable, due to a C–C bond length of about 0.145 nm. The atomic surface inside GO gets distorted and twisted when an external force is applied. The structure is stable and maintained due to the absence of reorganization and change in the alignment of carbon atoms. There is no energy dispersion in GO after an electron jump in the inner orbit, due to interference of external atoms and lattice defects. This uniqueness in the lattice structure contributes excellent properties to graphene oxide. Recent research has proved that there are a lot of methods for the preparation of graphene oxide from graphite powder; however the principal preparation methods include liquid phase stripping, mechanical stripping, redox methods, chemical vapor deposition, and epitaxial growth.

14.2 Synthesis of Graphene Oxide Nanoparticles

Methods being used for GO synthesis are due to the intercalation of layered graphite, as it permits the atoms of dynamic metals and some oxidizing agents to disrupt the horizontal strata of carbon of the crystalline graphite. This causes the aggregation of the interlayer distance, and the surfaces of layers are changed by the chemical cementing of the functional groups. Finally, a suitable oxidizing agent disassembles graphite crystals in detached single-layer carbon with a modified surface.

The renowned procedure for the synthesis of GO (Staudenmaier, Brodie, Hummer-Offman) includes a reaction between concentrated acids (sulfur and/or nitric) and potassium chlorate. The current procedure by Hummer-Offman requires a mixture of $KMnO_4/NaNO_3$ and concentrated H_2SO_4 for the complete oxidation of graphite into GO. This method provides a good yield in a short time compared to the previously described methods. The GO is synthesized (Fig. 14.2a) as an aqueous suspension of particles with yellowish coloration.

It is very important to understand that the chemical oxidation process of graphite essentially requires small (<10 μm) sized particles of the parent material. For an irreversible chemical reaction, intensive mixing or even ultra-sonication may be needed. Due to these factors, procurement of large-sized GO flakes becomes difficult. So, the average size of particles in commercial graphite powder gives GO that does not exceed some microns.

Hence, the achievement of large-sized GO particles remains a challenge for biotechnological applications and detection in living systems. For this reason, some recent researchers suggest improved methods of graphene oxide preparation targeting to achieve large- and ultra-large-sized GO particles. The idea behind improved

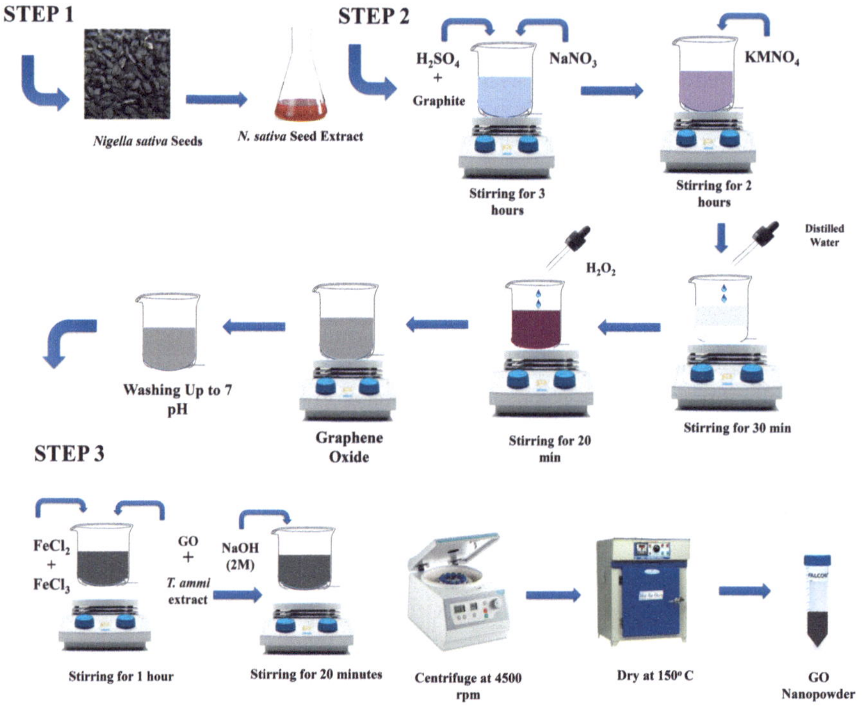

Fig. 14.2 Steps for green synthesis of graphene oxide nanoparticles

methods is to divide the GO synthesis into two steps: (1) soft chemical surface peeling of primary graphite powder into graphene flakes in liquid media and (2) oxidation of the synthesized product under moderate conditions. In controlled conditions, the interaction among crystalline graphite powder, chromium trioxide, and hydrogen peroxide results in an effective exfoliation of graphite under normal temperature. The precise process of synthesis helps to sustain the integrity of the large graphene flakes. Later, the oxidation of graphene flakes at a relatively low concentration of $KMNO_4$ allows the synthesis of GO flakes with a large and uniform area avoiding the formation of oxidation defects. These improved methods for GO synthesis are mainly based on Hummer's oxidation and exfoliation of graphite. The pilot-scale production technology of the graphene family should be addressed by researchers.

14.3 Green Synthesis of Graphene Oxide Nanoparticles

As far as the green synthesis of graphene oxide is concerned currently, electrochemical (EC) procedures have been worked out to synthesize graphene sheets due to their environmental-friendly properties, high efficacy, and cost-effectiveness. A number of plants have been worked out by researchers for the green synthesis of GO (Table 14.1).

Table 14.1 Green synthesis of graphene oxide nanoparticles and its modified forms

Source material	Type	Method of synthesis	Reference
Ocimum sanctum	Reduced GO Pb/cu bimetallic	Modified Hummer's	Sultana et al. (2020)
Vernonia amygdalina	Reduced GO	Improved Marcano	Meka Chufa et al. (2021)
Ascorbic acid	Porous spherical reduced GO	Modified hummer-Offman	Gao et al. (2020)
Gallic acid	GO-silver nanocomposites	Modified Hummer's	Bao et al. (2022)
Oil palm leaves	Adsorbant reduced GO	Catalytic acid spray	Fathy et al. (2019)
Green tea extract	Reduced GO	rGO reduced electrode	Vatandost et al. (2020)
Lemon juice	Reduced GO	Modified Hummer's	Mahiuddin and Ochiai (2021)
Lantana camara	Nanocrystalline reduced GO	Modified Hummer's	Medha et al. (2017)
Hibiscus sabdariffa	Reduced GO	Modified Hummer's	Alshamsi et al. (2020)
Nigella sativa	Reduced GO	Modified Hummer's	Mirza et al. (2022)
Vitis vinifera	Reduced GO	Modified Hummer's	Ansari et al. (2018)
Asparagus racemosus	Reduced GO	Hummer's	Paul et al. (2021)
Tecoma stans	Reduced GO	Modified Hummer's	Mahmoud et al. (2022a, b)
Trachyspermum ammi	Reduced GO	Modified Hummer's	Rafiq et al. (2023)
Citrus sinensis	Photocatalytic reduced GO	Modified Hummer's	Thanapackiam et al. (2022)
Cellulose	Reduced GO	Chemical exfoliation	Said et al. (2023)
Cinnamomum verum + *Zingiber officinale* + sesame oil	Multilayered GO	Solvothermal	Khaleghi Bousari et al. (2023)
Tea leaves + ascorbic acid	Reduced GO	Modified Hummer's	Zaib and Kyung (2022)
Domestic green waste	Reduced GO	Chemical exfoliation	Joshi et al. (2023)
Mussaenda glabrata	Reduced GO	Modified Hummer's	Parvathi et al. (2022a, b)
Allium cepa	Reduced GO	Modified Hummer's	Parvathi et al. (2023)
Artemisia vulgaris	Reduced GO	Modified Hummer's	Tamang et al. (2023)

(continued)

Table 14.1 (continued)

Source material	Type	Method of synthesis	Reference
Aloe vera	Reduced GO	Chemical exfoliation	Ramanathan et al. (2017)
Banana + mango leaves + potato extract	Reduced GO	Modified Hummer's	Sinha et al. (2022)
Aqueous sources	Reduced GO	Multiple methods	Parvathi et al. (2022a, b)
Plectranthus amboinicus	Reduced GO	Modified Hummer's	Dominic et al. (2022)
Salvia spinosa	Reduced GO	Modified Hummer's	Yang et al. (2021)
Ziziphus jujube	Reduced GO	Modified Hummer's	Mahmoud et al. (2022a, b)
Pomelo grandis + *Tamarindus indica*	Reduced GO	Hummer's	Panicker and Sahu (2021)
Euphorbia heterophylla	Reduced GO	Modified Hummer's	Lingaraju et al. (2019)
Tridax procumbens	Reduced G nanosheets	Modified Hummer's	Thiyagarajulu et al. (2020)
Acalypha indica + *Raphanus sativus*	Reduced GO	Hydrothermal	Punniyakotti et al. (2021)
Clinacanthus nutans	Reduced GO	Hummer's	Perumal et al. (2022)
Gallnut	Functionalized few-layer graphene–silver nanocomposites	Chemical exfoliation	Bu et al. (2022)
Crocus sativus	Graphene oxide encapsulated selenium nanoparticles	Chemical exfoliation	Naemi and Meshkini (2023)
Pleurotus sajor-caju	RGO nanosheets	Hummer's	Manimaran et al. (2022)
Sugarcane bagasse extract	Reduced GO	Modified Hummer's	Li et al. (2018)
Anisomeles malabarica	Reduced GO silver nanocomposites	Modified Hummer's	Rakkimuthu et al. (2022)
Pelargonium graveolens	Nanosilver-decorated graphene oxide sheets	Modified Hummer's	Rodríguez-González et al. (2016)
Citrullus colocynthis	Graphene nanosheets	Modified Hummer's	Zhu et al. (2017)
Atriplex halimus	Reduced GO and nano-zero valent iron composite	Modified Hummer's	Abdelfatah et al. (2022)

14.4 Characterization of Graphene Oxide Nanoparticles

The synthesized GO and RGO solutions are characterized through a couple of characterization methods, e.g., X-ray diffraction (XRD), UV-visible (UV–vis), Fourier transform infrared (FT-IR) spectroscopy, scanning electron microscopy (SEM), Raman spectroscopy (Xu et al. 2008), atomic force microscopy (AFM) (Yin et al. 2015), X-ray photoelectron spectroscopy (XPS), transmission electron microscopy (TEM) (Chen et al. 2017; Li et al. 2009), dynamic light scattering (DLS), NMR spectroscopy, IR spectroscopy, electrical conductivity, thermal conductivity, etc. According to the absorption spectra, the principal spectrum of GO has a robust absorption peak at 233 nm, credited to the shift of the C–C conjugated aromatic spheres and weak incorporation at 305 nm due to changes in the C=O bond. UV–vis absorption is due to the shift of the electrons at carbon (π–π^* transition) pi bonds, displaying an absorption peak of monolayer GO at ~230 nm. Contrastingly, the absorption peak between 250 and 270 nm can be seen in the case of pristine graphene. SEM imaging helps to detect the graphene folds, impurities, and discontinuities during the GO synthesis process. Research shows that the thickness of graphene sheets remains around 2 nm encapsulated by polyethylene terephthalate. The thickness of the graphene ranges between 1.57 nm and 50 nm as PET interacts with the graphene (Figs. 14.3 and 14.4).

These special optical, electronic, and mechanical characteristics of graphene structure propose that graphene oxide can be used as an auspicious next-generation antimicrobial material because of the various oxygen-containing functional groups, for example, carbonyl, epoxide, hydroxyl, and carboxyl groups. Current research has exposed that GO has an extensive variety of bactericidal activity and has significant inhibiting properties against a broad range of bacteria (Bu et al. 2022). It has been revealed that GO can deactivate a wide range of bacteria, such as *Xanthomonas oryzae* pv. *oryzae*, copper-resistant *Ralstonia solanacearum*, *Xanthomonas campestris* pv. *undulosa*, and *Pseudomonas syringae*. Moreover, GO has been proven as a strong inhibitor of the growing colony and germinating spores of various plant fungal pathogenic fungi, such as *Fusarium poae*, *Fusarium graminearum*, and *Fusarium oxysporum*. This high antimicrobial efficacy of GO against phytopathogens makes it an encouraging entity to suppress plant-pathogenic interactions.

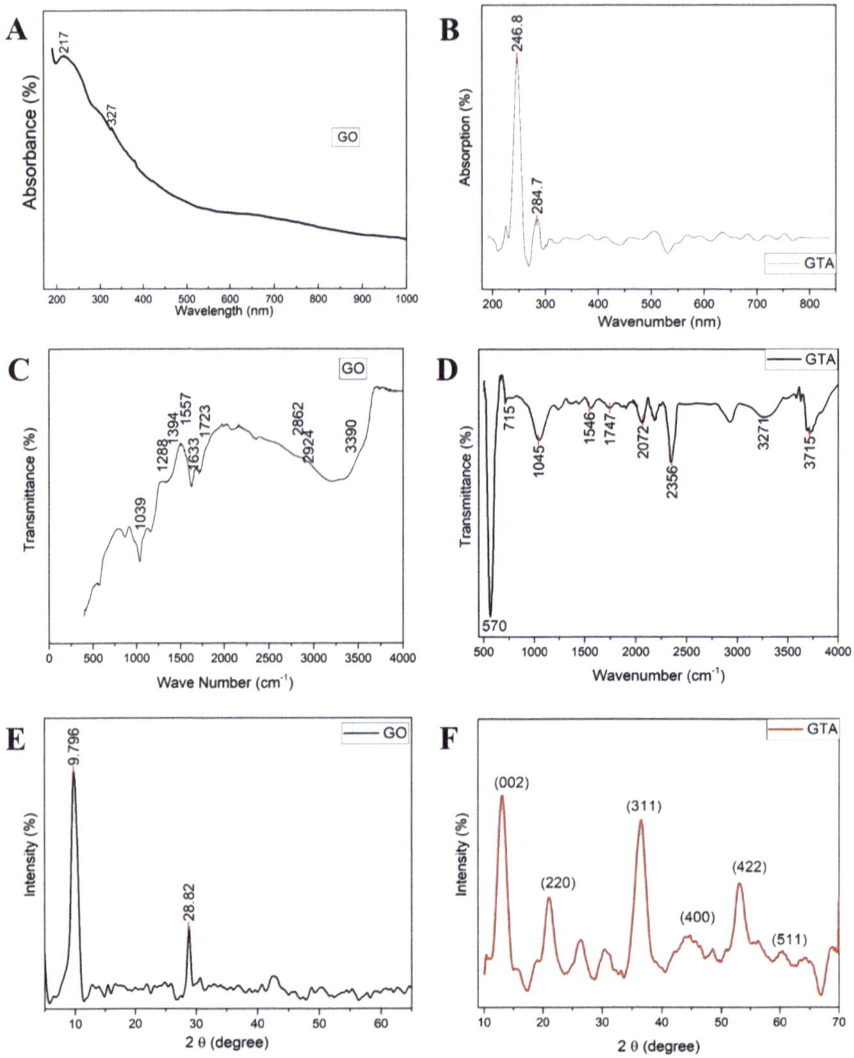

Fig. 14.3 (**a**) UV–vis analysis of synthesized graphene oxide; (**b**) UV–vis analysis of green synthesized graphene oxide nanoparticles; (**c**) FTIR spectrum of synthesized graphene oxide; (**d**) FTIR spectrum of green synthesized graphene oxide nanoparticles; (**e**) XRD pattern of synthesized graphene oxide; (**f**) XRD pattern of green synthesized graphene oxide nanoparticles

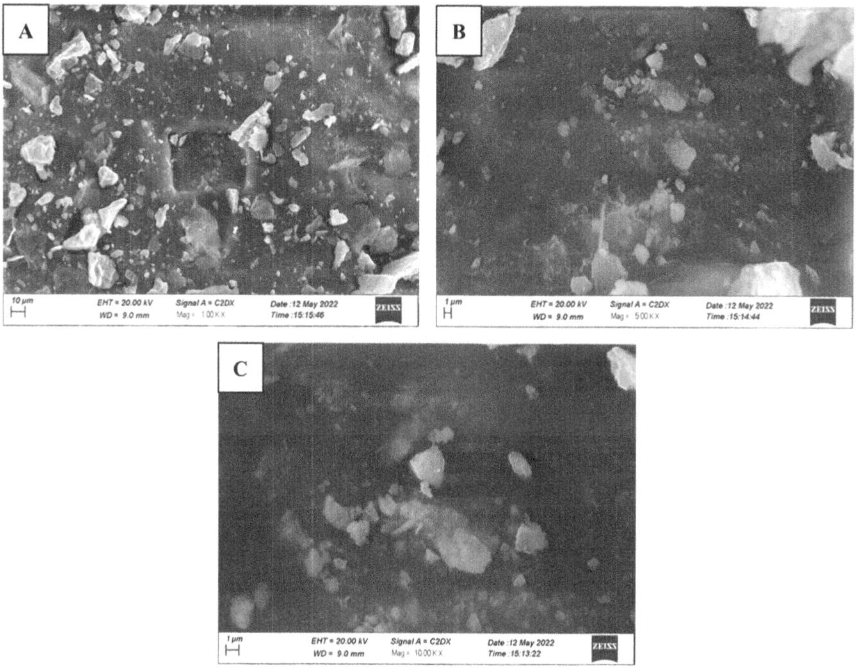

Fig. 14.4 SEM images of graphene oxide nanoparticles under different magnifications (**a**) = 1 KX; (**b**) 5 KX; (**c**) 10 KX

14.5 Graphene Oxide Loaded with Copper Oxide Nanoparticles as an Antibacterial Agent against *Pseudomonas Syringae Pv.* Tomato (Li et al. 2018)

Bacterial speck resulting from the infection of *Pseudomonas syringae pv. Tomato* (Pst) is among the most significant diseases of tomatoes. Pathogen infecting at the seedling stage could cause more than 75% of total quality and yield losses in the tomato drop. Huge applications of chemical bactericides to manage plant diseases have caused adverse resistance in pathogen races, environmental pollution, and serious risks to human health. A novel bactericidal was synthesized from CuO nanoparticles loaded on sheets of graphene oxide as GO–Cu NPs. The results revealed that the Go-loaded copper NPs showed 16 times more bactericidal activity than the synthetic bactericide, Kocide 3000. Additional investigations revealed that the GO–Cu nanocomposite damages the cell, increases the reactive oxygen species level, and decreases the DNA content in Pst cells. In vivo tests showed that GO–Cu NPs at concentrations of 4 and 8 mg mL^{-1} pointedly reduced the disease severity, i.e., less than 25% with no phytotoxicity to the crop.

Both copper biocide and streptomycin are considered highly efficient and frequently used biocidal agents for the management of bacterial diseases. Nevertheless, prolonged practice of these antibiotics has induced resistance to phytopathogen. Moreover, deposits of these chemicals in food and soil are harmful to the consumer as well as the environment. Hence, it is crucial to develop an alternative to these bactericides to protect tomato crops from Pst infection.

Chen et al. (2010) observed that GO showed an enormous antimicrobial activity toward phytopathogenic fungi and bacteria. Ocsoy et al. (2013) synthesized DNA-mediated Ag NPs on GO that showed effective control against bacterial spots on tomatoes. In another study, it was presented that the GO-silver nanocomposite wire had improved bactericidal activity along with improved cell compatibility. Additionally, Zhang et al. (2015) described that grapheme oxide had a significant effect on the promotion of seed germination and an improvement in seedling growth in tomatoes was also noted. Therefore, GO was proved to be a suitable host material to be loaded with Cu NPs.

14.6 Reduced Graphene Oxide Nanosheet-Decorated Copper Oxide Nanoparticles: A Potent Antifungal Nanocomposite Against Fusarium Root Rot and Wilt Diseases of Tomato and Pepper Plants (El Abeid et al. 2020)

Environment-friendly nanotechnological applications in crop protection require complete background knowledge of the life cycle of the plant, the toxic consequences of nanoparticles, and the underlying mechanism against the selected pathogens. In another research work, the efficacious synthesis of rGO-CuO NPs was proved through characterization, and its fungicidal activity was also successfully checked against *Fusarium oxysporum* wild strains affecting pepper and tomato plants. An in vitro relative analysis showed that the synthesized nanocomposite had significantly high fungicidal activity at a very low dose, i.e., even 1 mg L^{-1} than 2.5 g L^{-1} of frequently used fungicide, i.e., Kocide 2000. Some more studies suggest that rGO-CuO nanocomposite may generate small lesions on the fungal cell membranes leading cells to death. The results of an in vivo research showed that only 1 mg L^{-1} from the nanocomposite decreased the severeness of wilting and root rot diseases by less than 5% for pepper and tomato plants besides phytotoxicity up to 70 days (El-Abeid et al. 2020).

Moreover, some researchers have studied detailed antimicrobial activity on phytopathogenic fungi and bacteria for GO (Karthik et al. 2017). Similarly, Ocsoy et al. (2013) synthesized DNA-directed Ag NPs loaded on graphene oxide surfaces, for effective control of bacterial spots in tomatoes. Additionally, it was described that GO had significantly positive effects on seed germination along with healthy tomato seedling development. The presence of reactive hydrophilic oxygen functional groups makes GO scatters in water (Dreyer et al. 2010), considerably reducing or inhibiting the *Fusarium* infection and also delaying the onset of symptoms of *Fusarium* wilt to the level that the disease does not even significantly appear.

Additional alternative potential mechanisms may be the upregulation of host, defense-related genes by the action of GO-CuO NPs, and helping pepper and tomato plants to cope with infection; however, the exact pathway by which rGO-CuO NPs manage the fungal infections is not known. However, it is sturdily recommended that rGO-CuO NPs can cross the root cells and move systematically in the other parts of the treated plants (Elmer and White 2016), hence inducing systemic resistance.

14.7 Synergistic Antifungal Activity of Graphene Oxide Nanoparticles and Fungicides Against Fusarium Head Blight In Vitro and In Vivo (Wang et al. 2021)

Resistance to antimicrobial agents has become a censorious factor in plant and pathogen relationship for disease expansion, and this renders great barriers against disease. There is a necessity of time to explore new antimicrobials. A mixed application of antimicrobials with a couple of antifungal pathways had been established as an assuring method to control plant pathogens. Graphene oxide (GO) has become a rapidly emergent and extremely trustable antimicrobial against different phytopathogens in the field of agriculture. In another observation, the inhibitory effects of fungicides (difenoconazole, cyproconazol, and mancozeb) combined with GO against *Fusarium graminearum* were examined in vivo and in planta. Their findings exposed that the combination of fungicides with GO had substantial combined inhibitory activity on the fungal spore germination, mycelial growth, and fresh biomass, relative to only fungicides. In the field analysis, GO combined fungicides significantly reduced the disease occurrence and disease severity, showing an expressively enhanced control efficacy against *F. graminearum*.

14.8 Reduced Graphene Oxide-Based Nanometal Composite Comprising Copper and Silver Nanoparticles Protects Tomatoes and Peppers against *Xanthomonas Euvesicatoria* Infection (Bytešníková et al. 2022)

Xanthomonas euvesicatoria (Xe) belongs to the genera *Xanthomonas*, causing bacterial spot disease and infecting pepper and tomato plants, hence resulting in alarming economic deficits. Conventional antibacterials lead to resistance to pathogens, environmental pollution, and hazards to human health. Nanocomposite of Ag and Cu loaded on reduced graphene oxide (rGO) was created and analyzed for its antibacterial properties; nanocomposites were applied at concentrations of 50 and 500 μg mL^{-1}, to keep safe pepper and tomato plants in vivo. Both the host plants showed a significant decline in disease severity after treatment with the nanocomposite (16.7% and 15.6%, respectively) than treatment with the conventional antibacterial Kocide® 2000. The synthesized nanocomposite was determined as an efficient bactericide and had strong potential for crop disease control without any phytotoxic impression.

14.9 Conclusion

Very few elements are known to show such antagonistic and diverse effects on plants as GO nanoparticles, and hence it is difficult to draw an integrated description of the mode of action of GO NPs from the available literature. A huge range of effects have been attributed to GO NPs application in different research, caused by the physical and chemical properties of GO NPs, variations among plant species, concentration of NPs, time of exposure, and mode of application. Among the given factors, the physicochemical properties of GO NPs may be the prime important reason for this diversified behavior. In spite of everything, the GO NPs applied in several studies generally are synthesized from unlike sources, and minute changes in their functional groups lead to significant changes in their function. The chemical and physical characteristics of GO NPs are easily affected by the substrates on which they act. Prospects suggest that GO NPs should be categorized by their precise responses for their unified applications in agricultural research. It has also been proposed that the creation of GO NPs should be standardized on priority. The absence of a definite standard will affect the comparative experimentation on the biological effects of different types of GO NPs under the same set of conditions and may make the conclusions more objectionable.

The available literature predicts that GO NPs may be absorbed by the roots and delivered to the above-ground organs by crossing cell membranes and entering into different organelles. In most cases, the current observations of the antimicrobial efficacy of GO NPs applications are examined on membrane systems of the cell, root absorption capacity, photosynthetic activity of the plant, and last but not least oxidative stress. GO NPs may also have an indirect effect on the growth of the plant by modifying the microbial population and enzyme activity in the soil. Alteration in genetic expression in response to GO NPs treatment predominantly consists of (1) transcription factors, (2) factors for growth regulation, (3) genes related to nutrition metabolism, (4) hormone-related genes, (5)genes encoding for chloroplast proteins, and 6) stress-related genes. Nevertheless, the roles of all these genes in response to GO NPs application require more research.

References

Abdelfatah AM, El-Maghrabi N, Mahmoud AED, Fawzy M (2022) Synergetic effect of green synthesized reduced graphene oxide and nano-zero valent iron composite for the removal of doxycycline antibiotic from water. Sci Rep 12:19372

Alshamsi HA, Ali SK, Altaa SHA (2020) Green synthesis and characterization of reduced graphene oxide (RGO) using *Hibiscus sabdariffa* L extract and its solubility property. In: Journal of physics: conference series, vol 1664. IOP Publishing, p 12058

Ansari MZ, Lone MN, Sajid S, Siddiqui WA (2018) Novel green synthesis of graphene layers using zante currants and graphene oxide. Orient J Chem 34:2832–2837

Bao Y, Tian C, Yu H, He J, Song K, Guo J et al (2022) In situ green synthesis of graphene oxide-silver nanoparticles composite with using gallic acid. Front Chem 10:905781

Bu Y, Kushwaha A, Goswami L, Kim B-S (2022) Green production of functionalized few-layer graphene–silver nanocomposites using gallnut extract for antibacterial application. Micromachines 13:1232

Byteśníková Z, Pečenka J, Tekielska D, Kiss T, Švec P, Ridošková A, Bezdička P, Pekárková J, Eichmeier A, Pokluda R. Adam V (2022) Reduced graphene oxide-based nanometal-composite containing copper and silver nanoparticles protect tomato and pepper against Xanthomonas euvesicatoria infection. Chem Biol Technol Agric 9:p84.

Chen L, Yu H, Zhong J, Song L, Wu J, Su W (2017) Graphene field emitters: a review of fabrication, characterization and properties. Mater Sci Eng B 220:44–58

Chen W, Yan L, Bangal PR (2010) Preparation of graphene by the rapid and mild thermal reduction of graphene oxide induced by microwaves. Carbon N Y 48:1146–1152

Dominic RM, Punniyakotti P, Balan B, Angaiah S (2022) Green synthesis of reduced graphene oxide using *Plectranthus amboinicus* leaf extract and its supercapacitive performance. Bull Mater Sci 45:1–8

Dreyer DR, Park S, Bielawski CW, Ruoff RS (2010) The chemistry of graphene oxide. Chem Soc Rev 39:228–240

El-Abeid SE, Ahmed Y, Daròs JA, Mohamed MA (2020) Reduced graphene oxide nanosheet-decorated copper oxide nanoparticles: a potent antifungal nanocomposite against fusarium root rot and wilt diseases of tomato and pepper plants. Nanomaterials 10:1001

Elmer WH, White JC (2016) The use of metallic oxide nanoparticles to enhance growth of tomatoes and eggplants in disease infested soil or soilless medium. Environ Sci Nano 3:10721079

Fathy M, Hosny R, Keshawy M, Gaffer A (2019) Green synthesis of graphene oxide from oil palm leaves as novel adsorbent for removal of cu (II) ions from synthetic wastewater. Graphene Technol 4:33–40

Gao M, Li X, Qi D, Lin J (2020) Green synthesis of porous spherical reduced graphene oxide and its application in immobilized pectinase. ACS Omega 5:32706–32714

Joshi R, De Adhikari A, Dey A, Lahiri I (2023) Green reduction of graphene oxide as a substitute of acidic reducing agents for supercapacitor applications. MaterSci Eng B 287:116128

Karthik C, Ramkumar VS, Pugazhendhi A, Gopalakrishnan K, Arulselvi PI (2017) Biosorption and biotransformation of Cr (VI) by novel Cellulosimicrobium funkei strain AR6. J Taiwan Inst Chem Eng 70:282–290

Khaleghi Bousari A, Mohammadi S, Mohammadi A (2023) The effects of cinnamon, ginger and sesame oils on in-situ solvothermal reduction of multi-layered grapheme oxide in epoxy to improve hydrophobicity and corrosion resistance. Fullerenes Nanotub Carbon Nanostruct:1–20

Li B, Jin X, Lin J, Chen Z (2018) Green reduction of graphene oxide by sugarcane bagasse extract and its application for the removal of cadmium in aqueous solution. J Clean Prod 189:128–134

Li X, Cai W, Jung IH, An JH, Yang D, Velamakanni A et al (2009) Synthesis, characterization, and properties of large-area graphene films. ECS Trans 19:41

Lingaraju K, Naika HR, Nagaraju G, Nagabhushana H (2019) Biocompatible synthesis of reduced graphene oxide from *Euphorbia heterophylla* (L.) and their in-vitro cytotoxicity against human cancer cell lines. Biotechnol Rep 24:e00376

Mahiuddin M, Ochiai B (2021) Lemon juice assisted green synthesis of reduced graphene oxide and its application for adsorption of methylene blue. Technologies 9:96

Mahmoud AED, El-Maghrabi N, Hosny M, Fawzy M (2022a) Biogenic synthesis of reduced graphene oxide from *Ziziphus spina-christi* (Christ's thorn jujube) extracts for catalytic, antimicrobial, and antioxidant potentialities. Environ Sci Pollut Res 29:89772–89787

Mahmoud AED, Hosny M, El-Maghrabi N, Fawzy M (2022b) Facile synthesis of reduced graphene oxide by *Tecoma stans* extracts for efficient removal of Ni (II) from water: batch experiments and response surface methodology. Sustain Environ Res 32:1–16

Manimaran K, Prakash DG, Kumar S, Bogeshwaran K, Dahlous KA, Mohammed AAA et al (2022) Biological approaches of reduced graphene oxide (rGO) nanosheets using *Pleurotus sajor caju* extract and its in vitro pharmaceutical applications. Biomass Convers Biorefinery:1–11

Medha G, Sharmila C, Anil G (2017) Green synthesis and characterization of nanocrystalline graphene oxide. Int Res J Sci Eng 1:29–34

Meka Chufa B, Abdisa Gonfa B, Yohannes Anshebo T, Adam Workneh G (2021) A novel and simplest green synthesis method of reduced graphene oxide using methanol extracted *Vernonia Amygdalina*: large-scale production. Adv Condens Matter Phys 2021:1–10

Mirza FS, Aftab ZEH, Ali MD, Aftab A, Anjum T, Rafiq H, Li G (2022) Green synthesis and application of GO nanoparticles to augment growth parameters and yield in mungbean (Vigna radiata L.). Front. Plant Sci 13:1040037

Naemi S, Meshkini A (2023) Phytosynthesis of graphene oxide encapsulated selenium nanoparticles using *crocus Sativus* petals' extract and evaluation of their bioactivity. J Drug Deliv Sci Technol 81:104286

Ocsoy I, Paret ML, Ocsoy MA, Kunwar S, Chen T, You M, Tan W (2013) Nanotechnology in plant disease management: DNA-directed silver nanoparticles on graphene oxide as an antibacterial against Xanthomonas perforans. ACS Nano 7:8972–8980

Panicker NJ, Sahu PP (2021) Green reduction of graphene oxide using phytochemicals extracted from *pomelo Grandis* and *Tamarindus indica* and its supercapacitor applications. J Mater Sci Mater Electron 32:15265–15278

Parvathi E, Akshaya CV, Dilraj N, Arjun G, Deepak NK (2023) Green synthesis of reduced graphene oxide by shallots. Mater Today Proc

Parvathi E, Arjun G, Akshaya CV, Dilraj N, Deepak NK (2022a) Novel *Mussaenda glabrata* leaves extract for facile green synthesis of reduced graphene oxide with enhanced rhodamine B dye removal efficiency. Sustain Chem Pharm 30:100868

Parvathi E, Dilraj N, Akshaya CV, Deepak NK (2022b) A review on graphene-based adsorbents for the remediation of toxic heavy metals from aqueous sources. Int J Environ Sci Technol 20:1–28

Paul S, Samanta A, Sarkar S, Ghosh C, Nandi DK (2021) Green synthesis of reduced graphene oxide using root extracts of Asparagus racemosus. Curr Nanosci 17:772–778

Perumal D, Albert EL, Saad N, Hin TYY, Zawawi RM, Teh HF et al (2022) Fabrication and characterization of *Clinacanthus nutans* mediated reduced graphene oxide using a green approach. Crystals 12:1539

Punniyakotti P, Aruliah R, Angaiah S (2021) Facile synthesis of reduced grapheme oxide using *Acalypha indica* and *Raphanus sativus* extracts and their in vitro cytotoxicity activity against human breast (MCF-7) and lung (A549) cancer cell lines. 3 Biotech 11:1–11

Rafiq H, Akram W, Anjum T, Mirza FS, Yousuf Z, Li G (2023) A biological evaluation and molecular docking insight on green synthesized graphene oxide nanoparticles mediated growth promotion in mungbean. Sci Hortic (Amsterdam) 318:112097

Rakkimuthu R, Aarthi S, Neelamathi E, Sathishkumar P, Anandakumar AM, Sowmiya D (2022) Green synthesis of reduced graphene oxide silver nanocomposite using *Anisomeles malabarica* (l.) r. br. Leaf extract and its antibacterial activity. Rasayan J Chem 15:417

Ramanathan S, Elanthamilan E, Obadiah A, Durairaj A, Merlin JP, Ramasundaram S et al (2017) *Aloe vera* (L.) Burm. f. Extract reduced graphene oxide for supercapacitor application. J Mater Sci Mater Electron 28:16648–16657

Rodríguez-González C, Velázquez-Villalba P, Salas P, Castaño VM (2016) Green synthesis of nanosilver-decorated graphene oxide sheets. IET Nanobiotechnol 10:301–307

Said M, Ulfa M, Rachmat A, Desnelli D, Hariani PL (2023) Synthesis of reduced graphene oxide from cellulose and its applications for methylene blue adsorption. Solid State Phenom 345:153–170

Sinha S, Andia KC, Devi NA, Swain BP (2022) Structural, optical and electrochemical properties of banana, mango leaves and potato extracts reduced graphene oxide. Bull Mater Sci 45:226

Sultana S, Mech SD, Hussain FL, Pahari P, Borah G, Gogoi PK (2020) Green synthesis of graphene oxide (GO)-anchored Pd/cu bimetallic nanoparticles using *Ocimum sanctum* as bioreductant: an efficient heterogeneous catalyst for the Sonogashira cross-coupling reaction. RSC Adv 10:23108–23120

Tamang S, Rai S, Mondal MK, Bhattacharyya NK, Swain BP, Biswas J (2023) Microwave-assisted reduction of graphene oxide using *Artemisia vulgaris* extract for supercapacitor application. J Mater Sci Mater Electron 34:575

Thanapackiam SP, Selvarajan P, Dhinakar KG, Veeraputhiran V (2022) Biological reduction of graphene oxide using *Citrus sinensis* L. extract and its nano-structural photocatalytic-antibacterial performances. Adv Nat Sci Nanosci Nanotechnol 13:35016

Thiyagarajulu N, Arumugam S, Narayanan AL, Mathivanan T, Renuka RR (2020) Green synthesis of reduced graphene nanosheets using leaf extract of *Tridax procumbens* and its potential in vitro biological activities. Biointerface Res Appl Chem 11:9975–9984

Vatandost E, Ghorbani-HasanSaraei A, Chekin F, Raeisi SN, Shahidi S-A (2020) Green tea extract assisted green synthesis of reduced graphene oxide: application for highly sensitive electrochemical detection of sunset yellow in food products. Food Chem X 6:100085

Wang X, Peng F, Cheng C, Chen L, Shi X, Gao X, Li J (2021) Synergistic antifungal activity of graphene oxide and fungicides against Fusarium head blight in vitro and in vivo. Nanomaterials 11:2393

Xu C, Wang X, Zhu J (2008) Graphene—metal particle nanocomposites. J Phys Chem C 112:19841–19845

Yang J, Xia X, He K, Zhang M, Qin S, Luo M et al (2021) Green synthesis of reduced graphene oxide (RGO) using the plant extract of *Salvia spinosa* and evaluation of photothermal effect on pancreatic cancer cells. J Mol Struct 1245:131064

Yin PT, Shah S, Chhowalla M, Lee K-B (2015) Design, synthesis, and characterization of graphene–nanoparticle hybrid materials for bioapplications. Chem Rev 115:2483–2531

Zaib Q, Kyung D (2022) Optimized removal of hexavalent chromium from water using spent tea leaves treated with ascorbic acid. Sci Rep 12:8845

Zhang M, Gao B, Chen J, Li Y (2015) Effects of graphene on seed germination and seedling growth. J Nano Res 17:1–8

Zhu X, Xu X, Liu F, Jin J, Liu L, Zhi Y et al (2017) Green synthesis of grapheme nanosheets and their in vitro cytotoxicity against human prostate cancer (DU 145) cell lines. Nanomater Nanotechnol 7:1847980417702794

Chapter 15
The Role of Green Synthesized Nanoparticles in Biotic Stress Resistance in Vegetables

Maria Ehsan, Naveed Iqbal Raja, Zia-ur-Rehman Mashwani, Noor Fatima, Fozia Abasi, Feroza Hamid Wattoo, and Muhammad Iqbal

Abstract The primary factor that restricts plant development as well as growth is biotic stress. Creating biotic stress-resistant biotech cultivars can boost agricultural output and profitability. Increasing their produce and improving their nutritional security can give farmers more power. Through the use of genes and biotechnological techniques, resistant transgenic crops are altering agriculture. However, it is crucial to assess the results of fresh biotic stresses that have emerged recently and to come up with inventive solutions for their management. With numerous uses in agriculture and the food chain, nanotechnology has recently gained appeal as a tactic to boost plant product development. Biotic stressors refer to negative factors instigated by pathogenic insects, fungi, viruses, bacteria, or nematodes that result in economic losses by decreasing the shelf life, the quality of products, or possible yields. The use of nanotechnology can be utilized in a variety of ways to boost plant production, preserve plants, and store, package, and transport commodities from agriculture. It can also provide safe, efficient, and environmentally friendly methods for managing the spread of plant diseases. The use of nanoparticles to protect plants is one of the most recently developed methods for effectively managing microorganisms, particularly fungal and bacterial diseases. The best examples of nanopar-

M. Ehsan · N. I. Raja (✉) · Zia-ur-Rehman Mashwani · F. Abasi
Department of Botany, PMAS-Arid Agriculture University, Rawalpindi, Pakistan
e-mail: drnaveedraja@uaar.edu.pk

N. Fatima
Department of Botany, Lahore College for Women University, Lahore, Pakistan

F. H. Wattoo
University Institute of Biochemistry and Biotechnology, PMAS-Arid Agriculture University, Rawalpindi, Pakistan

M. Iqbal
Department of Botany, University of Chakwal, Chakwal, Pakistan

M. Khan, J.-T. Chen (eds.), *Nanoparticles in Plant Biotic Stress Management*,
https://doi.org/10.1007/978-981-97-0851-2_15

ticles with remarkable antibacterial action against multiple plant diseases are formed from silver, copper, silicon, and chitosan. A variety of nanoparticle types have been brought to the field of battling plant diseases. This chapter explains how nanobiotechnology can be applied to control biotic stressors and safeguard vegetables.

Keywords Nanotechnology · Disease stress · Agriculture · Nanomaterials

15.1 Introduction

There will likely be 10 billion people on the earth in 2050, 800 million of whom will lack access to food and 653 million will be underweight. As a result, meeting the demand for food around the world will be quite challenging (Giuntoli et al. 2023). Due to current scientific developments and disease management strategies, there won't be enough food to go around by 2050, and the First Green Revolution, which had a substantial impact on yield and food supply, has lately stagnated (Katz-Rosene et al. 2023). To satisfy the population's need for food, a Second Green Revolution is necessary. To control diseases of plants and stop phytopathogen-disease-causing organisms' establishment before and after crop harvest, the farming system mostly uses chemicals (Richard et al. 2022). Farmers follow numerous strategies to lessen the plant disease effects. When used excessively, commercial fungicides, herbicides, and pesticides, typically used for the treatment of plant diseases, have adverse effects on the health of humans and ecology (Gill and Garg 2014). Numerous commercial pesticide utilization results in increased disease and insect resistance, a decrease in nitrogen fixation, and harmful pesticide bioaccumulation (Ali et al. 2021).

An example of a synthetic commercial pesticide is DDT. However, to humans, it was later discovered as genetically toxic, producing endocrine problems. It was widely employed in agriculture to prevent plant infections (Sabarwal et al. 2018). Commercial fungicides, herbicides, and pesticides are among the chemicals that are widely and improperly used, and their use pollutes the water and soil and has a negative influence on both the environment and human health (Jayakumar et al. 2023). Chemical pesticides are being used more frequently, which has increased disease and insect resistance, decreased nitrogen fixation, and resulted in the bioaccumulation of harmful chemicals (Saini 2023). Due to their interactions with pathogens in soil/plant systems, these substances significantly contribute to the eradication of illnesses, particularly by creating a soil that is disease-suppressive (Hussain et al. 2023). Plant defense responses, which are mediated by SMs generated by certain bacteria, result in systemic resistance to pathogen invasion (Adedayo and Babalola 2023). There are currently over 1300 nanomaterials, and they can be used in many different ways. Nanoparticles can enter through stomatal pores and pass through the plasma membrane since the plasma membrane is permeable to macromolecules as tiny as 3.5–20 nm. However, the presence of stomata at the lower surface makes it difficult for nanoparticles to penetrate (Wang et al. 2023).

At every stage, including pre-harvest preparation, postharvest storage, transit, sale, and consumption in residences and commercial establishments, fresh produce is susceptible to pathogen contamination (Alegbeleye et al. 2022). Some of the pollution before harvest season include the use of wastewater in irrigation, soil, pesticides, dust, fungicides, compost, insects, and excrement from wild and household faunae (Piližota 2023). Postharvest pollution can also be caused by harvesting tools, trucks, cold water, rinsing water, pests, transportation packaging, and human handling. Infections are reduced or completely eradicated by the application of pesticides, while resistance to infections is increased by the distribution of fertilizer (Macieira et al. 2023). Since typical insecticide and manure preparations frequently comprehend dynamic constituents with less water solubility and restrictive yield access, cultivators are frequently required to utilize bigger measurements or amounts to control infections and achieve a sufficient output (Tudi et al. 2023). For agriculture to be sustainable, current nutrition sources must be modified, and new nutrient delivery methods must be created. Delivering a concentrated supply of plant nutrients with greater effectiveness, application, and absorption while reducing losses is now possible, thanks to nanotechnology (Yadav et al. 2023).

By delivering novel tackles to enhance the capacity of plants for nutrient absorption, resistance from disease at the molecular level and cure, and fast discovery of disease, nanotechnology has the potential for food transformation and farming businesses (Pramanik et al. 2020). Potential applications of nanotechnology in various fields have been highlighted in Fig. 15.1. Using environmentally friendly alternative energy sources, filters, and catalysts may also provide ecological protection indirectly by lowering contamination and getting rid of poisons that are already there. To find out how nanotechnology might be used to prevent disease, many studies have been done (Beni and Jabbari 2022). However, there hasn't been a single thorough analysis of the part that nanomaterials play in protecting fruits and vegetables from diseases. Scientific research on nanomaterials' role in preventing illness spread in vegetables is scant and poorly defined (Vithanage et al. 2023). This is true even though these new techniques for processing fruits and vegetables have numerous positive consequences, such as disease prevention, an increase in the amount of time

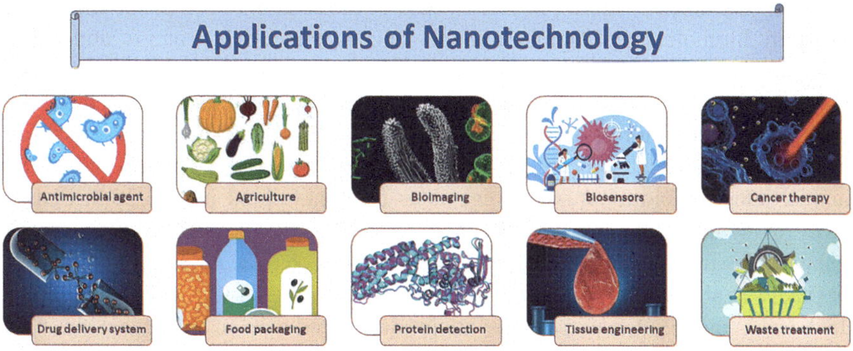

Fig. 15.1 Potential applications of nanotechnology in various fields

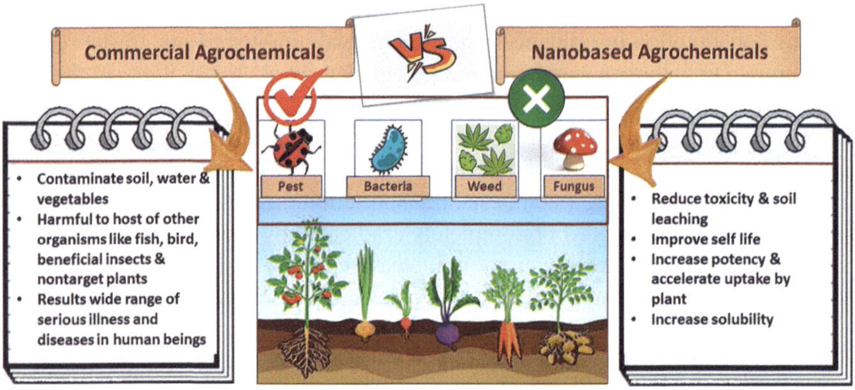

Fig. 15.2 Comparison of commercial and nano-based agrochemicals against biotic stress in vegetables

they may be preserved, nontoxic nano-packing, and other advantages that follow the "green" processing philosophy (Jalal et al. 2023). A comparison of commercial and nano-based agrochemicals on vegetables against biotic stress is highlighted in Fig. 15.2.

15.2 A Greener Way of Preparing Nanoparticles

To produce food, scientists actively attempt to develop farming techniques that are both resource and environmentally friendly. New technologies, like bionanotechnology, are crucial to achieve maintainable environmentally favorable nutrition production. Since it is more affordable and environmentally responsible than physical and chemical processes, green nanoparticle production has gained interest (Jha and Jha 2023). The green production of nanoparticles can be achieved by replacing traditional methods with living organisms. Using plants instead of chemicals to make nanoparticles is more preferred (Devi et al. 2023). To aid in the metal ion reduction, biomolecules like alkaloids, flavonoids, phenolics, proteins, and terpenoids may be utilized for the environmentally friendly creation of nanomaterials. Due to their environmentally friendly nature, using live creatures to make nanoparticles is becoming more and more popular (Dhaka et al. 2023) compared to using chemical or physical processes. The advantages of using green synthesized nanoparticles for biotic stress in vegetables are highlighted in Fig. 15.3. Plant's ability to synthesize nanoparticles on a massive scale, producing more robust nanoparticles in a variety of forms and sizes, has attracted the attention of researchers (Sadiq et al. 2023). A lot of metal oxides such TiO_2, Ag_2O, CuO, Au_2O, CeO_2, MnO, CaO, and Fe_3O_4 nanoparticles are produced and used in significant amounts (Bordiwala 2023).

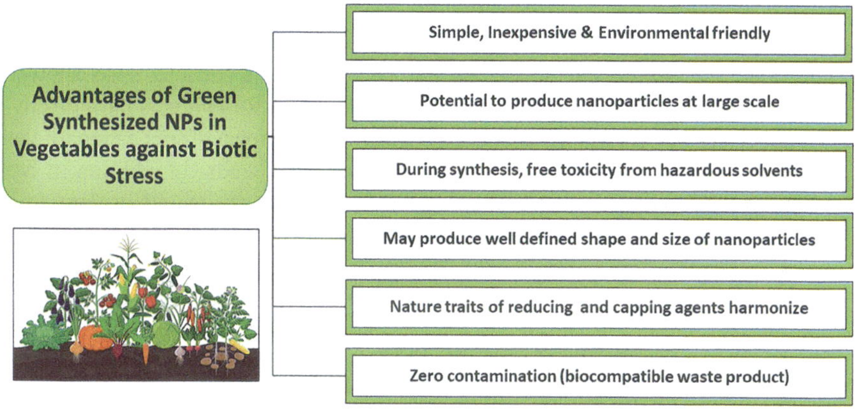

Fig. 15.3 A greener way of preparing nanoparticles for biotic stress in vegetables is beneficial

A sustainable and environmentally friendly process for creating nanoparticles has been developed using a range of plants. For instance, *Trifolium pratense* flower extract has been used to create ZnO nanoparticles that exhibit a wide range of antibacterial activity, including activity against *Pseudomonas aeruginosa* (Adraoui et al. 2023). Other plant species, such as *M. oleifera* and *A. linearis*, are employed for zinc oxide nanoparticle synthesis (Mirza et al. 2023). As evidence of their potential for use in medicine, silver nanoparticles (Ag NPs) generated from *Sageretia thea* have demonstrated excellent bactericidal effects in contradiction to both Gram-positive (like *S. aureus*) and Gram-negative (like *E. coli*) bacteria (Bruna et al. 2021). The usage of bio-synthesized nanoparticles (NPs) may have significant advantages for agriculture. Using the extract of the plant, nanoparticle preparation is more cost-effective, eco-friendly, and beneficial than using synthesis at a higher scale (Alsaiari et al. 2023). Although employing viruses to create nanoparticles is somewhat unusual, researchers have successfully created Au and Ag NPs using tobacco mosaic virus (TMV), both recombinant and wild-type (Roy et al. 2023).

FeO and Fe-Pt nanoparticles can be produced using the cowpea and brome mosaic virus. Nanotechnology offers a more environmentally friendly method of synthesizing nanoparticles than physical and chemical methods, making it potentially a green technology (Zahmanova et al. 2023). With an emphasis on clean, environmentally friendly methods, bionanotechnology can become a "green" expertise. The anti-pesticidal activity of metallic oxide-based nanoparticles contrary to numerous vegetable diseases has been documented in several researches (Namasivayam et al. 2023). For instance, *Alternaria citri*, which causes citrus black rot, was sensitive to biogenically produced ZnO nanoparticles made from lemon peel extract. Similar to this, ZnO nanoparticles biosynthesized from *Cinnamomum camphora* leaf extract were ineffective against *A. alternata*, resulting in early blight disease in *S. lycopersicum* (Munir et al. 2023). In addition to acting as fungicides against *Aspergillus terreus*, *Sclerotium sclerotia*, *Fusarium solani*, and *Fusarium oxysporum*, *Penicillium chrysogenum*-produced ZnO nanoparticles also showed activity against these other fungi (Cruz-Luna et al. 2023). Researchers have discovered a novel strategy for plant protection in which photoactivated ZnO nanoparticles can destroy *Escherichia coli* and *F. oxysporum*-infected seeds. In addition, it has been shown that photoactivated ZnO nanoparticles increase crop yield, prolong fruit shelf life, and decrease *B. cinerea* occurrences in strawberries (Decelis et al. 2017; Kalia et al. 2020).

15.3 Bionanotechnology for Management of Plant Diseases

Nanotechnology has been used in applications that enhance plant health and handle disease control difficulties. For plant disease management, NPscan be used as transporters, protectors, and for RNAi-based plant defense (Dong et al. 2023). Nanoparticles have the potential to lessen biotic stress in agriculture by augmenting plant defense mechanisms, improving plant nutrition, and lowering pathogen infections. Nanotechnology has been used to increase disease resistance in plants by using nanoparticles to regulate diseases and improve plant growth (Bala et al. 2023). Nanotechnology in disease stress management is illustrated in Fig. 15.4. Nanoparticles have been included in disease prevention strategies such as bactericides, fungicides, and nanofertilizers to enhance plant health. Because of their antibacterial properties, biological molecules and NPs could employed in a range of areas to eliminate bacteria, fungi, and yeast that cause diseases (Humbal and Pathak 2023).

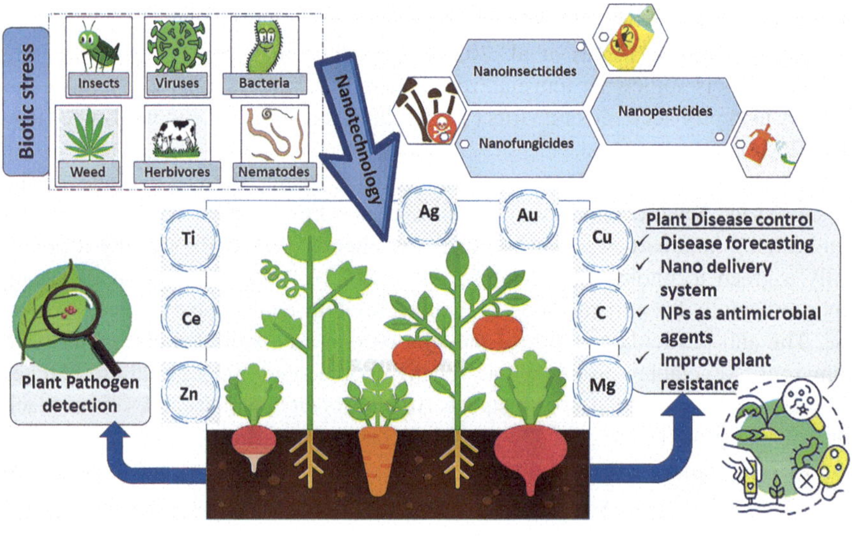

Fig. 15.4 Nanotechnology in the management of diseases in vegetables

15.3.1 Nanopesticides

On the market, numerous pesticides could be utilized for the elimination, or stop the spread, of pests, but they can accumulate in food and be dangerous for the environment and people's health. Nanotechnology is a promising alternative to conventional pesticides for managing plant diseases since it provides effective and ecologically acceptable methods (Naseer et al. 2023). Nanoparticles can be used as nano-pesticides, nano-bactericides, nano-fungicides, and nano-insecticides for the controlled release of pesticides in agriculture. Pesticides can accumulate in food and get into the food chain, putting public and ecological health in jeopardy (Dhoundiyal et al. 2023). Bioremediation and the use of biopesticides are the two environmentally friendly management strategies that are now necessary due to pesticide misuse, exposure, and adverse effects. Biochemical pesticides, which are naturally occurring compounds that control pests by nontoxic procedures, are less dangerous than chemical pesticides (Fenibo et al. 2022). Nanopesticides have two advantages over conventional pesticides: smaller active component particles and the ability to encapsulate the active ingredient to make it responsive to the environment. As pesticide technology advances, some registrants are utilizing nanoparticles to meet their objectives (Paradva and Kalla 2023).

Green nanotechnology focuses on environmentally responsible techniques to create nanoparticles, which can reduce the need for hazardous chemicals in farming. Nanoparticles can be used in agriculture to increase plant nutrition and growth, diagnose diseases, manage pests, and practice sustainable agriculture (Anandhi et al. 2020). Use of silica nanoparticles have been suggested as an effective water-loving distribution method for a measured discharge of pesticides. Both the deterrence and treatment of vegetable illnesses have demonstrated amazing potential for nanoparticles (Munir et al. 2023). Chitosan nanomaterials (ChNMs) positively affect the plants' innate immune responses. Treatment with chitosan nanoparticles boosts the natural immune system of plants by triggering defensive enzyme responses, upregulating defense genes and antioxidant enzymes, and enhancing the accumulation of nitric oxide (NO) and phenolics. Chitosan nanoparticles can be successfully used as a phytosanitary and disease-controlling agent in sustainable organic cultivation (Mukarram et al. 2023).

Sulfur nanoparticles (SNPs) have been employed as a green nano-pesticide to fight against phytopathogens like *Venturia inaequalis* and *Fusarium solani*. It has also been demonstrated that *A. nerii* insects are vulnerable to the insecticidal effects of silver nanoparticles (Hazarika et al. 2022). These nanoparticles provide promising crop plant pest management solutions. It is feasible to produce nanoparticles such as gold nanoparticles and iron oxide nanoparticles and employ them as vegetable fertilizers and pesticides (Hu and Xianyu 2021). They can also be employed as drug-delivery particles in human beings (Nallasamy et al. 2023). These nanomaterials can release therapeutic agents or active ingredients to treat various stresses on crop plants (Hyder et al. 2023).

15.3.2 Nano-Fungicides

Due to their potent antibacterial properties, metal nanoparticles can be used as fungicides and growth promoters in the cultivation of plants. Silver nanoparticles are used regularly to treat a variety of plant illnesses, as they cause less harm to people and faunae as compared to commercial antifungals (Malandrakis et al. 2022). Fungicides containing copper have the potential to produce extremely reactive hydroxyl radicals that could harm biomolecules. Chitosan nanoparticles can be used in sustainable organic farming to control illness and uphold phytosanitary conditions since they have a positive impact on plants' innate immune systems (Saritha et al. 2022). Metal nanoparticles have been employed successfully in agriculture to control phytopathogenic fungi, offering a potential way to lessen biotic stress in plants (Sonawane et al. 2022). Copper and zinc oxide nanoparticles have demonstrated potential in agriculture as fungicides and bactericides. Numerous plant diseases caused by bacteria, fungi, and oomycetes have been combated using copper nanoparticles (Jabri et al. 2022).

The use of copper nanoparticles may be advantageous for the agroecosystem because they did not affect advantageous bacteria. ZnO-NPs have been utilized in many biological applications, like antimicrobial agents (antifungals and antibacterials) in farming and food (Yuan et al. 2023). They have been found to promote ROS production results in cell death, enhance the growth and metabolism of plant, and activate the plant defense system. Due to characteristics including an increased ratio of surface and volume and tiny size, allowing effective contact with microorganisms, zinc oxide nanoparticles have demonstrated a more favored microbicidal activity than bulk zinc particles (Chandrasekaran et al. 2022).

The antifungal effectiveness of Au-NPs from *Abelmoschus esculentus* has been demonstrated against *P. graminis*, *C. albicans*, *A. flavus*, and *A. niger* (Upadhyay et al. 2022). The conventional well diffusion method also demonstrated the fungicidal potential of gold nanoparticles against *A. flavus*, *A. niger*, *C. albicans*, *P. graministritci*, and *P. graminis*. Au-NPs had the largest inhibitory zone when compared to other NPs (Dutta et al. 2021). Similar antifungal action was seen when *Magnaporthe oryzae*, *P. digitatum*, *Sclerotium rolfsii*, *Rhizoctonia solani*, *Botrytis cinerea*, *Colletotrichum graminicola*, and *Colletotrichum musae* were exposed to CuO-NPs (Bind et al. 2023). Additional metallic oxide-based nanoparticles, including ZnO, MgO, TiO_2, and Si, exhibit fungicidal potential against a variety of fungi (Salem and Fouda 2021).

The antifungal activities of MgO-NPs were ineffective for *Phomopsis vexans* (Rabea et al. 2023). On *Alternaria solani*, silica nanoparticles have antifungal activity. *Fusarium oxysporum* and *Fusarium radicis-lycopersici* were successfully repressed by SiO_2-NPs (Albalawi et al. 2022). Additionally, the antifungal action of Si-Ag-NPs was felt by *Magnaporthe grisea*, *Xanthomonas campestris*, *Colletotrichum gloeosporioides*, *Botrytis cinerea*, *Pythium ultimum*, *Rhizoctonia solani*, and *Pseudomonas syringae* (Salam et al. 2022). Similar to this, ZnO-NPs have demonstrated remarkable antifungal potential against *A. niger*, *P. expansum*, *A. alternata*, *F. oxysporum*, and *B. cinerea* (Yassin et al. 2023).

15.3.3 Nano-Bactericides

Various metal nanoparticles like Ag, Cu, TiO_2, and ZnO were extensively studied for their antiviral and antibacterial activities (Shweta and Sood 2023). For instance, the antibacterial potential of silver nanoparticles has been studied against *S. aureus*, *K. pneumonia*, *E. coli*, and *B. subtilis*. *E. coli*, *B. subtilis*, and *P. aeruginosa*, three dangerous food-borne pathogens, were repressed by silver nanoparticles. *P. syringae* was reportedly decreased by silver-chitosan composite as well as silver NPs (Parmar et al. 2022). Gold nanoparticles demonstrated antibacterial activity against *E. coli*. Other metal oxide nanoparticles (NPs) are ineffective against a variety of dangerous bacterial species (Rana et al. 2023). These include ZnO-NPs for the treatment of *Ralstonia solanacearum*, *E. coli*, *Candida albicans*, *P. aeruginosa*, *Aspergillus flavus*, and *Aspergillus niger*. When applied to plant follicles before and after being injected with the bacteria, zinc oxide nanoparticles, whether singly or coupled with nickel, iron, manganese, or copper metals, stopped *Pantoea ananatis* disease spread in maize. Additionally, *M. incognita*- and *P. syringae*-causing bacterial blight infections of pea plants were successfully managed by ZnO-NPs (Pena et al. 2022).

Increased rhizospheric microbial diversity, decreased incidence of *Ralstonia solanacearum*-related illnesses, and improved tomato plant development and antioxidant response are all benefits of adding ZnO-NPs to the soil (Kalaba et al. 2021). Also, in tomatoes, zinc oxide nanoparticles, produced by the flower extract of *Matricaria chamomilla*, were discovered to have antibacterial potential against *R. solanacearum* and to lessen wilting disease. *Aspergillus niger*, *Candida albicans*, *Bacillus subtilis*, *P. aeruginosa*, *Streptomyces sannanensis*, and zinc oxide nanoparticles produced by the peel of *Citrus medica* were also successful strains (Tortella et al. 2021). Using greener zinc oxide nanoparticles produced from *T. harzianum*, *T. reesei*, co-culture, and the strain Sx3 of *Paenibacillus polymyxa*, *oryzae*, which causes bacterial leaf blight in rice, was suppressed (Zaki et al. 2021). Plants with foliar spraying experienced faster growth and fewer bacterial leaf blight infections. Infection on the sugar beet, apple scab disease brought on by *Venturia inaequalis*, *Fusarium* wilting in potato and tomato brought on by *F. solani*, bacterial blight on geraniums, and leaf spotting on poinsettias brought on by *Pelargonium* and *Xanthomonas* pv. *poinsettiicola* all appeared to be suppressed by TiO_2-NPs (Parveen and Siddiqui 2022).

15.3.4 Nano-Nematicides

About 20% of agricultural losses are attributable to plant-parasitic nematodes. Nematodes hinder the growth of crops, especially in tropical and subtropical areas. Over 100 identified species of *Meloidogyne* are the most harmful root-knot nematodes (RKNs) (Baronia et al. 2020). RKNs are thought to be responsible for a $100 billion loss per year globally. RKNs are challenging to manage due to their wide host range and robust reproductive capacity. Crop rotation, biocatalysis, keeping land fallow, breeding resistant plant types, and other conventional techniques are used to control nematodes (Abrantes et al. 2023). Nematodes are employed to get rid of huge plant-parasitic nematodes despite being extremely toxic and damaging to the environment. NPs have been demonstrated to be efficient nematicides against a variety of parasitic plant nematodes because of their multi-site manner of action. For instance, toxic and artificial nematicides are less successful at killing nematodes than Au- and Ag-NPs (Sabry 2019).

Expanding applications of commonly available nanotechnology materials are being used to fight plant diseases brought on by RKNs like *M. incognita*. However, it was shown that *M. incognita* J2 and her eggs were unaffected by silicon-carbide nanoparticles at 172 ppm concentration and 50 nm size (Nazir et al. 2019). However, SiC nanoparticles had a significant impact on the development of first-stage *C. elegans* larvae (Fabiyi et al. 2020). Additionally, ZnO-NPs have antibacterial qualities and are effective against a variety of bacterial, fungal, and nematode-related plant diseases. TiO_2-NPs are also efficient against viruses and nematodes (Rudramurthy and Swamy 2018). In tomato, titanium oxide shows antinematicidal potential against *M. incognita*. Furthermore, TiO_2-NPs had an insecticidal effect on tomato plants that were infected with *Bactericera cockerelli* (Sulc). Ninety-three percent of the insects were wiped out by TiO_2-NPs delivered at doses greater than 100 ppm after 24 h (Zinovieva et al. 2023).

The rate of photosynthetic activity is accelerated when tomato plants are treated with TiO_2-NPs. Numerous physiological and metabolic processes in plants depend on silicon, a key element. SiO_2 nanoparticles enhance seed germination and seedling growth of *Agropyron elongatum* (Wahab et al. 2023). Nano-SiO_2 can be used to make fertilizers that are good for crops and prevent fertilizer loss through delayed and regulated release, enabling controlled and timely delivery. The *Cucurbita pepo* plant displayed improved seed germination following treatment with nano-SiO_2. The ability of tomato plants to create oxygen is increased when nano-Si is applied (Bairwa et al. 2023).

Copper is an important component for the growth and metabolism of plants and is an essential component for the efficient functioning of several plant enzymes. Electronics, catalysts, gas sensors, and antibacterial agents all use copper nanoparticles (Cu-NP). The photosynthesis of the *Spinacia oleracea* plant was improved by CuO-NPs. The antioxidant system of the *Lactuca sativa* was improved by $Cu(OH)_2$NPs (Manjula et al. 2022). The *Zea mays* plant benefits from $Mg(OH)_2$NP treatment in terms of germination and seedling development. The antioxidant system of *Raphanus sativus* is strengthened by MgO-NPs. It has been demonstrated that MgO-NPs can increase tomato plant resistance and aid in the development of the tobacco plant (Vaishnavi et al. 2023). AgNPs increase biomass and development of the *Vigna sinensis*. Apple trees are shielded by chitosan nanoparticles after harvest (Pattanayak et al. 2023). In the *Robusta coffee*, chitosan NPs promote plant development (Maluin et al. 2023).

15.4 Disease Stress

A few examples of physical barriers are the cell wall, stratum corneum, and wax that viruses can effectively penetrate to infect the host. To strengthen the mechanical strength of plants, silica is deposited in the stratum corneum and leaf epidermis (Sun et al. 2023). After silicon treatment, rice has fewer *P. oryzae* penetrative appressorial sites because the thick silicon coating that results on the grain's surface limits or delays pathogen penetration. To boost the leaf epidermal's resistance to soft rot (*Dickeya dadantii*), SNPs develop a layer of silica (Du et al. 2022). To lower the frequency of *P. syringae* infections in *Arabidopsis*, SNPs may cause SAR. By lowering ROS generation and NO levels, SNP therapy may decrease the negative effects of infection. SNPs increase plants' resistance to diseases, yet it is not yet apparent how this reaction is mediated at the cellular level (El-Shetehy et al. 2021). Nanotechnology as a futuristic approach for combating biotic stress in vegetables is illustrated in Fig. 15.5.

A crop's resistance to a variety of illnesses, such as nematocidal (*Meloidogyne incognita*), bacterial (*Pseudomona dadantii*), and fungal diseases (*A. niger*, *A. solani*, and *F. oxysporum*) can be significantly improved by SNPs. SNPs increase maize's resistance to *A. niger* and *F. oxysporum* (Kralova and Jampilek 2023). By controlling the salicylic acid signaling system, SNPs cause *Arabidopsis* to develop a protective resistance to *P. syringae*. Downregulation of RANI1, CSD1, PPO, and PAO in *Citrullus vulgaris* roots and root-dipping actions with mesoporous SNPs decrease 40% of disease harshness in watermelon (Wang et al. 2022).

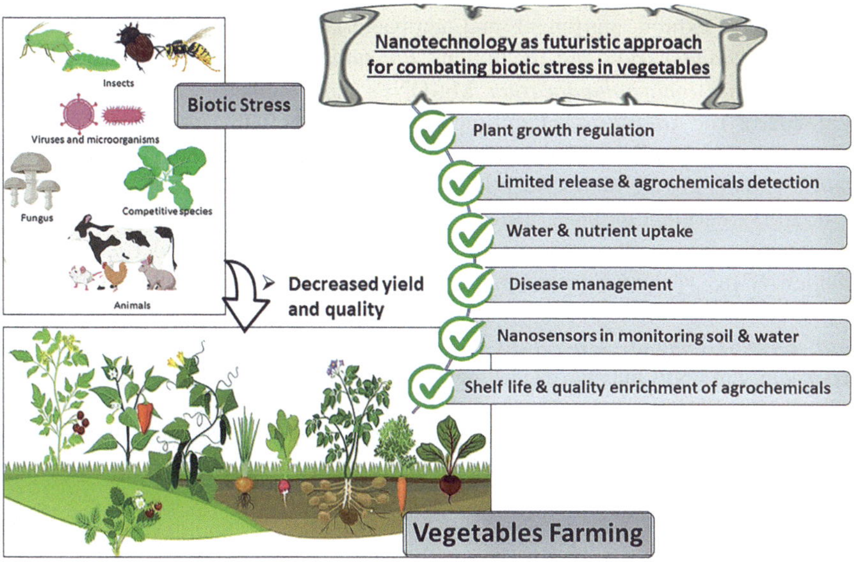

Fig. 15.5 Nanotechnology as a futuristic approach for combating biotic stress in vegetables

Nanomaterials can be used to get rid of bacteria. Antibacterial and antifungal activities are present in silver nanoparticles (Paul and Roychoudhury 2021). It has been discovered that silver nanoparticles have antimicrobial potential against *E. coli*. Both Gram-positive and Gram-negative pathogens are susceptible to AgNPs' antibacterial effects. *Bacillus subtilis*, *E. coli*, and *Pseudomonas aeruginosa* are examples of food-borne pathogens that AgNPs exhibit antibacterial action against (Singh et al. 2023). *Penicillium digitatum*, *Alternaria citri*, and *Alternaria alternata* fungi are all susceptible to the antifungal effects of AgNPs. The fungi that are resistant to the silver nanoparticle's antifungal effects include *Alternaria alternata*, *Macrophomina phaseolina*, *Botrytis cinerea*, *Sclerotinia sclerotiorum*, *Curvularia lunata*, and *Rhizoctonia solani* (Ahmed et al. 2023).

Silver-chitosan and silver nanoparticle composites are ineffective against *Pseudomonas syringae* pv. *syringae*. The fungi *Alternaria alternata* and *Macrophomina phaseolina* are resistant to the antifungal effects of chitosan NPs (Dos Santos et al. 2021). Chitosan NPs exhibit antifungal activity against *Aspergillus niger* and *Fusarium solani*. Against the *E. coli* bacterium, AuNPs exhibit antibacterial effectiveness. Against *E. coli* bacteria, ZnO and AuNPs have antibacterial activity (Madkhali 2023). In contrast to macro gold, the *Salmonella typhimurium* bacterium was adversely affected by the gold nanoparticles. Using the conventional well diffusion method, the AuNP's fungicidal potential was evaluated against *Candida albicans*, *Aspergillus niger*, *Puccinia graminis*, and *Aspergillus niger*. Against *P. graminis* and *C. albicans*, the AuNPs showed the broadest zone of inhibition (Mobed et al. 2021). *Botrytis cinerea*, *Colletotrichum graminicola*, *Rhizoctonia solani*, *Colletotrichum musae*, *Magnaporthe oryzae*, *Penicillium digitatum*, and *Sclerotium rolfsii* are all susceptible to the antifungal properties of CuO-NPs (Khan et al. 2021).

MgO-NPs exhibit antifungal and antibacterial action against the bacteria *Ralstonia solanacearum* and the fungus *Phomopsis vexans*, respectively. When exposed to *Ralstonia solanacearum*, Mg-ONPs show antibacterial activity (Rabea et al. 2023). The *Alternaria solani* fungus is susceptible to the antifungal effects of silica nanoparticles. *Fusarium oxysporum* f. sp. *lycopersici* and *F. oxysporum* f. sp. *radicis-lycopersici* are both susceptible to SiO_2-NPs' antifungal properties (Munir et al. 2023). *Pythium ultimum*, *Magnaporthe grisea*, *Colletotrichum gloeosporioides*, *Botrytis cinerea*, *Rhizoctonia solani*, *Pseudomonas syringae*, and *Xanthomonas campestris* pv. *vesicatoria* are among the fungi against which nano-Si-Ag has shown antifungal efficacy (Ramezani et al. 2019). On the mold, *Alternaria alternata*, *Botrytis cinerea*, *Aspergillus niger*, *Fusarium oxysporum*, and *Penicillium expansum*, ZnO-NPs may have antifungal properties. *Aspergillus flavus*, *Aspergillus niger*, *Aspergillus fumigates*, *Fusarium culmorum*, and *F. oxysporum* are all susceptible to ZnO-NPs' antifungal properties (Slavin and Bach 2022). *Aspergillus flavus*, *Trichoderma harzianum*, *Aspergillus nidulans*, and *Rhizopus stolonifer* are four fungi that ZnO-NPs may be able to inhibit (Daniel et al. 2023).

ZnO nanoparticles were produced using *Aeromonas hydrophila*, and their effectiveness against pathogenic bacteria like *P. aeruginosa* and fungi like *C. albicans*, *A. flavus*, and *A. niger* was evaluated (Gauba et al. 2023). *Fusarium oxysporum* f. sp. *radicis-lycopersici* and *Fusarium oxysporum* f. sp. *lycopersici* are both susceptible to TiO_2-NPs' antifungal properties. TiO_2-NPs are effective against *Cercospora beticola* as antifungals (Farahat et al. 2023). The worm *Meloidogyne incognita* is subjected to nematicidal effects by AccoTiO_2 nanoparticles. Because they can taint stored food and kill crops, pests are a serious problem in agriculture because they reduce the quality of the food produced and promote the spread of plant diseases (Danish et al. 2021). Chemicals can be distributed by SNPs, release control is improved, and insects can be harmed. Despite having strong antifeedant effects, many compounds have drawbacks like less shelf life, photo capability, limited accessibility, and usage challenges (Wang et al. 2022). However, using pesticides with SNPs added significantly increases their efficacy. SNPs and plant extracts can be combined to improve the insecticidal effectiveness and shelf life of these substances (Qasim et al. 2022).

Helicoverpa armigera larvae's growth was reduced by the trypsin inhibitor and SNPs. *Spodoptera littoralis* (Bosid.), *Sitophilus oryzae* (L.), *Rhyzopertha dominica*, *Tribolium castaneum*, and *Oryzaephilus surinamensis* are all immediately poisonous when exposed to SNPs. Ninety percent of *Sitophilus oryzae* mortality is caused by SNPs (Sharma et al. 2023). The insect's cuticle acts as its initial line of defense against outside intruders. In the *stratum corneum*, SNPs can bind with cuticles, dissolve the waxy coating barrier, and absorb water from the body of the insect, causing dehydration and death. This is made possible by the SNPs' huge specific surface area. By abrading the wall and limiting stomata, silicon nanoparticle treatment raises the diamondback moth's (*Plutella xylostella* L.) death rate to 85% (El-Monairy et al. 2023). By changing the values of peats' life cycle features, such as mine quantity, existence rate, and puparium mass, silicon nanoparticles may also be able to regulate pests. As a result, SNPs show enormous promise for pest management and may eventually replace conventional insecticides (Rana et al. 2021). Hence, a brief advantage of using nanotechnology in vegetable farming is illustrated in Fig. 15.6.

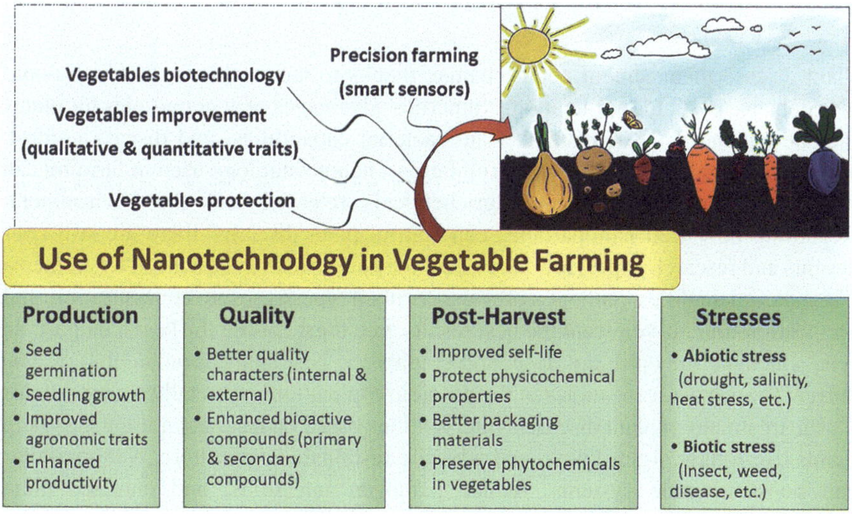

Fig. 15.6 Advantages of using nanotechnology in vegetable farming

15.5 Toxicity of Nanoparticles toward Humans

The toxicity of NPs, which may injure plants and other living things, is a significant problem in agriculture. Even though nanotechnology may be utilized in agriculture to make nanofertilizers, nano-herbicides, nano-fungicides, and nano-sensors, the detrimental influences of NPs on ecology and living beings are not well established (Altabbaa et al. 2023). Even though most fruits and vegetables are cleaned before eating, nanoparticles can diffuse and propagate past organic barriers, having unexpected consequences during touch with biological elements like genetic material, proteins, and cells (Yadav et al. 2023). Two significant possible toxicities linked to nanoparticles are genotoxicity and carcinogenic toxicity. Nanoparticles that are genotoxic and carcinogenic can alter and damage DNA, which could result in the occurrence of cancer (More et al. 2021). It has been demonstrated that nanoparticles pass through the intestinal lumen and enter the bloodstream, where they then go to other organs (Lee et al. 2023).

Carbon-containing nanomaterials, such as carbon nanotubes, are associated with cardiovascular issues because they promote arterial thrombosis, degrade aortal genetic material, promote platelet accretion, and elicit provocative responses (Lead et al. 2021). Depending on the organ, nanoparticle toxicity and reactions can differ significantly. To assess the toxicological effects on a range of organs, a complete knowledge of the relations and comebacks of living processes at the cellular, molecular, and physiological stages is crucial (Gurram et al. 2023). Nanoparticles' harmful impacts can change blood pressure and heart rate; increase the chance of cardiac strokes, coagulation, and myocardial infarctions; and decrease a person's life span in all of these ways (Smith and Edelman 2023).

15.6 Conclusion and Future Prospects

Plant disease management could change, thanks to nanotechnology, which could also enhance pesticide application. Numerous advantages of nanoparticles for plants include enhanced nutrient uptake, antibacterial capabilities, and disease control. When compared to conventional formulations, nanotechnology used in farming can decrease metals' number and agrichemicals released into the environment. Regarding how well nanoparticles can control plant diseases, there are still constraints and research gaps. How successful they are could be influenced by elements like material qualities, pathogen prevalence, plant type, exposure concentration, and application time. If you want the best results, you must choose the best nanoparticle type, calculate the optimum dose, and employ the best application technique. The current years have seen studies of the possible applications of metallic nanoparticles for the treatment of plant diseases. Interventions based on nanotechnology may help plants live longer, fight illness, and advance sustainable agriculture. Nanoparticles can boost immune systems, reduce pathogen infections, and enhance plant nutrition. To protect plants from bacterial and fungal crop diseases, nanoparticles' antibacterial characteristics can be used to make nano-pesticides.

All things considered, nanotechnology holds tremendous promise for improving plant health and disease control in agriculture, but more studies are required to properly comprehend and maximize its applications. Therefore, widespread use of these nanoparticles could result in metal pollution of the soil. A focused, efficient delivery mechanism must be created, and the dose must be carefully examined before use. If the objective of reducing the ecological influence of nanoparticles as agrochemicals is to be achieved, standardizing techniques for evaluating the physicochemical properties of nanoparticles will be crucial. Before releasing nano-pesticides for widespread agricultural use, it is crucial to assess their economic viability, societal acceptance, and regulatory compliance. Without these comparisons, it is challenging to make specific claims about the efficacy and economic viability of nano-enabled agrochemicals. Given the low profit margins associated with farming and nutrition manufacture, novel techniques must be just as efficient and cost-effective as those now in use. The development and implementation of nano-based farming techniques will play a vital part in the achievement and maintenance of global food safety and security.

References

Abrantes I, Almeida MT, Conceição IL, Esteves I, Maleita C (2023) Nematodes of potato and their management. In: Potato production worldwide. Academic Press, pp 213–240

Adedayo AA, Babalola OO (2023) Fungi that promote plant growth in the rhizosphere boost crop growth. J Fungi 9(2):239

Adraoui I, Mamouni R, Saffaj N, Achemchem F (2023) Eco-friendly synthesis of zinc oxide nanoparticles using saffron extract and their photocatalytic and antibacterial activities. J Mater Res 38:1–11

Ahmed A, Usman M, Ji Z, Rafiq M, Yu B, Shen Y, Cong H (2023) Nature-inspired biogenic synthesis of silver nanoparticles for antibacterial applications. Mater Today Chem 27:101339

Albalawi MA, Abdelaziz AM, Attia MS, Saied E, Elganzory HH, Hashem AH (2022) Mycosynthesis of silica nanoparticles using aspergillus Niger: control of Alternaria solani causing early blight disease, induction of innate immunity and reducing of oxidative stress in eggplant. Antioxidants 11(12):2323

Alegbeleye O, Odeyemi OA, Strateva M, Stratev D (2022) Microbial spoilage of vegetables, fruits and cereals. Appl Food Res 2(1):100122

Ali S, Ullah MI, Sajjad A, Shakeel Q, Hussain A (2021) Environmental and health effects of pesticide residues. In: Sustainable agriculture reviews 48: pesticide occurrence, analysis and remediation Vol. 2 analysis, pp 311–336

Alsaiari NS, Alzahrani FM, Amari A, Osman H, Harharah HN, Elboughdiri N, Tahoon MA (2023) Plant and microbial approaches as green methods for the synthesis of nanomaterials: synthesis, applications, and future perspectives. Molecules 28(1):463

Altabbaa S, Mann NA, Chauhan N, Utkarsh K, Thakur N, Mahmoud GAE (2023) Era connecting nanotechnology with agricultural sustainability: issues and challenges. Nanotechnol Environ Eng 8(2):481–498

Anandhi S, Saminathan VR, Yasotha P, Saravanan PT, Rajanbabu V (2020) Nano-pesticides in pest management. J Entomol Zool Stud 8:685–690

Bairwa P, Kumar N, Devra V, Abd-Elsalam KA (2023) Nano-biofertilizers synthesis and applications in agroecosystems. Agrochemicals 2(1):118–134
Bala M, Bansal SK, Fatima F (2023) Nanotechnology: a boon for agriculture. Mater Today Proc 73:267–270
Baronia R, Kumar P, Singh SP, Walia RK (2020) Silver nanoparticles as a potential nematicide against. J Nematol 52(1):1–9
Beni AA, Jabbari H (2022) Nanomaterials for environmental applications. Results Eng 15:100467
Bind S, Bind S, Chandra D (2023) Microbial nanotechnology: a green approach towards sustainable agriculture. In: Unravelling plant-microbe synergy. Academic Press, pp 195–211
Bordiwala RV (2023) Green synthesis and applications of metal nanoparticles. A review article. Results Chem 100832
Bruna T, Maldonado-Bravo F, Jara P, Caro N (2021) Silver nanoparticles and their antibacterial applications. Int J Mol Sci 22(13):7202
Chandrasekaran S, Anusuya S, Anbazhagan V (2022) Anticancer, anti-diabetic, antimicrobial activity of zinc oxide nanoparticles: a comparative analysis. J Mol Struct 1263:133139
Cruz-Luna AR, Vásquez-López A, Rojas-Chávez H, Valdés-Madrigal MA, Cruz-Martínez H, Medina DI (2023) Engineered metal oxide nanoparticles as fungicides for plant disease control. Plan Theory 12(13):2461
Daniel AI, Keyster M, Klein A (2023) Biogenic zinc oxide nanoparticles: a viable agricultural tool to control plant pathogenic fungi and its potential effects on soil and plants. Sci Total Environ:165483
Danish M, Altaf M, Robab MI, Shahid M, Manoharadas S, Hussain SA, Shaikh H (2021) Green synthesized silver nanoparticles mitigate biotic stress induced by Meloidogyne incognita in Trachyspermum ammi (L.) by improving growth, biochemical, and antioxidant enzyme activities. ACS Omega 6(17):11389–11403
Decelis S, Sardella D, Triganza T, Brincat JP, Gatt R, Valdramidis VP (2017) Assessing the antifungal efficiency of filters coated with zinc oxide nanoparticles. R Soc Open Sci 4(5):161032
Devi D, Julkapli NM, Sagadevan S, Johan MR (2023) Eco-friendly green synthesis approach and evaluation of environmental and biological applications of iron oxide nanoparticles. Inorg Chem Commun:110700
Dhaka A, Mali SC, Sharma S, Trivedi R (2023) A review on biological synthesis of silver nanoparticles and their potential applications. Results Chem:101108
Dhoundiyal S, Kaur A, Alam MA, Sharma A (2023) The future of nanopesticides, nanoherbicides, and nanofertilizers. In: Nanopesticides, nanoherbicides, and nanofertilizers. Formulations and applications, 1st edn. CRC Press
Dong BR, Jiang R, Chen JF, Xiao Y, Lv ZY, Chen WS (2023) Strategic nanoparticle-mediated plant disease resistance. Crit Rev Biotechnol 43(1):22–37
Dos Santos EMP, Martins CCB, de Oliveira Santos JV, da Silva WRC, Silva SBC, Pelagio-Flores MA et al (2021) Silver nanoparticles–chitosan composites activity against resistant bacteria: tolerance and biofilm inhibition. J Nanopart Res 23(8):196
Du J, Liu B, Zhao T, Xu X, Lin H, Ji Y et al (2022) Silica nanoparticles protect rice against biotic and abiotic stresses. J Nanobiotechnol 20(1):1–18
Dutta P, Das G, Boruah S, Kumari A, Mahanta M, Yasin A et al (2021) Nanoparticles as nanopriming agent for antifungal and antibacterial activity against plant pathogens. Biol Forum Int J 13:476–482
El-Monairy OM, Ibrahim AWA, Zedan MH, Abdel-Meguid AD (2023) Insecticidal activity of silica, zinc and copper nanoparticles against the German cockroach, Blattella germanica (L.) (Dictyoptera: Blattellidae). Egypt Acad J Biol Sci A Entomol 16(1):15–26
El-Shetehy M, Moradi A, Maceroni M, Reinhardt D, Petri-Fink A, Rothen-Rutishauser B et al (2021) Silica nanoparticles enhance disease resistance in Arabidopsis plants. Nat Nanotechnol 16(3):344–353
Fabiyi OA, Alabi RO, Ansari RA (2020) Nanoparticles' synthesis and their application in the management of phytonematodes: an overview. In: Management of phytonematodes: recent advances and future challenges, pp 125–140

Farahat GA, Sehsah MD, Salama NH (2023) Environment, biodiversity & soil security

Fenibo EO, Ijoma GN, Nurmahomed W, Matambo T (2022) The potential and green chemistry attributes of biopesticides for sustainable agriculture. Sustain For 14(21):14417

Gauba A, Hari SK, Ramamoorthy V, Vellasamy S, Govindan G, Arasu MV (2023) The versatility of green synthesized zinc oxide nanoparticles in sustainable agriculture: a review on metal-microbe interaction that rewards agriculture. Physiol Mol Plant Pathol:102023

Gill HK, Garg H (2014) Pesticide: environmental impacts and management strategies. Pesticides-Toxic Aspects 8(187):10–5772

Giuntoli J, Oliver T, Kallis G, Ramcilovic-Suominen S, Monbiot G (2023) Exploring new visions for a sustainable bioeconomy

Gurram S, Jha DK, Shah DS, Amin PD, Moravkar KK, Pardeshi CV (2023) Nanomaterials toxicology: an overview. In: Nanomaterial-based drug delivery systems: therapeutic and theranostic applications, pp 327–368

Hazarika A, Yadav M, Yadav DK, Yadav HS (2022) An overview of the role of nanoparticles in sustainable agriculture. Biocatal Agric Biotechnol 43:102399

Hu J, Xianyu Y (2021) When nano meets plants: a review on the interplay between nanoparticles and plants. Nano Today 38:101143

Humbal A, Pathak B (2023) Application of nanotechnology in plant growth and diseases management: tool for sustainable agriculture. In: Agricultural and environmental nanotechnology: novel technologies and their ecological impact. Singapore, Springer Nature Singapore, pp 145–168

Hussain M, Shakoor N, Adeel M, Ahmad MA, Zhou H, Zhang Z et al (2023) Nano-enabled plant microbiome engineering for disease resistance. Nano Today 48:101752

Hyder S, Ul-Nisa M, Shahid H, Gohar F, Gondal AS, Riaz N et al (2023) Recent trends and perspectives in the application of metal and metal oxide nanomaterials for sustainable agriculture. Plant Physiol Biochem:107960

Jabri HA, Saleem MH, Rizwan M, Hussain I, Usman K, Alsafran M (2022) Zinc oxide nanoparticles and their biosynthesis: overview. Life 12(4):594

Jalal H, Giammarco M, Lanzoni L, Akram MZ, Mammi LM, Vignola G et al (2023) Potential of fruits and vegetable by-products as an alternative feed source for sustainable ruminant nutrition and production: a review. Agriculture 13(2):286

Jayakumar M, Sruthi P, Surendran U, Safari Z, Ahammed SJ (2023) The impact of agricultural chemicals on the environment and human health in India: a review. Knowledge-Based Eng Sci 4(2):60–72

Jha, S. K., & Jha, A. (2023). Sustainable utilization of renewable plant–based material for the green synthesis of metal nanoparticles

Kalaba MH, Moghannem SA, El-Hawary AS, Radwan AA, Sharaf MH, Shaban AS (2021) Green synthesized ZnO nanoparticles mediated by Streptomyces plicatus: characterizations, antimicrobial and nematicidal activities and cytogenetic effects. Plan Theory 10(9):1760

Kalia A, Abd-Elsalam KA, Kuca K (2020) Zinc-based nanomaterials for diagnosis and management of plant diseases: ecological safety and future prospects. J Fungi 6(4):222

Katz-Rosene R, Heffernan A, Arora A (2023) Protein pluralism and food systems transition: a review of sustainable protein meta-narratives. World Dev 161:106121

Khan M, Khan AU, Hasan MA, Yadav KK, Pinto MM, Malik N et al (2021) Agro-nanotechnology as an emerging field: a novel sustainable approach for improving plant growth by reducing biotic stress. Appl Sci 11(5):2282

Kralova K, Jampilek J (2023) Applications of nanomaterials in plant disease management and protection. In: Nanotechnology in agriculture and agroecosystems. Elsevier, pp 239–296

Lead JR, Doak SH, Clift MJ (eds) (2021) Nanotoxicology in humans and the environment. Springer

Lee EJ, Krassin ZL, Abaci HE, Mahler GJ, Esch MB (2023) Pumped and pumpless microphysiological systems to study (nano) therapeutics. Wiley Interdiscip Rev: Nanomed Nanobiotechnol 15:e1911

Macieira A, Fernandes VC, Delerue-Matos C, Teixeira P (2023) A global perspective of synthetic agrochemicals in local farmers' markets. In: One health implications of agrochemicals and their sustainable alternatives. Singapore, Springer Nature Singapore, pp 461–483

Madkhali OA (2023) A comprehensive review on potential applications of metallic nanoparticles as antifungal therapies to combat human fungal diseases. Saudi Pharm J:101733

Malandrakis AA, Kavroulakis N, Chrysikopoulos CV (2022) Metal nanoparticles against fungicide resistance: alternatives or partners? Pest Manag Sci 78(10):3953–3956

Maluin FN, Hussein MZ, Wayayok A, Nik Ibrahim NNL, Hashim N (2023) Chitosan nanoparticles as a sustainable alternative nutrient formulation in hydroponically grown Brassica rapa subsp. chinensis (L.) Hanelt microgreen and its adult vegetable. Arch Agron Soil Sci 69(12):2401–2412

Manjula NG, Sarma G, Shilpa BM, Suresh Kumar K (2022) Environmental applications of green engineered copper nanoparticles. In: Phytonanotechnology. Singapore, Springer Nature Singapore, pp 255–276

Mirza S, Hussaini AA, Öztürk G, Turgut M, Öztürk T, Tugay O et al (2023) Photocatalytic and antibacterial activities of ZnO nanoparticles synthesized from Lupinus albus and Lupinus pilosus plant extracts via green synthesis approach. Inorg Chem Commun 155:111124

Mobed A, Hasanzadeh M, Seidi F (2021) Anti-bacterial activity of gold nanocomposites as a new nanomaterial weapon to combat photogenic agents: recent advances and challenges. RSC Adv 11(55):34688–34698

More SL, Kovochich M, Lyons-Darden T, Taylor M, Schulte AM, Madl AK (2021) Review and evaluation of the potential health effects of oxidic nickel nanoparticles. Nanomaterials 11(3):642

Mukarram M, Ali J, Dadkhah-Aghdash H, Kurjak D, Kačík F, Ďurkovič J (2023) Chitosan-induced biotic stress tolerance and crosstalk with phytohormones, antioxidants, and other signalling molecules. Front. Plant Sci 14

Munir N, Gulzar W, Abideen Z, Hancook JT, El-Keblawy A, Radicetti E (2023) Nanotechnology improves disease resistance in plants for food security: applications and challenges. Biocatal Agric Biotechnol:102781

Nallasamy L, Murugavelu GS, Ganesh S, Nandhakumar PK, Krishnamoorthy D, Chandrasekaran S, Balan L (2023) Green nanotechnology revolution in biomedical application and treatments. In: Nanovaccinology: clinical application of nanostructured materials research to translational medicine. Springer International Publishing, Cham, pp 181–191

Namasivayam SKR, Francis AL, Kavisri M, Alharbi NS, Thiruvengadam M, Moovendhan M (2023) Biocompatible nanoscale silica particles fabricated from aminopropyltriethoxysilane functionalized brick ash induced versatile pesticidal activity. Environ Res 117090

Naseer MM, Rafeeq H, Afzal M, BiBi M, Rehman A, Fatima T (2023) Eco-friendly nanopesticides: a comprehensive review of their applications and environmental impacts. Hamdard Medicus 66(2)

Nazir K, Mukhtar T, Javed H (2019) In vitro effectiveness of silver nanoparticles against root-knot nematode (Meloidogyne incognita). Pak J Zool 51(6)

Paradva KC, Kalla S (2023) Nanopesticides: a review on current research and future perspective. ChemistrySelect 8(26):e202300756

Parmar S, Kaur H, Singh J, Matharu AS, Ramakrishna S, Bechelany M (2022) Recent advances in green synthesis of ag NPs for extenuating antimicrobial resistance. Nanomaterials 12(7):1115

Parveen A, Siddiqui ZA (2022) Foliar spray and seed priming of titanium dioxide nanoparticles and their impact on the growth of tomato, defense enzymes and some bacterial and fungal diseases. Arch Phytopathol Plant Protect 55(5):527–548

Pattanayak SP, Bose P, Sunita P (2023) Interaction between nanoparticles and phytopathogens. In: Agricultural and environmental nanotechnology: novel technologies and their ecological impact. Singapore, Springer Nature Singapore, pp 169–220

Paul A, Roychoudhury A (2021) Go green to protect plants: repurposing the antimicrobial activity of biosynthesized silver nanoparticles to combat phytopathogens. Nanotechnol Environ Eng 6(1):10

Pena GA, Cárdenas MA, Monge MDP, Yerkovich N, Planes GA, Chulze SN (2022) Reduction of fusarium proliferatum growth and fumonisin accumulation by ZnO nanoparticles both on a maize based medium and irradiated maize grains. Int J Food Microbiol 363:109510
Piližota V (2023) Fruits and vegetables (including herbs). In: Food safety management. Academic Press, pp 235–268
Pramanik P, Krishnan P, Maity A, Mridha N, Mukherjee A, Rai V (2020) Application of nanotechnology in agriculture. Environ Nanotechnol 4:317–348
Qasim M, Akhtar W, Haseeb M, Sajjad H, Rasheed M (2022) Potential role of nanoparticles in plants protection. Life Sci J 19(2):31–38
Rabea A, Naeem E, Balabel NM, Daigham GE (2023) Management of potato brown rot disease using chemically synthesized CuO-NPs and MgO-NPs. Bot Studies 64(1):1–20
Ramezani M, Ramezani F, Gerami M (2019) Nanoparticles in pest incidences and plant disease control. In: Nanotechnology for agriculture: crop production and protection, pp 233–272
Rana RA, Siddiqui MN, Skalicky M, Brestic M, Hossain A, Kayesh E et al (2021) Prospects of nanotechnology in improving the productivity and quality of horticultural crops. Horticulturae 7(10):332
Rana A, Pathak S, Lim DK, Kim SK, Srivastava R, Sharma SN, Verma R (2023) Recent advancements in plant-and microbe-mediated synthesis of metal and metal oxide nanomaterials and their emerging antimicrobial applications. ACS Appl Nano Mater 6(10):8106–8134
Richard B, Qi A, Fitt BD (2022) Control of crop diseases through integrated crop management to deliver climate-smart farming systems for low-and high-input crop production. Plant Pathol 71(1):187–206
Roy S, Das A, Chakraborty T, Bhattacharya B (2023) Nanotechnology-based regenerative approaches. In: Nanostructured materials for tissue engineering. Elsevier, pp 181–280
Rudramurthy GR, Swamy MK (2018) Potential applications of engineered nanoparticles in medicine and biology: an update. J Biol Inorg Chem 23:1185–1204
Sabarwal A, Kumar K, Singh RP (2018) Hazardous effects of chemical pesticides on human health–cancer and other associated disorders. Environ Toxicol Pharmacol 63:103–114
Sabry AKH (2019) Role of nanotechnology applications in plant-parasitic nematode control. Nanobiotechnol Appl Plant Protect 2:223–240
Sadiq MU, Shah A, Haleem A, Shah SM, Shah I (2023) Eucalyptus globulus mediated green synthesis of environmentally benign metal based nanostructures: a review. Nanomaterials (Basel) 13(13):2019
Saini MP (2023) Pesticide hazards on human health: assessing exposure, risks, and mitigation strategies. J Healthcare Life-Sci Res 2(7):91–105
Salam A, Afridi MS, Javed MA, Saleem A, Hafeez A, Khan AR et al (2022) Nano-priming against abiotic stress: a way forward towards sustainable agriculture. Sustain For 14(22):14880
Salem SS, Fouda A (2021) Green synthesis of metallic nanoparticles and their prospective biotechnological applications: an overview. Biol Trace Elem Res 199:344–370
Saritha GNG, Anju T, Kumar A (2022) Nanotechnology-big impact: how nanotechnology is changing the future of agriculture? J Agric Food Res:100457
Sharma U, Sanjta S, Sharma SK (2023) Nanopesticides: an approach towards precision insect-pest management in agriculture. J Entomol Res 47(2):288–294
Shweta, Sood S (2023) A review on nanotechnological advancements in tomato: a model plant. J Hortic Sci Biotechnol 98:1–17
Singh C, Anand SK, Upadhyay R, Pandey N, Kumar P, Singh D et al (2023) Green synthesis of silver nanoparticles by root extract of Premna integrifolia L. and evaluation of its cytotoxic and antibacterial activity. Mater Chem Phys 297:127413
Slavin YN, Bach H (2022) Mechanisms of antifungal properties of metal nanoparticles. Nanomaterials (Basel) 12(24):4470
Smith BR, Edelman ER (2023) Nanomedicines for cardiovascular disease. Nat Cardiovasc Res 2(4):351–367
Sonawane H, Shelke D, Chambhare M, Dixit N, Math S, Sen S et al (2022) Fungi-derived agriculturally important nanoparticles and their application in crop stress management–prospects and environmental risks. Environ Res 212:113543

Sun T, Li C, Li X, Song H, Su B, You H et al (2023) Pharmaceutical nanotechnology. In: Nanomedicine. Singapore, Springer Nature Singapore, pp 179–283
Tortella G, Rubilar O, Fincheira P, Pieretti JC, Duran P, Lourenço IM, Seabra AB (2021) Bactericidal and virucidal activities of biogenic metal-based nanoparticles: advances and perspectives. Antibiotics 10(7):783
Tudi M, Yang L, Wang L, Lv J, Gu L, Li H et al (2023) Environmental and human health hazards from chlorpyrifos, pymetrozine and avermectin application in China under a climate change scenario: a comprehensive review. Agriculture 13(9):1683
Upadhyay P, Agarwal S, Upadhyay S (2022) Recent applications of natural polymers in the formulation of nanogel. Curr Drug Ther 17(5):334–348
Vaishnavi PA, Prakash NU, Jayanthi M, Suganthi M, Thenmozhi M (2023) Impact of metal oxide nanoparticles against biotic stress in plants. In: Nanometal oxides in horticulture and agronomy. Academic Press, pp 101–140
Vithanage M, Zhang X, Gunarathne V, Zhu Y, Herath L, Peiris K et al (2023) Plant nanobionics: fortifying food security via engineered plant productivity. Environ Res:115934
Wahab A, Munir A, Saleem MH, AbdulRaheem MI, Aziz H, Mfarrej MFB, Abdi G (2023) Interactions of metal-based engineered nanoparticles with plants: an overview of the state of current knowledge, research progress, and prospects. J Plant Growth Regul 42:1–21
Wang L, Ning C, Pan T, Cai K (2022) Role of silica nanoparticles in abiotic and biotic stress tolerance in plants: a review. Int J Mol Sci 23(4):1947
Wang X, Xie H, Wang P, Yin H (2023) Nanoparticles in plants: uptake, transport and physiological activity in leaf and root. Materials 16(8):3097
Yadav A, Yadav K, Abd-Elsalam KA (2023) Exploring the potential of Nanofertilizers for a sustainable agriculture. Plant Nano Biol:100044
Yassin MT, Elgorban AM, Al-Askar AA, Sholkamy EN, Ameen F, Maniah K (2023) Synergistic anticandidal activities of greenly synthesized ZnO nanomaterials with commercial antifungal agents against candidal infections. Micromachines 14(1):209
Yuan Z, Li Y, He Y, Qian K, Zhang Y (2023) Differential analysis of three copper-based nanomaterials with different morphologies to suppress Alternaria alternata and safety evaluation. Int J Mol Sci 24(11):9673
Zahmanova G, Aljabali AA, Takova K, Toneva V, Tambuwala MM, Andonov AP et al (2023) The plant viruses and molecular farming: how beneficial they might be for human and animal health? Int J Mol Sci 24(2):1533
Zaki SA, Ouf SA, Albarakaty FM, Habeb MM, Aly AA, Abd-Elsalam KA (2021) Trichoderma harzianum-mediated ZnO nanoparticles: a green tool for controlling soil-borne pathogens in cotton. J Fungi 7(11):952
Zinovieva SV, Udalova ZV, Khasanova OS (2023) Nanomaterials in plant protection against parasitic nemates. Uspehi Sovremennoj Biol 143(3):278–299

Chapter 16
Recent Advancements and Applications of Nano-agrochemicals Against Plant Fungal Diseases

Jaganathan Sakthi Yazhini Preetha, Duraisampath Sriram, Paramasivam Premasudha, Ramesh Namdeo Pudake, Arumugam Vijaya Anand, and Muthukrishnan Arun

Abstract Biotic stress in plants is considered one of the main environmental constraints which ultimately affects crop performance. Biotic agents such as pests, viruses, insects, fungi, and nematodes cause the malnourishment of plants by depriving them of nutrient content when interacting. It also affects photosynthetic activity and eventually makes the plants die in serious conditions. Globally, it is reported that 70–80% of plant diseases are caused by fungal organisms. Plants respond to fungal infections by the innate and systemic defense mechanisms. Although the plant has a natural defense against stress factors, fungal infections could affect the entire plant system and spread rapidly through spores in many cases. Therefore, effective and eco-friendly strategies are needed to be developed to mitigate fungal infection in plants. Nano-agricultural technology has become a promising strategy to control both resilient and persistent fungal infections in plants. Recent advancements made with nanoparticles to control plant fungal diseases have a better ability to prevent infection, maintain sustainable production, reduce environmental pollution, and have lesser toxic effects compared to conventional chemi-

J. S. Y. Preetha · M. Arun (✉)
Department of Biotechnology, Bharathiar University, Coimbatore, Tamil Nadu, India
e-mail: arun@buc.edu.in

D. Sriram · P. Premasudha
Department of Nano Science and Technology, Bharathiar University, Coimbatore, Tamil Nadu, India

R. N. Pudake
Amity Institute of Nanotechnology, Amity University Uttar Pradesh, Sector 125, Noida, Uttar Pradesh, India

A. V. Anand
Department of Human Genetics and Molecular Biology, Bharathiar University, Coimbatore, Tamil Nadu, India

M. Khan, J.-T. Chen (eds.), *Nanoparticles in Plant Biotic Stress Management*,
https://doi.org/10.1007/978-981-97-0851-2_16

cal methods in agriculture. The physicochemical properties of the engineered nanomaterials, specifically to the targeted pathogen, enable the plants to counteract and survive the infection. The nano-sized structure and high surface area of the synthesized nano-fungicides ensure the efficiency and availability of antifungal agents to the crops. This chapter reveals the advantages and applications of nanomaterials and their mode of action to reduce pathogenic fungal infection in plants.

Keywords Biotic stress · Nano-agricultural technology · Fungal infection · Antifungal efficiency

16.1 Introduction

Agricultural production is affected by numerous environmental stress factors. Among them, abiotic and biotic stress factors account for huge economic losses in the agricultural sector. Abiotic stress factors include salt, drought, temperature, heavy metals, and waterlogging, and biotic stress factors include nematodes, fungi, bacteria, weeds, insects, and pests (Umar 2016). Biotic stress in plants affects its growth stage including seed germination, plant growth, and fruit setting or yield and massively affects crop productivity (Huseynova 2014). Pathogens attack the plants in three ways: (i) necrotrophic (entire plant tissue is affected), (ii) biotrophic (pathogen acts as parasite or symbiont to survive and keeps the plant cells alive), and (iii) hemibiotrophic (initially keeps the host plant alive likely in the biotrophic phase and kills the plant in the second phase) (Sergeant and Renaut 2010). Biotic stress accounts for overall 68% of crop loss worldwide. Among the biotic stress, bacterial infections account for 16% and fungal pathogens account for 70–80% of yield loss (Mansoor et al. 2021). Many types of fungal strains are responsible for affecting the plant performance. Some of the important strains that cause heavy loss in crop production are *Alternaria* spp., *Colletotrichum* spp., *Aspergillus* spp., *Pythium* spp., *Leveillula taurica*, *Fusarium oxysporum*, *Cercospora* spp., and *Phytophthora* spp. (Hussain and Usman 2019). It is important to mitigate the adverse effects of fungal pathogens caused in plants to maintain crop productivity for sustaining global food demand. The usage of fungicides results in aquatic pollution and badly affects a broad range of aquatic organisms due to their toxic effects (Zubrod et al. 2019). Nanotechnology is an emerging new-era technology that has potential innovations to limit toxic effects and mitigate detrimental effects caused by biotic and abiotic stress factors (Almoneafy et al. 2023; Hassanisaadi et al. 2022). Research on nanoparticles and nano-based formulations has been proven to reduce the pathogenesis of fungal infection in plant systems (Lopez-Moya et al. 2019; González-Merino et al. 2021). Generally, nanoparticles are classified into three categories: organic-, inorganic-, and carbon-based. Organic-based nanomaterials comprise dendrimers, micelles, and liposomes. Inorganic forms of nanomaterials are divided into metal- and metal-oxide-based nanoparticles. Carbon-based nanoparticles comprise fullerenes, graphene, carbon nanotubes, carbon nanofiber, and carbon black (Ealias and Saravanakumar 2017). This chapter summarizes the types of nanomaterials and nanomaterial-based formulations in diminishing the fungal attack in plant systems (Fig. 16.1).

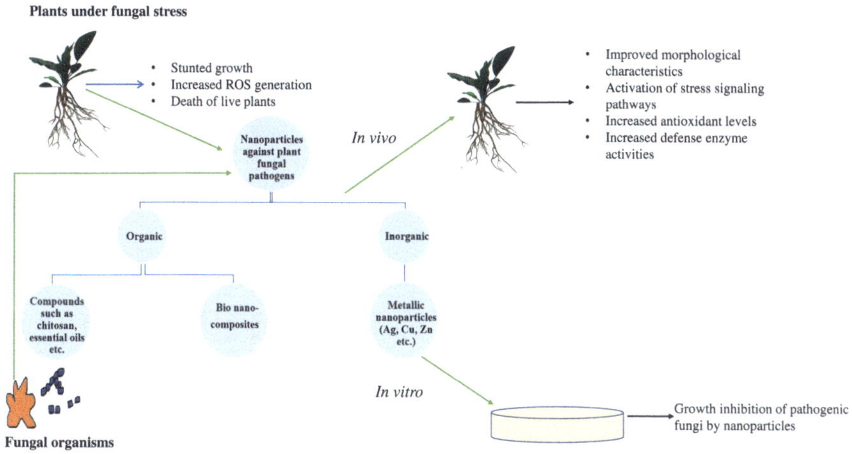

Fig. 16.1 Application of nanoparticles against plant fungal pathogens

16.2 Nano-chitosan and Its Implications on Fungal Stress

Several studies reported the role of chitosan as an antifungal agent against chitosan-sensitive fungi in infected plants. In chitosan-sensitive fungi, chitosan permeabilizes the plasma membrane as an energy-dependent process via ATP (adenine triphosphate) production and induces the generation of ROS (reactive oxygen species). This permeabilization of chitosan also blocks the electron transport chain and disturbs the antifungal activity response of chitosan-sensitive fungi (Lopez-Moya et al. 2019). Application of chitosan nanoparticles against *Pyricularia grisea* (blast fungus) in rice significantly suppressed the fungal growth in leaves (Manikandan and Sathiyabama 2016). Similarly, foliar application of chitosan nanoparticles (CNPs) suppressed the leaf spot disease caused by *Alternaria alternata* by enhancing defense-related enzymes (peroxidase, polyphenol oxidase, L-phenylalanine, β-1,3-glucanase activity), flavonoids, and total phenols and also augmented nitric oxide in chili plants (Sarkar et al. 2022). The effectiveness of CNPs was assessed both in vivo and in vitro against *Aspergillus flavus* in mung bean using an exogenous priming approach and growth inhibition test in PDA medium. The results showcased that CNPs were effective in combating the pathogenic fungus *Aspergillus flavus* in both in vivo and in vitro assessments (Sen et al. 2022). In maize, chitosan, nano-chitosan, and commercial fungicides were tested for their efficiency in reducing the severity of vascular disease caused by *Harpophora maydis* fungal pathogen. This study concludes that chitosan and nano-chitosan are more efficient than commercial fungicides in eradicating the fungal infection in maize (Hassan et al. 2022). Chitosan formulated with organic and inorganic compounds is also proven to have significant antifungal properties (Table 16.1).

Table 16.1 Application of nanoparticles in controlling plant pathogenic fungi

Nanoparticle	Synthesis method	Size (nm)	Shape	Fungal species	Evaluation method	References
Chitosan	Ionic gelation method	20–50	Spherical	*Pyricularia grisea*	In vitro	Manikandan and Sathiyabama (2016)
Chitosan	Ionotropic gelation method	<100	Spherical	*Alternaria alternata*	In vivo	Sen et al. (2022)
Chitosan	Ionic gelation method	–	Spherical	*Aspergillus flavus*	In vitro and in vivo	Hassan et al. (2022)
Chitosan	Ionotropic gelation method	65–80	–	*Harpophora maydis*	In vitro and in vivo	Chouhan et al. (2022)
Chitosan	Ionotropic gelation method	50	Spherical	Spores and sclerotia (*R. solani*, *F. solani*, *F. oxysporum*, *M. phaseolina*, *A. solani*, *Sclerotium rolfsii*, *Phytophthora infestans*)	In vitro	El-Mohamedy et al. (2019)
Chitosan-coupled copper	Ionic gelation method	163.8	–	*Rhizoctonia solani* and *Pythium aphanidermatum*	In vivo	Vanti et al. (2020)
Chitosan-coated silver nanoparticles	–	33.26	Spherical	*Fusarium oxysporum*	In vitro and in vivo	Gowda and Sriram (2023)
Chitosan-coated silver nanocomposites	–	4	Spherical	*Colletotrichum truncatum*	In vitro and in vivo	Kalagatur et al. (2018)
Chitosan loaded with protocatechuic acid	Ionic gelation method	30–35	–	*Pyricularia oryzae*	In vitro	Pham et al. (2019)
Clove oil-encapsulated chitosan	Droplet formation and solidification	129–148	–	*Aspergillus niger*	In vitro	Hasheminejad et al. (2019)
Nanoemulsion of essential oils (cinnamon, thyme, tea tree, and manuka)	Low-energy and high-energy emulsification methods	<30	Droplets	*Fusarium culmorum*, *Phytophthora cactorum*	In vitro	Hassanin et al. (2017)

Nanoemulsion of essential oils (clove, black seed, and lemon)	–	82–117	Droplets	*Botrytis cinerea*	In vivo	Ziedan et al. (2022)
Nanoemulsion of essential oils (clove and lemongrass)	Low-energy emulsification method	76.73	Spherical-shaped droplets	*Fusarium oxysporum*	In vivo	Sharma et al. (2018)
Ag NPs	–	5	–	*Alternaria alternata*, *Pestalotiopsis microspora*, *Diaporthe actinidiae*, and *Botryosphaeria dothidea*	In vitro	Kim et al. (2012)
Ag-NPs	Biogenic	9.39	Spherical	*Alternaria alternata*, *F. oxysporum*, *Aspergillus niger*, *Penicillium digitatum*	In vitro	Ahmed and El (2020)
Ag-NPs	Green synthesis	10–32	Spherical	*M. phaseolina*, *Alternaria alternata*, *F. oxysporum*	In vitro	Bahrami-Teimoori et al. (2017)
Ag-NPs	Green synthesis	77.45 and 63.20	Spherical	*Pestalotiopsis mangiferae*	In vitro	Rizwana et al. (2022)
Ag-NPs	Green synthesis	8–42	Spheroidal	*Alternaria alternata*, *Pestalotiopsis mangiferae*, *M. phaseolina*, *Colletotrichum musae*	In vitro	Borkow (2005)
Ag-NPs	Green synthesis	8–14	Spherical	*Alternaria solani*	In vitro	Mostafa et al. (2021)
Ag-NPs	Green synthesis	19.8–44.9	–	*Fusarium graminearum*	In vitro	Ibrahim et al. (2020)
Ag-NPs	–	<100	–	*Fusarium culmorum*, *Alternaria alternata*	In vitro	Ganash et al. (2018)
Ag-NPs	Chemical and biological method	50, 25	Spherical	*Alternaria solani*, *Corynespora cassiicola*, *Fusarium* spp.	In vitro and in vivo	Tyagi et al. (2020)
Ag-NPs	Biogenic	3–10	Spherical	*Fusarium* and *Alternaria* spp.	In vitro	Win et al. (2020)

(continued)

Table 16.1 (continued)

Nanoparticle	Synthesis method	Size (nm)	Shape	Fungal species	Evaluation method	References
Ag-NPs	Chemical reduction	10	Spherical	*Alternaria alternata*, *Alternaria citri*, *Penicillium digitatum*	In vitro	Salaheldin (2016)
Ag-NPs	Green synthesis	–	–	*Fusarium oxysporum*	In vitro	Singh et al. (2019)
Cu-NPs	Green synthesis	8	Crystal	*Fusarium oxysporum*, *Fusarium solani*, *Neofusicoccum* sp.	In vitro	Ibarra-Laclette et al. (2022)
Cu-NPs	Chemical reduction method	20–50	Spherical	*Fusarium* sp.	In vitro	Van et al. (2016)
Cu-NPs	Chemical method	5–10	Spherical	*Fusicladium oleagineum*, *Colletotrichum acutatum*	In vitro	Ntasiou et al. (2021)
Cu-NPs	Chemical method	25.5–25.8	Poly crystalline cubic	*Fusarium solani*	In vivo	Kamel et al. (2022)
Cu-NPs	Green chemistry	28	Spherical	*Aspergillus niger*, *Fusarium culmorum*	In vitro	Vanathi et al. (2016)
Cu-NPs	–	<50	–	*Alternaria tenuissima*, *Alternaria alternata*, *F. equiseti*, *F. oxysporum*, *F. solani*	In vitro	Yagoub et al. (2022)
Cu-NPs	Chemical reduction	26.5	Spherical	*F. oxysporum*	In vitro	Pariona et al. (2019)
ZnO-CuO-NPs	Mycosynthesis	54.18	Semispherical	*F. oxysporum*	In vitro	Gaber et al. (2023)
ZnO-NPs	Green synthesis	27	Spherical	*Aspergillus flavus*, *Aspergillus niger*	In vitro	Rajiv et al. (2013)
ZnO-NPs	Biogenic method	94	Hexagonal	*Fusarium oxysporum*	In vivo	Ezealigo et al. (2021)
IO-NPs	Green synthesis	20–80	Crystalline	*Alternaria mali*, *Botryosphaeria dothidea*, *Diplodia seriata*	In vitro	Ahmad et al. (2017)

16.2.1 Chitosan Nanoparticles Formulated with Inorganic Compounds Against Fungal Stress

Chitosan can be formulated with metals to form nanocomposites to fight against fungal infection (Chouhan et al. 2022). Chitosan is combined with nickel to form inorganic nanocomposites and tested against root rot disease caused by *Fusarium solani* in wheat. The results demonstrated that chitosan-nickel nanocomposites were effective in the complete eradication of root rot disease occurrence in wheat compared to commercial fungicides and nanochitosan. Similarly, nano-formulated chitosan, chitosan-saponin, and copper chitosan (Cu chitosan) were studied to evaluate their antifungal activity on phytopathogenic fungi *Alternaria alternata*, *Macrophomina phaseolina*, and *Rhizoctonia solani*. This study demonstrated that Cu chitosan and CNPs were more efficient in controlling fungal growth in in vitro studies (Saharan et al. 2013). Similar to this, Cu-CNPs were also studied against *Curvularia* leaf spot disease in maize (Choudhary et al. 2017). Cu-CNPs enhanced the antioxidants, and defense enzymes such as *polyphenol oxidase* and *phenylalanine ammonia lyase*, and also controlled the leaf spot disease. Chitosan is also used as a stabilizing agent in the nano-conversion of metallic silver. The synthesized silver nanoparticles coated with chitosan (As-Cs NPs) compound were tested for their activity against *Fusarium oxysporum* (wilt disease) in tomatoes. The maximum inhibition of 70% growth of *Fusarium oxysporum* strain in the germinated tomato seedlings was observed (Encinas Basurto et al. 2020). This study proves that As-Cs NPs can be used as a potential antifungal agent to suppress the fungal growth in the germination stage of plant seedlings. Likewise, in vivo and in vitro assessment of green synthesized chitosan-silver nanocomposites against *Colletotrichum truncatum* resulted in growth inhibition of the pathogens in both cases to prevent anthracnose disease in chilies (Gowda and Sriram 2023). Thus, the inorganic compounds (metal and metal oxide) formulated with chitosan as a coating agent have significant potential to control fungal infection in plants.

16.2.2 Organic Forms of Chitosan Nanoparticles Against Fungal Stress

Chitosan nanoparticles encapsulated with the essential oil of *Cymbopogon martinii* were treated against *Fusarium graminearum* (Kalagatur et al. 2018). Here, CNPs act as a carrier to prevent the instability of volatile essential oils. The result showed efficient antifungal and anti-mycotoxin activity of nano-chitosan encapsulated essential oil against *Fusarium* head blight disease. Likewise, CNPs encapsulated with thyme and clove essential oils were also tested on controlling fungal infection (*Botrytis cinerea* and *Penicillium expansum*) in apple fruits (Soppelsa et al. 2023). The thyme oil formulated with chitosan showed maximum inhibition of *Botrytis cinerea* in apples by 48%. Furthermore, the clove oil formulated with chitosan showed maximum inhibition of *Penicillium expansum* by 62%. Thus, these findings showed the potential of CNPs as an antifungal agent when formulated with organic compounds.

16.3 Essential Oil-Based Nanoformulation on Controlling Fungal Stress

Essential oils (EOs) are intricate blends of volatile compounds sourced from plants, and they fall within the category of secondary metabolites, primarily consisting of terpenes and phenolics (Arraiza et al. 2010). These EOs serve as nontoxic yet effective antimicrobial agents (Puškárová et al. 2017). A multitude of studies have affirmed the efficacy of EOs as antifungal agents against pathogenic fungi (Puškárová et al. 2017; Parikh et al. 2021; Tian et al. 2022). The mechanism of action involves disruption of fungal cell walls composed of chitin, inhibition of cell wall formation, impeding efflux pumps, and disruption of mitochondrial function (Nazzaro et al. 2017). The continuous usage of synthetic fungicides has led to the emergence of fungicide-resistant strains. In response, the application of nano-formulated EOs presents a promising alternative to synthetic fungicides, addressing concerns about environmental toxicity and the development of resistant strains. Nanoemulsions of oils from cinnamon, thyme, tea tree, and manuka (ranging from 0.5% to 2.5%) were synthesized using the low-energy phase inversion composition method (PIC) and ultrasonication. These nano-formulated EOs were then assessed against various fungal strains, including *Scopulariopsis brevicaulis*, *Fusarium culmorum* (Smith) Saccardo, *Microsporum gypseum*, *Trichophyton mentagrophytes*, *Aspergillus niger*, filamentous fungi, and plant pathogenic fungi (Schroeter), specifically *Phytophthora cactorum* using agar test plates. The findings of this study demonstrated that nano-formulated essential oils produced through ultrasonication exhibited remarkable antifungal activity against the fungal strains, surpassing the efficacy of pure oils and macroemulsions (Miastkowska et al. 2020). Similarly, nanoemulsion EOs derived from *Ocimum basilicum* L., *Majorana hortensis Moench.*, *Mentha piperita* L., *Menta spicata* L., and *Thymus vulgaris* L., synthesized using ultrasonication with water and nonionic surfactant, showcased greater antifungal activity compared to emulsions when tested against a *Fusarium oxysporum* strain isolated from geranium plants (Hassanin et al. 2017).

16.4 Metal or Metal Oxide Nanoparticles on Controlling Fungal Stress

Studies have highlighted the capacity of metal and metal-based nanoparticles to bolster plant growth, enhance resistance to both biotic and abiotic stresses, and elevate the plant's antioxidant and ROS scavenging systems. Additionally, these nanoparticles have been shown to regulate antioxidant defence system, endogenous hormone regulation, and transcriptional regulation of stress-related genes (Tortella et al. 2023; Tripathi et al. 2022). When fungal organisms are exposed to metallic nanoparticles, several effects are observed, including cell wall shrinkage, cell aggregation, changes in organelle arrangement, and distortion of inner membranes (Slavin and Bach 2022). This highlights the role of metallic nanoparticles, such as Ag, Cu, Fe, and Zn, as antifungal agents in plants (Table 16.1).

16.4.1 Silver Nanoparticles

Silver nanoparticles (Ag-NPs) find a wide range of applications in fields such as pharmaceuticals, biomaterial synthesis, and antimicrobial agents (Khan et al. 2021). The recent years have seen significant research on the efficacy of Ag-NPs against phyto-fungal pathogens. One notable study by Li et al. (2022) investigated the effects of 5 nm Ag-NPs on postharvest rot diseases caused by various strains of *Alternaria alternata*, *Pestalotiopsis microspora*, *Diaporthe actinidiae*, and *Botryosphaeria dothidea*. This investigation shed light on the molecular mechanisms through which Ag-NPs counteract fungal pathogens, operating at physiological, cytological, and transcriptomic levels. The outcome of this experiment showcased several key findings, notably, Ag-NPs led to an increase in cell membrane permeability, resulting in the leakage of intracellular substances within the fungi. Microscopic analysis unveiled observable effects, including cell shrinkage, distortion, and vacuolization within hyphal cells. Transcriptomic analysis further revealed significant variations in the expression of genes associated with diverse metabolic and catabolic pathways. The synthesis of Ag-NPs can be achieved through two primary methods: chemical synthesis and a more environmentally conscious approach (biogenic). The nano-sized colloidal solution of silver particles also played a significant role as an antifungal agent against various plant pathogenic fungi (Kim et al. 2012). Here, 50,000 μl ml^{-1} silver particles (7–25 nm) showed effective inhibition of most fungal pathogens at 100 ppm in the PDA (potato dextrose agar) medium used in this study. Green synthesis of silver nanoparticles using plant extracts is generally regarded as a safer, cost-effective, and less toxic alternative (Mansoor et al. 2021). The antimicrobial potential of Ag-NPs produced through green synthesis was explored in a study (Khan et al. 2021), wherein the authors demonstrated the efficacy of green-synthesized Ag-NPs (with a size of 55 nm) against *Fusarium oxysporum* fungus, as well as bacterial and nematode pathogens. Similarly, Ag-NPs synthesized from moringa leaf and flower extracts showed antifungal activity on *Pestalotiopsis mangiferae* (Jenish et al. 2022). Green-synthesized Ag-NPs from *O. majorana* were also witnessed to damage the conidia and hyphae morphological structures of fungal organisms in an in vitro study(Rizwana et al. 2022). Antifungal studies of Ag nanoparticles synthesized by chemical and biogenic methods are listed in Table 16.1.

16.4.2 Copper Nanoparticles

Copper nanoparticles (Cu-NPs) have potent biocidal applications as antibacterial, antiviral, and antifungal agents (Borkow 2005). Cu-NPs possess remarkable antimicrobial properties to control plant pathogens and are considerably cheap when compared to gold and silver nanoparticles in agricultural applications (Truong et al. 2023). Cu-NPs have been tested against many plant fungal pathogens synthesized through different approaches which are listed in Table 16.1. Among them, the antifungal effect of green-synthesized Cu-NPs against *Fusarium solani*, *Fusarium oxysporum*, and *Neofosicoccum* sp. was studied (Pariona et al. 2019). The results showed that different concentrations (0.1–1.0 mg L^{-1}) of aqueous-stabilized Cu-NPs inhibited mycelium growth in the PDA medium by affecting morphology, color, texture, form, and density. The molecular events involved in the antifungal activity of Cu-NPs in *Fusarium kuroshium* strain were demonstrated (Ibarra-Laclette et al. 2022). In this experiment, the morphology, differential gene expression, and gene ontology of *F. kuroshium* against Cu-NPs were validated. The Cu-NP-treated strains counteracted the toxicity of copper ions through reduction, internalization, and intracellular movement mechanisms. However, these mechanisms are compromised by the reduced ergosterol synthesis in *F. kuroshium* due to Cu-NPs' exposure. This leads to autophagy, apoptosis, and cell death of fungal pathogens. Likewise, Cu-NPs' antifungal efficacy was tested against plant fungal pathogens isolated from infected plants grown under arid regions. The results of the study demonstrated that maximum inhibitory potential was observed on *Alternaria tenuissima*, *Alternaria alternata*, *F. equiseti*, *F. oxysporum*, and *F. solani* isolates (Kasana et al. 2020). Thus, these significant studies show that Cu-NPs could be used as an alternative to conventional fungicides to reduce pathogenic fungal infection in plants.

16.4.3 Zinc Nanoparticles

Zinc (Zn) and zinc oxide (ZnO) nanoparticles have indeed been established to possess robust antimicrobial attributes (Yagoub et al. 2022). ZnO nanoparticles (ZnO-NPs) infiltrate the surfaces of microorganisms, disrupting their metabolic processes and resulting in their demise. The efficacy of this process is entirely contingent upon the size, configuration, and concentration of the ZnO-NPs (Siddiqi et al. 2018). In vitro and in vivo studies on biogenically synthesized ZnO-NPs were evaluated against *Fusarium* wilt disease in *Solanum melongena*. The application of ZnO-NPs yielded noteworthy benefits in vivo, ameliorating the detrimental effects of *Fusarium* strain-induced wilt disease in *S. melongena* plants. This enhancement was evident in terms of morphological attributes, photosynthetic pigments, total soluble carbohydrates, proteins, isozymes, phenolic content, and antioxidant activities. In vitro assessments revealed that a concentration of 1000 μg ml^{-1} of ZnO-NPs completely inhibited the growth of *F. oxysporum* in the PDA medium. This study underscores the veracity of ZnO-NP's effectiveness as a potent antifungal agent, thereby presenting an alternative to chemical fungicides in agricultural contexts (Abdelaziz et al. 2022). Further validation of ZnO-NPs' application as an antifungal agent can be found in Table 16.1.

16.4.4 Iron Oxide Nanoparticles

Iron oxide nanoparticles (IO-NPs) play a crucial role in biocidal applications owing to their superparamagnetic properties (Ezealigo et al. 2021). The antimicrobial characteristics of IO-NPs involve direct interaction with the cell surface, leading to the disruption of the envelope and perturbation of metabolic processes in microorganisms (Prodan et al. 2013). In vivo and in vitro investigations into green-synthesized IO-NPs (20 nm) against the plant pathogenic fungus *F. oxysporum* were studied (Ashraf et al. 2022). In this, the utilization of IO-NPs demonstrated a positive modulation of plant responses. This was manifested through enhanced growth, heightened antioxidant activities, and upregulated expression of defense-related genes (pathogenesis-related (PR) proteins and phenylalanine ammonia-lyase (PAL) as a countermeasure to wilt disease caused *by F. oxysporum* in tomato plants. Furthermore, an in vitro evaluation of IO-NPs (ranging from 0.01 to 15 μg L^{-1}) in a PDA medium against *F. oxysporum* exhibited remarkable inhibitory effects, with concentrations of 10–15 μg L^{-1} effectively suppressing 96–99% of mycelial growth. Consequently, this investigation underscores the potential utility of IO-NPs as antifungal agents in agriculture, mitigating losses attributable to fungal pathogens in plants. Likewise, numerous other studies corroborate the role of IO-NPs as antifungal agents, as summarized in Table 16.1.

16.5 Conclusion and Future Perspectives

In this section, the current advancement in diverse categories and uses of nanostructured compounds in the regulation of fungal pathogens affecting plants has been examined. Up to this point, substantial research has validated both natural sources like chitosan, and essential oils, and synthetic materials such as metal-based nanoparticles, as highly promising alternatives to conventional chemical fungicides. The nanoparticles, synthesized using various techniques including chemical and biological methods, also assume a pivotal role in influencing the efficacy of the compound by virtue of their specific size and shape. Further investigation is necessary to assess the real-world impact under field conditions (in vivo), and an in-depth exploration is warranted to uncover the molecular mechanisms that underlie the interaction between nanoparticles and the disintegration of fungal pathogens.

References

Abdelaziz AM, Salem SS, Khalil AMA, El-Wakil DA, Fouda HM, Hashem AH (2022) Potential of biosynthesized zinc oxide nanoparticles to control Fusarium wilt disease in eggplant (Solanum melongena) and promote plant growth. Biometals 35:601–616. https://doi.org/10.1007/s10534-022-00391-8

Ahmad H, Rajagopal K, Shah AH, Bhat AH, Venugopal K (2017) Study of bio-fabrication of iron nanoparticles and their fungicidal property against phytopathogens of apple orchards. IET Nanobiotechnol 11:230–235. https://doi.org/10.1049/iet-nbt.2015.0061

Ahmed M, El Y (2020) ISSN: 2090-0872. 12:29–43

Almoneafy AAA, Algam SAE, Alhammadi ASA, Moustafa-Farag M, Moghalles MAA (2023) Role of fungi-mediated nanoparticles in mitigation of biotic and abiotic stresses in plants. In: Fungal cell factories for sustainable nanomaterials productions and agricultural applications. Elsevier, pp 601–633

Arraiza MP, Ren J, Cheng Y-L, Lee C-Y, Huang Y-L, Buckner CA, Lafrenie RM, Dénommée JA, Caswell JM, Want DA, Gan GG, Leong YC, Bee PC, Chin E, Teh AKH, Picco S, Villegas L, Tonelli F, Merlo M, Rigau J, Diaz D, Masuelli M, Korrapati S, Kurra P, Puttugunta S, Picco S, Villegas L, Tonelli F, Merlo M, Rigau J, Diaz D, Masuelli M, Tascilar M, de Jong FA, Verweij J, Mathijssen RHJ, Amin J, Sharif M, Gul N, Kadry S, Chakraborty C, Dutt V, Chandrasekaran S, García-Díaz V (2010) We are IntechOpen, the world's leading publisher of Open Access books Built by scientists, for scientists TOP 1%. INTECH 34:57–67

Ashraf H, Batool T, Anjum T, Illyas A, Li G, Naseem S, Riaz S (2022) Antifungal potential of green synthesized magnetite nanoparticles black coffee–magnetite nanoparticles against wilt infection by ameliorating enzymatic activity and gene expression in Solanum lycopersicum L. Front Microbiol 13:1–23. https://doi.org/10.3389/fmicb.2022.754292

Bahrami-Teimoori B, Nikparast Y, Hojatianfar M, Akhlaghi M, Ghorbani R, Pourianfar HR (2017) Characterisation and antifungal activity of silver nanoparticles biologically synthesised by Amaranthus retroflexus leaf extract. J Exp Nanosci 12:129–139. https://doi.org/10.1080/17458080.2017.1279355

Borkow G (2005) Production aspects of organic photovoltaics and their impact on the commercialization of devices. MRS Bull 30:50–52. https://doi.org/10.1557/mrs2005.10

Choudhary RC, Kumaraswamy RV, Kumari S, Sharma SS, Pal A, Raliya R, Biswas P, Saharan V (2017) Cu-chitosan nanoparticle boost defense responses and plant growth in maize (Zea mays L.). Sci Rep 7:1–11. https://doi.org/10.1038/s41598-017-08571-0

Chouhan D, Dutta A, Kumar A, Mandal P, Choudhuri C (2022) Application of nickel chitosan nanoconjugate as an antifungal agent for combating Fusarium rot of wheat. Sci Rep 12:1–21. https://doi.org/10.1038/s41598-022-18670-2

Ealias AM, Saravanakumar MP (2017) A review on the classification, characterisation, synthesis of nanoparticles and their application. IOP Conf Ser Mater Sci Eng 263. https://doi.org/10.1088/1757-899X/263/3/032019

El-Mohamedy RSR, El-Aziz MEA, Kamel S (2019) Antifungal activity of chitosan nanoparticles against some plant pathogenic fungi in vitro. Agric Eng Int CIGR J 21:201–209

Encinas Basurto DA, Alvarez Carvajal F, Armenta Calderon A, Gonzalez Soto TE, Esquer Miranda E, Juarez Onofre J, Mendez Ibarra R (2020) Silver nanoparticles coated with chitosan against Fusarium oxysporum causing the tomato wilt. Biotecnia 22:73–80. https://doi.org/10.18633/biotecnia.v22i3.952

Ezealigo US, Ezealigo BN, Aisida SO, Ezema FI (2021) Iron oxide nanoparticles in biological systems: antibacterial and toxicology perspective. JCIS Open 4:100027. https://doi.org/10.1016/j.jciso.2021.100027

Gaber SE, Hashem AH, El-Sayyad GS, Attia MS (2023) Antifungal activity of myco-synthesized bimetallic ZnO–CuO nanoparticles against fungal plant pathogen Fusarium oxysporum. Biomass Convers Biorefinery. https://doi.org/10.1007/s13399-023-04550-w

Ganash M, Abdel Ghany TM, Omar AM (2018) Morphological and biomolecules dynamics of phytopathogenic fungi under stress of silver nanoparticles. BioNanoScience 8:566–573. https://doi.org/10.1007/s12668-018-0510-y

González-Merino AM, Hernández-Juárez A, Betancourt-Galindo R, Ochoa-Fuentes YM, Valdez-Aguilar LA, Limón-Corona ML (2021) Antifungal activity of zinc oxide nanoparticles in Fusarium oxysporum-Solanum lycopersicum pathosystem under controlled conditions. J Phytopathol 169:533–544. https://doi.org/10.1111/jph.13023

Gowda S, Sriram S (2023) Green synthesis of chitosan silver nanocomposites and their antifungal activity against Colletotrichum truncatum causing anthracnose in chillies. Plant Nano Biol 5:100041. https://doi.org/10.1016/j.plana.2023.100041

Hasheminejad N, Khodaiyan F, Safari M (2019) Improving the antifungal activity of clove essential oil encapsulated by chitosan nanoparticles. Food Chem 275:113–122. https://doi.org/10.1016/j.foodchem.2018.09.085

Hassan EO, Shoala T, Attia AMF, Badr OAM, Mahmoud SYM, Farrag ESH, El-Fiki IAI (2022) Chitosan and nano-chitosan for management of Harpophora maydis: approaches for investigating antifungal activity, pathogenicity, maize-resistant lines, and molecular diagnosis of plant infection. J Fungi 8. https://doi.org/10.3390/jof8050509

Hassanin M, Abd-El-Sayed M, Abdallah M (2017) Antifungal activity of some essential oil emulsions and nanoemulsions against fusarium oxysporum pathogen affecting cumin and geranium plants. Sci J Flowers Ornam Plants 4:245–258. https://doi.org/10.21608/sjfop.2017.11326

Hassanisaadi M, Barani M, Rahdar A, Heidary M, Thysiadou A, Kyzas GZ (2022) Role of agrochemical-based nanomaterials in plants: biotic and abiotic stress with germination improvement of seeds. Springer Netherlands

Huseynova I (2014) Improvement of crops in the era of climatic changes. Improv Crop Era Clim Chang 2:1–368. https://doi.org/10.1007/978-1-4614-8824-8

Hussain F, Usman F (2019) Fungal biotic stresses in plants and its control strategy. Abiotic Biot Stress Plants. https://doi.org/10.5772/intechopen.83406

Ibarra-Laclette E, Blaz J, Pérez-Torres CA, Villafán E, Lamelas A, Rosas-Saito G, Ibarra-Juárez LA, García-Ávila CD, Martínez-Enriquez AI, Pariona N (2022) Antifungal effect of copper nanoparticles against Fusarium kuroshium, an obligate symbiont of Euwallacea kuroshio Ambrosia Beetle. J Fungi 8. https://doi.org/10.3390/jof8040347

Ibrahim E, Zhang M, Zhang Y, Hossain A, Qiu W, Chen Y, Wang Y, Wu W, Sun G, Li B (2020) Green-synthesization of silver nanoparticles using endophytic bacteria isolated from garlic and its antifungal activity against wheat fusarium head blight pathogen fusarium graminearum. Nanomaterials 10(2):219. https://doi.org/10.3390/nano10020219

Jenish A, Ranjani S, Hemalatha S (2022) Moringa oleifera nanoparticles demonstrate antifungal activity against plant pathogenic fungi. Appl Biochem Biotechnol 194:4959–4970. https://doi.org/10.1007/s12010-022-04007-2

Kalagatur NK, Nirmal Ghosh OS, Sundararaj N, Mudili V (2018) Antifungal activity of chitosan nanoparticles encapsulated with Cymbopogon martinii essential oil on plant pathogenic fungi Fusarium graminearum. Front Pharmacol 9:1–13. https://doi.org/10.3389/fphar.2018.00610

Kamel SM, Elgobashy SF, Omara RI, Derbalah AS, Abdelfatah M, El-Shaer A, Al-Askar AA, Abdelkhalek A, Abd-Elsalam KA, Essa T, Kamran M, Elsharkawy MM (2022) Antifungal activity of copper oxide nanoparticles against root rot disease in cucumber. J Fungi 8:1–18. https://doi.org/10.3390/jof8090911

Kasana RC, Singh SK, Kumar P (2020) Antifungal activity of copper oxide nanoparticles against the fungal pathogens isolated from arid environment of Jodhpur. Int J Curr Microbiol Appl Sci 9:901–909. https://doi.org/10.20546/ijcmas.2020.911.108

Khan M, Khan AU, Bogdanchikova N, Garibo D (2021) Antibacterial and antifungal studies of biosynthesized silver nanoparticles against plant parasitic nematode meloidogyne incognita, plant pathogens ralstonia solanacearum and fusarium oxysporum. Molecules 26. https://doi.org/10.3390/molecules26092462

Kim SW, Jung JH, Lamsal K, Kim YS, Min JS, Lee YS (2012) Antifungal effects of silver nanoparticles (AgNPs) against various plant pathogenic fungi. Mycobiology 40:53–58. https://doi.org/10.5941/MYCO.2012.40.1.053

Li L, Pan H, Deng L, Qian G, Wang Z, Li W, Zhong C (2022) The antifungal activity and mechanism of silver nanoparticles against four pathogens causing kiwifruit post-harvest rot. Front Microbiol 13:1–16. https://doi.org/10.3389/fmicb.2022.988633

Lopez-Moya F, Suarez-Fernandez M, Lopez-Llorca LV (2019) Molecular mechanisms of chitosan interactions with fungi and plants. Int J Mol Sci 20. https://doi.org/10.3390/ijms20020332

Manikandan A, Sathiyabama M (2016) Preparation of Chitosan nanoparticles and its effect on detached rice leaves infected with Pyricularia grisea. Int J Biol Macromol 84:58–61. https://doi.org/10.1016/j.ijbiomac.2015.11.083

Mansoor S, Zahoor I, Baba TR, Padder SA, Bhat ZA, Koul AM, Jiang L (2021) Fabrication of silver nanoparticles against fungal pathogens. Front Nanotechnol 3:1–12. https://doi.org/10.3389/fnano.2021.679358

Miastkowska M, Michalczyk A, Figacz K, Sikora E (2020) Nanoformulations as a modern form of biofungicide. J Environ Heal Sci Eng 18:119–128. https://doi.org/10.1007/s40201-020-00445-4

Mostafa YS, Alamri SA, Alrumman SA, Hashem M, Baka ZA (2021) Green synthesis of silver nanoparticles using pomegranate and orange peel extracts and their antifungal activity against alternaria solani, the causal agent of early blight disease of tomato. Plan Theory 10:1–18. https://doi.org/10.3390/plants10112363

Nazzaro F, Fratianni F, Coppola R, De Feo V (2017) Essential oils and antifungal activity. Pharmaceuticals 10:1–20. https://doi.org/10.3390/ph10040086

Ntasiou P, Kerou AK, Karamanidou T, Vlachou A, Tziros GT, Tsouknidas A, Karaoglanidis GS (2021) Synthesis and characterization of novel copper nanoparticles for the control of leaf spot and anthracnose diseases of olive. Nanomaterials 11:1–19. https://doi.org/10.3390/nano11071667

Parikh L, Agindotan BO, Burrows ME (2021) Antifungal activity of plant-derived essential oils on pathogens of pulse crops. Plant Dis 105:1692–1701. https://doi.org/10.1094/PDIS-06-20-1401-RE

Pariona N, Mtz-Enriquez AI, Sánchez-Rangel D, Carrión G, Paraguay-Delgado F, Rosas-Saito G (2019) Green-synthesized copper nanoparticles as a potential antifungal against plant pathogens. RSC Adv 9:18835–18843. https://doi.org/10.1039/c9ra03110c

Pham TT, Nguyen TH, Thi TV, Nguyen TT, Le TD, Vo DMH, Nguyen DH, Nguyen CK, Nguyen DC, Nguyen TT, Bach LG (2019) Investigation of chitosan nanoparticles loaded with protocatechuic acid (PCA) for the resistance of Pyricularia oryzae Fungus against rice blast. Polymers (Basel) 11. https://doi.org/10.3390/POLYM11010177

Prodan AM, Iconaru SL, Chifiriuc CM, Bleotu C, Ciobanu CS, Motelica-Heino M, Sizaret S, Predoi D (2013) Magnetic properties and biological activity evaluation of iron oxide nanoparticles. J Nanomater 2013. https://doi.org/10.1155/2013/893970

Puškárová A, Bučková M, Kraková L, Pangallo D, Kozics K (2017) The antibacterial and antifungal activity of six essential oils and their cyto/genotoxicity to human HEL 12469 cells. Sci Rep 7:1–11. https://doi.org/10.1038/s41598-017-08673-9

Rajiv P, Rajeshwari S, Venckatesh R (2013) Bio-Fabrication of zinc oxide nanoparticles using leaf extract of Parthenium hysterophorus L. and its size-dependent antifungal activity against plant fungal pathogens. Spectrochim Acta Part A Mol Biomol Spectrosc 112:384–387. https://doi.org/10.1016/j.saa.2013.04.072

Rizwana H, Alzahrani T, Alwahibi MS, Aljowaie RM, Aldehaish HA, Alsaggabi NS, Ramadan R (2022) Phytofabrication of silver nanoparticles and their potent antifungal activity against phytopathogenic fungi. Processes 10. https://doi.org/10.3390/pr10122558

Saharan V, Mehrotra A, Khatik R, Rawal P, Sharma SS, Pal A (2013) Synthesis of chitosan based nanoparticles and their in vitro evaluation against phytopathogenic fungi. Int J Biol Macromol 62:677–683. https://doi.org/10.1016/j.ijbiomac.2013.10.012

Salaheldin TA (2016) Silver nanoparticles as a potent fungicide for citrus phytopathogenic fungi. J Nanomed Res 3. https://doi.org/10.15406/jnmr.2016.03.00065

Sarkar A, Chakraborty N, Acharya K (2022) Chitosan nanoparticles mitigate Alternaria leaf spot disease of chilli in nitric oxide dependent way. Plant Physiol Biochem 180:64–73. https://doi.org/10.1016/j.plaphy.2022.03.038

Sen SK, Tripathi R, Mandal P, Choudhuri C (2022) Assessment of antifungal activity of nano-chitosan against mung bean seed borne pathogen Aspergillus flavus through solid matrix priming. South Afr J Bot 150:372–386. https://doi.org/10.1016/j.sajb.2022.06.016

Sergeant K, Renaut J (2010) Plant biotic stress and proteomics. Curr Proteomics 7:275–297. https://doi.org/10.2174/157016410793611765

Sharma A, Sharma NK, Srivastava A, Kataria A, Dubey S, Sharma S, Kundu B (2018) Clove and lemongrass oil based non-ionic nanoemulsion for suppressing the growth of plant pathogenic Fusarium oxysporum f.sp. lycopersici. Ind Crop Prod 123:353–362. https://doi.org/10.1016/j.indcrop.2018.06.077

Siddiqi KS, Ur Rahman A, Tajuddin HA (2018) Properties of zinc oxide nanoparticles and their activity against microbes. Nanoscale Res Lett 13. https://doi.org/10.1186/s11671-018-2532-3

Singh R, Gupta AK, Patade VY, Balakrishna G, Pandey HK, Singh A (2019) Synthesis of silver nanoparticles using extract of Ocimum kilimandscharicum and its antimicrobial activity against plant pathogens. SN Appl Sci 1:1–12. https://doi.org/10.1007/s42452-019-1703-x

Slavin YN, Bach H (2022) Mechanisms of antifungal properties of metal nanoparticles. Nanomaterials 12. https://doi.org/10.3390/nano12244470

Soppelsa S, Van Hemelrijck W, Bylemans D, Andreotti C (2023) Essential oils and chitosan applications to protect apples against postharvest diseases and to extend shelf life. Agronomy 13. https://doi.org/10.3390/agronomy13030822

Tian F, Woo SY, Lee SY, Park SB, Zheng Y, Chun HS (2022) Antifungal activity of essential oil and plant-derived natural compounds against Aspergillus flavus. Antibiotics 11. https://doi.org/10.3390/antibiotics11121727

Tortella G, Rubilar O, Pieretti JC, Fincheira P, de Melo SB, Fernández-Baldo MA, Benavides-Mendoza A, Seabra AB (2023) Nanoparticles as a promising strategy to mitigate biotic stress in agriculture. Antibiotics 12. https://doi.org/10.3390/antibiotics12020338

Tripathi D, Singh M, Pandey-Rai S (2022) Crosstalk of nanoparticles and phytohormones regulate plant growth and metabolism under abiotic and biotic stress. Plant Stress 6. https://doi.org/10.1016/j.stress.2022.100107

Truong HT, Nguyen LCT, Quang Le L (2023) Synthesis and antifungal activity of copper nanoparticles against Fusarium oxysporum pathogen of plants. Mater Res Express 10. https://doi.org/10.1088/2053-1591/acdb34

Tyagi PK, Mishra R, Khan F, Gupta D, Gola D (2020) Antifungal effects of silver nanoparticles against various plant pathogenic fungi and its safety evaluation on Drosophila melanogaster. Biointerface Res Appl Chem 10:6587–6596. https://doi.org/10.33263/BRIAC106.65876596

Umar OB (2016) We are IntechOpen, the world's leading publisher of Open Access books Built by scientists, for scientists TOP 1%. INTECH, p 13

Van VP, Nguyen HT, Cao TM, Van HL (2016) Fusarium antifungal activities of copper nanoparticles synthesized by a chemical reduction method. J Nanomater 2016. https://doi.org/10.1155/2016/1957612

Vanathi P, Rajiv P, Sivaraj R (2016) Synthesis and characterization of Eichhornia-mediated copper oxide nanoparticles and assessing their antifungal activity against plant pathogens. Bull Mater Sci 39:1165–1170. https://doi.org/10.1007/s12034-016-1276-x

Vanti GL, Masaphy S, Kurjogi M, Chakrasali S, Nargund VB (2020) Synthesis and application of chitosan-copper nanoparticles on damping off causing plant pathogenic fungi. Int J Biol Macromol 156:1387–1395. https://doi.org/10.1016/j.ijbiomac.2019.11.179

Win TT, Khan S, Fu PC (2020) Fungus- (Alternaria sp.) mediated silver nanoparticles synthesis, characterization, and screening of antifungal activity against some phytopathogens. J Nanotechnol 2020. https://doi.org/10.1155/2020/8828878

Yagoub AEGA, Al-Shammari GM, Al-Harbi LN, Subash-Babu P, Elsayim R, Mohammed MA, Yahya MA, Fattiny SZA (2022) Antimicrobial properties of zinc oxide nanoparticles synthesized from Lavandula pubescens shoot methanol extract. Appl Sci 12. https://doi.org/10.3390/app122211613
Ziedan E-SHE, Saad MM, El-Kafrawy AA, Sahab AF, Mossa A-TH (2022) Evaluation of essential oils nanoemulsions formulations on Botrytis cinerea growth, pathology and grey mould incidence on cucumber fruits. Bull Natl Res Cent 46. https://doi.org/10.1186/s42269-022-00765-5
Zubrod JP, Bundschuh M, Arts G, Brühl CA, Imfeld G, Knäbel A, Payraudeau S, Rasmussen JJ, Rohr J, Scharmüller A, Smalling K, Stehle S, Schulz R, Schäfer RB (2019) Fungicides: an overlooked pesticide class? Environ Sci Technol 53:3347–3365. https://doi.org/10.1021/acs.est.8b04392

Chapter 17
The Beneficial Effect of Nanoparticles on Plant Growth and Their Role in Alleviating the Biotic Stress

Aiman Parveen and Zaki A. Siddiqui

Abstract In order to meet the need for food, the agriculture industry is crucial. It struggles with issues such as a lack of space, infections, and shifting agroclimatic conditions. To overcome these obstacles and boost output, technology must be embraced. Nanotechnology, which encompasses the manufacture, manipulation, and utilization of materials from less than a micron to individual atoms, is a viable strategy to overcome these difficulties. Similar to how nanoparticles have historically been created by chemical processes, current developments have made it possible to incorporate materials effectively. This review will look at how plants and microbes create nanoparticles and how that affects sustainable agriculture.

Keywords Nanoparticles · Plants · Pathogens · Pest

17.1 Introduction

A promising area of multidisciplinary research is nanotechnology. Professor Taniguchi and colleagues (Taniguchi et al. 1974) are credited for popularizing the term "nanotechnology" (Bulovic et al. 2004; Khan et al. 2021a). Advanced instrumentation to precisely isolate and characterize nanoparticles has been made possible by advances in nanotechnology (Adams et al. 2005; Bonnell and Huey 2001). Due to their unique characteristics, nanoparticles are used in a variety of industries, including engineering, medicine, agriculture, and electronics. Nanoparticles (NPs) are materials with a size of less than 100 nm (Adams et al. 2005; Bonnell and Huey 2001; Thomas et al. 2012). In terms of both their fundamental features and

A. Parveen (✉) · Z. A. Siddiqui
Department of Botany, Aligarh Muslim University, Aligarh, Uttar Pradesh, India

M. Khan, J.-T. Chen (eds.), *Nanoparticles in Plant Biotic Stress Management*,
https://doi.org/10.1007/978-981-97-0851-2_17

physicochemical properties, NPs are different from the corresponding bulk substance. Organic, inorganic, metal, and hybrid nanoparticles with novel optical, physical, and biological features are actively being created by biologists and chemists (Adams et al. 2005; Bonnell and Huey 2001; Thomas et al. 2012).

Crop diseases result in substantial yield loss and are a constant source of conflict for crop growers. A disease is a specific aberrant condition that can afflict a plant in whole or part, characterized by a structural or functional problem. Plant pathology is the study of causes, symptoms, and management of plant diseases. Pathogens are the organisms responsible for plant diseases, and to sustain the proper food supply by farmers around the world, these plant diseases are to be managed.

The use of nanotechnology has become increasingly important in a variety of scientific sectors, including medicine, engineering, pharmaceuticals, agriculture, and the food industry (Gul et al. 2014). Most developing nations depend heavily on agriculture to feed directly and indirectly their people. According to estimates, there will be eight billion people on the earth by 2025 and nine billion by 2050 (Jo et al. 2009).

Nanoparticles exhibit unique qualities like particle shape, size, and distribution (Kaviya et al. 2011). The physicochemical qualities of nanoparticles that improve their antibacterial, electrical, magnetic, catalytic, and optical properties are due to the large surface area (Kaviya et al. 2011; Catauro et al. 2005).

Uses of nanotechnology include the creation of insect-resistant plants through the transfer of genes or DNA via nanoparticles, food storage, and extended product shelf life. The advancement of biomass-to-fuel production technologies is predicted to be accelerated by nanotechnology. According to experts, the concerns for the land, water, and fisheries must be balanced against the potential use of nanotechnology for crop production, food, and aquaculture (Kaviya et al. 2011; Catauro et al. 2005; Chhipa 2017).

Agricultural output must rise to feed the fast expanding global population. Nanotechnology offers fresh agrochemicals and innovative delivery systems to boost crop output while allegedly lowering pesticide usage. Nano-formulations of agrochemicals to apply pesticides and fertilizers to improve crop yields are some of the applications of nanotechnology in agriculture (Fig. 17.1) (Chhipa 2017).

17.2 Nanoparticles: Synthesis and Impact on Plant Growth and Phytopathogens

17.2.1 Silver Nanoparticles: Biosynthesis and Impact on Plant Growth and Phytopathogens

The most common and environmentally benign way to create nanoparticles is through the green synthesis of plant extracts (Table 17.1). Plant extracts are normally distributed, conveniently available, much safer to handle, and also operate as a source of various metabolites, thus this has the advantage.

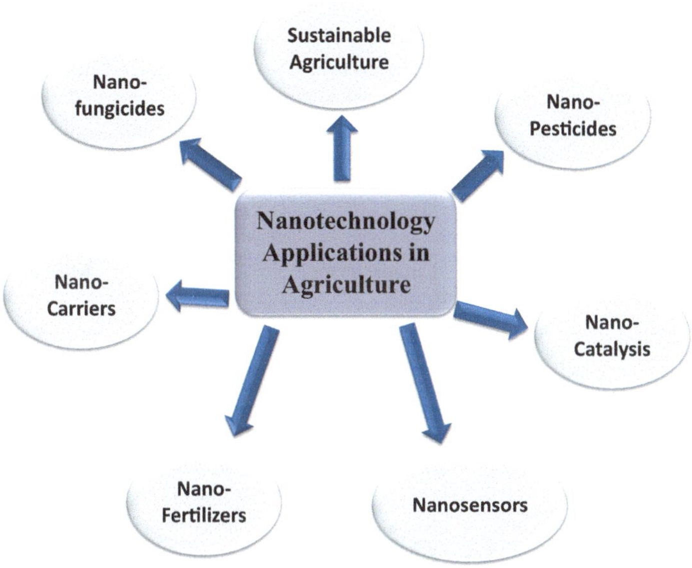

Fig. 17.1 The diverse applications of nanotechnology in agriculture

Additionally, the production of NPs using plant-mediated means is inexpensive and requires a single step that does not involve the use of highly toxic chemicals, high pressure, energy, or temperature (Ankamwar et al. 2005). Methanol was used to extract seed powder from *Gloriosa superba* seeds. Methanol seed extract and 1 M silver nitrate solution were combined, and the mixture was left to sit at room temperature for 15 h. As silver nanoparticles (AgNPs) develop, the color changes from yellowish brown to dark brown (Saradhadevi et al. 2017). Plants such as *Gloriosa superba* (Saradhadevi et al. 2017), *Pistacia vera* (Khan et al. 2021b), strawberry (Khan et al. 2021c), and *Trigonella foenum-graecum* (Khan et al. 2019) were used for the green synthesis of AgNPs (Table 17.1).

Prokaryotes such as bacteria have also been used to synthesize AgNPs. The simplicity of handling and straightforward genetic modification of bacteria make them an advantageous choice for the manufacturing of NPs. Contrary to intracellular production, extracellular synthesis of NPs utilizing cell filtrate is advantageous. Abo-State and Partila (2018) confirmed that AgNPs are produced extracellularly by the four bacterial species *Pseudomonas aeruginosa* (MAM-42), *Bacillus cereus* (MAM-I.11), *Achromobacter xylosoxidans* (MAM-29), and *Ochrobactrum* sp. (MAM-C9).

About 650 different types of microbes can be killed by AgNPs, an inorganic antimicrobial agent that is nontoxic, safe, and effective (Jo et al. 2009). According to Jiang et al. (2004), silver is well known for having an inhibitory effect on a variety of bacterial strains and pathogens that are frequently encountered in industrial

Table 17.1 Biosynthesis of nanoparticles

Nanoparticles	Size	Shape	Plant	Reference
Magnesium oxide nanoparticles (MgO NPs)	100 nm	Spherical to hexagonal	Strawberries	Khan et al. (2022b)
MgO NPs	27 and 68 nm	Irregular shapes	*Annona squamosa*	Sharma et al. (2020)
MgO NPs	100 nm	Spherical	Lemon fruit	Ahamad et al. (2023a)
Zinc oxide nanoparticles (ZnO NPs)	20 nm	Circular shape	Almond	Khan et al. (2022c)
ZnO/Fe_3O_4 nanocomposites	45, 35, and 60 nm	Rod-shaped particles of ZnO and cubic-shaped structures of Fe_3O_4	*Callistemon viminalis*	Dwivedi et al. (2021)
Silver nanoparticles (Ag NPs)	20 nm	Irregular shapes	Pistachio (*Pistacia vera*)	Khan et al. (2021b)
Ag NPs	55 nm	Rectangular	Strawberry	Khan et al. (2021c)
Ag NPs	20–25 nm	Spherical	*Trigonella*	Khan et al. (2019)
Pt NPs	15–19 nm	–	*Diospyros kaki*	Song et al. (2010)
Ag and Cu NPs	18 nm and 10.5 nm	_	*Euphorbia ceae latex*	Patil et al. (2012)
Au NPs	10–30 nm	Face-centered cubic	*Vitex negundo* L.	Zargar et al. (2011)
Indium oxide (In_2O_3) NPs	5–50 nm	Spherical	*Aloe vera*	Maensiri et al. (2008)

and medicinal operations (Table 17.2). Pesticides made chemically can be replaced with AgNPs without experiencing toxicity issues (Jo et al. 2009). It was made clear that AgNPs are effective and inhibit the fungus *Aspergillus flavus* (Al-Othman et al. 2014). At 200 and 175 ppm AgNPs, 100% fungal growth suppression was achieved, while at 150 ppm, 95% inhibition was recorded. According to Al-Othman et al. (2014), the inclusion of AgNPs resulted in a 95.5–81.1% decrease in the formation of mycotoxins.

Numerous scientific studies looked into the use of nano-silver for improving plant production. Sriram and Pandidurai (2017) found that *Zea mays* L. grows more quickly in vitro when exposed to a tiny quantity of AgNPs. Shoot and root growth in *Zea mays* was significantly increased at a 40 ppm concentration of silver NPs. Similar to other concentrations, the 60 ppm concentration of AgNPs increased seed germination.

Researchers have looked into the impact of 20-nm-diameter AgNPs on *Trigonella foenum-graecum* (fenugreek) seedlings (Hojjat 2015). At a concentration of 10 g/mL of AgNPs, the maximum seed germination (76.11%), speed of germination (4.10), root length (76.94 mm), root fresh weight (2.78), and root dry weight (1.20) were noted. According to El-Batal et al. (2016), selenium and silver NPs increased the plant's vegetative and chemical parameters and also reduced the severity of potato disease. In this case, AgNPs were created utilizing a biological method and cell-free supernatant from *Trichoderma viride*. Additional chemical synthesis was also carried out using polyvinylpyrrolidone (PVP) by gamma irradiation. The antibacterial activity of the NPs was tested in the field against *Alternaria solani*, the causal organism of the early blight potato. The plant's biochemical, vegetative, and disease severity parameters were determined. The severity of the disease was significantly reduced (8.39%) when the NPs 150 g/mL + K_2SO_4 (2%) were applied topically compared to untreated plants (86.17%), and it also led to a substantial rise in the physiological parameters.

17.2.2 *Zinc Oxide Nanoparticles: Biosynthesis and Impact on Plant Growth and Phytopathogens*

Zinc oxide nanoparticles (ZnO-NPs) were used to treat eggplant by Khan and Siddiqui (2018), and they observed that foliar application of ZnO-NPs results in the greatest increase in plant growth. According to Awasthi et al. (2017), ZnO-NPs have a beneficial impact on seed germination on *Triticum aestivum*. Venkatachalam et al. (2017) observed that ZnO NPs increase the amount of photosynthetic pigment in the *Leucaena leucocephala*. Treatment of *Sesamum indicum* plant with ZnO-NPs caused high levels of chlorophyll-a, chlorophyll-b, and total chlorophyll (Narendhran et al. 2016).

According to Jamdagni et al. (2016), *Alternaria alternata*, *Botrytis cinerea*, *Aspergillus niger*, *Fusarium oxysporum*, and *Penicillium expansum* are only a few of the fungi that ZnO-NPs have shown to have promising antifungal effect against. According to Navale et al. (2015), ZnO-NPs have shown promising antifungal activity against the fungi *Aspergillus flavus* and *A. fumigatus*. According to Rajiv et al. (2013), it has been shown that ZnO-NPs have antifungal action against *Aspergillus flavus*, *A. niger*, *A. fumigatus*, *Fusarium culmorum*, and *A. oxysporum*. ZnO-NPs were efficient against *Aspergillus flavus*, *Trichoderma harzianum*, *A. nidulans*, and *Rhizopus stolonifer* (Gunalan et al. 2012). ZnO-NPs' antifungal efficacy was demonstrated together with their interaction with a biocontrol bacterium to inhibit the growth of *Fusarium graminearum* by Dimkpa et al. (2013). *F. graminearum* and other fungal plant infections cause significant global economic losses in cereal crops, but a few effective control strategies may be adopted for their management. There is a possibility of using ZnO-NPs and biocontrol microorganisms for fungal disease control (Table 17.2).

By using *Aeromonas hydrophila* to produce ZnO-NPs, Jayaseelan et al. (2012) reported their activity against *Pseudomonas aeruginosa* as well as fungi, *Candida albicans*, *Aspergillus niger*, and *A. flavus*. The replicating bacterium *A. hydrophila* was used as an eco-friendly reducing and capping agent in a low-cost and simple

Table 17.2 Nanoparticles in plant disease management

Nanoparticles	Pathogen	Effect	Reference
Ag_2O NPs	*Xanthomonas campestris* pv. *vesicatoria, Pseudomonas syringae* pv. *tomato, Ralstonia solanacearum, Pectobacterium carotovorum subsp. carotovorum, Fusarium oxysporum f.sp. lycopersici* and *Alternaria solani*	0.10 and 0.20 g L^{-1} Ag_2O NPs concentration inhibited the growth of tested pathogens	Parveen and Siddiqui (2023)
TiO_2 NPs	*X. campestris* pv. *vesicatoria, P. syringae* pv. *tomato, R. solanacearum, P. carotovorum* subsp. *carotovorum, F. oxysporum* f.sp. *lycopersici* and *A. solani*	0.10 and 0.20 mL L^{-1} TiO_2 NPs concentration was inhibitory to the growth of all pathogens	Parveen and Siddiqui (2022a)
ZrO_2 NPs	*X. campestris* pv. *vesicatoria, P. syringae* pv. *tomato, R. solanacearum, P.carotovorum* subsp. *carotovorum, F. oxysporum* f.sp. *lycopersici,* and *A. solani*	0.10 and 0.20 g L^{-1} ZrO_2 NPs concentration inhibited the growth of all pathogens	Parveen and Siddiqui (2022b)
ZnO NPs	*X. campestris* pv. *vesicatoria, P. syringae* pv. *tomato,R. solanacearum, P. carotovorum* subsp. *carotovorum, F. oxysporum* f.sp. *lycopersici, and A. solani*	0.10 and 0.20 mL^{-1} ZnO NPs concentration restrained the growth of all pathogens	Parveen and Siddiqui (2021)
Graphene oxide (GO) and ZnO NPs	*P. carotovorum, X. campestris* pv. *carotae, Meloidogyne javanica, Alternaria dauci* and *Fusarium solani*	Two concentrations, 0.05 and 0.10 mg mL^{-1} effectively controlled the pest and pathogens	Siddiqui et al. (2019b)
SiO_2 NPs	*X. campestris* pv. *vesicatoria, Pseudomonas syringae pv. tomato,R. solanacearum, Pectobacterium carotovorum subsp. carotovorum, F. oxysporum f. sp. lycopersici* and *Alternaria solani*	SiO_2 NPs at 0.10 and 0.20 g L^{-1} concentration was inhibitory to the pathogens	Parveen and Siddiqui (2022c)
Si	*Fusarium oxysporum* f.sp. *vasinfectum*	Physical and biochemical resistance induced in Si-treated cotton plants against the pathogen	Whan et al. (2016)
TiO_2 NPs	*Rhizoctonia solani, P. betavasculorum* and *M. incognita*	100 and 200 mg L^{-1} TiO_2 NPs reduced the growth of all pathogens	Khan and Siddiqui (2021)
Rhizophagus irregularis and SiO_2 NPs	*M. incognita, Alternaria dauci* and *R. solani*	Improved the growth of carrot plants and reduced the growth of pathogens	Ahamad et al. (2023b)

ZrO_2 NPs and AM fungus	Wilt disease complex	Increase in photosynthetic pigments in bottle gourd	Siddiqui and Khan (2023)
Pseudomonas putida, ZnO NPs and *Rhizobium leguminosarum*	*M. incognita* and *Pseudomonas syringae* pv. *pisi*	Improved growth of infected pea plants	Kashyap and Siddiqui (2023)
ZnO NPs and *Rhizobium leguminosarum*	Blight disease complex	Enhanced pea plant growth in infected plants	Kashyap and Siddiqui (2022)
SiO_2 NPs with *Rhizobium leguminosarum*	Bacterial blight disease complex	Growth in pea plants was improved in infected plants	Kashyap and Siddiqui (2021)
SiO_2 NPs and *Rhizobium ciceri*	*Rhizoctonia solani*, *Fusarium solani* and *M. incognita*	Improvement in growth of infected chickpea plants and reduction in disease severity	Siddiqui and Aziz (2020)
SiO_2 NPs	Disease complex of *M. incognita*, *P. betavasculorum* and *R. solani*	Severity of disease complex decreased and increase in growth of beetroot plants	Khan and Siddiqui (2020)
SiO_2 NPs	*P. carotovorum, X. campestris* pv. *carotae, R. solani, F.solani* and *A. dauci*	Disease management and plant growth improvement in carrot	Siddiqui et al. (2020)
ZnO NPs	*Alternaria alternata, F. oxysporum* f.sp. *lentis, Xanthomonas axonopodis* pv. *phaseoli, P. syringae* pv. *syringae* and *M. incognita*	Disease indices were reduced and plant growth was improved in lentil with ZnO NPs spray	Siddiqui et al. (2018)
Graphene oxide NPs	*M. incognita* and *Macrophomina phaseolina*	Growth was increased and disease severity got reduced in infected lentil plants	Guroo et al. (2016)

approach for the production of ZnO-NPs. The highest zone of inhibition against *P. aeruginosa* and *A. flavus* was demonstrated by ZnO-NPs (25 g/mL).

17.2.3 Silicon Dioxide Nanoparticles: Impact on Plant Growth and Phytopathogens

For many physiological and metabolic processes in plants, silicon is a crucial element that is necessary (Bao et al. 2004). *Agropyron elongatum* seedling development and germination could be enhanced by silicon dioxide (SiO_2) NPs (Azimi et al. 2014). Through delayed and controlled release, nano-SiO_2 can be utilized to create fertilizers that are effective for crops, reduce fertilizer loss, and enable controlled and timely delivery (Nair et al. 2010). In *Cucurbita pepo*, seed germination was enhanced after treatment with nano-SiO_2 (Siddiqui et al. 2014). Haghighi and Pessarakli (2012) applied nano-Si to tomato plants boosting their ability to produce oxygen. Spraying of SiO_2 NPs improved the photosynthetic pigments chlorophyll a and b, plant length, shoot dry weight, and root dry weight of eggplant. Disease severity in eggplant caused by pathogens such as *Phomopsis vexans*, *Ralstonia solanacearum*, and *Meloidogyne incognita* was also reduced (Khan et al. 2022a).

17.2.4 Titanium Dioxide Nanoparticles: Impact on Plant Growth and Phytopathogens

About 3000 tons of titanium dioxide (TiO_2) NPs is generated annually, and more than 50% of them are used in personal care products, as reported by Keller and Lazareva (2013). TiO_2 application to crops encourages plant growth factors and may speed up photosynthetic activity. TiO_2 and ZnO NPs were employed by Siddiqui et al. (2019a) on the beetroot plant. They discovered that both NPs enhanced plant growth, increased chlorophyll and carotenoid content, and enhanced plant superoxide dismutase (SOD), catalase (CAT), H_2O_2, and proline content. The performance of ZnO NPs was found better than TiO_2 NPs. According to Raliya et al. (2015), treatment with TiO_2 NPs enhanced the shoot length of *Vigna radiata*. Lawre and Raskar (2014) found that TiO_2 NPs at a lower concentration improved seed germination and seedling growth in onion plants. TiO_2 NPs were discovered to have a beneficial effect on *Triticum aestivum* (Rafique et al. 2014). Mahmoodzadeh et al. (2013) observed a beneficial effect of TiO_2 NPs on the germination of *Brassica napus* seeds. According to Qi et al. (2013), the application of TiO_2 NPs increased the rate of photosynthetic activity in tomato plants.

17.2.5 Copper Oxide Nanoparticles: Impact on Plant Growth and Phytopathogens

Copper is a crucial element for the development and growth in plants. Numerous plant enzymes depend heavily on copper for proper function. Electronics, catalysts, gas sensors, and antibacterial agents all use copper nanoparticles (Cu NPs) (Kasana et al. 2017). According to Wang et al. (2019), copper oxide (CuO) NPs improved photosynthesis in *Spinacia oleracea*. $Cu(OH)_2$ NPs improved *Lactuca sativa's* antioxidant system, according to Zhao et al. (2014). Fan et al. (2020) found that Cu composites had antibacterial action against *Xanthomonas euvesicatoria*. According to Huang et al. (2015), due to the antifungal capabilities of CuO NPs, *Botrytis cinerea, Colletotrichum graminicola, Rhizoctonia solani, Colletotrichum musae, Magnaporthe oryzae, Penicillium digitatum*, and *Sclerotium rolfsii* are susceptible. The antifungal activity of CuO and Cu_2O NPs against *Phytophthora infestans* was demonstrated by Giannousi et al. (2013).

17.2.6 Gold Nanoparticles: Impact on Plant Growth and Phytopathogens

Jayaseelan et al. (2013) reported that *Abelmoschus esculentus* seed aqueous extract and its antifungal properties were used to create gold nanoparticles (Au NPs). They discussed the antifungal activity of *A. esculentus* seed aqueous extract in the manufacturing of Au NPs. The antifungal activity of Au NPs was tested using the conventional well diffusion method against *Candida albicans, Aspergillus flavus*, and *Puccinia graminis tritici*. With regard to *C. albicans* (18 mm) and *P. graminis* (17 mm), the Au NPs created the largest zones of inhibition. The results suggest that synthesized Au NPs may have antifungal properties. Au NPs have a lot of potential for producing significant antifungal effectiveness, which opens the door to developing drugs to treat fungal illnesses. The plant-mediated biosynthesis of some nanoparticles and their impact on plant pathogens has been listed in Tables 17.1 and 17.2, respectively.

17.3 Conclusions

Research indicates that current agricultural techniques are insufficient to meet food demand for the future, and there is an urgent need for an increase in global food production. The use of chemical-based pesticides in agriculture has a great adverse effect on the environment. According to estimates, around 192.8 Mt. of chemical fertilizers was utilized globally in 2016–2017. Environmental contamination brought on by the usage of harmful pesticides has adversely affected both wildlife and flora. This has resulted in resistance to pathogens and pests against fungicides and pesticides. As a result, there is an urgent need to develop a strategy that is less harmful to the environment and which may boost agricultural production to meet our food demand.

Therefore, it is important to utilize hazardous chemical pesticides and fungicides as little as possible. The use of nanotechnology is useful to produce materials with nanostructures that may carry and deliver fertilizers and pesticides. These nanomaterials may improve nutrient availability and low environmental contamination and can be used to create "smart fertilizers." This may serve as a substitute that can lessen yield losses and increase crop productivity. With the help of this cutting-edge technology, hazardous chemicals and pesticides that pollute groundwater, the environment, and soil can be used less frequently.

References

Abo-State MAM, Partila AM (2018) Production of silver nanoparticles (AgNPs) by certain bacterial strains and their characterization. Novel Res Microbiol J 1(2):19–32

Adams F, Van Vaeck L, Barrett R (2005) Advanced analytical techniques: platform for nano materials science. Spectrochim Acta B At Spectrosc 60(1):13–26

Ahamad L, Azmat AL, Khan M, Farid O, Mahboob A (2023a) Exploring the nano-fungicidal efficacy of green synthesized magnesium oxide nanoparticles (MgO NPs) on the development, physiology, and infection of carrot (*Daucus carota* L.) with *Alternaria* leaf blight (ALB): Molecular docking. J Integr Agric. https://doi.org/10.1016/j.jia.2023.02.034

Ahamad L, Siddiqui ZA, Hashem A, Abd-Allah EF (2023b) Use of AM fungus *Rhizophagus irregularis* and silicon dioxide nanoparticles for the management of *Meloidogyne incognita*, *Alternaria dauci* and *Rhizoctonia solani* and growth of carrot. Arch Phytopathol Plant Prot. https://doi.org/10.1080/03235408.2023.2207958

Al-Othman MR, Abd El-Aziz ARM, Mahmoud MA, Eifan SA, El-Shikh MA, Majrashi M (2014) Application of silver nanoparticles as antifungal and anti aflatoxin B1 produced by *Aspergillus flavus*. Dig J Nanomater Biostruct 9(1):151–157

Ankamwar B, Ahmad A, Sastry M (2005) Biosynthesis of gold and silver nanoparticles using *Emblica officinalis* fruit extract, their phase transfer and transmetallation in an organic solution. J Nanosci Nanotechnol 5:1665–1671

Awasthi A, Bansal S, Jangir LK, Awasthi G, Awasthi KK, Awasthi K (2017) Effect of ZnO nanoparticles on germination of *Triticum aestivum* seeds. Macromol Symp 376:1700043

Azimi R, Borzelabad MJ, Feizi H, Azimi A (2014) Interaction of SiO_2 nanoparticles with seed prechilling on germination and early seedling growth of tall wheatgrass (*Agropyron elongatum* L.). Pol J Chem Technol 16:25–29

Bao SL, Chun HL, Li JF, Shu CQ, Min Y (2004) Effect of TMS (nanostructured silicon dioxide) on growth of Changbai larch seedlings. J For Res 15:138–140

Bonnell DA, Huey BD (2001) Basic principles of scanning probe microscopy. In: Bonnell DA (ed) Scanning probe microscopy and spectroscopy: theory, techniques, and applications, vol 123(39). Wiley-VCH, New York, pp 9725–9725

Bulovic V, Mandell A, Perlman A (2004) Molecular memory device. US 20050116256 A1

Catauro M, Raucci MG, De Gaaetano FD, Marotta A (2005) Sol-gel processing of drug delivery materials and release kinetics. J Mater Sci Mater Med 16(3):261–265

Chhipa H (2017) Nanofertilizers and nanopesticides for agriculture. Env Chem Lett 15(1):15–22

Dimkpa CO, McLean JE, Britt DW, Anderson AJ (2013) Antifungal activity of ZnO nanoparticles and their interactive effect with a biocontrol bacterium on growth antagonism of the plant pathogen *Fusarium graminearum*. Biometals 26:913–924

Dwivedi P, Jatrana I, Khan AU, Khan AA, Satiya H, Khan M, Moon IS, Alam M (2021) Photoremediation of methylene blue by biosynthesized ZnO/Fe_3O_4 nanocomposites using *Callistemon viminalis* leaves aqueous extract: a comparative study. Nanotechnol Rev 10(1):1912–1925

El-Batal AI, Sidkey NM, Ismail AA, Arafa RA, Fathy RM (2016) Impact of silver and selenium nanoparticles synthesized by gamma irradiation and their physiological response on early blight disease of potato. J Chem Pharm Res 8(4):934–951

Fan Q, Liao YY, Kunwar S, Da Silva S, Young M, Santra S, Paret ML (2020) Antibacterial effect of copper composites against *Xanthomonas euvesicatoria*. Crop Prot 139:105366

Giannousi K, Avramidis I, Dendrinou-Samara C (2013) Synthesis, characterization and evaluation of copper based nanoparticles as agrochemicals against *Phytophthora infestans*. RSC Adv 3:21743–21752

Gul HT, Saeed S, Khan FZA, Manzoor SA (2014) Potential of nanotechnology in agriculture and crop protection: a review. Appl Sci Bus Econ 1(2):23–28

Gunalan S, Sivaraj R, Rajendran V (2012) Green synthesized ZnO nanoparticles against bacterial and fungal pathogens. Prog Nat Sci Mater 22:693–700

Guroo JA, Khan M, Ahmad A, Azam A, Siddiqui ZA (2016) Management of *Meloidogyne incognita* and *Macrophomina phaseolina* by graphene oxide on lens culinaris. Acta Phytopath Entom Hung 51:43–56

Haghighi M, Afifipour Z, Mozafarian M (2012) The effect of N-Si on tomato seed germination under salinity levels. J Biol Environ Sci 6:87–90

Hojjat SS (2015) Impact of silver nanoparticles on germinated fenugreek seed. Int J Agric Crop Sci 8:627–630

Huang S, Wang L, Liu L, Hou Y, Li L (2015) Nanotechnology in agriculture, livestock, and aquaculture in China. A review. Agron Sustain Dev 35:369–400

Jamdagni P, Khatri P, Rana JS (2016) Green synthesis of zinc oxide nanoparticles using flower extract of *Nyctanthes arbor-tristis* and their antifungal activity. J King Saud Univ Sci. https://doi.org/10.1016/j.jksus.2016.10.002

Jayaseelan C, Rahuman AA, Kirthi AV, Marimuthu S, Santhoshkumar T, Bagavan A, Rao KVB (2012) Novel microbial route to synthesize ZnO nanoparticles using *Aeromonas hydrophila* and their activity against pathogenic bacteria and fungi. Spectrochim Acta Part A Mol Biomol Spectrosc 90:78–84. https://doi.org/10.1016/j.saa.2012.01.006

Jayaseelana C, Ramkumar R, Rahumana AA, Perumal P (2013) Green synthesis of gold nanoparticles using seed aqueous extract of *Abelmoschus esculentus* and its antifungal activity. Indust Crops Prod 45:423–429

Jiang H, Manolache S, Wong ACL, Denes FS (2004) Plasma-enhanced deposition of silver nanoparticles onto polymer and metal surfaces for the generation of antimicrobial characteristics. J Appl Polym Sci 93(3):1411–1422

Jo YK, Kim BH, Jung G (2009) Antifungal activity of silver ions and nanoparticles on phytogenic fungi. Plant Dis 93:1037–1043

Kasana RC, Panwar NR, Kaul RK, Kumar P (2017) Biosynthesis and effects of copper nanoparticles on plants. Environ Chem Lett 15:233–240

Kashyap D, Siddiqui ZA (2021) Effect of silicon dioxide nanoparticles and *Rhizobium leguminosarum* alone and in combination on the growth and bacterial blight disease complex of pea caused by *Meloidogyne incognita* and *Pseudomonas syringae* pv. *pisi*. Arch Phytopathol Plant Protect 9-10:499–515. https://doi.org/10.1080/03235408.2020.184330629

Kashyap D, Siddiqui ZA (2022) Effect of Zinc oxide nanoparticles (ZnO NPs) and *Rhizobium leguminosarum* alone and in combination on the growth, photosynthetic pigments and on the blight disease complex of pea. Gesunde Pflanzen 74:29–40

Kashyap D, Siddiqui ZA (2023) Effects of *Pseudomonas putida* and zinc oxide nanoparticles (ZnO NPs) with *Rhizobium leguminosarum* on the management of *Meloidogyne incognita* and *Pseudomonas syringae* pv. *pisi* on pea. Acta Phytopath Entom Hung. https://doi.org/10.1556/038.2022.00139

Kaviya S, Santhanalakshmi J, Viswanathan B (2011) Green synthesis of silver nanoparticles using *Polyalthia longifolia* leaf extract along with D-sorbitol: study of antibacterial activity. J Nanotechnol 2011:1–5

Keller AA, Lazareva A (2013) Predicted releases of engineered nanomaterials: from global to regional to local. Environ Sci Technol Lett 1:65–70

Khan M, Siddiqui ZA (2018) Zinc oxide nanoparticles for the management of *Ralstonia solanacearum*, *Phomopsis vexans* and *Meloidogyne incognita* incited disease complex of eggplant. Indian Phytopathol 71:355–364

Khan MR, Siddiqui ZA (2020) Use of silicon dioxide nanoparticles for the management of *Meloidogyne incognita*, *Pectobacterium betavasculorum* and *Rhizoctonia solani* disease complex of beetroot (*Beta vulgaris* L.). Sci Hort 265:109211. https://doi.org/10.1016/j.scienta.2020.109211

Khan MR, Siddiqui ZA (2021) Efficacy of titanium dioxide nanoparticles in the management of disease complex of beetroot (*Beta vulgaris* L.) caused by *Pectobacterium betavasculorum*, *Rhizoctonia solani*, and *Meloidogyne incognita*. Gesunde Pflanzen 73(4):445–464

Khan AU, Khan M, Khan MM (2019) Antifungal and antibacterial assay by silver nanoparticles synthesized from aqueous leaf extract of *Trigonella foenum-graecum*. BioNanoScience 9:597–605

Khan M, Khan AU, Hasan MA, Yadav KK, Pinto MM, Malik N, Yadav VK, Khan AH, Islam S, Sharma GK (2021a) Agro-nanotechnology as an emerging field: a novel sustainable approach for improving plant growth by reducing biotic stress. Appl Sci 11(5):2282

Khan M, Khan AU, Moon IS, Falimban R, Alam M (2021b) Synthesis of biogenic silver nanoparticles from the seed coat waste of pistachio (*Pistacia vera*) and their effect on the growth of eggplant. Nanotechnol Rev 10:1789–1800

Khan M, Khan AU, Bogdanchikova N, Garibo D (2021c) Antibacterial and antifungal studies of biosynthesized silver nanoparticles against plant parasitic nematode *Meloidogyne incognita*, plant pathogens *Ralstonia solanacearum* and *Fusarium oxysporum*. Molecules 26:2462

Khan M, Siddiqui ZA, Parveen A, Khan AA, Moon IS, Alam M (2022a) Elucidating the role of silicon dioxide and titanium dioxide nanoparticles in mitigating the disease of the eggplant caused by *Phomopsis vexans*, *Ralstonia solanacearum*, and root-knot nematode *Meloidogyne incognita*. Nanotechnol Rev 10:1789–1800

Khan AU, Khan M, Parveen A, Ansari S, Alam M (2022b) Effect of phyto-assisted synthesis of magnesium oxide nanoparticles (MgO-NPs) on bacteria and the root-knot nematode. Bioinorg Chem Appl. https://doi.org/10.1155/2022/3973841

Khan AU, Khan M, Malik N, Parveen A, Sharma P, Min K, Gupta M, Alam M (2022c) Screening of biosynthesized zinc oxide nanoparticles for their effect on *Daucus carota* pathogen and molecular docking. Microsc Res Tech 85(10):3365–3373. https://doi.org/10.1002/jemt.24191

Laware SL, Raskar S (2014) Effect of titanium dioxide nanoparticles on hydrolytic and antioxidant enzymes during seed germination in onion. Int J Curr Microbiol Appl Sci 3:749–760

Maensiri S, Laokul P, Klinkaewnarong J, Phokha S, Promarak V, Seraphin S (2008) Indium oxide (In_2O_3) nanoparticles using *Aloe vera* plant extract: synthesis and optical properties.J Optoelectron. Adv Mater 10:161–165

Mahmoodzadeh H, Nabavi M, Kashefi H (2013) Effect of nanoscale titanium dioxide particles on the germination and growth of Canola (*Brassica napus*). J Ornam Hortic Plant 3:25–32

Nair R, Varghese SH, Nair BG, Maekawa T, Yoshida Y, Kumar DS (2010) Nanoparticulate material delivery to plants. Plant Sci 179:154–163

Narendhran S, Rajiv P, Rajeshwari S (2016) Influence of zinc oxide nanoparticles on growth of *Sesamum indicum*, L. in zinc deficient soil. Int J Pharm Pharm Sci 8:365–371

Navale GR, Thripuranthaka M, Late DJ, Shinde SS (2015) Antimicrobial activity of ZnO Nanoparticles against pathogenic bacteria and fungi. JSM Nanotechnol Nanomed 3:10–33

Parveen A, Siddiqui ZA (2021) Zinc oxide nanoparticles affect growth, photosynthetic pigments, proline content and bacterial and fungal diseases of tomato. Arch Phytopathol Plant Protect 54(17–18):1519–1538

Parveen A, Siddiqui ZA (2022a) Foliar spray and seed priming of titanium dioxide nanoparticles and their impact on the growth of tomato, defense enzymes and some bacterial and fungal diseases. Arch Phytopathol Plant Protect 55(5):527–548

Parveen A, Siddiqui ZA (2022b) Zirconium dioxide nanoparticles affect growth, photosynthetic pigments, defense enzymes activities and mitigate severity of bacterial and fungal diseases of tomato. Gesunde Pflanzen 74(3):615–628. https://doi.org/10.1007/s10343-022-00636-z

Parveen A, Siddiqui ZA (2022c) Impact of silicon dioxide nanoparticles on growth, photosynthetic pigments, proline, activities of defense enzymes and some bacterial and fungal pathogens of tomato. Vegetos 35:83–93. https://doi.org/10.1007/s42535-021-00280-4

Parveen A, Siddiqui ZA (2023) Effect of silver oxide nanoparticles on growth, activities of defense enzymes and fungal and bacterial diseases of tomato. Gesunde Pflanzen 75(2):405–414

Patil SV, Borase HP, Patil CD, Salunke BK (2012) Biosynthesis of silver nanoparticles using latex from few euphorbian plants and their antimicrobial potential. Appl Biochem Biotechnol 167:776–790

Qi M, Liu Y, Li T (2013) Nano-TiO(2) improve the photosynthesis of tomato leaves under mild heat stress. Biol Trace Elem Res 156:323–328

Rafique R, Arshad M, Khokhar M, Qazi I, Hamza A, Virk N (2014) Growth response of wheat to titania nanoparticles application. NUST J Eng Sci 7:42–46

Rajiv P, Rajeshwari S, Venckatesh R (2013) Bio-fabrication of zinc oxide nanoparticles using leaf extract of *Parthenium hysterophorus* L. and its size-dependent antifungal activity against plant fungal pathogens. Spectrochim Acta A 112:384–387

Raliya R, Biswas P, Tarafdar J (2015) TiO_2 nanoparticle biosynthesis and its physiological effect on mung bean (*Vigna radiata* L.). Biotechnol Rep 5:22–26

Saradhadevi M, Gnanadesigan M, Kapildev G, Vasanth D (2017) Dataset on antitumor properties of silver nanoparticles from *Gloriosa superba* (L.) seed on *Dalton Lymphoma Ascites* (DLA) tumor: facile and biocompatible approach. Data Brief 14:524–530

Sharma SK, Khan AU, Khan M, Gupta M, Gehlot A, Park S, Alam M (2020) Biosynthesis of MgO nanoparticles using *Annona squamosa* seeds and its catalytic activity and antibacterial screening. Micro Nano Lett 15(1):30–34

Siddiqui ZA, Aziz S (2020) Management of *Rhizoctonia solani*, *Fusarium solani* and *Meloidogyne incognita* by Silicon dioxide nanoparticles and *Rhizobium ciceri* alone and in combination on chickpea. J Biosens Renew Sci 1(2) JBRS.MS.ID.000109

Siddiqui ZA, Khan MR, Aqib (2023) Comparative efficacy of Zirconium dioxide nanoparticles and AM fungus against wilt disease complex of bottle gourd and upregulation of biochemical and physiological processes. Intern J Veg Sci 29:243–260. https://doi.org/10.1080/19315260.2023.2187503

Siddiqui MH, Al-Whaibi MH, Faisal M, Al Sahli AA (2014) Nano-silicon dioxide mitigates the adverse effects of salt stress on *Cucurbita pepo* L. Environ Toxicol Chem 33:2429–2437

Siddiqui ZA, Khan A, Khan MR, Abd-allah EF (2018) Effects of zinc oxide nanoparticles (ZnO NPs)and some plant pathogens on the growth and nodulation of lentil (*Lens culinaris* Medik.). Acta Phytopath Entom Hung 53(2):195–211

Siddiqui ZA, Khan MR, Abdallah EF, Parveen A (2019a) Titanium dioxide and zinc oxide nanoparticles affect some bacterial diseases, and growth and physiological changes of beetroot. Int J Veg Sci 25:409–430

Siddiqui ZA, Parveen A, Ahmad L, Hashem A (2019b) Effects of graphene oxide and zinc oxide nanoparticles on growth, chlorophyll, carotenoids, proline contents and diseases of carrot. Sci Hortic 249:374–382

Siddiqui ZA, Hashmi A, Khan MR, Parveen A (2020) Management of bacteria *Pectobacterium carotovorum*, *Xanthomonas campestris* pv. *carotae*, and fungi *Rhizoctonia solani*, *Fusarium solani* and *Alternaria dauci* with silicon dioxide nanoparticles on carrot. Intern J Veg Sci 26(6):547–557. https://doi.org/10.1080/19315260.2019.1675843

Song JY, Kwon EY, Kim BS (2010) Biological synthesis of platinum nanoparticles using *Diospyros kaki* leaf extract. Bioprocess Biosyst Eng 33(1):159–164

Sriram T, Pandidurai V (2017) In vitro growth analysis of *Zea mays* L. using silver nanoparticles. Int J Pharma Bio Sci 8(3b):30–37

Taniguchi N, Arakawa C, Kobayashi T (1974) On the basic concept of nano-technology'. In: Proceedings of the international conference on production engineering, 1974–8, vol 2, pp 18–23

Thomas R, Jasim B, Mathew J, Radhakrishnan EK (2012) Extracellular synthesis of silver nanoparticles by endophytic *Bordetella* sp. isolated from *Piper nigrum* and its antibacterial activity analysis. Nano Biomed Eng 4:183–187

Venkatachalam P, Jayaraj M, Manikandan R, Geetha N, Rene ER, Sharma NC, Sahi SV (2017) Zinc oxide nanoparticles (ZnO NPs) alleviate heavy metal-induced toxicity in *Leucaena leucocephala* seedlings: a physiochemical analysis. Plant Physiol Biochem 110:59–69

Wang Y, Lin Y, Xu Y, Yin Y, Guo H, Du W (2019) Divergence in response of lettuce (var. ramosa Hort.) to copper oxide nanoparticles/microparticles as potential agricultural fertilizer. Environ Pollut Bioavailab 31:80–84

Whan JA, Dann EK, Aitken EA (2016) Effects of silicon treatment and inoculation with *Fusarium oxysporum* f. sp. *vasinfectum* on cellular defences in root tissues of two cotton cultivars. Ann Bot 118:219–226

Zargar M, Hamid AA, Bakar FA, Shamsudin MN, Shameli K, Jahanshiri F (2011) Green synthesis and antibacterial effect of silver nanoparticles using *Vitex negundo* L. Molecules 16:6667–6676

Zhao LJ, Peralta-Videa JR, Rico CM, Hernandez-Viezcas JA, Sun Y, Niu G, Servin A, Nunez JE, Duarte-Gardea M, Gardea-Torresdey JL (2014) CeO_2 and ZnO nanoparticles change the nutritional quality of cucumber (*Cucumis sativus*). J Agric Food Chem 62:2752–2759

Chapter 18
Nanomaterials as Effective Fertilizers and Fungicides in Plant Protection for Improving Crop Growth and Inducing Tolerance Against Phytopathogens

Masudulla Khan, Jelena Popović-Djordjević, and Jelena Katanić Stanković

Abstract Nanotechnology is utilized in agriculture to increase crop output, lower yield losses, and activate plant defense mechanisms against pests, phytopathogens, and other environmental threats. Global agriculture is dealing with a number of difficulties due to a changing climate, such as temperature changes, landslides, and growing food demand. Crop output can be increased, crop tolerance can be improved, and environmental pollution can be reduced with the use of nanotechnology. Because plants can absorb nanoparticles and because they affect physiological processes, nanoparticles (NPs) are frequently employed in agriculture. The reduction of stress caused by such NPs varies significantly depending on the number of NPs administered, the stage of plant growth, and environmental conditions. Stress, hormone control, and the activation of the antioxidant defense system are only a few examples of the physiological and biochemical processes. Stress, hormone regulation, and the activation of the antioxidant defense system are only a few examples of the physiological and biochemical processes in plants that can be modified and controlled by NPs.

Keywords Nanoparticles · Plant physiology · Phytohormone · Biotic stress

M. Khan (✉)
Botany Section, Women's College, Aligarh Muslim University, Aligarh, Uttar Pradesh, India

J. Popović-Djordjević
Department of Food Technology and Biochemistry, University of Belgrade—Faculty of Agriculture, Belgrade, Serbia
e-mail: jelenadj@agrif.bg.ac.rs

J. K. Stanković
Department of Science, University of Kragujevac, Institute for Information Technologies Kragujevac, Kragujevac, Serbia
e-mail: jkatanic@kg.ac.rs

M. Khan, J.-T. Chen (eds.), *Nanoparticles in Plant Biotic Stress Management*,
https://doi.org/10.1007/978-981-97-0851-2_18

18.1 Introduction

It is necessary to preserve the agricultural system for future generations since it is the foundation of a prosperous civilization. Agriculture's sustainable development has risen to the top of the agenda. The annual usage of agrochemicals worldwide is expected to be 50×10^6 t for phosphate-based fertilizers, 120×10^6 t for N-based fertilizers, and more than 2.6×10^6 t for insecticides in 2020 (Gilbertson et al. 2020; World Fertilizer Trends and Outlook 2017). In order to increase crop yields, conventional agricultural practices frequently over-apply agrochemicals, which include fertilizers and insecticides, to crops (Dolijanović et al. 2022). Currently, less than 10% of applied pesticides reach their intended targets, and more than 50% of applied N and 85% of applied P are not absorbed by crops.

Although overapplication is a practical on-farm practice in many regions of the world, more effective delivery systems will be required in the future due to the costs and availability of raw materials. Net agrochemical discharges into the environment contaminate the soils, surface waters, and groundwater, which have serious negative effects on human health and the environment (such as anoxic water bodies, ecotoxicity, biodiversity loss, and contaminated drinking water sources) (Gilbertson et al. 2020; World Fertilizer Trends and Outlook 2017; Atwood and Paisley-Jones 2017; Dolijanović et al. 2022; Elser and Platz 2015; Springmann et al. 2018; Lassaletta et al. 2014).

The greatest problem of all is meeting the demand for food because the human population is growing quickly (Khan et al. 2021a). Agricultural food production suffers a significant loss due to several biotic and abiotic causes. Plant parasitic nematodes, fungi, bacteria, and viruses, as well as other pathogens, are biotic elements that significantly reduce plant viability during crop production. Globally, 20–40% of all agricultural products are harmed by pest attacks, with plant diseases responsible for 16% of the yield loss (Khan et al. 2021a; Worrall et al. 2018). Several strategies to suppress pests and pathogens have been used to lessen the damage they cause. A ban on some crops was imposed by many nations in order to stop the spread of illness.

To boost production, a variety of chemical, biological, and other integrated pest management techniques are used. Despite all of this, illnesses continue to cause severe production loss in agriculture. Nanoparticles can be utilized in agriculture to enhance crop growth (Grillo et al. 2021). Nutrient intake, photosynthesis, nutrient transport, and high photosynthate accumulation are all improved by using nano-biofertilizers. As crop growth increases, plant quality and productivity are also rising. By encasing silver and gold nanoparticles in a biofertilizer, the usage of nano-biofertilizer in agricultural products has been found to greatly boost crop growth. When nanoparticles were used directly as fertilizer, growth and nutritional enhancement of the plant were seen in a variety of agricultural products (Dikshit et al. 2013; Mohammad Ghasemi et al. 2020; Moradi Rikabad et al. 2019). By encouraging the germination of the crops in the seedlings, PGPR increases the productivity of the plants. Additionally, some leguminous plants may be effective at boosting crop productivity (Rahman and Zhang 2018).

Continuous use of chemical fertilizers reduces the soil's fertility because it depletes the soil's yield content. The degradation of soil fertility is also significantly impacted by soil acidification (Rahman and Zhang 2018). Due to the significant correlations between low soil fertility and soil nutrient imbalance and decreased crop productivity, soil nutrient deficiencies, and decreased food nutritional value, farmers are in a difficult position (Monreal et al. 2016). Nano-biofertilizers offer a cost-effective, sustainable, and effective integrated nutrient management solution to address these drawbacks. They improve plants' ability to absorb and assimilate nutrients (Mohammad Ghasemi et al. 2020). By leaching or competing with other organisms, they also lower nutrient losses in the soil (Janmohammadi et al. 2016). The bioorganic elements of nano-biofertilizer (PGPR) aid in phosphate dissolution, nitrogen fixation, and soil nutrient richness restoration. The biofertilizer's nanoparticle coating aids in the synchronized slow release of nutrients in response to crop demand (Morales-Díaz et al. 2017).

Nanoparticles can be absorbed and transported by plants, which can be affected by a variety of factors including particle size, surface charge, concentration, exposure time, and plant species. In plants, two pathways for the uptake of nanoparticles and their transport occur (Ma et al. 2015). Transport of NPs in plants may occur from the root to the stem, leaf, or fruit through the xylem or phloem (Hernandez-Viezcas et al. 2013; Zhao et al. 2014). Root hairs, stomata, and leaf surfaces are the ways that nanoparticles can get into a plant's system. After entry into plants, the nanoparticles can either diffuse, bulk flow, or phloem load their way through the system. Previous studies have primarily used seed treatment, branch injection, root application, and foliar spraying to apply nanoparticles to plants. According to some studies, plants can absorb nanoparticles. However, limited direct evidence of nanoparticle transport in plants exists (Rico et al. 2011). To create the best nanoparticles for agricultural use, it is essential to comprehend how agricultural nanoparticles are absorbed and transported by plants (Grillo et al. 2021).

This chapter discusses the use of agricultural nanoparticles in plant protection and physiology as well as the uptake and transport of nanoparticles.

18.2 Applications of Nanotechnology in Plant Protection

The science of creating, modifying, and implementing materials at the molecular, atomic, or supramolecular level is known as nanotechnology. The National Nanotechnology Initiative defined nanotechnology as "the understanding and control of matter at dimensions between approximately 1 and 100 nm, where unique phenomena enable novel applications" in its general definition. In comparison to their bulk counterparts, nanoparticles exhibit special qualities such as chemical stability, enhanced thermal conductivity, catalytic reactivity, and biological activity.

This is because they have a high surface area-to-volume ratio (Khan et al. 2021a; Srećković et al. 2023; Mihailović et al. 2023).

Applications for nanoparticles include biomolecular detection, antibacterial action, and disease diagnosis (Khan et al. 2021a; Moradi Rikabad et al. 2019; Vuković et al. 2023). The most recent and promising field of research in the modern period is the fusion of biology and nanotechnology. The goal of bionanotechnology, which combines biotechnology and nanotechnology, is to provide environmentally benign and biologically based methods for the synthesis of nano-based materials (Mihailović et al. 2023; Srećković et al. 2021).

Nanomaterials can be utilized in the field of plant protection to transport pesticides to specific areas of the plant. Since nanomaterials are known to have antifungal, antibacterial, and antiviral activities, they can be used as a foliar spray in seed treatment and root application (Khan et al. 2021a; Vuković et al. 2023; Srećković et al. 2021).

Metallic oxides, metalloids, nonmetals, and carbon nanomaterials are among the NPs that have been shown to inhibit plant diseases. The most often researched and utilized NPs include iron, titanium oxide, gold, silver, zinc oxide, and copper (Khan et al. 2021a; Moradi Rikabad et al. 2019). Numerous phytopathogenic microbesP, including the fungi *Phomopsis vexans*, *Rhizoctonia solani*, and *Fusarium oxysporum*, phytopathogenic bacteria, and plant parasitic nematodes *Meloidogyne incognita*, *Ralstonia solanacearum*, and *Pectobacterium carotovorum* have been successfully combated by the various nanoparticles (Khan et al. 2022; Ahamad and Siddiqui 2021; Parveen and Siddiqui 2021).

It has been demonstrated that magnesium oxide nanoparticles (MgO-NPs) are efficient against the growth, physiology, and *Alternaria alternata* infection of the carrot (*Daucus carota* L.). In their study, Ahamad et al. reported this (Ahamad et al. 2023). MgO-NPs showed fungicidal activity against the fungus under study. Silicon dioxide (SiO_2-NPs) at 0.2 mg/L inhibits the growth of *Xanthomonas campestris*, *Pseudomonas syringae*, *Pectobacterium carotovorum*, *Ralstonia solanacearum*, *Fusarium oxysporum*, and *Alternaria solani* (Parveen and Siddiqui 2021). Additionally, SiO_2-NPs prevent the growth of the phytopathogenic fungus *Phomopsis vexans* and the bacteria *Ralstonia solanacearum* (Khan et al. 2022a). Plants gain strength from silicon nanoparticles to combat disease. Silicon has the ability to successfully develop host resistance (Rastogi et al. 2019; Rajput et al. 2021).

With a global yearly production of 33,400 tons, zinc oxide (ZnO) nanoparticles are the third most commonly used NPs and can treat plant diseases (Sabir et al. 2014; Bondarenko et al. 2013). With regard to the fungi *Aspergillus flavus*, *A. niger*, *Rhizopus oryzae*, *Trichophyton*, *Fusarium*, *Fonsecaea*, *Ramichloridium schulzeri*, and *Cladosporium*, George et al. (2014) screened the antifungal activity of ZnO and TiO_2 NPs. The test microorganisms were successfully neutralized by both nanoparticles. ZnO and TiO_2 NP application lessens illness brought on by phytopathogenic microorganisms (Siddiqui et al. 2018). Silver nanoparticles (NPs) have antibacterial properties and may be employed to treat diseases (Figs. 18.1 and 18.2) (Kaur et al. 2019). According to Namburi et al. (2021), biogenic AgNPs had antibacterial efficacy against the bacterium that causes rice's bacterial leaf blight, *Xanthomonas oryzae* pv. *oryzae*. The inhibitory impact of nanoparticles on plant pathogens is shown in Table 18.1.

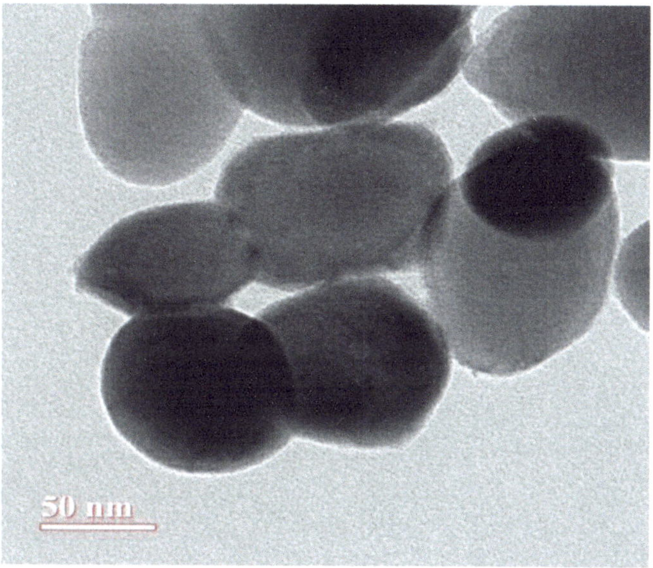

Fig. 18.1 TEM image of synthesized AgNPs (Taken from Khan et al. 2022, *Biomolecules*. 12(2), 74)

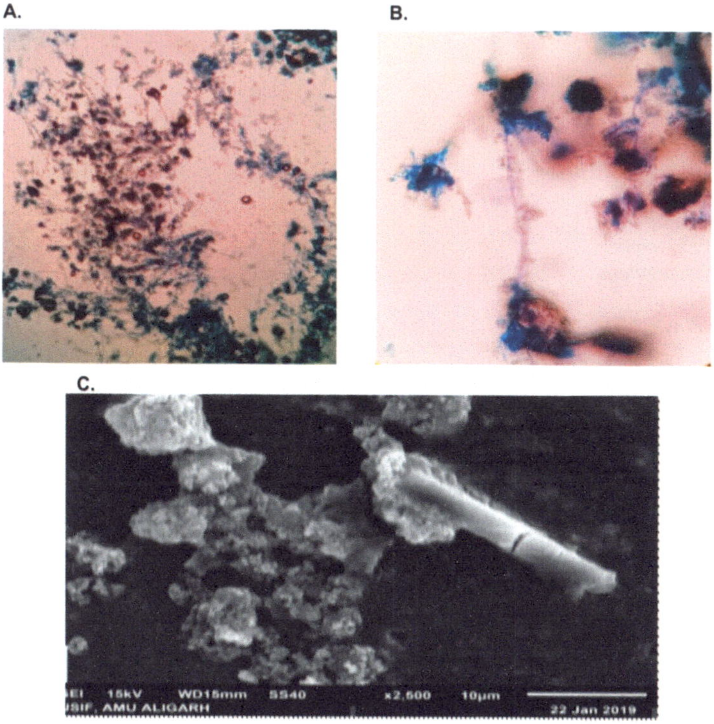

Fig. 18.2 (**a**, **b**). Light microscope images of *A. flavus* fungal mycelium treated with AgNPs; (**c**). SEM image showing disturbed fungal mycelium treated with biosynthesized AgNPs (Taken from Khan et al. 2022, *Biomolecules*. 12(2), 74)

Table 18.1 Inhibitory impact of nanoparticles on plant pathogens

Tested pathogen	Size	Concentration	Nanoparticle	Effect	Reference
Botrytis cinerea, *Colletotrichum graminicola*, *Colletotrichum musae*, *Magnaporthe oryzae*, *Penicillium digitatum*, *Rhizoctonia solani*, *Sclerotium rolfsii*	20 nm	250, 500, and 750 mg kg^{-1}	CuO	In vitro inhibitory effect	Huang et al. (2015)
Phytophthora infestans	11–55 nm	35, 54, 110, and 224 g hl^{-1}	CuO and Cu_2O	In vivo (foliar) inhibitory effect	Giannousi et al. (2013)
Pactobacterium carotovorum, *P. vexans*, *M. incognita*	100 nm	100 μg/ml	CuO	In vitro inhibitory effect	Khan et al. (2023)
Alternaria alternata, *A. brassicicola*, *A. solani*, *Fusarium oxysporum* f.sp. *cucumerinum*, *F. oxysporum* f. sp. *lycopersici*, *F. oxysporum*, *F. solani*, *Fusarium* sp., *Botrytis cinerea*, *Cladosporium cucumerinum*, *Corynespora cassiicola*, *Cylindrocarpon destructans*, *Didymella bryoniae*, *Glomerellacingulata*, *Monosporascus cannonballus*, *Pythium aphanidermatum*, *P. spinosum*, *Stemphylium Lycopersici*	7–25 nm	10,000–50,000 μl mL^{-1}	Ag	In vitro inhibitory effect	Kim et al. (2012)
Fusarium culmorum	~20 nm	0.12–10 ppm	Ag	Inhibitory effect under In vitro	Kasprowicz et al. (2010)
Alternaria alternata, *Sclerotinia sclerotiorum*, *Macrophomina phaseolina*, *Rhizoctonia solani*, *B. cinerea*, and *Curvularia lunata*	10–50 nm	5, 10, and 15 mg L^{-1}	Ag	In vitro inhibitory effect	Krishnaraj et al. (2012)

Table 18.1 (continued)

Tested pathogen	Size	Concentration	Nanoparticle	Effect	Reference
Colletotrichum sp.	4–8 nm	10, 30, 50, 100 ppm	Ag	In vivo (pepper plant) and in vitro	Lamsal et al. (2011)
Meloidogyne incognita, Ralstonia solanacearum and Fusarium oxysporum	100 nm	100 ppm	AgNPs	Inhibitory effect	Khan et al. (2021b)
Aspergillus flavus, A. niger, A. fumigatus, Fusarium culmorum, and F. oxysporum	27 ± 5 (spherical) and 84 ± 2 (hexagonal)	25, 50, 100 μg mL^{-1}	ZnO	In vitro	Rajiv et al. (2013)

18.3 Plant Hormones and Pathogenesis-Related Proteins

The key defense mechanism by which host plants become resistant is when signaling molecules like jasmonic acid (JA), salicylic acid (SA), and ethylene (ET) activate the proper defense responses. Auxin (IAA), gibberellin (GA), cytokinin (CK), abscisic acid (ABA), brassinosteroids (BRs), and strigolactone (SL) are other plant growth-regulating hormones that have the potential to control defensive reactions. Nanoparticles might alter the hormonal balance of plants. After being exposed to CuO-NPs, *Arabidopsis thaliana* showed increased levels of SA and JA as well as the transcription of genes involved in defensive signaling, according to Landa et al. (2017). Again, in tomato plants under salt stress, the combination of polyvinyl alcohol hydrogels (Cs-PVA) and Cu-NPs causes overexpression of the JA gene (Hernández-Hernández et al. 2018).

While the JA level remained unchanged, Cai et al. (2020) found increased SA and ABA levels in *N. benthamiana*. The aforementioned findings thus imply that the specific interactions between NPs and plants, timing, and application doses all affect how a particular plant hormone is expressed. When *Capsicum annuum* L. was exposed to Ag-NPs, it was discovered that the expression of other hormones, such as CK, also increased (Vinković et al. 2017). At moderate and high concentrations, ZnO-NPs may also function as a stressor molecule that upregulates plant defenses and the level of the stress hormones SA and ABA. They also stimulate cytokinin production (Landa et al. 2017). Numerous methods are used by plants to control their growth and their defenses against diseases like viruses. Tobacco plant infected with TMV upon exposure to TiO_2 NPs and Fe_2O_3 alters the level of phytohormones such as ABA, zeatin riboside, and brassinoid (Hao et al. 2016).

Furthermore, it was found that ethylene signaling-dependent biochemical processes in plants are under the control of ZnO-NPs. A considerable rise in ethylene-related genes was also discovered in cucumbers treated with Ag2S-NPs (Wang et al.

2016). Expression of genes connected to ethylene significantly increased in the presence of ZnO-NPs. Following seed preparation with FeNPs, Kasote et al. (2019) research on watermelon seedlings showed that JA-mediated defense responses were activated. It has been found that stress-coping JA signaling is enhanced in some circumstances with extra abiotic and biotic restrictions and low nanoparticle concentration, which ultimately activates defense responses and has negative consequences on plants (Hernández-Hernández et al. 2018). Additionally, it was revealed that applying copper sulfide nanoparticles to infected plants during the fungal infection of *O. sativa* brought on by *G. fujikuroi* boosted the synthesis of jasmonic acid (Shang et al. 2020).

18.4 Nanomaterials in Crop Growth, Seed Germination, and Quality Enrichment

The creation of low-cost nanomaterials and their use in promoting increased seed germination, plant growth, and environmental adaption are all a part of the emerging field of study known as nanoscience. Numerous research have shown that the usage of nanomaterials has a good effect on seed germination, development, and plant growth. Maize, barley, soybean, maize, tomato, wheat, garlic, and peanut seeds all germinate better when multi-walled carbon nanotubes (MWCNTs) are added (Joshi et al. 2018). Similar to this, seed germination is aided by the application of nano-zeolite, TiO_2, and SiO_2 to cultivated plants (Manjaiah et al. 2019). According to another research, nanomaterials may activate the enzymatic system, boosting the capacity to absorb and use water, which, by penetrating the seed coat, enhances seedling growth and germination (Lu et al. 2002). Additionally, studies with TiO_2, FeO, MWCNTs, ZnFeCu-oxide, ZnO, and hydroxyfluorene nanomaterials have demonstrated that they promote crop development and growth as well as crop quality in wheat, onion, potato, mung bean, tomato, soybean, spinach, saffron, and peanut crop types (Moradi Rikabad et al. 2019; Gao et al. 2011). After being treated with fertilizer containing nano-iron, the *Dracocephalum moldavica* plant's essential oil concentration improved (Chen et al. 2015). Depending on the type of plant, application method, dose, and type of nanoparticle, nanoparticles can have a variety of effects on plant growth (Liu et al. 2015). The dosage and size of nanoparticles significantly affect their behavior, toxicity, and reactivity. Tomato seeds treated with carbon nanotubes sprout more quickly and generate more biomass. The seeds germinate more quickly due to increased water uptake by the seeds and carbon nanotubes' ability to pierce the dense seed coat (Rizwan et al. 2019). Nanoparticles in plant growth improvement are shown in Table 18.2.

Table 18.2 Nanoparticles in plant growth improvement

Nanoparticles	Concentration	Plant	Effect	Reference
TiO_2	25 and 50mg/L	*Saffron (Crocus sativus* L.)	Increase of total anthocyanins and malondialdehyde (MDA) in saffron leaves, and total phenolics and total flavonoids, radical scavenging activity ((DPPH) in saffron stigmas	Moradi Rikabad et al. (2019)
SiO_2 and TiO_2 NPs	0.2 mg/L	*Solanum melongena*	Increase photosynthetic pigments, growth parameters	Khan et al. (2022a)
Zn and FeO NPs	100 mg/L	Wheat	Increased the wheat growth	Rizwan et al. (2019)
TiO_2-NPs	10 mg/L	*Vigna radiata* L.	An improvement was observed in shoot length and root length	Raliya et al. (2015)
ZnO-NPs	600 mg/L	*Vigna mungo*	Improve germination and seedling vigor	Raja et al. (2019)

Table 18.3 Nanoparticle as carriers of insecticides

Nanoparticle	Insecticide	Target pest	Reference
Chitosan and ZnO	Azadirachtin	Groundnut bruchid (*C. serratus*)	Jenne et al. (2018)
Chitosan/TPP	Nicotine	House fly (*M. domestica*)	Yang et al. (2018)
Dendrimers	Thiamethoxam	Cotton bollworm (*H. armigera*)	Yang et al. (2018)
Chitosan	Azadirachtin	Tobacco cutworm (*S. litura*) culture	Lu et al. (2013)

18.5 Nanoparticles in the Delivery of Insecticides

Insecticide delivery by nanoparticles was first attempted in the early 2000s. *Helicoverpa armigera*, *Spodoptera litura*, and *Tetranychus urticae* are the most often targeted pests (Jenne et al. 2018; Yang et al. 2009, 2018). Nanoparticles can make insecticides more soluble and less harmful. Less water-soluble insecticides have been successfully loaded into porous silica (Jenne et al. 2018; Wang et al. 2014) and chitosan (Feng and Peng 2012). According to some studies, nanoparticle carriers may be used to make active pesticides more soluble, as indicated in Table 18.3. Another problem is pesticide volatilization, which can be effectively dealt with by encapsulating volatile active ingredients inside nanoparticle carriers. For instance, red flour beetles (*Tribolium castaneum*) died at an 80% rate when polyethylene glycol (PEG) was used to encapsulate garlic essential oil (Yang et al. 2009). Nanoparticles as carriers of insecticides are shown in Table 18.3.

Table 18.4 Some specific studies using nanoparticles in herbicide delivery

Target Pest	Nanoparticle	Herbicide	Reference
Potato	Poly (lactic-co-glycolic acid)	Atrazine	Schnoor et al. (2018)
Rice; *D. sanguinalis; A. thaliana*	Polymer	Metolachlor	Tong et al. (2017)
Maize	Nanosized rice husk	2,4-D	Chidambaram (2016)
B. pilosa	Chitosan/ tripolyphosphate/alginate	Imazapic and Imazapyr combined	Maruyama et al. (2016)
E. crus-galli	Chitosan	Diuron	Yu et al. (2015)
Mustard (*Brassica* sp.); Maize	Tripolyphosphate/chitosan	Paraquat	Grillo et al. (2014)

18.6 Nanoparticles in the Delivery of Herbicides

The main goal of nano-carrier-based herbicide delivery is to reduce the damage that herbicides do to the environment. Herbicides made using nanoparticles come in a wide range of varieties, for instance, nano-sized tubular halloysite, amino-activated iron (II, III) oxide magnetic NPs, hollow core-shell manganese carbonate, and nano-sized rice husks (Chidambaram 2016). During genotoxicity tests on *Allium cepa* cell cultures, simazine, ametryn, or atrazine-containing polymeric nanoparticles were marginally less hazardous than the herbicide alone (Grillo et al. 2012). According to Silva et al. (2010), paraquat-loaded chitosan/alginate NPs in soil had less herbicide leakage than paraquat alone. In addition, the release rate lasted 2 h longer than with just herbicide. Some specific studies using nanoparticles in herbicide delivery are shown in Table 18.4.

18.7 Nanofertilizers

The misuse of hazardous chemicals in agricultural production has ruined the characteristics of the soil, decreased the amount of essential nutrients, and damaged the environment. The extensive use of chemicals as fertilizers damages the environment, lowers the quality of the soil, and leaves residues in the soil, on agricultural goods, and on crop plants. A novel method called nanofertilizers uses nanoparticles to control the release of nutrients and make them more efficient and affordable (Mohammad Ghasemi et al. 2020). Advanced fertilizers called nanofertilizers were created using nanotechnology to offer a sustainable and eco-friendly way to fertilize crops. At least half of the particles in nanofertilizers are smaller than 100 nm.

They are made to give the plant a steady stream of nutrients in a controlled way, providing a gradual release. Nanomaterials are effective at holding and releasing nutrients because they have a high surface area-to-volume ratio. Additionally, using nanofertilizers can improve environmental sustainability by lowering runoff and nutrient leaching into the environment. They can also increase the effectiveness of fertilizer use, which increases crop yields and lowers the overall cost of fertilizer application. In regions where conventional fertilizers are inefficient or useless, nanofertilizers are very helpful. With fewer negative effects on the environment, nanofertilizers can offer a more effective and economical approach to fertilizing crops. They are the result of a promising new technology that can support meeting the rising demand for food while enhancing the sustainability of agriculture. Given their usage of nanomaterials, nanofertilizers currently have drawbacks, such as greater production costs and potential safety and environmental issues. Additionally, more research is required to fully comprehend their long-term effects on soil health, crop development, and the environment. Nanofertilizers are a crucial component of plant health products and have the potential to increase a plant's resilience to disease. Fusarium wilt in chrysanthemums, Fusarium crown, root rot in asparagus, Verticillium wilt in brinjal, red root rot in tea, Fusarium wilt in tomato, etc. have all been studied in detail (Elmer and White 2018; Elmer et al. 2018). Some NPs can directly enhance plant development in addition to acting as nanofertilizers (Pandey et al. 2021; Bala et al. 2019).

18.8 Conclusion and Future Remarks

Knowing how to regulate phytohormone synthesis and signaling under stressful conditions is crucial for the future growth and sustainability of plant production. It discusses how nanoparticles impact plant growth as well as how they are absorbed and transported by plants. Nanoparticles may interfere with a number of hormone production and signaling pathways when plant cells are under stress. Targets for nanoparticle absorption by plant leaves include plant roots, primary roots, damaged areas, stomata, and pores of the stratum corneum of the stem cell wall. These nanoparticles are transported inside the plant body by both symplastic and apoplastic pathways, as well as by xylem and phloem, which are employed to transmit them between various plant tissues.

The absorption and transport process is influenced by the physical characteristics of nanoparticles, including size, chemical composition, surface charge, and surface modification. Additionally, nanoparticles have a variety of effects on plant seed germination and growth when they are moving inside a plant. The ability of certain nanoparticles to harm plants at high concentrations while promoting plant development at low quantities has been demonstrated. The genotoxic effects of nanoparticles, such as chromosome aberration and the development of micronuclei, will lead to alterations in plant hormone concentrations and even changes in gene expression in plants. Although the harmful effects of nanoparticles on plants have been well confirmed, nothing is known about the mechanism underlying plant toxicity.

Additionally, NPs can harm plants and their cells, impairing the strength of their cell walls and causing membrane damage as well as cytoplasm contraction. The quantity of NPs exposed is the main determining factor. Many NP species have been proven to stimulate plant growth at lower concentrations, whereas plants are likely to suffer at greater concentrations of NPs. Therefore, the type of nanoparticles employed, their concentration, and timing of application rely on how bio-based natural polymer nanoparticles are used as an emerging technology to increase agricultural productivity. It should also be further researched in order to maximize its interaction in a given setting. These technologies consequently support predicted plant culture and output.

In conclusion, research to date emphasizes the unstoppable trend of rising food demands, which calls for a corresponding change in agricultural methods. The current approaches cannot satisfy the expanding needs of humanity. The widespread use of chemical pesticides in agriculture has had negative environmental effects that are severe and long-lasting. This necessity necessitates the development of alternate strategies that demonstrate less environmental toxicity while concurrently fulfilling rising food requirements. According to statistics, chemical fertilizers were used in about 192.8 million metric tons during 2016 and 2017. Given their intrinsic toxicity, it becomes necessary to maximize the application of chemical pesticides and fungicides. Nanotechnology, known for its amazing powers, has produced nanostructured materials that are essential delivery systems for fertilizers and pesticides.

Intelligent fertilizers can be developed using nanomaterials, which can also increase nutrient availability and reduce environmental contamination. An encouraging alternate route is provided by the developing field of nanotechnology, which has the power to reduce crop illnesses and increase productivity. Previous research has confirmed that nanoparticles have a beneficial effect on agricultural plants, supporting its promise as a game-changing technology. The use of harmful chemicals and pesticides, which are renowned for damaging soil, ecosystems, and groundwater, may be reduced as a result of this breakthrough. In order to scale up and standardize this technology for widespread use, more research is absolutely necessary. Unwise use of chemical agents and pesticides has always resulted in environmental contamination, taking a noticeable toll on the fauna and flora. Additionally, the growing need for inventing fresh remedies is further highlighted by the infections' and pests' growing resistance to fungicides and insecticides.

Acknowledgments Authors JPDj and JKS acknowledge the Ministry of Science, Technological Development, and Innovation of the Republic of Serbia for financial support (Grant No. 451-03-47/2023-01/200116).

References

Ahamad L, Siddiqui ZA (2021) Effects of silicon dioxide, zinc oxide and titanium dioxide nanoparticles on *Meloidogyne incognita, Alternaria dauci and Rhizoctonia solani* disease complex of carrot. Exp Parasitol 230:108176

Ahamad L, Azmat AL, Khan M, Farid O, Mahboob A (2023) Exploring the nano-fungicidal efficacy of green synthesized magnesium oxide nanoparticles (MgO NPs) on the development,

physiology, and infection of carrot (*Daucus carota* L.) with Alternaria leaf blight (ALB): molecular docking. J Integr Agric 10:3069–3080. https://doi.org/10.1016/j.jia.2023.02.034
Atwood D, Paisley-Jones C (2017) Pesticides industry sales and usage: 2008–2012 market estimates. United States Environmental Protection Agency
Bala R, Kalia A, Dhaliwal SS (2019) Evaluation of efficacy of ZnO nanoparticles as remedial zinc nanofertilizer for rice. J Soil Sci Plant Nutr 19:379–389
Bondarenko O, Juganson K, Ivask A, Kasemets K, Mortimer M, Kahru A (2013) Toxicity of Ag, CuO and ZnO nanoparticles to selected environmentally relevant test organisms and mammalian cells in vitro: a critical review. Arch Toxicol 87(7):1181–1200
Cai L, Cai L, Jia H, Liu C, Wang D, Sun X (2020) Foliar exposure of Fe_3O_4 nanoparticles on *Nicotiana benthamiana*: evidence for nanoparticles uptake, plant growth promoter and defense response elicitor against plant virus. J Hazard Mater 393:122415
Chen G, Qiu J, Liu Y, Jiang R, Cai S, Liu Y, Zhu F, Zeng F, Luan T, Ouyang G (2015) Carbon nanotubes act as contaminant carriers and translocate within plants. Sci Rep. 5 art. No: 15682
Chidambaram R (2016) Application of rice husk nanosorbents containing 2, 4-dichlorophenoxyacetic acid herbicide to control weeds and reduce leaching from soil. J Taiwan Inst Chem Eng 63:318–326
Dikshit A, Shukla SK, Mishra RK (2013) Exploring nanomaterials with PGPR in current agricultural scenario. Lap Lambert Academic Publishing, Saarbrücken. ISBN 978-3-659-36774-8
Dolijanović Ž, Nikolić SR, Dragicevic V, Mutić J, Šeremešić S, Jovović Z, Popović DJ (2022) Mineral composition of soil and the wheat grain in intensive and conservation cropping systems. Agronomy 12(6):1321. https://doi.org/10.3390/agronomy12061321
Elmer W, White JC (2018) The future of nanotechnology in plant pathology. Annu Rev Phytopathol 56:111–133
Elmer W, Ma C, White J (2018) Nanoparticles for plant disease management. Curr Opin Environ Sci Health 6:66–70
Elser JJ, Platz MS (2015) Closing the human phosphorus cycle. National Science Foundation
Feng BH, Peng LF (2012) Synthesis and characterization of carboxymethyl chitosan carrying ricinoleic functions as an emulsifier for azadirachtin. Carbohydr Polym 88:576–582
Gao J, Wang Y, Folta KM, Krishna V, Bai W, Indeglia P, Georgieva A, Nakamura H, Koopman B, Moudgil B (2011) Polyhydroxy fullerenes (fullerols or fullerenols): beneficial effects on growth and lifespan in diverse biological models. PLoS One 6:e19976
George SA, Raj MS, Solomon D, Roselin P (2014) Comparative study of the anti-fungal activity of zinc oxide and titanium dioxide nano and bulk particles with anti-fungals against fungi isolated from infected skin and dandruff flakes. Res Rev J Microbiol Biotechnol 3(3):23–30
Giannousi K, Avramidis I, Dendrinou-Samara C (2013) Synthesis, characterization and evaluation of copper based nanoparticles as agrochemicals against *Phytophthora infestans*. RSC Adv 3(44):21743–21752
Gilbertson LM, Pourzahedi L, Laughton S, Gao X, Zimmerman JB, Theis TL, Westerhoff P, Lowry GV (2020) Guiding the design space for nanotechnology to advance sustainable crop production. Nat Nanotechnol 15(9):801–810
Grillo R, dos Santos NZP, Maruyama CR, Rosa AH, de Lima R, Fraceto LF (2012) Poly (ε-caprolactone) nanocapsules as carrier systems for herbicides: physico-chemical characterization and genotoxicity evaluation. J Hazard Mater 231:1–9
Grillo R, Rosa A, Fraceto LF (2014) Poly (ε-caprolactone) nanocapsules carrying the herbicide atrazine: Effect of chitosan-coating agent on physico-chemical stability and herbicide release profile. Int J Environ Sci Technol 11:1691–1700
Grillo R, Mattos BD, Antunes DR, Forini MML, Monikh FA, Rojas OJ (2021) Foliage adhesion and interactions with particulate delivery systems for plant nanobionics and intelligent agriculture. Nano Today 37:101078
Hao Y, Yu F, Lv R et al (2016) Carbon nanotubes filled with different ferromagnetic alloys affect the growth and development of rice seedlings by changing the C: n ratio and plant hormones concentrations. PLoS One 11(6):e0157264

Hernández-Hernández H, Juárez-Maldonado A, Benavides-Mendoza A, Ortega-Ortiz H, Cadenas-Pliego G, Sánchez-Aspeytia D, González-Morales S (2018) Chitosan-PVA and copper nanoparticles improve growth and overexpress the SOD and JA genes in tomato plants under salt stress. Agronomy 8(9):175

Hernandez-Viezcas JA, Castillo-Michel H, Andrews JC, Cotte M, Rico C, Peralta- Videa JR, Gardea-Torresdey JL (2013) In situ synchrotron X-ray fluorescence mapping and speciation of CeO_2 and ZnO nanoparticles in soil cultivated soy-bean (*Glycine max*). ACS Nano 7:1415–1423

Huang S, Wang L, Liu L, Hou Y, Li L (2015) Nanotechnology in agriculture, livestock, and aquaculture in China. A review. Agron Sustain Dev 35(2):369–400

Janmohammadi M, Navid A, Segherloo AE, Sabaghnia N (2016) Impact of nano-chelated micronutrients and biological fertilizers on growth performance and grain yield of maize under deficit irrigation condition. Biologija 62:134–147

Jenne M, Kambham M, Tollamadugu NP, Karanam HP, Tirupati MK, Reddy Balam R, Shameer S, Yagireddy M (2018) The use of slow releasing nanoparticle encapsulated Azadirachtin formulations for the management of Caryedon serratus O.(groundnut bruchid). IET Nanobiotechnol 12(7):963–967

Joshi A, Kaur S, Dharamvir K, Nayyar H, Verma G (2018) Multi-walled carbon nanotubes applied through seed-priming influence early germination, root hair, growth and yield of bread wheat (Triticum aestivum L.). J Sci Food Agric 98:3148–3160

Kasote DM, Lee JH, Jayaprakasha GK, Patil BS (2019) Seed priming with iron oxide nanoparticles modulate antioxidant potential and defense-linked hormones in watermelon seedlings. ACS Sustain Chem Eng 7(5):5142–5151

Kasprowicz MJ, Kozioł M, Gorczyca A (2010) The effect of silver nanoparticles on phytopathogenic spores of *Fusarium culmorum*. Can J Microbiol 56(3):247–253

Kaur J, Singh J, Rawat M (2019) An efficient and blistering reduction of 4-nitrophenol by green synthesized silver nanoparticles. SN Appl Sci 1(9):1–6

Khan M, Khan AU, Hasan MA, Yadav KK (2021a) Agro-nanotechnology as an emerging field: a novel sustainable approach for improving plant growth by reducing biotic stress. Applied Sciences 11:2282

Khan M, Khan AU, Bogdanchikova N, Garibo D (2021b) Antibacterial and antifungal studies of biosynthesized silver nanoparticles against plant parasitic nematode Meloidogyne incognita, plant pathogens Ralstonia solanacearum and Fusarium oxysporum. Molecules 26(9):2462

Khan M, Khan AU, Rafatullah M, Alam M, Bogdanchikova N, Garibo D (2022) Search for effective approaches to fight microorganisms causing high losses in agriculture: application of P. lilacinum metabolites and mycosynthesised silver nanoparticles. Biomolecules 12(2):74

Khan M, Siddiqui ZA, Parveen A, Khan AA, Moon IS, Alam M (2022a) Elucidating the role of silicon dioxide and titanium dioxide nanoparticles in mitigating the disease of the eggplant caused by Phomopsis vexans, Ralstonia solanacearum, and root-knot nematode *Meloidogyne incognita*. Nanotechnol Rev 11(1):1606–1619

Khan M, Khan AA, Parveen A, Min K, Yadav VK, Khan AU, Alam M (2023) Mitigating the growth of plant pathogenic bacterium, fungi, and nematode by using plant-mediated synthesis of copper oxide nanoparticles (CuO NPs). Green Chem Lett Rev 16(1):2177520. https://doi.org/10.1080/17518253.2023.2177520

Kim SW, Jung JH, Lamsal K, Kim YS, Min JS, Lee YS (2012) Antifungal effects of silver nanoparticles (AgNPs) against various plant pathogenic fungi. Mycobiology 40(1):53–58

Krishnaraj C, Ramachandran R, Mohan K, Kalaichelvan P (2012) Optimization for rapid synthesis of silver nanoparticles and its effect on phytopathogenic fungi. Spectrochim Acta A 93:95–99

Lamsal K, Kim SW, Jung JH, Kim YS, Kim KS, Lee YS (2011) Application of silver nanoparticles for the control of *Colletotrichum* species in vitro and pepper anthracnose disease in field. Mycobiology 39(3):194–199

Landa P, Dytrych P, Prerostova S, Petrova S, Vankova R, Vanek T (2017) Transcriptomic response of Arabidopsis thaliana exposed to CuO nanoparticles, bulk material, and ionic copper. Environ Sci Technol 51(18):10814–10824

Lassaletta L, Billen G, Grizzetti B, Anglade J, Garnier J (2014) 50 year trends in nitrogen use efficiency of world cropping systems: the relationship between yield and nitrogen input to cropland. Environ Res Lett 9:105011

Liu X, Wang F, Shi Z, Tong R, Shi X (2015) Bioavailability of Zn in ZnO nanoparticle-spiked soil and the implications to maize plants. J Nanopart Res 17:11

Lu C, Zhang C, Wen J, Wu G, Tao MX (2002) Research of the effect of nanometer materials on germination and growth enhancement of Glycine max and its mechanism. Soybean Sci 21:168–171

Lu W, Lu ML, Zhang QP et al (2013) Octahydrogenated retinoic acid-conjugated glycol chitosan nanoparticles as a novel carrier of azadirachtin: synthesis, characterization, and in vitro evaluation. J Polym Sci A Polym Chem 51:3932–3940

Ma CX, White JC, Dhankher OP, Xing B (2015) Metal-based nanotoxicity and detoxification pathways in higher plants. Environ Sci Technol 49(12):7109–7122

Manjaiah KM, Mukhopadhyay R, Paul R, Datta SC, Kumararaja P, Sarkar B (2019) Clay minerals and zeolites for environmentally sustainable agriculture. In: Mariano M, Binoy S, Alessio L (eds) Modified clay and zeolite nanocomposite materials. Elsevier, Amsterdam, The Netherlands, pp 309–329

Maruyama CR, Guilger M, Pascoli M, Bileshy-José N, Abhilash P, Fraceto LF, de Lima R (2016) Nanoparticles based on chitosan as carriers for the combined herbicides imazapic and imazapyr. Sci Rep:6

Mihailović VB, Srećković N, Nedić ZP, Dimitrijević S, Matić M, Obradović A, Selaković D, Rosić G, Katanić Stanković JS (2023) Green synthesis of silver nanoparticles using *Salvia verticillata* and *Filipendula ulmaria* extracts: optimization of synthesis, biological activities, and catalytic properties. Molecules 28(2):808. https://doi.org/10.3390/molecules28020808

Mohammad Ghasemi V, Siavash Moghaddam S, Rahimi A, Pourakbar L, Popović-Djordjević J (2020) Winter cultivation and nano fertilizers improve yield components and antioxidant traits of dragon's head (*Lallemantia iberica* (M.B.) Fischer & Meyer). Plants 9:252. https://doi.org/10.3390/plants9020252

Monreal CM, De Rosa M, Mallubhotla SC, Bindraban PS, Dimkpa C (2016) Nanotechnologies for increasing the crop use efficiency of fertilizer-micronutrients. Biol Fertil Soils 52:423–437

Moradi Rikabad M, Pourakbar L, Moghaddam SS, Popović-Djordjević J (2019) The impact of TiO_2 nanoparticles on growth, chemical and antioxidant traits of saffron (*Crocus sativus* L.) exposed to UV-B stress. Ind Crop Prod 137:137–143. https://doi.org/10.1016/j.indcrop.2019.05.017

Morales-Díaz AB, Ortega-Ortíz H, Juárez-Maldonado A, Cadenas-Pliego G, González-Morales S, Benavides-Mendoza A (2017) Application of nanoelements in plant nutrition and its impact in ecosystems. Adv Nat Sci Nanosci Nanotechnol 8:013001. (13 pp)

Namburi KR, Kora AJ, Chetukuri A, Kota VSMK (2021) Biogenic silver nanoparticles as an antibacterial agent against bacterial leaf blight causing rice phytopathogen *Xanthomonas oryzae* pv. *oryzae*. Bioprocess Biosyst Eng:1–14

Pandey S, Giri VP, Tripathi A, Bajpai R, Sharma D, Bahadur L, Mishra A (2021) Interaction, fate and risks associated with nanomaterials as fertilizers and pesticides. In: Advances in nanofertilizers and nano-pesticides in agriculture. Woodhead Publishing, pp 229–248

Parveen A, Siddiqui ZA (2021) Impact of silicon dioxide nanoparticles on growth, photosynthetic pigments, proline, activities of defense enzymes and some bacterial and fungal pathogens of tomato. Vegetos:1–11

Rahman KM, Zhang D (2018) Effects of fertilizer broadcasting on the excessive use of inorganic fertilizers and environmental sustainability. Sustainability 10(3):759

Raja K, Sowmya R, Sudhagar R, Moorthy PS, Govindaraju K, Subramanian KS (2019) Biogenic ZnO and Cu nanoparticles to improve seed germination quality in black gram (*Vigna mungo*). Mater Lett 235:164–167

Rajiv P, Rajeshwari S, Venckatesh R (2013) Bio-Fabrication of zinc oxide nanoparticles using leaf extract of *Parthenium hysterophorus* L. and its size-dependent antifungal activity against plant fungal pathogens. Spectrochim Acta A 112:384–387

Rajput VD, Minkina T, Feizi M, Kumari A, Khan M, Mandzhieva S et al (2021) Effects of silicon and silicon-based nanoparticles on rhizosphere microbiome, plant stress and growth. Biology 10:791

Raliya R, Biswas P, Tarafdar J (2015) TiO_2 nanoparticle biosynthesis and its physiological effect on mung bean (*Vigna radiata* L.). Biotechnol Rep 5:22–26

Rastogi A, Tripathi DK, Yadav S, Chauhan DK, Zivcak M, Ghorbanpour M, El-Sheery NI, Brestic M (2019) Application of silicon nanoparticles in agriculture. 3 Biotech 9:90

Rico CM, Majumdar S, Duarte-Gardea M, Peralta-Videa JR, Gardea-Torresdey JL (2011) Interaction of nanoparticles with edible plants and their possible implications in the food chain. J Agric Food Chem 59:3485–3498

Rizwan M, Ali S, Ali B et al (2019) Zinc and iron oxide nanoparticles improved the plant growth and reduced the oxidative stress and cadmium concentration in wheat. Chemosphere 214:269–277. https://doi.org/10.1016/J.CHEMOSPHERE.2018.09.120

Sabir S, Arshad M, Chaudhari SK (2014) Zinc oxide nanoparticles for revolutionizing agriculture: synthesis and applications. Sci World J 2014:1–8

Schnoor B, Elhendawy A, Joseph S, Putman M, Chacón-Cerdas R, Flores-Mora D, Bravo-Moraga F, Gonzalez-Nilo F, Salvador-Morales C (2018) Engineering atrazine loaded poly (lactic-co-glycolic acid) nanoparticles to ameliorate environmental challenges. J Agric Food Chem 66(30):7889–7898

Shang H, Ma C, Li C, White JC, Polubesova T, Chefetz B, Xing B (2020) Copper sulfide nanoparticles suppress *Gibberella fujikuroi* infection in rice (*Oryza sativa* L.) by multiple mechanisms: contact-mortality, nutritional modulation and phytohormone regulation. Environ Sci Nano 7(9):2632–2643

Siddiqui ZA, Khan MR, Abd-Allah EF, Parveen A (2018) Titanium dioxide and zinc oxide nanoparticles affect some bacterial diseases, and growth and physiological changes of beetroot. Int J Vegetable Sci 25(5):409–430

Silva MDS, Cocenza DS, De Melo NFS, Grillo R, Rosa AH, Fraceto LF (2010) Alginate nanoparticles as a controlled release system for clomazone herbicide. Quim Nova 33(9)

Springmann M et al (2018) Options for keeping the food system within environmental limits. Nature 562:519–525

Srećković NZ, Nedić ZP, Liberti D, Monti DM, Mihailović NR, Katanić Stanković JS, Dimitrijević S, Mihailović VB (2021) Application potential of biogenically synthesized silver nanoparticles using *Lythrum salicaria* L. extracts as pharmaceuticals and catalysts for organic pollutant degradation. RSC Adv 11:35585. https://doi.org/10.1039/d1ra05570d

Srećković NZ, Nedić ZP, Monti DM, D'Elia L, Dimitrijević SB, Mihailović NR, Katanić Stanković JS, Mihailović VB (2023) Biosynthesis of silver nanoparticles using *Salvia pratensis* L. aerial part and root extracts: bioactivity, biocompatibility, and catalytic potential. Molecules 28:1387. https://doi.org/10.3390/molecules28031387

Tong Y, Wu Y, Zhao C, Xu Y, Lu J, Xiang S, Zong F, Wu X (2017) Polymeric nanoparticles as a metolachlor carrier: water-based formulation for hydrophobic pesticides and absorption by plants. J Agric Food Chem 65:7371–7378

Vinković T, Novák O, Strnad M, Goessler W, Jurašin DD, Parađiković N, Vrček IV (2017) Cytokinin response in pepper plants (*Capsicum annuum* L.) exposed to silver nanoparticles. Environ Res 156:10–18

Vuković S, Popović-Djordjević JB, Kostić AŽ, Pantelić ND, Srećković N, Akram M, Laila U, Katanić Stanković JS (2023) Allium species in the Balkan Region—major metabolites, antioxidant and antimicrobial properties. Horticulturae 9(3):408. https://doi.org/10.3390/horticulturae9030408

Wang Y, Cui H, Sun C et al (2014) Construction and evaluation of controlled-release delivery system of Abamectin using porous silica nanoparticles as carriers. Nanoscale Res Lett 9:2490

Wang X, Yang X, Chen S, Li Q, Wang W, Hou C, Gao X, Wang L, Wang S (2016) Corrigendum on zinc oxide nanoparticles affect biomass accumulation and photosynthesis in Arabidopsis. Front Plant Sci 7:559

World Fertilizer Trends and Outlook to 2020: Summary Report (Food and Agriculture Organization of the United Nations, 2017.

Worrall EA, Hamid A, Mody KT, Mitter N, Pappu HR (2018) Nanotechnology for plant disease management. Agronomy 8(12):285

Yang FL, Li XG, Zhu F et al (2009) Structural characterization of nanoparticles loaded with garlic essential oil and their insecticidal activity against *Tribolium castaneum* (Herbst) (Coleoptera: Tenebrionidae). J Agric Food Chem 57(21):10156–10162

Yang Y, Cheng J, Garamus VM et al (2018) Preparation of an environmentally friendly formulation of the insecticide nicotine hydrochloride through encapsulation in chitosan/tripolyphosphate nanoparticles. J Agric Food Chem 66(5):1067–1074

Yu Z, Sun X, Song H, Wang W, Ye Z, Shi L, Ding K (2015) Glutathione-responsive carboxymethyl chitosan nanoparticles for controlled release of herbicides. Mater Sci Appl 6:591–604

Zhao L, Peralta-Videa JR, Rico CM, Hernandez-Viezcas JA, Sun Y, Niu G, Servin A, Nunez JE, Duarte-Gardea M, Gardea-Torresdey JL (2014) CeO_2 and ZnO nanoparticles change the nutritional qualities of cucumber (*Cucumis sativus*). J Agric Food Chem 62:2752–2759

Chapter 19
Nanoparticle-Enhanced Plant Defense Mechanisms Harnessed by Nanotechnology for Sustainable Crop Protection

Dali Vilma Francis, Anam Asif, and Zienab Fawzy Reiad Ahmed

Abstract Nanotechnology stands as a transformative force in modern agriculture, providing innovative solutions to fortify plant defense mechanisms against diverse biotic stressors. This comprehensive exploration delves into how nanoparticles contribute to bolstering plant immunity and resilience. Amid population growth and climate change, the imperative of ensuring food security while minimizing the environmental impact of agriculture has never been more critical. Biotic stressors, ranging from pests to pathogens and herbivores, pose significant threats to crop yields. Nanoparticles, with their distinctive physicochemical properties, emerge as promising tools to address these challenges. The review explores the underlying mechanisms, various nanoparticle types, applications, and the potential benefits of employing nanoparticles for plant defense. Understanding the distinctive characteristics of nanoparticle types—such as metal nanoparticles, metal oxide nanoparticles, and carbon-based nanoparticles—is pivotal. Size, shape, and surface chemistry play crucial roles in tailoring nanoparticles for specific plant defense applications. This article provides a comprehensive overview of these essential nanoparticle properties. The intricate and multifaceted mechanisms underpinning nanoparticle-induced plant defense responses are explored. Nanoparticles act as triggers for signal transduction pathways, generate reactive oxygen species (ROS), and modulate the expression of defense-related genes. This exploration offers insights into the molecular interactions between nanoparticles and plants. Practical applications of nanoparticles in pest and pathogen management take center stage in this review. Numerous case studies and experiments underscore the efficacy of nanoparticles in controlling various biotic stressors, including insect pests, fungal pathogens, and

D. V. Francis · A. Asif · Z. F. R. Ahmed (✉)
Integrative Agriculture Department, College of Agriculture and Veterinary Medicine, United Arab Emirates University, Al Ain, United Arab Emirates
e-mail: dali_f@uaeu.ac.ae; anam_a@uaeu.ac.ae; zienab.ahmed@uaeu.ac.ae

M. Khan, J.-T. Chen (eds.), *Nanoparticles in Plant Biotic Stress Management*,
https://doi.org/10.1007/978-981-97-0851-2_19

nematodes. Moreover, the impact of nanoparticle interventions on beneficial plant-microbe interactions is examined, highlighting their potential to enhance symbiotic relationships while combating pathogens. Environmental considerations and safety aspects linked to nanoparticle use in agriculture are thoroughly addressed. Potential risks and benefits are evaluated, alongside regulatory aspects governing nanoparticle applications in crop protection. The article concludes by discussing future prospects and challenges, offering insights into the commercial viability and scalability of nanoparticle-enhanced plant defense mechanisms. This comprehensive review encapsulates the exciting possibilities of nanotechnology in revolutionizing sustainable crop protection. It emphasizes the significance of environmentally friendly strategies in modern agriculture, serving as a valuable resource for researchers, policymakers, and stakeholders seeking innovative solutions to address biotic stress challenges in food production.

Keywords Biotic stressors · Climate change · Herbivores · Nanomaterials · Pathogens

19.1 Introduction

In a world grappling with burgeoning populations and the escalating impacts of climate change, the imperative to ensure sustainable food production has never been more urgent. With estimates projecting a need to increase food production by up to 70% to feed a projected 9.2 billion people by 2050, the challenges are monumental (Francis et al. 2020). Agriculture, caught in the delicate balance between meeting escalating food demands and mitigating the environmental toll of conventional practices, faces disruptions from climate change, traditional farming methods, and relentless biotic stressors (Beckford 2018).

Traditional agricultural practices, marked by intensive pesticide use and monocultures, have provided short-term gains but at a significant environmental cost, contributing to pollution, soil degradation, and biodiversity loss. Biotic stressors, including pests, pathogens, and herbivores, add complexity, threatening crop yields and global food security (Gomiero et al. 2011).

Recognizing the need for environmental stewardship in agriculture, sustainable crop protection has emerged as a pivotal discourse. This paradigm shift seeks innovative solutions that harmonize environmental, economic, and social dimensions (Anderson et al. 2019). In response to this imperative, nanotechnology emerges as a transformative force in agriculture (Kim et al. 2018). Nanoparticles, with their unique characteristics and precision, offer a nuanced approach to fortifying plant defense mechanisms, presenting an alternative path toward sustainable crop protection (Wahab et al. 2023). Incorporating nanoparticles into plant defense mechanisms represents a shift in perspective, moving away from conventional practices (Mittal et al. 2020). Nanoparticles, operating at the nanoscale, bring efficiency and customization previously unattainable, allowing tailored solutions for specific biotic stressors (Elhawat et al. 2018). Their adaptability positions them as versatile tools capable of addressing a spectrum of challenges faced by modern agriculture.

This review is significant for its in-depth exploration of nanoparticle-enhanced plant defense mechanisms, unraveling the scientific complexities of this innovative approach. Positioned as a guiding influence in the dynamic landscape of agricultural innovation, it makes a meaningful contribution to the broader scientific discourse. Beyond theoretical insights, the review delves into practical applications, addressing environmental considerations and safety aspects associated with nanoparticle use in crop protection. Its importance lies in its capacity to enlighten decision-makers, influence future research trajectories, and inspire responsible strategies for nanoparticle-mediated plant defense. Ultimately, the review aspires to catalyze change, fostering a deep understanding of nanotechnology's pivotal role in reshaping the future of global food security.

19.2 Nanoparticle Applications in Agriculture

Nanoparticles, defined as particles with dimensions ranging from 1 to 100 nm, exhibit unique physicochemical properties owing to their size and increased surface area (Khan et al. 2019). In agriculture, nanoparticles find application in various domains due to their versatility (Dhiman et al. 2021). They are classified based on composition, structure, and origin, with common categories including metal nanoparticles (e.g., silver, gold, and copper), metal oxide nanoparticles (e.g., zinc oxide, titanium dioxide), and carbon-based nanoparticles (e.g., carbon nanotubes, graphene) (Sengul and Asmatulu 2020). Each category possesses distinct characteristics that influence their behavior and interactions within agricultural systems (Chen 2018).

19.2.1 Nanoparticles as Delivery Systems for Agrochemicals

In the ever-evolving field of agricultural innovation, nanoparticles play a pivotal role as carriers for agrochemicals (Singh et al. 2021a). Engineered nanoparticles can encapsulate pesticides, herbicides, and fertilizers, providing targeted and controlled delivery to crops, thereby amplifying the efficacy of agrochemicals while championing environmental sustainability (Singh et al. 2021b). Nanoparticle encapsulation improves fungicide efficacy, reduces environmental impact, and optimizes resource utilization (Kumar et al. 2019). Other contributions showcase how nanoparticles as fertilizers result in improved crop nutrient uptake, enhanced yields, and decreased environmental nutrient runoff (Iqbal et al. 2019a). Additionally, research on nanoparticle-mediated pesticide delivery systems reveals promising outcomes in pest management, highlighting their potential to minimize overuse and reduce soil and water contamination (Banerjee et al. 2022). This collective body of research underscores the multifaceted potential of nanoparticle-mediated delivery systems to revolutionize agricultural practices, making them not only more efficient and targeted but also more environmentally sustainable.

19.2.2 Nanoparticles in Plant Growth Promotion

Plant growth promotion by nanoparticles stands as a pivotal contributor, showcasing their potential to revolutionize agricultural practices. Research examples highlight the efficacy of specific nanoparticles, such as silicon and zinc oxide, in enhancing various aspects of plant growth and development (Elsheery et al. 2020). Notably, silicon nanoparticles have demonstrated plant growth-promoting properties by enhancing stress tolerance, facilitating improved nutrient uptake, and bolstering plant resistance to pathogens. This multifaceted impact positions silicon nanoparticles as valuable tools for cultivating resilient and robust crops, especially under challenging environmental conditions (Akhtar et al. 2021).

Similarly, zinc oxide nanoparticles have been identified as influential agents in promoting plant growth. Studies illustrate their capacity to enhance photosynthesis, increase nutrient absorption, and stimulate root development (Thounaojam et al. 2021). Table 19.1 demonstrates how different nanoparticles enhance plant growth parameters. It provides insights into their application modes, concentrations, and the associated plant species. These findings underscore the diverse ways in which nanoparticles can positively influence crucial physiological processes within plants, ultimately contributing to improved crop yields and overall plant resilience.

The application of these nanoparticles, whether directly to plant roots or as foliar sprays, further exemplifies their potential to transform agricultural practices (Grillo et al. 2021). By targeting specific aspects of plant growth and development, nanoparticles offer a precise and efficient means to enhance crop performance. This targeted application showcases the versatility of nanoparticles in addressing key challenges in modern agriculture, from nutrient management to stress resilience, ultimately contributing to the sustainable intensification of crop production.

19.2.3 Nanoparticles for Heavy Metal (HM) Stress Mitigation

Nanoparticles stand at the forefront of transformative solutions for soil remediation, offering innovative approaches to combat soil pollution and alleviate nutrient depletion. Research examples have spotlighted the significant potential of metal nanoparticles, specifically iron and titanium oxide nanoparticles, in contributing to soil remediation endeavors (Kumar et al. 2021). Studies reveal the remarkable capability of these nanoparticles to facilitate the degradation of organic pollutants, employing intricate processes such as adsorption and reduction (Gusain et al. 2019). This results in the immobilization of heavy metals, effectively reducing soil

Table 19.1 Improved plant growth parameters utilizing diverse nanoparticles, along with their application modes, concentrations, and specific plant species

Nanoparticle	Plant species	Mode of application	Concentration of nanoparticle	Reference
Cu	Maize	Root	69.4 µM	Van Nguyen et al. (2022)
Cu	Pigeon pea	Root	20 ppm	Shende et al. (2017)
CuO	*Brassica pekinensis*	Seed	10–1000 mg/L	Wang et al. (2020)
CuO	*Oryza sativa*	Seed	250 mg/L	Yang et al. (2020)
CuO	*Lactuca sativa*	Seed	0.2–80 µg/mL	Pelegrino et al. (2020)
CuO, ZnO, Mn_2O_3	Chrysanthemum	Foliar	0.5–1 mg/L	Elmer et al. (2021)
CuO, ZnO	*Amaranthus hybridus*	Seed priming, foliar, and root	0.6–0.24 µM	Francis et al. (2022)
ZnO	*Oryza sativa*	Foliar	50–100 mg/L	Ali et al. (2019)
ZnO	*Gossypium hirsutum*	Root	25–200 mg/L	Venkatachalam et al. (2017)
ZnO	Tomato	Root	2–16 mg/L	Faizan et al. (2018)
ZnO, FeO	Wheat	Seeds	5–100 mg/L	Rizwan et al. (2019)
Fe_2O_3	*Citrus maxima*	Root	20–100 mg/L	Li et al. (2018)
Fe_2O_3	Alfalfa	seed	50 mg/L	Yang et al. (2023)
Fe_2O_3	Wheat	Root	200 and 500 mg/L	Feng et al. (2022)
FeO	Flax	Seed priming	25–100 ppm	Waqas Mazhar et al. (2022)
SiO, FeO	Wheat	Root	25–100 mg/kg	Manzoor et al. (2023)
Ag	Tomato	Root	100 ppm	Zakharova et al. (2017)
Ag	Fenugreek	Foliar	20–60 mg/L	Sadak (2019)
Ag	*Allium cepa*	Root	5–100 µg/mL	Casillas-Figueroa et al. (2020)
Ag	Tomato	Root	10–30 mg/L	Noori et al. (2020)
Ag	Wheat	Root	10–30 mg/L	Matras et al. (2023)
MgO	Mustard	Seed	10–150 mg/L	Gautam et al. (2023)
MgO	Rice	Root	150 mg/L	Song et al. (2023)
Al_2O_3	Lettuce	Root	0.4–2 mg/mL	Hayes et al. (2020)
NiO-Al_2O_3	*Nigella arvensis*	Root	50–100 mg/L	Chahardoli et al. (2020)
MnO_2	*Phaseolus vulgaris*	Foliar	10–40 ppm	Salama et al. (2022)
Mn_2O_3	*Artemisia annua*	Seed, foliar	25–100 mg/L	Salehi et al. (2023)
FeO, MnO, ZnO	Cabbage & *Andean lupin*	Root	270 ppm	Murgueitio-Herrera et al. (2022)

(continued)

Table 19.1 (continued)

Nanoparticle	Plant species	Mode of application	Concentration of nanoparticle	Reference
TiO_2	Wheat	Root	40 mg/L	Gülmezoğlu et al. (2020)
TiO_2	*Vigna radiata*	Seed	25–250 μg/mL	Sunny et al. (2022)
TiO_2	*Hordeum vulgare*	Seed	25-50 μg/mL	Ma et al. (2023)
Au, chitosan	*Vigna unguiculata*	Root	100 μg/mL	Panichikkal and Krishnankutty (2022)
Au	Wheat	Seed	5–50 μg/L	Venzhik et al. (2022)
Si	Cucumber	Root	100–400 mg/kg	Alsaeedi et al. (2019)
Si	Sugar beet	Foliar	12.5 mg/L	Alharbi et al. (2022)

contaminants and playing a crucial role in the restoration of soil health (Rajput et al. 2022). Zinc oxide nanoparticles (ZnO NPs) have also shown efficacy in mitigating HM stress. Combined application of ZnO NPs and biochar effectively relieved Cd stress, outperforming individual treatments. ZnO NPs demonstrated their ability to reduce Cd accumulation in rice shoots, with the combined application of ZnO NPs and biochar being more efficient. Additionally, ZnO NPs were found to alleviate drought stress in sorghum by increasing the absorption of nitrogen and potassium (Zhou et al. 2020).

Metal nanoparticles, particularly in the context of soil remediation, demonstrate their dual impact by not only addressing pollution challenges but also by enhancing nutrient availability (Perea Velez et al. 2021). The immobilization of heavy metals not only mitigates environmental risks associated with contamination but also enhances the soil's nutrient profile, fostering conditions conducive to robust plant growth and overall ecosystem health (Mani and Kumar 2014). Non-metal-based nanoparticles, such as selenium (Se) and silicon (Si), have shown significant potential in reducing HM stress. Se-containing fertilizer application enhanced plant stress resistance, with foliar spraying of Se NPs proving effective in reducing Cd content in grains and increasing yield in rice (Zhou et al. 2020). Si, although not an essential element, has been linked to enhanced plant growth and stress resistance. Si NPs, particularly in foliar applications, effectively decreased Cd accumulation in grains and increased the yield in wheat. The combined application of Se NPs and Si NPs further reduced Cd and Pb in rice grains (Hussain et al. 2020).

Hydroxyapatite (HAP) NPs, a phosphorus-containing amendment, demonstrated effectiveness in remediating HM-contaminated soils. In *Brassica chinensis* L., HAP NPs enhanced biomass, chlorophyll, and antioxidant enzyme activities while decreasing Cd and malondialdehyde (MDA) contents in plant shoots (Alhammad et al. 2023). HAP NPs also showed promise in reducing Cd content in rice plants and immobilizing Pb in Cd and Pb compound stress conditions (Wu et al. 2023a).

Furthermore, the adaptability of nanoparticles is exemplified through their engineered applications to augment soil structure and water retention. Research indicates that nanoparticles can be precisely tailored to specific soil properties, leading to improvements in physical structure and water-holding capacity (Denisse et al.

2021). This engineered approach contributes to the creation of a more favorable soil environment, promoting optimal conditions for sustained plant growth. A study demonstrated that both soil and foliar application of Fe_2O_3 NPs to wheat under Cd stress resulted in reduced leaf electrolyte leakage, decreased Cd content in grains, increased antioxidant enzyme activity, and enhanced dry weight of wheat. The foliar application of Fe NPs was preferred over soil application due to potential factors affecting soil absorption, such as pH and mineral interactions. In composite stress conditions, combining Fe NPs with biochar further improved the alleviating effects on Cd stress in rice (El-Saadony et al. 2022).

Integrating metal nanoparticles into soil remediation is a sophisticated and multifaceted strategy, addressing soil pollution, immobilizing heavy metals, and enhancing soil structure. This approach, using nanoparticles as powerful tools, not only advances soil remediation techniques but also underscores the potential for environmentally friendly and sustainable soil management. The collective findings indicate that both metal-based and non-metal-based nanoparticles show significant promise in alleviating heavy metal stress in plants. However, further research is crucial to understand the intricate interplay of nanoparticles with soil, plant, and stress conditions, elucidating potential ecological risks and refining application strategies.

19.2.4 Nanoparticles in Pest and Pathogen Control

Nanoparticles play a crucial role in the agricultural sector by offering innovative solutions in pest and pathogen management (Yadav and Yadav 2018). The unique physicochemical characteristics of metal nanoparticles contribute to their effectiveness in disrupting the cellular structures of pathogens, thereby inhibiting their growth and proliferation (Khezerlou et al. 2018). Particularly, metal nanoparticles, such as silver (Sharma et al. 2018) and copper (Rai et al. 2018) nanoparticles, have demonstrated significant potential in combating agricultural pests and pathogens. Noteworthy examples include the use of chitosan, a derivative of chitin, with strong insecticidal activity against *Spodoptera* spp. The chitosan nanoparticle-coated fungal metabolite (CNPCFM) demonstrated superior efficacy compared to uncoated counterparts, indicating its potential in managing pests (Alfy et al. 2020). AgNPs were utilized in combination with essential oils, like *Aristolochia indica*, resulting in significant antifeedant and larvicidal activities against pests like *Helicoverpa armigera* (Dangi and Verma 2021). In the context of postharvest pest management, nanoparticles like nanostructured alumina, silica nanoparticles (SNPs), and polyethylene glycol-coated nanoparticles loaded with garlic essential oil have demonstrated insecticidal activity against pests like *Sitophilus* spp (Kitherian 2017). and *Callosobruchus* spp. These nanoparticles offer innovative alternatives to conventional insecticides, addressing issues related to residual toxicity and insect resistance. Furthermore, the development of DNA-tagged nanogold, DNA-tagged CdS, nano-TiO_2, and nano-Ag showcases the potential of nanotechnology in creating smart delivery systems. These nanomaterials were effective against *Spodoptera litura* larvae (Chakravarthy et al. 2012), demonstrating controlled release mechanisms and targeted pest management.

Silver nanoparticles, copper nanoparticles, zinc oxide, and titanium dioxide have demonstrated significant efficacy against a range of plant pathogens (Panhwar et al. 2021), showcasing their versatility. For instance, silver nanoparticles exhibit potent antimicrobial properties, effectively reducing bacterial and fungal infections in crops. Copper oxide (Francis et al. 2022) nanoparticles have proven effective in managing fungal diseases threatening plant health. Zinc oxide and titanium dioxide nanoparticles, on the other hand, exhibit promise in controlling insect pests, thereby safeguarding crop yield.

Expanding the repertoire of nanoparticles, silicon oxide (SiO_2) nanoparticles have shown potential in disease management (Abdelrhim et al. 2021). Studies highlight their inhibitory effects against viruses such as tomato yellow leaf curl virus (TYLCV) and papaya ringspot virus (PRSV), indicating a role in bolstering plant defenses (Mishra et al. 2022). Iron oxide (Fe_2O_3) nanoparticles, initially recognized for bioremediation, have evolved into versatile agents. Derived from plant extracts, these nanoparticles exhibit strong antifungal and antibacterial activities. They inhibit fungal diseases in crops and demonstrate efficacy against bacterial strains such as *E. coli* and MRSA (Ansari and Asiri 2021). The antimicrobial mechanism involves accumulation in the pathogen's cytoplasm, leading to conformational changes in DNA and proteins, ultimately causing cell death. Magnesium oxide (MgO) nanoparticles, characterized by excellent properties, have proven effective against plant pathogens (Chen et al. 2020). The inhibitory effect is attributed to complex formations with the pathogen's cell wall, enhanced reactive oxygen species (ROS) production (Ahmed et al. 2018), and disruption of quorum sensing (Siddique et al. 2022).

In the realm of commercialization, nanoparticle-based agricultural products are gaining momentum globally. Patented products such as SIVANTO® prime and CLARIVA® offer effective solutions for controlling pests and soil bacteria (Ganguli 2019). Regulatory laws, such as those by the US Food and Drug Administration and European regulations, play a crucial role in ensuring the safety and proper use of nanoparticle-based products in agriculture (Kumari et al. 2023).

Overall, these studies and applications collectively emphasize the promising role of nanoparticles in fortifying plant defenses against pests and pathogens. The field is continually evolving, with ongoing research contributing to sustainable approaches for enhancing crop protection and productivity. This targeted approach not only enhances the efficiency of pest control but also reduces the reliance on traditional chemical pesticides, contributing to sustainable agricultural practices. Additionally, the use of nanoparticles in plant protection helps bolster the plant's natural defense mechanisms, making it more resilient to pest infestations (Kumar et al. 2019). Table 19.2 provides a summary of various nanoparticles and their documented antimicrobial effects against plant pathogens, while Table 19.3 comprises a list of nanoparticles exhibiting pesticidal effects on pests, impacting plant growth.

Table 19.2 The antimicrobial efficacy of various nanoparticles against plant pathogens

Nanoparticle	Plant pathogen	Reference
Cu, CuO	*Agrobacterium tumefaciens, Xanthomonas campestris, Xanthomonas axonopodis, Ralstonia solanacearum, Fusarium solani, Neofusicoccum* sp., *Fusarium oxysporum, Macrophomina phaseolina, Pectobacterium carotovorum, Fusarium culmorum, Aspergillus niger, Alternaria solani , Ralstonia solanacearum, Erwinia amylovora, Alternaria brassicae, Alternaria brassicicola*	Kashyap et al. (2023), Pariona et al. (2019), Kamel et al. (2022), Abou-Salem et al. (2022), Vanathi et al. (2016), El-Batal et al. (2020), Chen et al. (2019), Gaba et al. (2022)
ZnO	*Colletotrichum* sp., *Xanthomonas oryzae, Botrytis cinerea, Penicillium expansum, Fusarium oxysporum, Aspergillus flavus, Aspergillus niger, Bipolaris oryzae, Sphaerulina oryzina, Burkholderia glumae* and B. gladioli, *Alternaria solani*	T-Thienprasert et al. (2021), Abdallah et al. (2020), He et al. (2011), González-Merino et al. (2021), Rajiv et al. (2013), Cheema et al. (2022), Ogunyemi et al. (2019), Ahmed et al. (2021a)
FeO, Fe_2O_3	*Rhizopus oryzae, Ralstonia solanacearum, Xanthomonas oryzae*	Saqib et al. (2019), Alam et al. (2019), Ahmed et al. (2022)
Ag	*Alternaria alternata, Botrytis cinerea, Rhizopus* sp., *Aspergillus* sp., *Xanthomonas oryzae, Alternaria solani, Fusarium graminearum, Rhizoctonia solani, Erwinia carotovora,* and *Alternaria solani*	Malandrakis et al. (2019), Medda et al. (2015), Mostafa et al. (2021), Ibrahim et al. (2020), Das and Dutta (2021), Namburi et al. (2021), Ahmed et al. (2020), Abbas et al. (2015)
Ag-SnO	*Xanthomonas oryzae*	Elizabeth Sunny and Kumar Shanmugam (2021)
Au	*Puccinia graminis tritici, Aspergillus flavus, Aspergillus niger, Sclerotinia sclerotiorum, Botrytis cinerea, Fragaria x ananassa, Rhizoctonia solani, Xanthomonas oryzae*	Jayaseelan et al. (2013), Elgorban (2016), Soltani Nejad et al. (2022)
TiO_2	*Dickeya dadantii, Ustilago tritici,* Xanthomonas sp. strain Xr-1	Hossain et al. (2019), Irshad et al. (2020), Paret et al. (2013)
Al_2O_3	*Fusarium oxysporum*	Shenashen et al. (2017)
CeO_2	*Ustilago tritici*	Alotaibi et al. (2023)
ZrO_2	*Pestalotiopsis* versicolor	Ahmed et al. (2021b)

This dual role of nanoparticles, addressing both pests and pathogens, highlights their versatility in promoting crop health and productivity. As agricultural challenges evolve, the integration of nanoparticle-based strategies provides a sophisticated and sustainable means of managing pest and pathogen pressures, contributing to the overall resilience and sustainability of agricultural systems (Chaud et al. 2021).

Table 19.3 The pesticidal efficacy of various nanoparticles against pests

Nanoparticles	Pests	References
$CaCO_3$	*Bactrocera dorsalis*	Hua et al. (2015)
CdS, Ag and TiO_2	*Spodoptera litura*	Chakravarthy et al. (2012)
Ag and Zn	*Aphis nerii Rouhani*	Rouhani et al. (2012)
$CaCO_3$	*Olive fruit Fly Bactrocera oleae*	Fenjan (2020)
Ag	*Spodoptera litura* and *Achaea janata*	Yasur and Usha Rani (2015)
Se	*Spodoptera litura*	Arunthirumeni et al. (2022)
Ag	*Bifenthrin Lygus hesperus* and *Acheta domesticus*	Pathipati and Kanuparthi (2021)
Ag	*Spodoptera litura*	Bharani and Namasivayam (2017)
Fe	*Spodoptera litura*	Xu et al. (2020)
Au	*Spodoptera litura*	Chakravarthy (2012)
Ag and Zn	*Spodoptera littoralis*	Ibrahim and Ali (2018)
ZnO	*Aphis gossypii* and *Spodoptera litura*	Elmasry (2021)
ZnO	*Anopheles stephensi*	Amuthavalli et al. (2021)
Ag	*Aedes aegypti* and *Anopheles stephensi*	Borase et al. (2014)
ZnO	*Culex quinquefasciatus* larvae	Yazhiniprabha et al. (2019)
Ag	*Aedes albopictus* larvae and pupae	Ga'al et al. (2018)
Ag	*Anopheles stephensi* Larvae	Nalini et al. (2017)

19.2.5 Nanoparticles for Drought Tolerance

Exploring the multifaceted role of NPs in fortifying plant defense mechanisms against drought stress, the focus encompasses critical aspects such as membrane stability, nutrient uptake, photosynthetic efficiency, osmolyte accumulation, hormonal crosstalk, phenolic compound accumulation, ROS detoxification, expression of stress-responsive genes, ultrastructural changes, and the overall impact on plant growth, yield, and quality under drought conditions (Rasheed et al. 2022).

NPs, notably ZnO and TiO_2, emerge as pivotal contributors to mitigating the adverse effects of drought stress on cellular membranes (Ramadan et al. 2022). They play a key role in reducing reactive oxygen species (ROS) production, lipid peroxidation, and malondialdehyde (MDA) accumulation, thereby upholding membrane stability and fostering improved plant-water relationships. Additionally, NPs positively influence the dynamics of nutrient uptake, translocation, and allocation, alleviating deficiencies induced by drought stress. In terms of photosynthesis, NPs enhance chlorophyll synthesis, chlorophyll fluorescence, and chlorophyllase activity, thereby contributing to sustained photosynthetic efficiency. Furthermore, NPs modulate osmolyte accumulation, including proline, sugars, total proteins, glycine betaine, and phenolic compounds, acting as effective combatants against drought-induced oxidative stress (Ghani et al. 2022).

The hormonal crosstalk induced by NPs, involving indole acetic acid (IAA) and gibberellins (GA), emerges as a key factor promoting plant growth under drought conditions. Phenolic compounds, pivotal for antioxidant activities, exhibit increased accumulation with NP application, leading to a reduction in oxidative damage. NPs play a crucial role in detoxifying reactive oxygen species (ROS) and fortifying antioxidant defense mechanisms, including the enhancement of enzyme activities such as ascorbate peroxidase (APX), catalase (CAT), and superoxide dismutase (SOD) (Rasheed et al. 2022).

Moreover, NPs influence the upregulation of stress-responsive genes, transcription factors, and proteins associated with drought tolerance. Their impact extends to ultrastructural changes in plant anatomy, positively influencing physiological and biochemical mechanisms. Importantly, NPs contribute to enhanced growth, yield, and quality parameters of crops even under drought-stress conditions.

19.2.6 Nanoparticles for Salinity Tolerance

The impact of nanoparticles (NPs) on the biochemical, physiological, and phytotoxic aspects of plants under salinity stress is a multifaceted and complex phenomenon. In terms of the biochemical aspects, NPs have been observed to modulate the antioxidant defense systems in plants. Several studies have demonstrated that NPs, such as cobalt, iron, cesium, and cerium oxide, can act similarly to antioxidant enzymes like catalase (CAT) and peroxidase (POD), thereby enhancing the plant's ability to scavenge reactive oxygen species (ROS) generated under salinity stress. Additionally, certain NPs, including silver (Ag-NPs) and titanium dioxide (TiO_2-NPs), have been shown to improve the activity of antioxidant enzymes such as superoxide dismutase (SOD), CAT, and glutathione peroxidase (GPX), leading to increased plant growth and tolerance to salinity stress (Etesami et al. 2021).

Moreover, phytohormones play a crucial role in plant adaptation to environmental stresses, and NPs have been reported to influence the biosynthesis of these hormones (Kasote et al. 2019). For instance, the presence of Ag-NPs altered the levels of abscisic acid (ABA), ethylene, and gibberellin in rice plants (Saha et al. 2021), while the application of Ag-NPs on wheat plants increased the content of indole-3-butyric acid (IBA), 1-naphthalene acetic acid (NAA), and 6-benzylaminopurine (BAP) and decreased ABA levels, contributing to improved stress tolerance (Ismail and Abou-Zeid 2018). The regulation of compatible solutes, such as amino acids and soluble sugars, by NPs has also been associated with enhanced salinity tolerance in various plants.

In the physiological aspect, NPs have shown promise in improving photosynthesis, a process highly affected by salinity stress (Hameed et al. 2021). NPs, including manganese (Mn-NPs (Zhang et al. 2022a)), carbon dots (Li et al. 2019a), and selenium-NPs (Iqbal et al. 2022), have been reported to enhance chlorophyll content, photosynthetic efficiency, and the activity of enzymes involved in photosynthesis, such as ribulose-1,5-bisphosphate carboxylase/oxygenase (RuBisCO). These improvements contribute to increased photosynthetic capacity and, consequently, better plant growth under salinity stress.

Additionally, NPs have been found to regulate ion homeostasis, maintaining a higher potassium-to-sodium ratio, which is crucial for plant resistance to salinity stress (Wahid et al. 2020). This regulation helps improve osmotic potential and facilitates better plant growth under saline conditions. Furthermore, NPs contribute to the maintenance of plant water balance by influencing aquaporin expression and activity, resulting in increased water uptake and retention.

Despite these positive effects, it is essential to consider NPs' phytotoxicity. The concentration, type, size distribution, morphology, and surface charge of NPs can influence their toxicity. High concentrations of NPs may lead to oxidative stress in plant cells, triggering the production of antioxidant defenses. Therefore, careful consideration of the appropriate NP concentration is crucial to avoid adverse effects on plant growth and development (Etesami et al. 2021).

19.2.7 Nanoparticles for UV Stress Resilience

Nanoparticles play a crucial role in enhancing UV stress resilience in plants. Specifically, titanium dioxide nanoparticles (TiO_2-NPs) have demonstrated a protective effect in *Arabidopsis* seedlings exposed to heightened UV-B radiation. Acting as a type of "sunscreen," TiO_2-NPs contribute to improved physiological parameters and mitigate adverse impacts on photosynthesis. Their role involves activating the plant's antioxidant system, reducing reactive oxygen species, and physically shielding plants from UV-B radiation. Additionally, TiO_2-NPs increase flavonoid accumulation, enhancing the plant's tolerance to UV stress. This research highlights the potential of nanoparticles, such as TiO_2-NPs, as valuable tools for bolstering plant resilience under UV stress conditions. Continued exploration is essential to understand their mechanisms of action and potential biological toxicity, ensuring safe and effective application in agriculture (Wang et al. 2021).

19.2.8 Nanoparticles for Temperature Stress Adaptation

Nanoparticles play a pivotal role in temperature stress adaptation in plants, offering a promising avenue for enhancing resilience in the face of rising global temperatures. The multifaceted effects of nanoparticles, particularly highlighted by selenium nanoparticles (Se-NPs) in the context of sorghum under heat stress, underscore their potential as innovative tools for temperature stress mitigation (Djanaguiraman et al. 2018).

One of the key contributions of nanoparticles is their ability to safeguard the photosynthetic machinery of plants. Under high-temperature stress, chlorophyll content is often compromised, leading to a decline in photosynthetic efficiency.

Nanoparticles, such as Se-NPs, have demonstrated a protective role by mitigating chlorophyll loss and preserving thylakoid membrane integrity. This preservation is crucial for maintaining the functionality of the photosynthetic apparatus, which is otherwise susceptible to damage under elevated temperature conditions (Djanaguiraman et al. 2018).

Nanoparticles further act as potent antioxidants, countering the oxidative stress induced by high temperatures. Reactive oxygen species (ROS), including superoxide radicals and hydrogen peroxide, tend to accumulate under heat stress, causing cellular damage (Sachdev et al. 2021). Nanoparticles contribute to reducing ROS levels, thereby preventing oxidative damage to cellular components (Flores-López et al. 2019). The application of Se-NPs, for instance, has shown a significant decrease in superoxide radical, hydrogen peroxide, and malondialdehyde (MDA) levels, indicative of their role in maintaining cellular redox balance (Djanaguiraman et al. 2018).

The enzymatic antioxidant defense system, comprising superoxide dismutase (SOD), catalase (CAT), peroxidase (POX), and glutathione peroxidase (GPX), is bolstered by nanoparticles, particularly under heat stress conditions. This augmentation of antioxidant enzyme activity helps plants efficiently neutralize ROS, enabling them to cope with temperature-induced oxidative challenges (Iqbal et al. 2019b).

Beyond cellular and physiological aspects, nanoparticles contribute to the reproductive success of plants under heat stress. Parameters such as pollen germination percentage, seed set percentage, and overall seed yield are positively influenced by nanoparticle applications (Yadav et al. 2022). This indicates that nanoparticles not only protect essential physiological processes but also play a role in sustaining plant reproductive functions, crucial for overall crop productivity.

Nanoparticles emerge as versatile tools for temperature stress adaptation in plants. Their protective effects on photosynthetic machinery, antioxidant properties, and positive impact on reproductive parameters collectively position nanoparticles as valuable assets in the pursuit of climate-resilient agriculture. As research in this field progresses, further exploration of nanoparticle-mediated metabolic pathways and genetic responses promises to unlock additional insights into their role in temperature stress adaptation, paving the way for sustainable and innovative agricultural practices.

19.2.9 Nanoparticles in Symbiotic Interactions

Nanoparticles exert a significant influence on various symbiotic interactions within the intricate web of plant-microbe relationships. These effects span across mycorrhizal associations, nitrogen-fixing symbiosis, and rhizobial symbiosis, showcasing the diverse impacts of NPs on key processes in agroecosystems.

In mycorrhizal associations, where plants form mutually beneficial relationships with arbuscular mycorrhizal fungi (AMF), NPs have been reported to exhibit complex and context-dependent effects. Some studies suggest that certain NPs, such as SiO_2NPs, TiO_2NPs, and CeO_2NPs, may promote AMF colonization and enhance the physiological functions of mycorrhizae. The presence of weakly toxic NPs like FeNPs, SiNPs, and TiO_2NPs, and even highly toxic AgNPs, has been associated with positive effects on AMF infection rates under specific conditions. The feedback of NPs to mycorrhizal structures is influenced by factors including experimental conditions, types of nanomaterials, and the plant-AMF species involved (Wang et al. 2022a).

Moving to nitrogen-fixing symbiosis, where leguminous plants form nodules with nitrogen-fixing bacteria, NPs have demonstrated varying impacts on nodulation processes and nitrogen fixation. The effects of NPs in this context depend on factors such as NP types, concentrations, and experimental conditions. Understanding these nuances is essential for optimizing the benefits of nitrogen-fixing symbiosis, a pivotal process for enhancing soil fertility and plant nutrition (De Souza-Torres et al. 2021).

In the case of rhizobial symbiosis, NPs play a role in modulating the efficiency of nitrogen fixation in leguminous plants. The interactions between NPs and rhizobia bacteria can influence nitrogen fixation processes, with the outcomes shaped by NP characteristics, concentrations, and the specific legume-rhizobia system under consideration (De Souza-Torres et al. 2021).

Considering mycorrhizal associations, nitrogen-fixing symbiosis, and rhizobial symbiosis collectively, the impact of NPs on these symbiotic interactions contributes to our evolving understanding of nanoparticle applications in agroecosystems. As research progresses, continued exploration of the intricate interdependencies between NPs, plants, and symbiotic microorganisms will be essential to harness the positive aspects while mitigating potential negative consequences. This holistic approach is crucial for the development of sustainable and resilient agricultural practices that leverage the benefits of nanotechnology in symbiotic plant-microbe relationships.

19.3 Natural Defense Mechanisms in Plants

Plants, as resilient organisms intricately connected to their environments, have evolved a repertoire of sophisticated defense mechanisms to navigate the challenges posed by climatic conditions and interactions with living organisms. This section delves into the multifaceted world of natural defense mechanisms employed by plants, shedding light on their ingenious strategies to thrive amidst diverse stressors.

19.3.1 Physical Barriers

Plants employ a sophisticated range of physical defenses to fortify themselves against environmental threats. The cuticle, comparable to human skin, forms an initial defense by creating a protective barrier against external toxins. Specialized structures like waxes and plant hairs enhance this defense, contributing to overall resilience (Lamalakshmi Devi et al. 2017).

The structural integrity of epidermal cell walls is crucial, hindering pathogen attachment. Various physical barriers, including the cell wall, cuticle, trichomes, thorns, bark, sclerenchyma tissue, prickles, and spines, undergo specific modifications (Jh 2015). The cell wall acts as a rigid shield, preventing pathogen entry. The cuticle, derived from the cell wall, inhibits water loss and pathogen entry, delaying fungal growth (Hückelhoven 2007).

Thorns shield against primary herbivores, while the bark and sclerenchyma tissue provide solid protection. Epidermal modifications like prickles induce pain or injury, and spines, as leaf modifications, hinder pathogen invasion. These modifications showcase the intricate and seamless strategies plants employ to bolster themselves against environmental challenges (Jh 2015).

19.3.2 Chemical Defenses

Plants have evolved intricate chemical defense mechanisms, employing a diverse array of secondary metabolites to deter potential threats. These chemicals serve as elicitors, inhibitors, repellents, toxins, or signals, collectively contributing to the overall defense strategies of plants. This chemical warfare, a result of the plant's evolutionary journey, plays a pivotal role in reducing the impact of pathogen feeding (Bennett and Wallsgrove 1994).

The chemical substances involved in plant defense may be consistently present or induced in response to an attack. They exhibit remarkable diversity in composition and concentration, which varies across individual plants, communities, developmental stages, and tissue types. Genetic variability and environmental factors further shape the intricate profile of secondary metabolites (Mithöfer and Boland 2012).

Terpenoids, a class of secondary metabolites, encompass compounds like resins and oils, showcasing potent antimicrobial and insecticidal properties. These compounds act as formidable defenders, hindering the accessibility of plant tissues to pathogens. On the other hand, phenolics, including flavonoids and tannins, demonstrate antioxidant properties, adding an extra layer of protection by making plant tissues less vulnerable to pathogenic intrusion (Bhatla 2018).

In specific plants, specialized transporters and membranes act as an additional line of defense, forming a protective barrier against potential harm caused by glycosides. This dynamic interplay of various chemical components highlights the sophistication of plant defense mechanisms. Terpenoids and phenolics, in particular, emerge as key players in fortifying plants against a myriad of environmental challenges (Agrawal et al. 2012).

19.3.3 Induced Systemic Resistance

Induced systemic resistance (ISR) is a remarkable phenomenon in which plants enhance their defense mechanisms against diseases by engaging in communication with beneficial microorganisms, particularly bacteria or fungi that contribute to plant growth. This intricate process stimulates unique defense mechanisms within the plant, showcasing the remarkable adaptability of plants to external challenges (Choudhary et al. 2007).

The induction and signaling of ISR are complex processes influenced by various factors, including the type of microbe, plant species, and the specific disease in question (Pieterse et al. 2014). Key components contributing to ISR are abundant in many microorganisms, such as lipopolysaccharides (LPS), flagellin, or peptidoglycan. These molecules serve as plant receptors, initiating fundamental immune responses like oxygen burst, calcium influx, and the activation of mitogen-activated protein kinase (MAPK) (Abdul Malik et al. 2020).

Beyond the direct interaction with the microorganisms, ISR is also influenced by gases released by these beneficial microbes, including acetoin, 2,3-butanediol, or dimethyl disulfide (Yu et al. 2022). These gases can disperse through the air and reach the plant, effectively altering the levels of jasmonic acid (JA) and ethylene (ET), two crucial signaling molecules involved in plant defense (Cofer et al. 2018). The modulation of these signaling pathways contributes to the induction of ISR, further highlighting the intricacies of the plant-microbe relationship.

ISR exemplifies the sophisticated ways in which plants can adapt and fortify themselves against diseases through strategic communication with beneficial microorganisms, demonstrating the dynamic and responsive nature of plant defense mechanisms (Khanna et al. 2021).

19.3.4 Plant Defense Response

Diving into the complex world of plant defenses reveals a remarkable array of protective mechanisms employed by plants. From robust physical barriers and intricate chemical warfare to the dynamic realm of induced systemic resistance, the exploration through physical, chemical, and induced defenses unveils a comprehensive tapestry of plant survival strategies. This section synthesizes these revelations, shedding light on the nuanced interplay between direct and indirect defense responses and showcasing the sophisticated adaptability of plants to environmental challenges. Figure 19.1 visually presents the various direct and indirect defense responses in plants, providing a comprehensive overview of their intricate survival strategies.

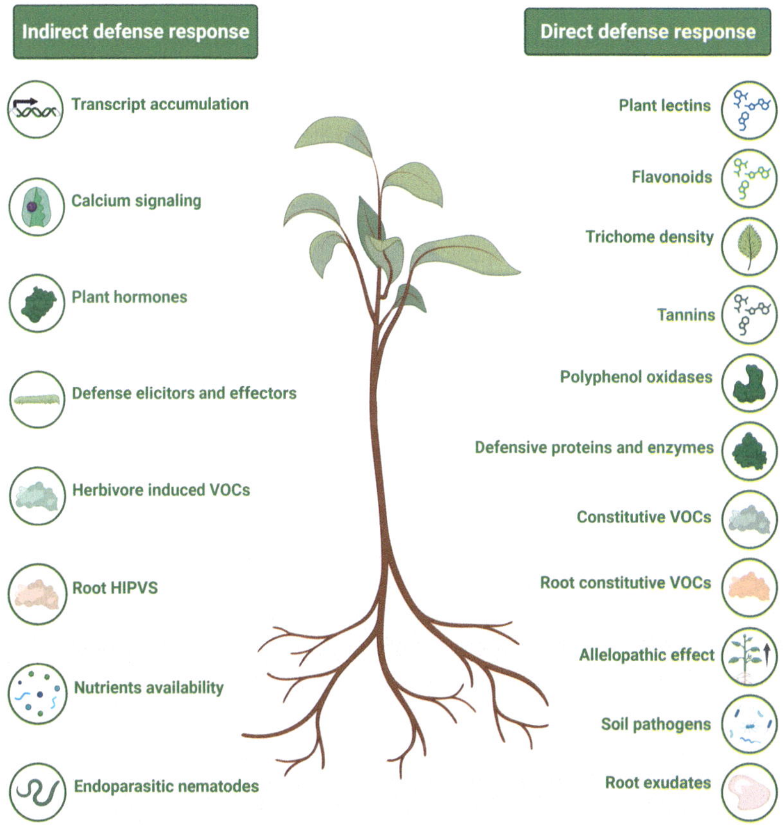

Fig. 19.1 The intricate interplay of direct and indirect defense responses in plants

Direct Defense Responses

Direct defense responses involve immediate and targeted actions by plants to fend off potential threats, such as pathogens or herbivores. These responses often include the production of specific chemicals, proteins, and physical structures that directly impede or deter the invaders (Ghosh et al. 2019).

Plant Lectin: Plant lectins are proteins that play a crucial role in obstructing pathogen growth. By binding to the surfaces of pathogens, these proteins act as a formidable defense mechanism. Their interference disrupts essential processes for the pathogen, impeding its ability to proliferate and cause harm to the plant. In essence, plant lectins serve as sentinels, directly targeting and neutralizing potential threats at the molecular level (Jain et al. 2022).

Flavonoids: Flavonoids, classified as secondary metabolites, function as biochemical warriors in the plant's defense arsenal. Endowed with antioxidant properties, these compounds hinder the development of pathogens. By mitigating oxidative stress within plant tissues, flavonoids not only protect the plant's cellular components but also create an inhospitable environment for invading pathogens (Shen et al. 2022). The intricate chemistry of flavonoids contributes to the plant's resilience by thwarting the progress of potential threats.

Trichome Density: Trichomes, the hair-like structures covering plant surfaces, constitute a physical defense against herbivores and pathogens. Their high density acts as a formidable barrier, physically impeding the movement and attachment of these invaders. The mechanical interference provided by trichomes serves as a first line of defense, deterring herbivores and preventing pathogens from establishing a foothold on the plant's surfaces (Hanley et al. 2007).

Tannins: Tannins are compounds known for their astringency and toxicity, making them potent deterrents against herbivores. These chemicals create an unpleasant and harmful environment for herbivores that attempt to feed on plant tissues. By inducing discomfort and potentially harmful effects, tannins act as a direct defense mechanism, dissuading herbivores from consuming the plant and minimizing the risk of damage (Manetas 2012).

Polyphenol Oxidases: Polyphenol oxidases are enzymes that contribute to the plant's defense by producing toxic compounds. When pathogens attempt to breach plant tissues, these enzymes catalyze the formation of toxic products, creating a hostile environment for the invaders. The toxic compounds generated by polyphenol oxidases directly fend off pathogens, serving as a biochemical weapon in the plant's intricate defense strategy (Fürstenberg-Hägg et al. 2013).

Defensive Proteins and Enzymes: Within plant defense, certain proteins and enzymes take on the role of direct defenders. Cyclophilin-like proteins, ribosome-inactivating proteins, thaumatin-like proteins, chitin-binding proteins, PR-1 proteins, LEA II proteins (Abdul Aziz et al. 2021), defensins/thionins, chitinase, lignin-forming peroxidase, and β-glucanases are among the key players in this intricate defense mechanism (Ferreira et al. 2007). These proteins and enzymes are designed to target and inhibit pathogenic invaders directly. By disrupting essential processes or structures within the pathogens, these defensive proteins and enzymes act as molecular guards, actively thwarting the progression of potential threats. Each of these components plays a specific role in plant defense, contributing to the overall robustness of the plant's immune response. Notably, high expression of cyclophilins has been found to be related to infection, underscoring the dynamic nature of plant defense responses (Singh et al. 2018). Additionally, a thaumatin-like protein of *Ocimum basilicum* has been identified as conferring tolerance to both fungal pathogens and abiotic stresses (Misra et al. 2016). Chitinase is involved in breaking down chitin, a major component of fungal cell walls (Sandhu et al. 2017), while defensins/thionins exhibit antimicrobial properties (Ji et al. 2015). Understanding the functions of these proteins provides insights into the sophisticated strategies plants employ to protect themselves from various pathogens.

Constitutive volatile organic compounds (VOCs): VOCs emitted by plants, known as constitutive VOCs, play a vital role in repelling herbivores and attracting predators. These compounds, released continuously by the plant, create an olfactory shield. They deter herbivores from feeding on the plant while simultaneously attracting natural enemies of the herbivores, establishing a harmonious ecological balance within the plant's surroundings (Effah et al. 2019).

Root constitutive VOCs: Below-ground, roots release constitutive VOCs that influence the soil microbiome and deter pests. These compounds, emanating from the roots, shape the rhizosphere—the region of soil influenced by root secretions. By influencing microbial interactions and creating an environment less conducive to pests, root-constitutive VOCs contribute to the plant's overall defense strategy (Russo et al. 2022).

Allelopathic Effect: Plants deploy an allelopathic effect by releasing chemicals to inhibit the growth of neighboring competing plants. This indirect competition for resources creates a strategic advantage, ensuring the plant's access to essential nutrients, water, and space. The allelopathic chemicals act as a preemptive defense mechanism, shaping the plant's immediate environment to its benefit (Arroyo et al. 2018).

Soil Pathogens: In a symbiotic twist, microorganisms in the soil can act as a line of defense against potential threats. These soil pathogens, while potentially harmful to other organisms, contribute to the plant's defense by outcompeting or directly antagonizing pathogenic invaders. The soil becomes a dynamic arena where beneficial microorganisms act as allies, fortifying the plant against potential harm (Ab Rahman et al. 2018).

Root Extrudates: Substances released by plant roots, known as root extrudates, play a crucial role in shaping the rhizosphere and influencing microbial interactions. These extrudates include a complex mix of chemicals that serve various functions, from nutrient mobilization to allelopathic effects. By sculpting the immediate soil environment, root extrudates actively participate in the plant's defense by fostering beneficial microbial communities and deterring potential threats (Patra and Mandal 2022).

Indirect Defense Responses

Indirect defense responses are more strategic and involve preparatory or systemic actions by plants to enhance their overall resilience. These responses often involve signaling pathways, hormonal regulation, and interactions with the environment to activate defenses (Lamalakshmi Devi et al. 2017). Examples include the following.

Transcript Accumulation: In anticipation of potential stressors, plants execute a strategic defense response by changing gene expression. This process, known as transcript accumulation, involves the buildup of specific messenger RNA (mRNA) molecules. By preloading the plant with the genetic information needed for defense, transcript accumulation enables a rapid and targeted molecular response when the plant encounters environmental challenges (MacRae 2010).

Calcium Signaling: A sophisticated defense mechanism in plants involves the intricate involvement of calcium ions in signaling cascades. Calcium signaling acts as a molecular conductor, initiating and coordinating defense responses. This regulatory network enables plants to transmit signals effectively, ensuring a swift and precise activation of defense mechanisms when needed. The orchestration of calcium ions highlights the dynamic nature of plant responses to environmental stressors (Chen et al. 2015).

Plant Hormones: Plant defense is finely tuned through the regulation of hormone levels. Hormones, such as jasmonic acid (JA) and salicylic acid (SA), play pivotal roles in modulating defense mechanisms. This intricate hormonal regulation allows plants to tailor their responses to specific threats, orchestrating a harmonized defense strategy. The dynamic interplay of plant hormones exemplifies the plant's ability to adapt its defense mechanisms based on the nature of the encountered stressors (Li et al. 2019b).

Defense Elicitors and Effectors: Molecules with the ability to induce or suppress immune responses are integral to plant defense. These defense elicitors and effectors act as molecular messengers, signaling the plant to activate or inhibit specific defense mechanisms. By responding to these molecular cues, plants can fine-tune their immune responses, ensuring a measured and targeted reaction to potential threats (Mishra et al. 2012).

Herbivore-Induced VOCs: In response to herbivore attacks, plants unleash a complex chemical defense strategy involving the release of volatile organic compounds (VOCs). These herbivore-induced VOCs serve as signaling molecules, broadcasting the presence of herbivores to attract natural predators. This indirect defense mechanism not only protects the plant from immediate threats but also contributes to the overall ecological balance in the plant's environment (Pinto-Zevallos et al. 2016).

Root HIPVs (Herbivore-Induced Plant Volatiles): Below the ground, plant roots deploy herbivore-induced plant volatiles (HIPVs) to influence the defenses of neighboring plants. These below-ground VOCs serve as chemical messengers, signaling nearby plants to prepare for potential herbivore attacks. Communication through root HIPVs exemplifies the interconnected nature of plant defenses, extending beyond individual plants to create a collaborative response within plant communities (Kigathi et al. 2009).

Nutrient Availability: The availability of nutrients plays a crucial role in shaping plant defense responses. Plants adjust their defense mechanisms based on nutrient levels, strategically allocating resources to maximize their resilience against potential threats. This dynamic interplay between nutrient availability and defense highlights the intricate balance that plants maintain to optimize their survival strategies in varying environmental conditions (Monson et al. 2022).

Endoparasitic Nematodes: Microscopic worms known as endoparasitic nematodes trigger intricate defense mechanisms in plants. These nematodes, while potentially harmful, stimulate the plant to activate specific defense responses. The plant's recognition of nematode presence initiates a cascade of molecular reactions, showcasing the plant's ability to sense and respond to the intricate challenges posed by microscopic invaders. The activation of defense mechanisms against endoparasitic nematodes underscores the adaptability of plants in confronting diverse stressors (Liu and Park 2018).

These indirect responses are often systemic, affecting not only the targeted area but also influencing other parts of the plant or even neighboring plants, showcasing the plant's ability to mount a coordinated and adaptive defense strategy.

19.3.5 Challenges in Enhancing Plant Defense Mechanisms

Elevating plant defense mechanisms introduces intricate challenges that necessitate a nuanced approach. The delicate equilibrium involved in allocating resources for the production and upkeep of physical and chemical defenses demands thoughtful consideration. Secondary metabolites, crucial for defense, wield significant influence over essential metabolic pathways by affecting carbon and nitrogen availability (Isah 2019). Achieving a harmonious balance between the advantages of defense and the associated costs becomes imperative, particularly in the context of fluctuating environmental conditions and diverse stress levels (Cipollini et al. 2017). A meticulous examination of these challenges serves as a comprehensive guide for formulating strategies that fortify plant defenses while safeguarding overall plant health.

19.3.6 The Need for Sustainable and Eco-Friendly Approaches

In the pursuit of sustainable agriculture, the imperative for eco-friendly approaches to crop protection becomes paramount. Balancing the demands of food production with environmental stewardship necessitates innovative and sustainable strategies. The exploration of natural defense mechanisms in plants serves as a foundational framework for considering novel approaches, such as the integration of nanoparticle-enhanced defenses (Chakraborty et al. 2023). This section delves into the pressing need for sustainable agricultural practices, emphasizing the pivotal role of eco-friendly approaches in ensuring the long-term health of both crops and the broader ecosystem.

19.4 Nanoparticle-Mediated Enhancement of Plant Defense

19.4.1 Mechanisms of Nanoparticle Entry into Plant Cells

The exploration of nanoparticle-mediated enhancement of plant defense delves into the intricate processes governing the entry of nanoparticles into plant cells, revealing a dynamic interplay between nanotechnology and plant biology. This comprehensive understanding is vital for unlocking the full potential of nanoparticles in

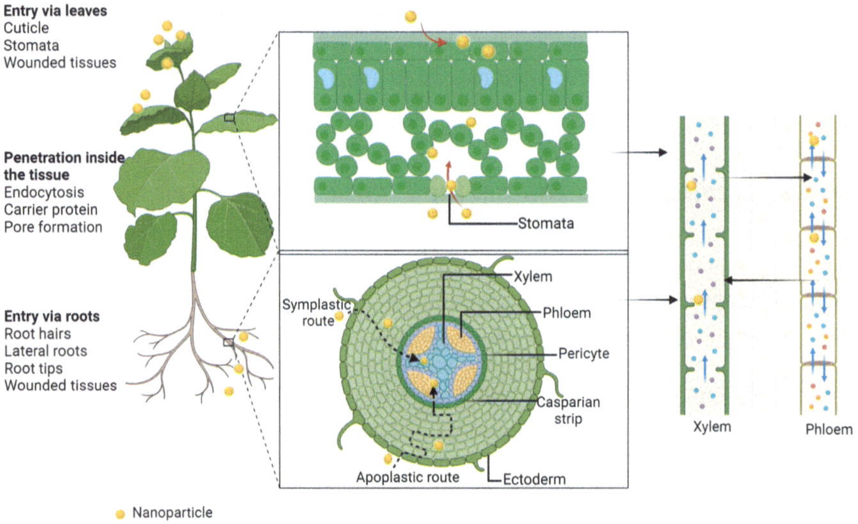

Fig. 19.2 The potential pathways through which nanoparticles enter the plant system

fortifying plant defense mechanisms. Research sheds light on diverse pathways and processes involved in nanoparticle entry into plant cells, emphasizing the significance of physicochemical properties.

Nanoparticles showcase remarkable versatility in gaining entry into plant tissues through various routes. Figure 19.2 illustrates potential routes for nanoparticles to enter the plant system, encompassing direct penetration of plant cell walls, absorption by roots from the soil, and foliar application (Su et al. 2019). The physicochemical attributes of nanoparticles, including size, shape, and surface charge, significantly influence their interaction with plant cell membranes (Hu et al. 2020). This impact on uptake by plant roots subsequently influences translocation and induces physiological effects.

Nanoparticle uptake in plants involves various entry points and routes. When applied to leaves, nanoparticles can enter through stomata, small openings on the leaf surface, or traverse the cuticle, the protective outer layer (Hong et al. 2021). Additionally, entry may be possible through wounds in leaves, presenting alternative access points (Fiol et al. 2021). Roots, the primary site for nanoparticle uptake from the soil, facilitate entry through root hairs, tips, lateral roots, and wounds, forming a crucial gateway for soil-applied nanoparticles (Dong et al. 2022).

Once in proximity to plant cells, nanoparticles may enter through different mechanisms. Endocytosis, a common mechanism for nanoparticle entry into plant cells, involves cells engulfing particles (Palocci et al. 2017). Delving into the intricacies of endocytosis in nanoparticle uptake provides insights into the molecular and cellular events facilitating the internalization of nanoparticles into plant cells. This mechanistic understanding is crucial for comprehending how nanoparticles navigate plant cell barriers to reach intracellular compartments where their effects manifest (Zhang et al. 2022b). Carrier proteins on the cell membrane may also facilitate nanoparticle entry, as well as the formation of pores (Beddoes et al. 2015). Understanding these cellular entry points sheds light on the interactions between nanoparticles and plant cells.

Furthermore, nanoparticles may navigate through the plant via apo plastic and symplastic routes (Avellan et al. 2021). The apoplastic route involves movement through the extracellular spaces, while the symplastic route traverses the interconnected cytoplasm of plant cells (Francis et al. 2020). As nanoparticles progress, they can enter the plant's vascular system, reaching the xylem and phloem. This systemic transport allows nanoparticles to distribute across various plant tissues, influencing physiological processes and interactions within the plant (Wang et al. 2023).

19.4.2 Activation of Plant Defense Responses by Nanoparticles

Reactive Oxygen Species (ROS) Generation

The induction of reactive oxygen species (ROS) generation within plant cells by nanoparticles represents a pivotal aspect of plant defense mechanisms (Yang et al. 2017). ROS, encompassing molecules such as superoxide radicals and hydrogen peroxide, assumes a dual role as both signaling molecules and effectors in plant defense responses (Marcec et al. 2019). This intricate interplay serves as a cornerstone for the activation of defense pathways against various biotic and abiotic stressors. Several research studies offer detailed insights into the dynamics of ROS generation facilitated by nanoparticles, emphasizing the significance of this process in bolstering plant defense. Going beyond their conventional function, the investigation into ROS dynamics unveils their intricate involvement in coordinating defense responses. ROS induced by nanoparticles acts as essential messengers, amplifying the plant's capacity to respond effectively to potential threats (Farooq et al. 2023). This exploration into the contributions of nanoparticles to signaling cascades provides valuable insights into the complex molecular dynamics of plant-nanoparticle interactions. These discoveries form the basis for leveraging nanomaterials to enhance sustainable crop protection, emphasizing the integral role of ROS in this process.

Understanding the intersection of ROS signaling and nanoparticle-induced defense responses in plants, the research reveals intricate molecular interactions and pathways (Sharma et al. 2019). The cascade initiated by nanoparticle-induced ROS production activates defense-related genes and proteins, enhancing the plant's ability to repel threats (Takehara et al. 2023). These findings significantly contribute to advancing nanotechnology in agriculture by deepening our understanding of molecular mechanisms in plant defense.

Hormonal Signaling Pathways

In plant defense mechanisms, hormonal signaling pathways play a pivotal role, with nanoparticles identified as key modulators. The modulation of plant hormones, including salicylic acid (SA), jasmonic acid (JA), and ethylene, represents a crucial aspect of how nanoparticles influence plant defense responses (Tripathi et al. 2022). Research has explored the impact of nanoparticles on salicylic acid signaling, revealing nuanced ways nanoparticles fine-tune plant defense mechanisms (García-Sánchez and Bernales 2015). Additionally, work has delved into jasmonic acid and ethylene signaling modulation, highlighting nanoparticles' dual regulatory role (Shang et al. 2020). These findings promise targeted strategies for sustainable crop protection in agriculture, emphasizing the intricate and environmentally responsive nature of nanoparticle-mediated hormonal modulation.

Induction of Defense-Related Genes

Nanoparticles serve as influential agents in activating defense-related genes within the intricate realm of plant defense. Prompting the initiation of specific genes, such as those encoding proteins engaged in diverse aspects of plant defense—such as pathogen recognition, synthesis of antimicrobial compounds, and fortification of cell walls—nanoparticles contribute to critical functions. The orchestrated upregulation of these defense-related genes, including those like PR1, PR5, and NPR1, enhances the plant's inherent ability to protect itself against various adversaries, such as pathogens and pests (Almutairi 2019).

Highlighting the importance of understanding molecular defense mechanisms, the study emphasizes limited but crucial evidence demonstrating nanoparticles' ability to activate molecular defense, increasing plant tolerance to environmental stress. It delves into common molecular defense mechanisms regulated by nanoparticles, focusing on stress-responsive genes such as WRKY, MYB, and MAPK, as well as nanoparticles-responsive genes (Salem et al. 2022). The involvement of hormones like ABA and SA, along with the repression of certain genes like ERF1 and JAZ, under different stress conditions is discussed (Yang et al. 2019).

High-throughput transcriptomic and proteomic technologies are recommended to uncover intricate molecular mechanisms and gene networks related to nanoparticle alleviation under stress. The ultimate goal is to achieve a deeper understanding of gene regulation during plant responses to stress for agricultural and biological advancements.

Modulation of Secondary Metabolites

Nanoparticles exert a notable influence on the modulation of secondary metabolites, playing a key role in enhancing a plant's resilience against stressors. Secondary metabolites, encompassing phenolics, flavonoids, and terpenoids, are pivotal in this defense mechanism (Jan et al. 2021). Research findings demonstrate that nanoparticles, such as MgO, act as novel elicitors, inducing the synthesis of secondary metabolites for valuable biosynthesis (Rezaei et al. 2019). TiO_2 nanoparticles contribute to this modulation by inducing enhanced production of proline, an osmotic regulator, while also mitigating cadmium toxicity in *Coriandrum sativum* (Sardar et al. 2022). This mitigation involves the modulation of antioxidant systems, stress markers, and the reduction of cadmium uptake. Additionally, silver nanoparticle treatment increases carnosic acid content in rosemary (Hadi Soltanabad et al. 2020). The nanoparticle-mediated modulation of secondary metabolites not only strengthens the plant's resistance to pests and pathogens but also enhances its adaptability to environmental challenges. Understanding this intricate interplay provides a valuable foundation for tailoring nanoparticle applications to optimize the production of specific secondary metabolites for targeted plant defense.

Nanoparticle-Mediated Regulation of Antioxidant Enzymes

An integral aspect of plant defense mechanisms involves the regulation of antioxidant enzymes. Nanoparticles have been identified as influential modulators in this regulatory network. Studies demonstrate that nanoparticles induce the activity of antioxidant enzymes such as superoxide dismutase, catalase, and peroxidase (Wu et al. 2023b). This heightened enzymatic activity serves as a defense strategy, mitigating oxidative stress and reinforcing the plant's capacity to combat various threats. Exploring the nexus between nanoparticles and antioxidant enzyme regulation provides critical insights into how nanotechnology can be harnessed to enhance the plant's antioxidative machinery, thereby bolstering its defense mechanisms in a targeted and sustainable manner.

Crosstalk Between Defense Signaling Pathways

A holistic understanding of nanoparticle-induced plant defense mechanisms involves unraveling the crosstalk between different signaling pathways (Jalil and Ansari 2019). Nanoparticles, through their multifaceted interactions, demonstrate the ability to orchestrate a coordinated response involving various signaling molecules (Almutairi 2019). This crosstalk extends beyond individual pathways, integrating hormonal signaling, ROS generation, and defense-related gene expression. Investigating this interconnected signaling network sheds light on how nanoparticles amplify the plant's ability to respond synergistically to diverse stressors (Adeleke et al. 2022). By deciphering the intricacies of crosstalk, researchers pave the way for the development of nanoparticle-based strategies that harness synergistic defense responses, presenting a paradigm shift in sustainable crop protection.

The exploration of nanoparticle-mediated enhancement of plant defense is a dynamic and interconnected field. From the mechanisms of nanoparticle entry into plant cells to the modulation of secondary metabolites, regulation of antioxidant enzymes, and crosstalk between defense signaling pathways, each facet contributes to a comprehensive understanding (Roy et al. 2022). This knowledge forms the basis for targeted and effective applications of nanoparticles in fortifying plant defense mechanisms, ensuring sustainable crop protection in the face of evolving environmental challenges.

19.5 Case Studies on Nanoparticle-Enhanced Plant Defense

In the realm of nanoparticle applications for plant defense, a groundbreaking study on biosynthesized silver nanoparticles (AgNP) sheds light on their remarkable dual impact within the *A. thaliana* and A. *brassicicola* pathosystem. This research not only demonstrates the effective countering of the pathogen *A. brassicicola* by AgNP but also reveals their ability to enhance the plant's immune responses through intricate molecular mechanisms. The study highlights SNP's role in mitigating biotic stress induced by the pathogen, specifically through the regulation of sugar and sterol metabolism. Additionally, AgNPs play a crucial role in initiating an early defense response in plants, thereby reducing the production of reactive oxygen species (ROS) and stress markers. The process of autophagy emerges as a potential contributor to this enhanced defense, as evidenced by the upregulation of autophagy-related proteins. The research also highlights AgNP's significant contribution to protein biogenesis, amino acid synthesis, and the subsequent production of antimicrobial compounds. Enhanced immunity and osmoregulation are observed through the upregulation of proteins involved in methionine biosynthesis and methyl esterase. In conclusion, this study unravels the multifaceted roles of silver nanoparticles in plant defense, presenting a promising avenue for their application in sustainable crop protection strategies (Kumari et al. 2020).

In a comprehensive exploration of Fe_3O_4NPs' impact on plant growth and the antioxidant system, the study revealed notable advancements in tobacco growth following foliar exposure to Fe_3O_4NPs. The investigation emphasizes the crucial function of salicylic acid (SA) communication in the defense mechanisms of plants, uncovering that Fe_3O_4NPs stimulate the increase of SA levels and induce the expression of SA-responsive genes (PR1 and PR2). This showcases the potential of Fe_3O_4NPs as stimulators of plant inherent immunity. Furthermore, the research examines Fe_3O_4NPs' direct deactivation effects against the tobacco mosaic virus (TMV) in both laboratory and living organism environments. Fe_3O_4NPs display direct toxicity toward TMV particles, leading to their aggregation and structural fracture. Significantly, the application of Fe_3O_4NPs on the leaves demonstrates an extraordinary capability to enhance the resistance of plants to TMV infection (Cai et al. 2020).

Another study highlights the ability of chitosan nanoparticles (CNPs) to enhance defense responses in tea plants, with CNP demonstrating superior efficacy compared to chitosan. The findings suggest that CNP could serve as a potent and cost-effective plant immune elicitor for disease management in organic cultivation. The study provides valuable insights into the cellular localization and mechanisms of action of chitosan nanoparticles in plant defense modulation (Chandra et al. 2015).

The investigation establishes that the application of silica nanoparticles (SNPs) to the leaves of plants significantly strengthens the mechanisms that protect against bacterial wilt caused by *Ralstonia solanacearum*. Additionally, the examination of the signaling of phytohormones reveals a positive correlation between the application of SNPs, the concentration of silicon, and the expression of genes related to salicylic acid (SA). SNPs effectively activate the systemic acquired resistance (SAR) pathway dependent on SA, resulting in an increased expression of genes related to SA (PR-1, PR-5, and PAL). This activation of the SAR pathway dependent on SA emphasizes the ability of SNPs to enhance the defense mechanisms of plants, presenting a promising and environmentally friendly strategy to enhance the resilience of plants against bacterial wilt (Wang et al. 2022b).

The examination of nanoparticle (NP) exposure's impact on the plant transcriptome revealed that, compared to other stress conditions, biotic stress had the most substantial effect, while saline or drought stresses had lesser impacts. The genetic resources of *Arabidopsis thaliana* enabled it to withstand NP-induced stress to a considerable extent. Notably, downregulation of gene expression was a predominant response to NPs, with implications for stress response control. The study identified common regulatory mechanisms in the *Arabidopsis* response to NPs, potentially preventing adverse effects on plant development. NP exposure significantly affected genes involved in the biotic stress response, leading to the downregulation of key components of pathogen-detection pathways. Furthermore, NP-induced repression targeted genes related to phosphate starvation and root hair development. The study underscores the complexity of plant responses to NP exposure, offering valuable insights for understanding and mitigating potential risks associated with nanomaterials (García-Sánchez and Bernales 2015).

19.6 Sustainable Crop Protection: Navigating Environmental, Economic, and Regulatory Dimensions

In agriculture, the utilization of nanoparticles introduces a dual challenge, necessitating a delicate balance between achieving sustainable crop protection and minimizing potential ecological repercussions (Agrawal et al. 2022). A primary ecological concern is nanoparticle persistence, prompting questions about their fate, transport, and potential bioaccumulation in soil and water systems, requiring a thorough examination of their ecological impact (Jogaiah et al. 2021). Researchers actively explore strategies to address nanoparticle persistence, such as modifying nanoparticles for controlled release and developing biodegradable nanoparticles. These approaches aim to balance the benefits of nanoparticles in agriculture with mitigating their potential ecological footprint. Integral to this endeavor are comprehensive environmental monitoring and risk assessments, including fate and transport studies that track nanoparticle movement and ecotoxicological assessments examining nanoparticle-induced toxicity and effects on biodiversity.

Sustainable crop protection involves implementing integrated crop management practices, allowing for practices that minimize environmental impacts while optimizing crop yields. Stakeholder collaboration and policy development play pivotal roles in balancing agricultural productivity with environmental stewardship (Davies and White 2012). As the agricultural sector explores the potential benefits of nanoparticle applications, ongoing research, technological advancements, and collaborative efforts are crucial. This collaborative approach ensures that innovations align with principles of environmental stewardship, contributing to long-term agricultural sustainability (Pe'er et al. 2020).

The economic feasibility and scalability of nanoparticle-based approaches are vital considerations for their adoption in agriculture (Gilbertson et al. 2020). Although initial costs associated with nanoparticle production, formulation, and application may pose challenges, long-term economic benefits, such as increased crop yields and reduced reliance on traditional agrochemicals, contribute to viability. Research and development efforts focus on optimizing production processes and developing cost-effective formulations, enhancing the scalability and accessibility of nanoparticle technologies for farmers (Acharya and Pal 2020).

19.7 Potential Risks and Concerns Associated with Nanoparticle Use in Agriculture

Navigating the promising landscape of nanoparticle applications in agriculture requires confronting potential risks and concerns entwined with their utilization. The physicochemical properties that render nanoparticles effective in various agricultural applications prompt inquiries into their environmental and health implications (Seleiman et al. 2020). Research has raised questions about the potential accumulation of nanoparticles in soil and water systems, with studies probing the bioavailability of nanoparticles to plants and organisms, shedding light on their potential interactions within ecosystems (Boxall et al. 2007). Apprehensions about the long-term ecological consequences of widespread nanoparticle use in agriculture, including unintended effects on nontarget organisms, have become focal points of scrutiny (Zhang and Goss 2022).

The mobility of nanoparticles in the environment raises concerns about their potential to leach into groundwater, affecting aquatic ecosystems (Dibyanshu et al. 2023). Ongoing research investigates pathways through which nanoparticles may move within soil and water systems, influencing broader environmental dynamics. The impact of nanoparticles on soil microbial communities and beneficial soil organisms is a critical area of ongoing research, with disruptions potentially having cascading effects on overall soil health, affecting plant growth and nutrient cycling (Rajput et al. 2018).

As the field of nanoparticle applications in agriculture advances, addressing these concerns becomes imperative. Establishing robust risk assessment protocols, developing comprehensive regulatory frameworks, and promoting responsible application practices are essential steps toward ensuring the safe and sustainable integration of nanoparticles into agricultural systems. Striking a balance between harnessing the innovative potential of nanoparticles and comprehensively understanding their ecological implications is pivotal for maximizing benefits while minimizing potential risks in the pursuit of sustainable agriculture.

19.8 Future Perspectives

In navigating the challenges of nanoparticle-enhanced plant defense, the future promises advancements through collaborative efforts. Ongoing research into the fate and transport of nanoparticles will deepen our understanding, fostering more effective and sustainable applications. Prioritizing the tailoring of nanoparticles for broad applicability and specificity across diverse plant species remains imperative, urging continuous refinement. The evolution of robust risk assessment protocols is essential to address ecological and human health concerns, ensuring responsible applications. Overcoming economic barriers, including production costs and scalability, requires concerted investment and innovation for widespread adoption.

19.9 Conclusion

Throughout this exploration of nanoparticle-enhanced plant defense, we have uncovered transformative potential in sustainable agriculture. The mechanisms fortifying plant defense responses, diverse applications in managing stressors, and crucial ecological and economic considerations lay the foundation for responsible implementation. The revolutionary potential of nanoparticle-enhanced plant defense cannot be overstated. Nanoparticles, with their precise .customization of plant immunity and activation of signaling pathways, offer an efficient approach to crop protection. Case studies underscored their effectiveness in pest and pathogen control, stress tolerance, and symbiotic interactions, presenting a promising technology for resilient and environmentally conscious agriculture. As agriculture addresses global food security and environmental impacts, nanoparticle-enhanced plant defense emerges as a beacon of hope. Integrating nanotechnology into sustainable practices opens frontiers for a resilient agricultural future. A collective call to action is essential. Researchers should delve deeper into nanoparticle interactions, optimizing properties and filling research gaps. Policymakers must establish regulatory frameworks balancing innovation with safety. Industry stakeholders should invest in scalability, ensuring economic viability worldwide. In unison, this collaborative effort can shape a future where sustainable agriculture is a reality.

Nanoparticle-enhanced plant defense showcases the potential of innovative technologies in addressing complex agricultural challenges. Now is the time for action, harnessing nanotechnology for a future where food security and environmental stewardship converge for the benefit of all generations.

References

Ab Rahman SF, Singh E, Pieterse CMJ, Schenk PM (2018) Emerging microbial biocontrol strategies for plant pathogens. Plant Sci 267:102–111

Abbas A, Naz SS, Syed SA (2015) Antimicrobial activity of silver nanoparticles (AgNPs) against Erwinia carotovora pv. carotovora and Alternaria solani. Int J Biosci 6:9–14

Abdallah Y et al (2020) Bioinspired green synthesis of chitosan and zinc oxide nanoparticles with strong antibacterial activity against rice pathogen Xanthomonas oryzae pv. oryzae. Molecules 25:4795

Abdelrhim A et al (2021) Silicon dioxide nanoparticles induce innate immune responses and activate antioxidant machinery in wheat against Rhizoctonia solani. Plants 10:2758

Abdul Aziz M, Sabeem M, Mullath SK, Brini F, Masmoudi K (2021) Plant Group II LEA proteins: intrinsically disordered structure for multiple functions in response to environmental stresses. Biomolecules 11:1662

Abdul Malik NA, Kumar IS, Nadarajah K (2020) Elicitor and Receptor molecules: orchestrators of plant defense and immunity. IJMS 21:963

Abou-Salem E, Ahmed AR, Elbagory M, Omara AE-D (2022) Efficacy of biological copper oxide nanoparticles on controlling damping-off disease and growth dynamics of sugar beet (Beta vulgaris L.) plants. Sustainability 14:12871

Acharya A, Pal PK (2020) Agriculture nanotechnology: translating research outcome to field applications by influencing environmental sustainability. NanoImpact 19:100232

Adeleke BS, Akinola SA, Adedayo AA, Glick BR, Babalola OO (2022) Synergistic relationship of endophyte-nanomaterials to alleviate abiotic stress in plants. Front Environ Sci 10:1015897

Agrawal AA, Petschenka G, Bingham RA, Weber MG, Rasmann S (2012) Toxic cardenolides: chemical ecology and coevolution of specialized plant–herbivore interactions. N Phytol 194:28–45

Agrawal S, Kumar V, Kumar S, Shahi SK (2022) Plant development and crop protection using phytonanotechnology: a new window for sustainable agriculture. Chemosphere 299:134465

Ahmed B, Hashmi A, Khan MS, Musarrat J (2018) ROS mediated destruction of cell membrane, growth and biofilms of human bacterial pathogens by stable metallic AgNPs functionalized from bell pepper extract and quercetin. Adv Powder Technol 29:1601–1616

Ahmed T et al (2020) Silver nanoparticles synthesized by using Bacillus cereus SZT1 ameliorated the damage of bacterial leaf blight pathogen in rice. Pathogens 9:160

Ahmed T et al (2021a) Bioinspired green synthesis of zinc oxide nanoparticles from a native bacillus cereus strain RNT6: characterization and antibacterial activity against rice panicle blight pathogens Burkholderia glumae and B. gladioli. Nanomaterials 11:884

Ahmed T et al (2021b) Green synthesis and characterization of zirconium oxide nanoparticles by using a native Enterobacter sp. and its antifungal activity against bayberry twig blight disease pathogen Pestalotiopsis versicolor. NanoImpact 21:100281

Ahmed T et al (2022) Bioengineered chitosan-iron nanocomposite controls bacterial leaf blight disease by modulating plant defense response and nutritional status of rice (Oryza sativa L.). Nano Today 45:101547

Akhtar N et al (2021) Synergistic effects of plant growth promoting rhizobacteria and silicon dioxide nano-particles for amelioration of drought stress in wheat. Plant Physiol Biochem 166:160–176

Alam T et al (2019) Biogenic synthesis of iron oxide nanoparticles via Skimmia laureola and their antibacterial efficacy against bacterial wilt pathogen Ralstonia solanacearum. Mater Sci Eng C 98:101–108

Alfy H, Soltan ELSAYED, Farag DA (2020) Impact of chitosan nanoparticles as insecticide and nematicide against Spodoptera littoralis, Locusta migratoria, and Meloidogyne incognita. Plant Cell Biotechnol Mol Biol 21:126–140

Alhammad BA, Ahmad A, Seleiman MF (2023) Nano-Hydroxyapatite and ZnO-NPs mitigate Pb stress in maize. Agronomy 13:1174

Alharbi K, Hafez E, Omara AE-D, Awadalla A, Nehela Y (2022) Plant growth promoting rhizobacteria and silica nanoparticles stimulate sugar beet resilience to irrigation with saline water in salt-affected soils. Plants 11:3117

Ali S et al (2019) Combined use of biochar and zinc oxide nanoparticle foliar spray improved the plant growth and decreased the cadmium accumulation in rice (Oryza sativa L.) plant. Environ Sci Pollut Res 26:11288–11299

Almutairi ZM (2019) Plant molecular defense mechanisms promoted by nanoparticles against environmental stresses. Int J Agric Biol 21(2):259–270. https://doi.org/10.17957/IJAB/15.0890

Alotaibi MO et al (2023) Effect of green synthesized cerium oxide nanoparticles on fungal disease of wheat plants: a field study. Chemosphere 339:139731

Alsaeedi A et al (2019) Silica nanoparticles boost growth and productivity of cucumber under water deficit and salinity stresses by balancing nutrients uptake. Plant Physiol Biochem 139:1–10

Amuthavalli P et al (2021) Zinc oxide nanoparticles using plant Lawsonia inermis and their mosquitocidal, antimicrobial, anticancer applications showing moderate side effects. Sci Rep 11:8837

Anderson CR, Bruil J, Chappell MJ, Kiss C, Pimbert MP (2019) From transition to domains of transformation: getting to sustainable and just food systems through agroecology. Sustainability 11:5272

Ansari MA, Asiri SMM (2021) Green synthesis, antimicrobial, antibiofilm and antitumor activities of superparamagnetic γ-Fe_2O_3 NPs and their molecular docking study with cell wall mannoproteins and peptidoglycan. Int J Biol Macromol 171:44–58

Arroyo AI et al (2018) Evidence for chemical interference effect of an allelopathic plant on neighboring plant species: a field study. PLoS One 13:e0193421

Arunthirumeni M, Veerammal V, Shivakumar MS (2022) Biocontrol efficacy of mycosynthesized selenium nanoparticle using Trichoderma sp. on insect pest Spodoptera litura. J Clust Sci 33:1645–1653

Avellan A et al (2021) Critical review: role of inorganic nanoparticle properties on their Foliar uptake and in planta translocation. Environ Sci Technol 55:13417–13431

Banerjee S, Mazumder S, Chatterjee D, Bose S, Majee SB (2022) Nanotechnology for cargo delivery with a special emphasis on pesticide, herbicide, and fertilizer. In: Nano-enabled agrochemicals in agriculture. Elsevier, pp 105–144. https://doi.org/10.1016/B978-0-323-91009-5.00002-1

Beckford FB (2018) Poverty and climate change: restoring a global biogeochemical equilibrium. Routledge. https://doi.org/10.4324/9780429437892

Beddoes CM, Case CP, Briscoe WH (2015) Understanding nanoparticle cellular entry: a physicochemical perspective. Adv Colloid Interface Sci 218:48–68

Bennett RN, Wallsgrove RM (1994) Secondary metabolites in plant defence mechanisms. N Phytol 127:617–633

Bharani RSA, Namasivayam SKR (2017) Biogenic silver nanoparticles mediated stress on developmental period and gut physiology of major lepidopteran pest Spodoptera litura (Fab.) (Lepidoptera: Noctuidae)—an eco-friendly approach of insect pest control. J Environ Chem Eng 5:453–467

Bhatla SC (2018) Secondary metabolites. In: Plant physiology, development and metabolism. Springer Singapore, pp 1099–1166. https://doi.org/10.1007/978-981-13-2023-1_33

Borase HP et al (2014) Mosquito larvicidal and silver nanoparticles synthesis potential of plant latex. J Entomol Acarol Res 46:59

Boxall AB, Tiede K, Chaudhry Q (2007) Engineered nanomaterials in soils and water: how do they behave and could they pose a risk to human health? Nanomedicine 2:919–927
Cai L et al (2020) Foliar exposure of Fe_3O_4 nanoparticles on Nicotiana benthamiana: evidence for nanoparticles uptake, plant growth promoter and defense response elicitor against plant virus. J Hazard Mater 393:122415
Casillas-Figueroa F et al (2020) Argovit™ silver nanoparticles effects on Allium cepa: plant growth promotion without cyto genotoxic damage. Nanomaterials 10:1386
Chahardoli A, Karimi N, Ma X, Qalekhani F (2020) Effects of engineered aluminum and nickel oxide nanoparticles on the growth and antioxidant defense systems of Nigella arvensis L. Sci Rep 10:3847
Chakraborty S, Bera D, Roy L, Ghosh CK (2023) Biomimetic and bioinspired nanostructures: recent developments and applications. In: Sen M, Mukherjee M (eds) Bioinspired and green synthesis of nanostructures. Wiley, pp 353–404. https://doi.org/10.1002/9781394174928.ch15
Chakravarthy AK (2012) DNA-tagged nano gold: a new tool for the control of the armyworm, Spodoptera litura Fab. (Lepidoptera: Noctuidae). Afr J Biotechnol 11
Chakravarthy et al (2012) Bio efficacy of inorganic nanoparticles CdS, Nano-Ag and Nano-TiO_2 against Spodoptera litura (Fabricius) (Lepidoptera: Noctuidae). Curr Biot 6:271–281
Chandra S, Chakraborty N, Dasgupta A, Sarkar J et al (2015) Chitosan nanoparticles: a positive modulator of innate immune responses in plants. Sci Rep
Chaud M et al (2021) Nanopesticides in agriculture: benefits and challenge in agricultural productivity, toxicological risks to human health and environment. Toxics 9:131
Cheema AI et al (2022) Antimicrobial activity of the biologically synthesized zinc oxide nanoparticles against important rice pathogens. Physiol Mol Biol Plants 28:1955–1967
Chen H (2018) Metal based nanoparticles in agricultural system: behavior, transport, and interaction with plants. Chem Spec Bioavailab 30:123–134
Chen J, Gutjahr C, Bleckmann A, Dresselhaus T (2015) Calcium signaling during reproduction and biotrophic fungal interactions in plants. Mol Plant 8:595–611
Chen J, Mao S, Xu Z, Ding W (2019) Various antibacterial mechanisms of biosynthesized copper oxide nanoparticles against soilborne *Ralstonia solanacearum*. RSC Adv 9:3788–3799
Chen J et al (2020) Comparative study on the fungicidal activity of metallic MgO nanoparticles and macroscale MgO against soilborne fungal phytopathogens. Front Microbiol 11:365
Choudhary DK, Prakash A, Johri BN (2007) Induced systemic resistance (ISR) in plants: mechanism of action. Indian J Microbiol 47:289–297
Cipollini D, Walters D, Voelckel C (2017) Costs of resistance in plants: from theory to evidence. In: Roberts JA (ed) Annual plant reviews online. Wiley, pp 263–307. https://doi.org/10.1002/9781119312994.apr0512
Cofer TM, Seidl-Adams I, Tumlinson JH (2018) From acetoin to (*Z*)-3-Hexen-1-ol: the diversity of volatile organic compounds that induce plant responses. J Agric Food Chem 66:11197–11208
Dangi K, Verma AK (2021) Efficient & eco-friendly smart nano-pesticides: emerging prospects for agriculture. Mater Today Proceed 45:3819–3824
Das A, Dutta P (2021) Antifungal activity of biogenically synthesized silver and gold nanoparticles against sheath blight of rice. J Nanosci Nanotechnol 21:3547–3555
Davies AL, White RM (2012) Collaboration in natural resource governance: reconciling stakeholder expectations in deer management in Scotland. J Environ Manage 112:160–169
De Souza-Torres A et al (2021) Fe_3O_4 nanoparticles and Rhizobium inoculation enhance nodulation, nitrogen fixation and growth of common bean plants grown in soil. Rhizosphere 17:100275
Denisse V-GJ et al (2021) Influence of nanoparticles on the physical, chemical, and biological properties of soils. In: Nanomaterials for soil remediation. Elsevier, pp 151–182. https://doi.org/10.1016/B978-0-12-822891-3.00008-6
Dhiman S, Yadav A, Debnath N, Das S (2021) Application of core/shell nanoparticles in smart farming: a paradigm shift for making the agriculture sector more sustainable. J Agric Food Chem 69:3267–3283

Dibyanshu K, Chhaya T, Raychoudhury T (2023) A review on the fate and transport behavior of engineered nanoparticles: possibility of becoming an emerging contaminant in the groundwater. Int J Environ Sci Technol 20:4649–4672

Djanaguiraman M, Belliraj N, Bossmann SH, Prasad PVV (2018) High-temperature stress alleviation by selenium nanoparticle treatment in grain sorghum. ACS Omega 3:2479–2491

Dong S et al (2022) Root hair apex is the key site for symplastic delivery of graphene into plants. Environ Sci Technol 56:12179–12189

Effah E, Holopainen JK, McCormick AC (2019) Potential roles of volatile organic compounds in plant competition. Perspect Plant Ecol Evol Syst 38:58–63

El-Batal AI, El-Sayyad GS, Mosallam FM, Fathy RM (2020) Penicillium chrysogenum-mediated mycogenic synthesis of copper oxide nanoparticles using gamma rays for in vitro antimicrobial activity against some plant pathogens. J Clust Sci 31:79–90

Elgorban A (2016) Extracellular synthesis of silver nanoparticles using Aspergillus versicolor and evaluation of their activity on plant pathogenic fungi. Mycosphere 7:844–852

Elhawat N et al (2018) Nanoparticle-associated phytotoxicity and abiotic stress under agroecosystems. In: Faisal M, Saquib Q, Alatar AA, Al-Khedhairy AA (eds) Phytotoxicity of nanoparticles. Springer International Publishing, pp 241–268. https://doi.org/10.1007/978-3-319-76708-6_10

Elizabeth Sunny N, Kumar Shanmugam V (2021) Anti- blight effect of green synthesized pure and Ag-doped tin oxide nanoparticles from Averrhoa bilimbi fruit extract towards Xanthomonas oryzae-the leaf blight pathogen of rice. Inorg Chem Commun 133:108866

Elmasry N (2021) Efficacy of zinc oxide nanoparticle adding to two neonicotinoids pesticides against Spodoptera litura (Lepidoptera: Noctuidae) and Aphis gossypii Glover. Egypt Acad J Biol Sci F Toxicol Pest Control 13:209–215

Elmer WH et al (2021) Foliar application of copper oxide nanoparticles suppresses fusarium wilt development on chrysanthemum. Environ Sci Technol 55:10805–10810

El-Saadony MT et al (2022) Role of nanoparticles in enhancing crop tolerance to abiotic stress: a comprehensive review. Front Plant Sci 13:946717

Elsheery NI, Helaly MN, El-Hoseiny HM, Alam-Eldein SM (2020) Zinc oxide and silicone nanoparticles to improve the resistance mechanism and annual productivity of salt-stressed mango trees. Agronomy 10:558

Etesami H, Fatemi H, Rizwan M (2021) Interactions of nanoparticles and salinity stress at physiological, biochemical and molecular levels in plants: a review. Ecotoxicol Environ Saf 225:112769

Faizan M, Faraz A, Yusuf M, Khan ST, Hayat S (2018) Zinc oxide nanoparticle-mediated changes in photosynthetic efficiency and antioxidant system of tomato plants. Photosynthetica 56:678–686

Farooq MA et al (2023) Phytotoxicity response and defense mechanisms of nanocomposites/mixture of nanoparticles. In: Husen A (ed) Nanomaterials and nanocomposites exposures to plants. Springer Nature Singapore, pp 43–58. https://doi.org/10.1007/978-981-99-2419-6_3

Feng Y et al (2022) Effects of iron oxide nanoparticles (Fe3O4) on growth, photosynthesis, antioxidant activity and distribution of mineral elements in wheat (Triticum aestivum) plants. Plants 11:1894

Fenjan, Saleh Falih (2020). Study on the effect of calcium carbonate nanoparticles as a new strategy to control the olive fruit fly *Bactrocera oleae* (Rossi) in olive orchards. http://hdl.handle.net/2067/50282.

Ferreira RB et al (2007) The role of plant defence proteins in fungal pathogenesis. Mol Plant Pathol 8:677–700

Fiol DF et al (2021) Nanotechnology in plants: recent advances and challenges. J Chem Technol Biotechnol 96:2095–2108

Flores-López LZ, Espinoza-Gómez H, Somanathan R (2019) Silver nanoparticles: electron transfer, reactive oxygen species, oxidative stress, beneficial and toxicological effects. Mini review. J Appl Toxicol 39:16–26

Francis DV, Sood N, Gokhale T (2020) Applications of metal nanoparticles in agriculture. In: Progress and prospects in nanoscience today. Nova Science Publishers
Francis DV, Sood N, Gokhale T (2022) Biogenic CuO and ZnO nanoparticles as nanofertilizers for sustainable growth of Amaranthus hybridus. Plants 11:2776
Fürstenberg-Hägg J, Zagrobelny M, Bak S (2013) Plant defense against insect herbivores. IJMS 14:10242–10297
Ga'al H et al (2018) Synthesis, characterization and efficacy of silver nanoparticles against Aedes albopictus larvae and pupae. Pest Biochem Physiol 144:49–56
Gaba S, Varma A, Goel A (2022) Protective and curative activity of biogenic copper oxide nanoparticles against Alternaria blight disease in oilseed crops: a review. J Plant Dis Prot 129:215–229
Ganguli P (2019) Intellectual property rights in nano-biopesticides. In: Abd-Elsalam KA, Prasad R (eds) Nanobiotechnology applications in plant protection. Springer International Publishing, pp 1–17. https://doi.org/10.1007/978-3-030-13296-5_1
García-Sánchez S, Bernales I (2015) Early response to nanoparticles in the Arabidopsis transcriptome compromises plant defence and root-hair development through salicylic acid signalling. BMC Dermatol. https://doi.org/10.1186/s12864-015-1530-4
Gautam A et al (2023) Magnesium oxide nanoparticles improved vegetative growth and enhanced productivity, biochemical potency and storage stability of harvested mustard seeds. Environ Res 229:116023
Ghani MI et al (2022) Foliar application of zinc oxide nanoparticles: an effective strategy to mitigate drought stress in cucumber seedling by modulating antioxidant defense system and osmolytes accumulation. Chemosphere 289:133202
Ghosh S, Malukani KK, Chandan RK, Sonti RV, Jha G (2019) How plants respond to pathogen attack: interaction and communication. In: Sopory S (ed) Sensory biology of plants. Springer Singapore, pp 537–568. https://doi.org/10.1007/978-981-13-8922-1_20
Gilbertson LM et al (2020) Guiding the design space for nanotechnology to advance sustainable crop production. Nat Nanotechnol 15:801–810
Gomiero T, Pimentel D, Paoletti MG (2011) Environmental impact of different agricultural management practices: conventional vs. organic agriculture. Crit Rev Plant Sci 30:95–124
González-Merino AM et al (2021) Antifungal activity of zinc oxide nanoparticles in *Fusarium oxysporum-Solanum lycopersicum* pathosystem under controlled conditions. J Phytopathol 169:533–544
Grillo R et al (2021) Foliage adhesion and interactions with particulate delivery systems for plant nanobionics and intelligent agriculture. Nano Today 37:101078
Gülmezoğlu N, Köleli N, Karakaya B, Dağhan H (2020) Impact of titanium dioxide nanoparticles (TiO_2-NPs) on growth and mineral nutrient uptake of wheat (Triticum vulgare L.). Biotech Studies 29:69–76
Gusain R, Gupta K, Joshi P, Khatri OP (2019) Adsorptive removal and photocatalytic degradation of organic pollutants using metal oxides and their composites: a comprehensive review. Adv Colloid Interface Sci 272:102009
Hadi Soltanabad M, Bagherieh-Najjar MB, Mianabadi M (2020) Carnosic Acid Content Increased by Silver Nanoparticle Treatment in Rosemary (Rosmarinus officinalis L.). Appl Biochem Biotechnol 191:482–495
Hameed A et al (2021) Effects of salinity stress on chloroplast structure and function. Cells 10:2023
Hanley ME, Lamont BB, Fairbanks MM, Rafferty CM (2007) Plant structural traits and their role in anti-herbivore defence. Perspect Plant Ecol Evol Syst 8:157–178
Hayes KL et al (2020) Effects, uptake, and translocation of aluminum oxide nanoparticles in lettuce: a comparison study to phytotoxic aluminum ions. Sci Total Environ 719:137393
He L, Liu Y, Mustapha A, Lin M (2011) Antifungal activity of zinc oxide nanoparticles against Botrytis cinerea and Penicillium expansum. Microbiol Res 166:207–215
Hong J et al (2021) Foliar application of nanoparticles: mechanisms of absorption, transfer, and multiple impacts. Environ Sci Nano 8:1196–1210

Hossain A et al (2019) Lemon-fruit-based green synthesis of zinc oxide nanoparticles and titanium dioxide nanoparticles against soft rot bacterial pathogen Dickeya dadantii. Biomolecules 9:863
Hu P et al (2020) Nanoparticle charge and size control foliar delivery efficiency to plant cells and organelles. ACS Nano 14:7970–7986
Hua K-H, Wang H-C, Chung R-S, Hsu J-C (2015) Calcium carbonate nanoparticles can enhance plant nutrition and insect pest tolerance. J Pestic Sci 40:208–213
Hückelhoven R (2007) Cell wall–associated mechanisms of disease resistance and susceptibility. Annu Rev Phytopathol 45:101–127
Hussain B et al (2020) Foliage application of selenium and silicon nanoparticles alleviates Cd and Pb toxicity in rice (Oryza sativa L.). Sci Total Environ 712:136497
Ibrahim AMA, Ali AM (2018) Silver and zinc oxide nanoparticles induce developmental and physiological changes in the larval and pupal stages of Spodoptera littoralis (Lepidoptera: Noctuidae). J Asia Pac Entomol 21:1373–1378
Ibrahim E et al (2020) Green-synthesization of silver nanoparticles using endophytic bacteria isolated from garlic and its antifungal activity against wheat fusarium head blight pathogen Fusarium graminearum. Nanomaterials 10:219
Iqbal M, Umar S, Mahmooduzzafar. (2019a) Nano-fertilization to enhance nutrient use efficiency and productivity of crop plants. In: Husen A, Iqbal M (eds) Nanomaterials and plant potential. Springer International Publishing, pp 473–505. https://doi.org/10.1007/978-3-030-05569-1_19
Iqbal M et al (2019b) Assessment of AgNPs exposure on physiological and biochemical changes and antioxidative defence system in wheat (*Triticum aestivum* L) under heat stress. IET Nanobiotechnol 13:230–236
Iqbal M et al (2022) Selenium and nano-selenium-mediated heat stress tolerance in plants. In: Hossain MA et al (eds) Selenium and nano-selenium in environmental stress management and crop quality improvement. Springer International Publishing, pp 149–171. https://doi.org/10.1007/978-3-031-07063-1_8
Irshad MA et al (2020) Synthesis and characterization of titanium dioxide nanoparticles by chemical and green methods and their antifungal activities against wheat rust. Chemosphere 258:127352
Isah T (2019) Stress and defense responses in plant secondary metabolites production. Biol Res 52:39
Ismail G, Abou-Zeid H (2018) The role of priming with biosynthesized silver nanoparticles in the response of Triticum aestivum L. to salt stress. Egypt J Bot
Jain M, Amera GM, Muthukumaran J, Singh AK (2022) Insights into biological role of plant defense proteins: a review. Biocatal Agric Biotechnol 40:102293
Jalil SU, Ansari MI (2019) Nanoparticles and abiotic stress tolerance in plants. In: Plant signaling molecules. Elsevier, pp 549–561. https://doi.org/10.1016/B978-0-12-816451-8.00034-4
Jan R, Asaf S, Numan M, Lubna, Kim K-M (2021) Plant secondary metabolite biosynthesis and transcriptional regulation in response to biotic and abiotic stress conditions. Agronomy 11:968
Jayaseelan C, Ramkumar R, Rahuman AA, Perumal P (2013) Green synthesis of gold nanoparticles using seed aqueous extract of Abelmoschus esculentus and its antifungal activity. Ind Crop Prod 45:423–429
Jh D (2015) An overview of plant immunity. J Plant Pathol Microbiol 6
Ji H et al (2015) The role of thionins in rice defence against root pathogens. Mol Plant Pathol 16:870–881
Jogaiah S et al (2021) Phytotoxicological effects of engineered nanoparticles: an emerging nanotoxicology. Sci Total Environ 801:149809
Kamel SM et al (2022) Antifungal activity of copper oxide nanoparticles against root rot disease in cucumber. JoF 8:911
Kashyap P, Shirkot P, Das R, Pandey H, Singh D (2023) Biosynthesis and characterization of copper nanoparticles from Stenotrophomonas maltophilia and its effect on plant pathogens and pesticide degradation. J Agric Food Res 13:100654

Kasote DM, Lee JHJ, Jayaprakasha GK, Patil BS (2019) Seed priming with iron oxide nanoparticles modulate antioxidant potential and defense-linked hormones in watermelon seedlings. ACS Sustain Chem Eng 7:5142–5151

Khan I, Saeed K, Khan I (2019) Nanoparticles: properties, applications and toxicities. Arab J Chem 12:908–931

Khanna K, Kohli SK, Ohri P, Bhardwaj R (2021) Plants-nematodes-microbes crosstalk within soil: a trade-off among friends or foes. Microbiol Res 248:126755

Khezerlou A, Alizadeh-Sani M, Azizi-Lalabadi M et al (2018) Nanoparticles and their antimicrobial properties against pathogens including bacteria, fungi, parasites and viruses. Microb Pathog 123:505–526

Kigathi RN et al (2009) Emission of volatile organic compounds after herbivory from Trifolium pratense (L.) under laboratory and field conditions. J Chem Ecol 35:1335–1348

Kim D et al (2018) Recent developments in nanotechnology transforming the agricultural sector: a transition replete with opportunities. J Sci Food Agric 98:849–864

Kitherian S (2017) Nano and bio-nanoparticles for insect control. Res J Nanosci Nanotechnol 7:1–9

Kumar S et al (2019) Nano-based smart pesticide formulations: emerging opportunities for agriculture. J Control Release 294:131–153

Kumar L, Ragunathan V, Chugh M, Bharadvaja N (2021) Nanomaterials for remediation of contaminants: a review. Environ Chem Lett 19:3139–3163

Kumari M et al (2020) Omics-based mechanistic insight into the role of bioengineered nanoparticles for biotic stress amelioration by modulating plant metabolic pathways. Front Bioeng Biotechnol 8:242

Kumari R et al (2023) Regulation and safety measures for nanotechnology-based agri-products. Front Genome Ed 5:1200987

Lamalakshmi Devi E et al (2017) Adaptation strategies and defence mechanisms of plants during environmental stress. In: Ghorbanpour M, Varma A (eds) Medicinal plants and environmental challenges. Springer International Publishing, pp 359–413. https://doi.org/10.1007/978-3-319-68717-9_20

Li J, Hu J, Xiao L, Wang Y, Wang X (2018) Interaction mechanisms between α-Fe_2O_3, γ-Fe_2O_3 and Fe_3O_4 nanoparticles and Citrus maxima seedlings. Sci Total Environ

Li H et al (2019a) Enhanced RuBisCO activity and promoted dicotyledons growth with degradable carbon dots. Nano Res 12:1585–1593

Li N, Han X, Feng D, Yuan D, Huang L-J (2019b) Signaling crosstalk between salicylic acid and ethylene/jasmonate in plant defense: do we understand what they are whispering? IJMS 20:671

Liu W, Park S-W (2018) Underground mystery: interactions between plant roots and parasitic nematodes. Curr Plant Biol 15:25–29

Ma J et al (2023) Bacillus mycoides PM35 in combination with titanium dioxide (TiO_2)-nanoparticles enhanced morpho-physio-biochemical attributes in Barley (Hordeum vulgare L.) under cadmium stress. Chemosphere 323:138224

MacRae TH (2010) Gene expression, metabolic regulation and stress tolerance during diapause. Cell Mol Life Sci 67:2405–2424

Malandrakis AA, Kavroulakis N, Chrysikopoulos CV (2019) Use of copper, silver and zinc nanoparticles against foliar and soil-borne plant pathogens. Sci Total Environ 670:292–299

Manetas Y (2012) The defence of a stationary organism. In: Alice in the land of plants. Springer, Berlin Heidelberg, pp 243–284. https://doi.org/10.1007/978-3-642-28338-3_7

Mani D, Kumar C (2014) Biotechnological advances in bioremediation of heavy metals contaminated ecosystems: an overview with special reference to phytoremediation. Int J Environ Sci Technol 11:843–872

Manzoor N et al (2023) Comparative efficacy of silicon and iron oxide nanoparticles towards improving the plant growth and mitigating arsenic toxicity in wheat (Triticum aestivum L.). Ecotoxicol Environ Saf 264:115382

Marcec MJ, Gilroy S, Poovaiah BW, Tanaka K (2019) Mutual interplay of Ca^{2+} and ROS signaling in plant immune response. Plant Sci 283:343–354

Matras E et al (2023) Silver nanoparticles affect wheat (Triticum aestivum L.) germination, seedling blight and yield. Funct Plant Biol 50:390–406
Medda S, Hajra A, Dey U, Bose P, Mondal NK (2015) Biosynthesis of silver nanoparticles from Aloe vera leaf extract and antifungal activity against Rhizopus sp. and Aspergillus sp. Appl Nanosci 5:875–880
Mishra AK, Sharma K, Misra RS (2012) Elicitor recognition, signal transduction and induced resistance in plants. J Plant Interact 7:95–120
Mishra P, Kumari S, Saini P, Meena RK (2022) Role of nanotechnology in management of plant viral diseases. Mater Today Proceed 69:1–10
Misra RC, Sandeep K, M., Kumar, S. & Ghosh, S. (2016) A thaumatin-like protein of Ocimum basilicum confers tolerance to fungal pathogen and abiotic stress in transgenic Arabidopsis. Sci Rep 6:25340
Mithöfer A, Boland W (2012) Plant defense against herbivores: chemical aspects. Annu Rev Plant Biol 63:431–450
Mittal D, Kaur G, Singh P, Yadav K, Ali SA (2020) Nanoparticle-based sustainable agriculture and food science: recent advances and future outlook. Front Nanotechnol 2:579954
Monson RK, Trowbridge AM, Lindroth RL, Lerdau MT (2022) Coordinated resource allocation to plant growth–defense tradeoffs. N Phytol 233:1051–1066
Mostafa YS, Alamri SA, Alrumman SA, Hashem M, Baka ZA (2021) Green synthesis of silver nanoparticles using pomegranate and orange peel extracts and their antifungal activity against Alternaria solani, the causal agent of early blight disease of tomato. Plants 10:2363
Murgueitio-Herrera E et al (2022) Synthesis of iron, zinc, and manganese nanofertilizers, using Andean blueberry extract, and their effect in the growth of cabbage and lupin plants. Nanomaterials 12:1921
Nalini M, Lena M, Sumathi P, Sundaravadivelan C (2017) Effect of phyto-synthesized silver nanoparticles on developmental stages of malaria vector, *Anopheles stephensi* and dengue vector, *Aedes aegypti*. Egypt J Basic Appl Sci 4:212–218
Namburi KR, Kora AJ, Chetukuri A, Kota VSMK (2021) Biogenic silver nanoparticles as an antibacterial agent against bacterial leaf blight causing rice phytopathogen Xanthomonas oryzae pv. oryzae. Bioprocess Biosyst Eng 44:1975–1988
Noori A, Ngo A, Gutierrez P, Theberge S, White JC (2020) Silver nanoparticle detection and accumulation in tomato (Lycopersicon esculentum). J Nanopart Res 22:131
Ogunyemi SO et al (2019) Green synthesis of zinc oxide nanoparticles using different plant extracts and their antibacterial activity against *Xanthomonas oryzae* pv. oryzae. Artif Cells Nanomed Biotechnol 47:341–352
Palocci C et al (2017) Endocytic pathways involved in PLGA nanoparticle uptake by grapevine cells and role of cell wall and membrane in size selection. Plant Cell Rep 36:1917–1928
Panhwar S et al (2021) Importance and analytical perspective of green synthetic strategies of copper, zinc, and titanium oxide nanoparticles and their applications in pathogens and environmental remediation. CAC 17:1169–1181
Panichikkal J, Krishnankutty RE (2022) Chitosan and gold nanoparticles supplementation for augmentation of Indole-3-acetic acid production by rhizospheric Pseudomonas aeruginosa and plant growth enhancement. Curr Microbiol 79:185
Paret ML, Palmateer AJ, Knox GW (2013) Evaluation of a light-activated nanoparticle formulation of titanium dioxide with zinc for management of bacterial leaf spot on Rosa 'Noare'. Horts 48:189–192
Pariona N et al (2019) Green-synthesized copper nanoparticles as a potential antifungal against plant pathogens. RSC Adv 9:18835–18843
Pathipati UR, Kanuparthi PL (2021) Silver nanoparticles for insect control: bioassays and mechanisms. In: Silver nanomaterials for agri-food applications. Elsevier, pp 471–494. https://doi.org/10.1016/B978-0-12-823528-7.00007-X

Patra D, Mandal S (2022) Non-rhizobia are the alternative sustainable solution for growth and development of the nonlegume plants. Biotechnol Genet Eng Rev:1–30. https://doi.org/10.1080/02648725.2022.2152623

Pe'er G et al (2020) Action needed for the EU Common Agricultural Policy to address sustainability challenges. People Nat 2:305–316

Pelegrino MT et al (2020) Effects of copper oxide nanoparticles on growth of lettuce (Lactuca sativa L.) seedlings and possible implications of nitric oxide in their antioxidative defense. Environ Monit Assess 192:232

Perea Velez YS, Carrillo-Gonzalez R, Gonzalez-Chavez MD (2021) Interaction of metal nanoparticles-plants-microorganisms in agriculture and soil remediation. J Nanopart Res 23(9):206

Pieterse CMJ et al (2014) Induced systemic resistance by beneficial microbes. Annu Rev Phytopathol 52:347–375

Pinto-Zevallos DM, Strapasson P, Zarbin PHG (2016) Herbivore-induced volatile organic compounds emitted by maize: electrophysiological responses in Spodoptera frugiperda females. Phytochem Lett 16:70–74

Rai M et al (2018) Copper and copper nanoparticles: role in management of insect-pests and pathogenic microbes. Nanotechnol Rev 7:303–315

Rajiv P, Rajeshwari S, Venckatesh R (2013) Bio-Fabrication of zinc oxide nanoparticles using leaf extract of Parthenium hysterophorus L. and its size-dependent antifungal activity against plant fungal pathogens. Spectrochim Acta Part A: Mol Biomol Spectrosc 112:384–387

Rajput VD et al (2018) Effect of nanoparticles on crops and soil microbial communities. J Soil Sediment 18:2179–2187

Rajput VD et al (2022) Nanotechnology in the restoration of polluted soil. Nanomaterials 12:769

Ramadan T, Sayed SA, Abd-Elaal AKA, Amro A (2022) The combined effect of water deficit stress and TiO_2 nanoparticles on cell membrane and antioxidant enzymes in Helianthus annuus L. Physiol Mol Biol Plants 28:391–409

Rasheed A et al (2022) The role of nanoparticles in plant biochemical, physiological, and molecular responses under drought stress: a review. Front Plant Sci 13:976179

Rezaei Z, Jafarirad S, Kosari-Nasab M (2019) Modulation of secondary metabolite profiles by biologically synthesized MgO/perlite nanocomposites in Melissa officinalis plant organ cultures. J Hazard Mater 380:120878

Rizwan M et al (2019) Zinc and iron oxide nanoparticles improved the plant growth and reduced the oxidative stress and cadmium concentration in wheat. Chemosphere 214:269–277

Rouhani M, Amin Samih M, Kalantari S (2012) Insecticied effect of silver and zinc nanoparticles against Aphis nerii Boyer of fonscolombe (Hemiptera: Aphididae). Chilean J Agric Res 72:590–594

Roy D et al (2022) Impact of CuO nanoparticles on maize: comparison with CuO bulk particles with special reference to oxidative stress damages and antioxidant defense status. Chemosphere 287:131911

Russo A, Pollastri S, Ruocco M, Monti MM, Loreto F (2022) Volatile organic compounds in the interaction between plants and beneficial microorganisms. J Plant Interact 17:840–852

Sachdev S, Ansari SA, Ansari MI, Fujita M, Hasanuzzaman M (2021) Abiotic stress and reactive oxygen species: generation, signaling, and defense mechanisms. Antioxidants 10:277

Sadak MS (2019) Impact of silver nanoparticles on plant growth, some biochemical aspects, and yield of fenugreek plant (Trigonella foenum-graecum). Bull Natl Res Cent 43:38

Saha I et al (2021) Silver-nanoparticle and abscisic acid modulate sub1A quantitative trait loci functioning towards submergence tolerance in rice (Oryza sativa L.). Environ Exp Bot 181:104276

Salama DM, Abd El-Aziz ME, Osman SA, Abd Elwahed MSA, Shaaban EA (2022) Foliar spraying of MnO_2-NPs and its effect on vegetative growth, production, genomic stability, and chemical quality of the common dry bean. Arab J Basic Appl Sci 29:26–39

Salehi H, Cheheregani Rad A, Raza A, Djalovic I, Prasad PVV (2023) The comparative effects of manganese nanoparticles and their counterparts (bulk and ionic) in Artemisia annua plants via seed priming and foliar application. Front Plant Sci 13:1098772

Salem D et al (2022) Nanobiotechnological approaches to enhance drought tolerance in catharanthus roseus plants using salicylic acid in bulk and nanoform. Molecules 27:5112

Sandhu JS, Sidhu MK, Yadav IS (2017) Control of fungal diseases in agricultural crops by Chitinase and glucanase transgenes. In: Lichtfouse E (ed) Sustainable agriculture reviews, vol 22. Springer International Publishing, pp 163–212

Saqib S et al (2019) Organometallic assembling of chitosan-Iron oxide nanoparticles with their antifungal evaluation against *Rhizopus oryzae*. Appl Org Chem 33:e5190

Sardar R, Ahmed S, Yasin NA (2022) Titanium dioxide nanoparticles mitigate cadmium toxicity in Coriandrum sativum L. through modulating antioxidant system, stress markers and reducing cadmium uptake. Environ Pollut 292:118373

Seleiman MF et al (2020) Nano-fertilization as an emerging fertilization technique: why can modern agriculture benefit from its use? Plants 10:2

Sengul AB, Asmatulu E (2020) Toxicity of metal and metal oxide nanoparticles: a review. Environ Chem Lett 18:1659–1683

Shang H et al (2020) Copper sulfide nanoparticles suppress *Gibberella fujikuroi* infection in rice (*Oryza sativa* L.) by multiple mechanisms: contact-mortality, nutritional modulation and phytohormone regulation. Environ Sci Nano 7:2632–2643

Sharma J, Singh VK, Kumar A, Shankarayan R, Mallubhotla S (2018) Role of silver nanoparticles in treatment of plant diseases. In: Patra JK, Das G, Shin H-S (eds) Microbial biotechnology. Springer Singapore, pp 435–454. https://doi.org/10.1007/978-981-10-7140-9_20

Sharma S, Singh VK, Kumar A, Mallubhotla S (2019) Effect of nanoparticles on oxidative damage and antioxidant defense system in plants. In: Roychoudhury A, Tripathi D (eds) Molecular plant abiotic stress. Wiley, pp 315–333. https://doi.org/10.1002/9781119463665.ch17

Shen N et al (2022) Plant flavonoids: classification, distribution, biosynthesis, and antioxidant activity. Food Chem 383:132531

Shenashen M, Derbalah A, Hamza A, Mohamed A, El Safty S (2017) Antifungal activity of fabricated mesoporous alumina nanoparticles against root rot disease of tomato caused by *Fusarium oxysporium*. Pest Manag Sci 73:1121–1126

Shende S, Rathod D, Gade A, Rai M (2017) Biogenic copper nanoparticles promote the growth of pigeon pea (*Cajanus cajan* L.). IET Nanobiotechnol 11:773–781

Siddique MH et al (2022) Ecofriendly phytosynthesized zirconium oxide nanoparticles as antibiofilm and quorum quenching agents against *Acinetobacter baumannii*. Drug Dev Ind Pharm 48:502–509

Singh K, Winter M, Zouhar M, Ryšánek P (2018) Cyclophilins: less studied proteins with critical roles in pathogenesis. Phytopathology 108:6–14

Singh H et al (2021a) Recent advances in the applications of nano-agrochemicals for sustainable agricultural development. Environ Sci: Processes Impacts 23:213–239

Singh RP, Handa R, Manchanda G (2021b) Nanoparticles in sustainable agriculture: an emerging opportunity. J Control Release 329:1234–1248

Soltani Nejad M, Samandari Najafabadi N, Aghighi S, Pakina E, Zargar M (2022) Evaluation of Phoma sp. Biomass as an endophytic fungus for synthesis of extracellular gold nanoparticles with antibacterial and antifungal properties. Molecules 27:1181

Song Y, Zheng C, Li S, Chen J, Jiang M (2023) Chitosan-magnesium oxide nanoparticles improve salinity tolerance in rice (*Oryza sativa* L.). ACS Appl Mater Interfaces 15:20649–20660

Su Y et al (2019) Delivery, uptake, fate, and transport of engineered nanoparticles in plants: a critical review and data analysis. Environ Sci Nano 6:2311–2331

Sunny NE et al (2022) Effect of green synthesized nano-titanium synthesized from Trachyspermum ammi extract on seed germination of Vigna radiate. Chemosphere 300:134600

Takehara Y et al (2023) Molecular analysis of MgO nanoparticle-induced immunity against Fusarium Wilt in tomato. IJMS 24:2941

Thounaojam TC, Meetei TT, Devi YB, Panda SK, Upadhyaya H (2021) Zinc oxide nanoparticles (ZnO-NPs): a promising nanoparticle in renovating plant science. Acta Physiol Plant 43:136

Tripathi D, Singh M, Pandey-Rai S (2022) Crosstalk of nanoparticles and phytohormones regulate plant growth and metabolism under abiotic and biotic stress. Plant Stress 6:100107

T-Thienprasert NP et al (2021) Large scale synthesis of green synthesized zinc oxide nanoparticles from banana peel extracts and their inhibitory effects against Colletotrichum sp., isolate KUFC 021, causal agent of anthracnose on dendrobium orchid. J Nanomater 2021:1–10
Van Nguyen D et al (2022) Copper nanoparticle application enhances plant growth and grain yield in maize under drought stress conditions. J Plant Growth Regul 41:364–375
Vanathi P, Rajiv P, Sivaraj R (2016) Synthesis and characterization of Eichhornia-mediated copper oxide nanoparticles and assessing their antifungal activity against plant pathogens. Bull Mater Sci 39:1165–1170
Venkatachalam P, Priyanka N, Manikandan K et al (2017) Enhanced plant growth promoting role of phycomolecules coated zinc oxide nanoparticles with P supplementation in cotton (Gossypium hirsutum L.). Plant Physiol Biochem 110:118–127
Venzhik Y, Deryabin A, Popov V, Dykman L, Moshkov I (2022) Gold nanoparticles as adaptogens increazing the freezing tolerance of wheat seedlings. Environ Sci Pollut Res 29:55235–55249
Wahab A et al (2023) Current knowledge, research progress, and future prospects of phyto-synthesized nanoparticles interactions with food crops under induced drought stress. Sustainability 15:14792
Wahid I et al (2020) Silver nanoparticle regulates salt tolerance in wheat through changes in ABA concentration, ion homeostasis, and defense systems. Biomolecules 10:1506
Wang W et al (2020) Impact of copper oxide nanoparticles on the germination, seedling growth, and physiological responses in Brassica pekinensis L. Environ Sci Pollut Res 27:31505–31515
Wang J et al (2021) Effects of TiO_2-NPs pretreatment on UV-B stress tolerance in Arabidopsis thaliana. Chemosphere 281:130809
Wang L, Yang D, Ma F, Wang G, You Y (2022a) Recent advances in responses of arbuscular mycorrhizal fungi—plant symbiosis to engineered nanoparticles. Chemosphere 286:131644
Wang L et al (2022b) Silica nanoparticles activate defense responses by reducing reactive oxygen species under Ralstonia solanacearum infection in tomato plants. NanoImpact 28:100418
Wang X, Xie H, Wang P, Yin H (2023) Nanoparticles in plants: uptake, transport and physiological activity in leaf and root. Materials 16:3097
Waqas Mazhar M et al (2022) Seed priming with iron oxide nanoparticles raises biomass production and agronomic profile of water-stressed flax plants. Agronomy 12:982
Wu H, Jiang X, Tong J, Wang J, Shi J (2023a) Effects of Fe_3O_4 nanoparticles and nano hydroxyapatite on Pb and Cd stressed rice (Oryza sativa L.) seedling. Chemosphere 329:138686
Wu S et al (2023b) Cichorium intybus L. is a potential Cd-accumulator for phytoremediation of agricultural soil with strong tolerance and detoxification to Cd. J Hazard Mater 451:131182
Xu J, Zhang K, Cuthbertson AGS, Du C, Ali S (2020) Toxicity and biological effects of Beauveria brongniartii Fe^0 nanoparticles against Spodoptera litura (Fabricius). Insects 11:895
Yadav A, Yadav K (2018) Nanoparticle-based plant disease management: tools for sustainable agriculture. In: Abd-Elsalam KA, Prasad R (eds) Nanobiotechnology applications in plant protection. Springer International Publishing, pp 29–61. https://doi.org/10.1007/978-3-319-91161-8_2
Yadav MR et al (2022) Impacts, tolerance, adaptation, and mitigation of heat stress on wheat under changing climates. IJMS 23:2838
Yang J, Cao W, Rui Y (2017) Interactions between nanoparticles and plants: phytotoxicity and defense mechanisms. J Plant Interact. https://doi.org/10.1080/17429145.2017.1310944
Yang J et al (2019) The crosstalks between jasmonic acid and other plant hormone signaling highlight the involvement of jasmonic acid as a core component in plant response to biotic and abiotic stresses. Front Plant Sci 10:1349
Yang Z et al (2020) Effects of copper oxide nanoparticles on the growth of rice (Oryza Sativa L.) seedlings and the relevant physiological responses. IJERPH 17:1260
Yang Y et al (2023) Enhanced germination and growth of alfalfa with seed presoaking and hydroponic culture in Fe_2O_3 magnetic nanoparticles. J Nanomater 2023:1–11
Yasur J, Usha Rani P (2015) Lepidopteran insect susceptibility to silver nanoparticles and measurement of changes in their growth, development and physiology. Chemosphere 124:92–102

Yazhiniprabha M, Vaseeharan B, Sonawane A, Behera A (2019) In vitro and In vivo toxicity assessment of phytofabricated ZnO nanoparticles showing bacteriostatic effect and larvicidal efficacy against Culex quinquefasciatus. J Photochem Photobiol B Biol 192:158–169

Yu Y et al (2022) Induced systemic resistance for improving plant immunity by beneficial microbes. Plants 11:386

Zakharova OV et al (2017) Sodium tallow amphopolycarboxyglycinate-stabilized silver nanoparticles suppress early and late blight of Solanum lycopersicum and stimulate the growth of tomato plants. BioNanoSci 7:692–702

Zhang Y, Goss GG (2022) Nanotechnology in agriculture: comparison of the toxicity between conventional and nano-based agrochemicals on non-target aquatic species. J Hazard Mater 439:129559

Zhang Y et al (2022a) Effects of metal nanoparticles and other preparative materials in the environment on plants: from the perspective of improving secondary metabolites. J Agric Food Chem 70:916–933

Zhang H et al (2022b) Nanoparticle cellular internalization is not required for RNA delivery to mature plant leaves. Nat Nanotechnol 17:197–205

Zhou P et al (2020) Application of nanoparticles alleviates heavy metals stress and promotes plant growth: an overview. Nanomaterials 11:26

SPRINGER NATURE

GPSR Compliance

The European Union's (EU) General Product Safety Regulation (GPSR) is a set of rules that requires consumer products to be safe and our obligations to ensure this.

If you have any concerns about our products, you can contact us on ProductSafety@springernature.com

In case Publisher is established outside the EU, the EU authorized representative is:

Springer Nature Customer Service Center GmbH
Europaplatz 3
69115 Heidelberg, Germany

The manufacturer's authorised representative in the EU is Springer Nature Customer Service Centre GmbH, Europaplatz 3, 69115 Heidelberg, Germany. If you have any concerns regarding our products, please contact ProductSafety@springernature.com

Printed and bound by CPI Group (UK) Ltd, Croydon, CR0 4YY
01/10/2026
02225859-0019